PRODUCT DEVELOPMENT AND MANAGEMENT BODY OF KNOWLEDGE

PRODUCT DEVELOPMENT AND MANAGEMENT BODY OF KNOWLEDGE

A Guidebook for Product Innovation Training and Certification

Third Edition

Allan Anderson
Chad McAllister
Ernie Harris

WILEY

Published by John Wiley & Sons, Inc., Hoboken, New Jersey.
Published simultaneously in Canada.

For general information on our other products and services or for technical support, please contact our Customer Care Department within the United States at (800) 762-2974, outside the United States at (317) 572-3993 or fax (317) 572-4002.

Wiley also publishes its books in a variety of electronic formats. Some content that appears in print may not be available in electronic formats. For more information about Wiley products, visit our web site at www.wiley.com.

Library of Congress Cataloging-in-Publication Data applied for:

Paperback ISBN: 9781119829942

Cover Design: Wiley
Cover Image: © aditya M S/Shutterstock

Set in 11/13 pts and Helvetica Neue LT Std by Straive, Chennai, India.

SKY10069735_031424

Contents

About the Authors

Allan M. Anderson, PhD, BTech (Hons), FNZIFST, NPDP

Allan is currently Emeritus Professor in Product Development in the School of Engineering and Advanced Technology at Massey University, New Zealand. He has extensive experience in product innovation over 40 years in general management, product management, consultancy, and academia. Research, teaching, and application of innovation management in a wide range of organizations internationally has provided Allan with a depth of knowledge and experience across the broad spectrum of practices, processes, and techniques that define product innovation management.

Allan was part of the establishment team for PDMA-NZ in 2007; president of PDMA-NZ until December 2011; was Vice President Asia-Pacific until 2015 and PDMA President and Chairman from 2016 to 2019; has NPDP certification; and is an accredited NPDP training provider. Allan's passion for product innovation education prompted his active involvement in the expansion of PDMA's New Product Development Professional Certification (NPDP) into new markets including China, India, and Indonesia. In 2016, he compiled the first edition of PDMA's Body of Knowledge (BoK) which laid the foundation for the second edition in 2019 and for this, the third edition.

Chad McAllister, PhD, PMP, NPDP, CIL

Chad is a recognized Top 40 Product Management Influencer and a Top 10 Innovation Blogger. He is the host of *Product Mastery Now,* a top-1% business podcast enjoyed by thousands of product managers, leaders, and innovators since 2014. Each week he provides a valuable discussion exploring topics that equip you for success. Find the podcast on your favorite podcast player by searching for *Product Mastery Now,* or at www.ProductMasteryNow.com.

With 30+ years of product experience, he helps product managers and leaders move toward product mastery, creating greater value for customers and their organizations. He frequently works with product teams, facilitating their move to higher performance. He also teaches product and innovation management graduate courses at Boston University, Colorado State University, University of Fredericton, and other universities. He holds two degrees in Electrical Engineering, a PhD in *Innovation*, and professional certifications from PMI (Project Management Professional), PDMA (New Product Development Professional), and AIPMM (Certified Innovation Leader). Connect with Chad via LinkedIn at https://www.linkedin.com/in/chadmcallister/.

Ernie Harris

For more than 30 years, Ernie has been involved in new product innovation in the software and services industry. Ernie was part of the establishment of the PDMA Tampa chapter in Florida and since joining PDMA in 2007 has gone on to serve in many capacities. Ernie has chaired PDMA's International Innovation conference, founded multiple chapters, and served as the Treasurer for the PDMA International Board of Directors. In 2019, he was elected President and Chairman of PDMA International where he helped build on the foundation of PDMA's international growth through education and certification. Working closely with the Board of Directors and many others, he also helped to expand PDMA into countries like Turkey, Egypt, and more.

Ernie holds the NPDP certification and sits on the International NPDP certification committee supporting the global NPDP exam. He has traveled on behalf of PDMA as a recognized innovation leader speaking on innovation management best practices in China.

Ernie currently serves as the President and Chief Operating Officer of a leading Third-Party Administrator offering employee benefits administration solutions to employers in the United States of all sizes.

Acknowledgments

PDMA's Body of Knowledge is founded on over 40 years of research and application of "best practices" in product innovation and product management. Those who have contributed to the evolution of PDMA's Body of Knowledge are far too many to list individually. We are reminded of Sir Isaac Newton's saying, "If I have seen further, it is by standing on the shoulders of giants." Thank you to all the PDMA "giants" who have made BoK3 a reality.

Following is specific acknowledgement to those who have been directly involved in the preparation of BoK3.

Chapter authors BoK 2nd edition
Stephen Atherton, NPDP
Jean-Jacques Verhaeghe, MBA-LS, PMP, NPDP
Carlos M Rodriguez, PhD, NPDP.
Karen Dworaczyk, BS (chemeng), NPDP
Teresa Jurgens-Kowal, PhD, PMP, NPDP
Jerry Fix, B.S., MBA, NPDP and Allan Anderson PhD, NPDP

Contributors the 1st, 2nd, or 3rd editions of the BoK
Aruna Shekar, PhD
Brenna Leever, BS Design
Brian Ottum, PhD
Chad McAllister, PhD, PMP, NPDP
Charles H Noble, PhD
Doug Collins, MBA
Ernie Harris, NPDP
Greg Coticchia, MBA
Jack Hsieh, NPDP
Jama L Bradley, PhD, PMP, NPDP
Jelena Spanjol, PhD
Lou Zheng, NPDP
Mark Adkins, NPDP
Mark Tunnicliffe, PhD
Martijn Antonisse, PhD, NPDP
Nikhil Kumtha, MBA, PMP
Peter Bradford, MSTC, NPDP
Sophie Xiao, PhD
Steve Johnson, Pragmatic CPM
Susan Burek, MS, NPDP
Vicky Zachareli, MSc, MBA

CASE STUDIES

Thanks to the following who have provided case studies for BoK3, presented in Appendices B and C:

Strategy: Global Dairy Co, Allan Anderson
Process: LeanMed, Mark Adkins
Design and Development: Stingray Digital Radio, Jessica Braddon-Parsons, Errol Chrystal, Afnan Kayed, Jacob Patrick
Market Research: OBO Hockey Equipment, Simon Barnett
Culture and Teams: Aobe Kickbox and rready®, Chad McAllister
Outstanding Corporate Innovator Awards: Sally E. Kay

Introduction

The Product Development and Management Association (PDMA) is proud to introduce the 3rd edition of its Body of Knowledge (BoK).

1 ABOUT PDMA

The Product Development and Management Association (PDMA) was founded in 1976. Founded in the USA, it has International Affiliates in Europe, South/Central America, and Asia/Pacific. It is the premier association worldwide for everyone involved in product innovation.

Membership includes practitioners, academics, and service providers and represents a broad cross-section of product and service industries, both "business-to-business" (B2B) and "business-to-consumer" (B2C). Members come from many industries, with a few including:

- Consumer goods
- Heavy machinery
- Information Technology
- Food and beverage
- Banking
- Healthcare
- Software
- Consultancy

Central to PDMA is its Body of Knowledge (BoK). This book is the guide to the PDMA BoK. PDMA also provides a professional qualification, the New Product Development Professional (NPDP) certification. Earning the certification demonstrates competency in the BoK concepts.

The BoK is founded on a broad spectrum of independent research, gathered by PDMA globally, from academic and practitioner sources, over the past 50 years. These include:

- ***The Journal of Product Innovation Management:*** an interdisciplinary, international journal that seeks to advance theoretical and managerial knowledge of innovation management and product development. The journal publishes original articles on organizations of all sizes (start-ups, small to medium-sized enterprises, and large corporations) and from the consumer, business-to-business, and policy domains.
- ***The PDMA Knowledge Hub (kHUB):*** a website of resources facilitating the creation and exchange of product management and development knowledge and best practices.
- ***The Outstanding Corporate Innovators Award (OCI):*** a rigorous assessment of the innovation capabilities of award applicants. The award-winning company or companies have created and captured long-term value through product and service innovation, demonstrating their innovation leadership.
- ***Research forums and Conferences:*** virtual and in-person meetings organized to facilitate the global exchange of ideas and practices among thought leaders.
- ***Publications covering a range of topics:*** PDMA explores current and emerging topics through a variety of publications. These are listed at the end of this introduction.

- ***Best Practice Surveys:*** these research studies started in 1997 and are conducted globally on a regular basis. The survey results identify the factors leading to high product innovation performance in organizations. Specifically, the results identify what is different between the *best*-performing organizations and the *rest*, lower-performing organizations.

The latest PDMA Best Practice Survey (Bsteiler & Noble, 2023) provides a compelling finding in support of the imperative for ongoing improvement in product innovation practices.

PDMA Best Practice Survey 2021 Finding

"The overall NPD [New Product Development] success rate (59.6%) has not changed materially in 30 years of PDMA's Best Practice surveys. Although many differences between the Best and the Rest are statistically significant, the absolute differences are not that large. Numerous new tools and practices such as stage-gate, concurrent, waterfall, and agile product development processes, Voice-of-the-Customer research, virtual team management, and electronic communication techniques and advanced digital development tools, have been implemented by firms, and they have improved the efficiency and effectiveness of NPD programs. Together, these results imply that all firms must continually evolve their NPD capabilities just to "stay in the game" as circumstances and the environment change."

2 ABOUT THIS BOOK

PDMA's 3rd edition of its Body of Knowledge (the BoK3) is designed to present a holistic view of Product Innovation Management, discussing individual components and their integration into the whole.

The Product Development Handbook, 4th edition (also published by PDMA) presents an Innovation Management (IM) Framework to help new product development managers identify those activities required to be a successful innovator (Markham, 2023). The IM Framework (Figure 1) was developed by the Center for Innovation Management Studies (CIMS). It describes a systematic way to think about managing innovation, breaking it down into competencies across several dimensions that can be learned, practiced, measured, and improved.

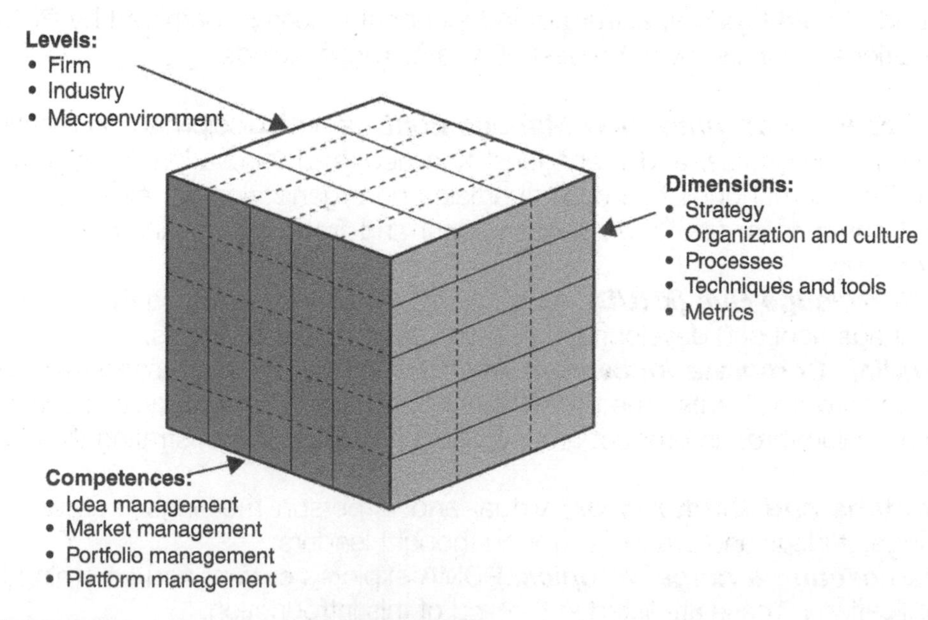

FIGURE 1 An innovation management framework.
Source: Markham, 2023 / John Wiley & Sons.

3 THE BOOK STRUCTURE

Inspired by the IM Framework, BoK3 provides a structured description of the key activities involved in product innovation management, comprising an "end to end" description of the product innovation: product strategy, portfolio management, product innovation processes, product design and development, market research, and organizational culture. All are presented in the context of an overall framework of product innovation management.

BoK3 is divided into seven chapters, diagrammatically presented in Figure 2. At the center of the figure is strategy, which defines the direction and goals for product innovation. This provides the basis for the product innovation portfolio, for the process leading to individual new products, and the consequent management of these products through their life cycles. Contributing at all levels of strategy, portfolio, and life cycle management are market research, specific tools for design and development, and people (culture, organization, leadership, and teams).

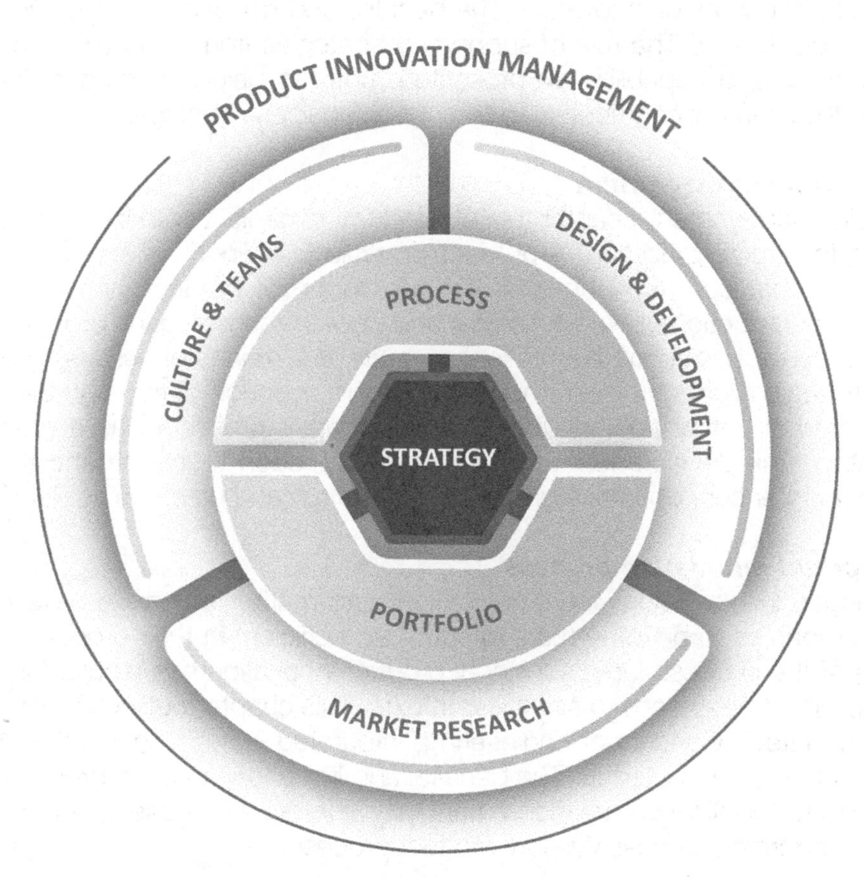

FIGURE 2 The seven chapters of PDMA's BoK3

A brief outline of each chapter is provided below:

Chapter 1 Product innovation management

The first chapter addresses the role of product innovation management. The first part defines product innovation and the key factors that lead to successful products. The breadth of the product innovation management role is discussed, together with the skill sets and knowledge required to manage across a range of disciplines and functions and how this differs from project management. Management of product innovation through the stages of the product life cycle – development, introduction, growth, maturity, decline, and retirement – is addressed with a focus on the product innovation strategies required at each stage. The second part focuses on the product life cycle and how product innovation is managed through this life cycle. While specific tools and techniques are discussed in other chapters, specific reference in this chapter is placed on road-mapping – product and technology, and light coverage of project management with reference to the Project Management Institute's PMBoK (2021).

Throughout BoK3, we emphasize the fundamentals for successful product innovation. There is no single recipe that can be used across all organizations. The third part of this chapter discusses the development and application of product innovation metrics, particularly how these form critical ingredients in organizational learning and continuous improvement.

Chapter 2 Strategy

The strategy chapter covers various strategies, from corporate, business, and functional strategies. An emphasis is placed on the innovation strategy, particularly as it sets out the framework and provides direction for product innovation. The benefits and limitations of specific innovation strategic frameworks are discussed. The role of supporting strategies from technology, marketing, platforms, intellectual property, and capability are presented, both as being directed by higher-level business strategy and in their mutual contribution to the overall business strategy.

Chapter 3 Portfolio management

The portfolio management chapter relates strategy to project selection, selecting the right innovation projects for the organization. A product portfolio is defined as the set of current and potential new products that can form the basis for a program of product innovation, including product improvement, cost reductions, line extensions, and new-to-the-company products.

Methods for project selection are presented, both as a means of assessing project potential and of achieving strategic alignment regarding individual project prioritization and balance across specific categories of product innovation. Portfolio management is presented as a cross-functional activity that encompasses the development of new products through to launch and the ongoing review of existing products to ensure optimal alignment with strategy and resource availability.

Chapter 4 Product innovation process

Rapid changes in technology, communication, and market demands have placed considerable pressure on companies to become more effective and efficient in their product innovation. A greater understanding of the success factors for new product innovation has resulted in the application of a range of new product processes to specific contexts. This chapter outlines many of these processes, including Stage-Gate®, Concurrent Engineering, Integrated Product Innovation, Hybrid Agile-Stage-Gate, Lean, Agile, and Lean Startup. The benefits and limitations of each process are discussed, and specific contexts for application are recommended. Reference is made to the tools and metrics that are required to underpin a successful new product process.

Chapter 5 Product design and development

This chapter focuses on the Design and Development stage of the product innovation process. The chapter is divided into two sections: Design (concept and embodiment design) and Development (specifications, usability testing, performance testing, endurance testing, quality assurance, design for manufacture and assembly, and design for sustainability). The evolution of the product through each of these phases is discussed, and specific tools are introduced together with their benefits and limitations. Some of these tools are applicable across a range of industries and products, while others are more specific in their application.

Chapter 6 Market research in product innovation

Market research is required to provide market-related information and data to underpin decision-making in all aspects of strategy development, portfolio management, product design and development, and life cycle management. The application of market research extends across the full cycle of product innovation, from initial idea generation to final product launch and post-launch reviews. This chapter covers a range of market research tools, including primary vs. secondary research, qualitative vs. quantitative, focus groups, customer site visits, ethnography, consumer panels, social media, big data, crowdsourcing, alpha and beta testing, biometric methods, multivariate techniques, and market testing. The benefits and limitations of each tool are discussed together with their potential for application at various stages of product innovation. Specific emphasis is placed on the accuracy and reliability of the various tools and their value in decision-making.

Chapter 7 Culture and teams

It is widely recognized that new product innovation cannot be successful through good processes alone. Success is dependent on people, the culture of the company, and the environment that is created to foster innovation. This chapter outlines the characteristics of an innovative culture. It also focuses on the requirements for a high-performing team and of team structures to support cross-functional teams in an innovative environment and in different project contexts, including virtual. Management roles and responsibilities at various levels and within different stages of product innovation are also discussed.

4 WHAT IS NEW IN BoK3?

New product innovation techniques and practices are continually being developed. PDMA is committed to keeping abreast of this through a periodic update of its BoK. This, the 3rd edition, succeeds the 2nd edition published in 2020. Following is a summary of changes made in this latest edition:

1. Although much of the concepts remain unchanged from previous editions, emphasis in BoK3 has been placed on the interrelationships between content across the various chapters, thus providing a greater appreciation of how individual components of the BoK work in concert to deliver improved product innovation performance. Examples and case studies have also been added to provide relevance and application of specific techniques and processes. This includes an appendix summarizing the practices of PDMA's Outstanding Corporate Innovators award winners (Appendices B and C).
2. *Chapter 1,* Product Innovation Management provides an overview of the Product Innovation Management role and a context for the remaining chapters of the BoK3. The basic content remains much the same, moved from Chapter 7 in BoK2, with greater details in some areas, including product life cycle management and benchmarking for continuous improvement. Feasibility analysis and financial valuation have been moved from this chapter to Chapter 3, Portfolio Management.
3. *Chapter 2* has added Porter's 5 forces to the Strategy Framework section, and the Digital Strategy and Open Innovation section have been expanded.
4. *Chapter 3*, Portfolio Management has been re-structured based on four goals of portfolio management: *value maximization, business strategy alignment, balance*, and *the right number of projects*. Although some may argue that *ideation* is not strictly part of portfolio management, we have included it at the beginning of Chapter 3 to demonstrate its relationship with and value to portfolio management. Chapter 3 also includes an extensive discussion of financial analysis as applied to selection of projects to be included in the product portfolio. Some new techniques have been added to this section – *Bang for Buck Index*, *Expected Commercial Value*, and *Options Pricing Theory*. Other new topics included in this chapter include the challenges of adapting traditional portfolio management to dynamic processes such as *agile* and hybrid *Agile-Stage-Gate*.

5. *Chapter 4*, Product Innovation Process, places greater emphasis on the *front end of innovation (FEI)* and the *product innovation charter (PIC)* in providing context at the start of the chapter. Discussion of most processes included in previous editions continues in BoK3 with the addition of benefits and limitations of each process. The sections on *design thinking*, *lean product innovation,* and *hybrid models* have been extended. A new section on *jobs to be done* has been added. The chapter concludes with a comparison of the various processes and a discussion of the question, "*is there a right process*?".
6. *Chapter 5*, Design and Development, although focusing on most of the tools discussed in previous editions, has been structured into two sections: *design* and *development*. This is intended to better describe the stage of application and the contribution of each tool to the design and development stages of product innovation.
7. *Chapter 6*, Market Research includes most of the tools as the previous edition, expanding sections including biometrics-based methods and multivariate techniques. Summaries of the strengths and weaknesses of most techniques have been added.
8. *Chapter 7*, Culture and Teams addresses similar topics as the previous edition with the addition of examples and expansion of the *virtual team* section.

5 WHO WILL BENEFIT FROM THIS BOOK?

PDMA's BoK provides a singular reference for anyone currently involved in or planning a career in product management and product innovation. It provides a proven framework for product innovation which can be applied to a wide cross-section of product and service industries at various levels of an organization, including and not limited to:

- Senior executives
- Product managers
- Brand managers
- Product owners
- Portfolio managers
- Program managers
- Project managers
- Business analysts
- Product designers
- Product developers
- Educators and trainers

6 HOW TO READ THIS BOOK?

The book is organized in a logical manner from the beginning to the end. You can read it straight through and be introduced to the topics (practices, concepts, tools, processes, and methodologies) that lead to successful product innovation. If you are new to product innovation or beginning a career that emphasizes innovation, you will gain a better understanding of how the topics relate to each other by reading the chapters in order. Perhaps unexpectedly, we suggest the same reading order for those moving into an innovation leadership role. Many practitioners have found the tremendous breadth of topics presented in BoK3 allows them to integrate their previous experiences while strengthening a strategic perspective they must have as leaders. The need for strategic thinking is both blatant and subtle throughout the chapters. For the strategic thinker, the chapters are organized like scaffolding, with later chapters building upon previous chapters.

However, suppose you already have product innovation knowledge and experience, and your immediate needs don't involve moving into a leadership role. In that case, you will find benefit in reading specific chapters to meet your needs. For example, if viable product ideas are failing to gain support and resourcing, the Strategy topics in Chapter 2 along with the project selection topics in Chapter 3 are essential to apply. If proper team formation or fostering an innovation culture are where current challenges are, skip to Chapter 7 to learn about both topics. So, if you have a clear immediate need and already possess a reasonable framework for product innovation, skip to the chapter(s) addressing your needs.

7 WHAT IS A PRODUCT?

A product is a term used to describe all goods, services, and knowledge comprising bundles of attributes (features, functions, benefits, and uses). A product can be tangible, as in the case of physical goods; intangible, as in the case of those associated with service benefits; or can be a combination of the two.

Products are classified into two categories: Consumer and Industrial. Each of these are divided into further categories as shown in Figure 3.

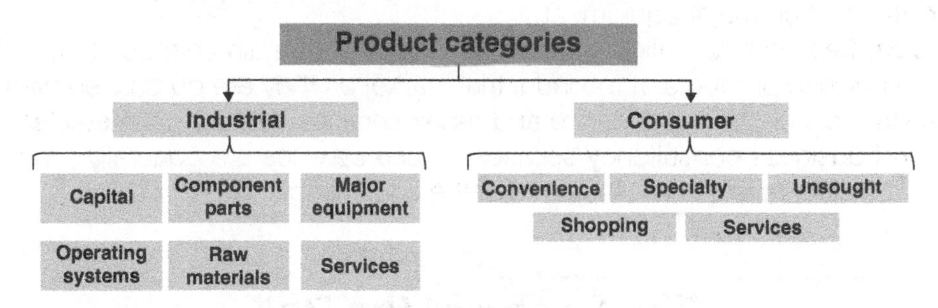

FIGURE 3 Types of products

Consumer Products

Consumer products can be further classified as convenience products, shopping products, specialty products, unsought products, and services:

- **Convenience products** are purchased by consumers repeatably and without much thought. Once consumers choose their convenience products, they typically keep to it unless they see a reason to switch — such as an interesting advertisement that compels them to try it or mere convenience at the checkout aisle.
- **Shopping products** include higher-end items like cars or houses or smaller items like clothing and electronics. Consumers typically spend more time conducting research, comparing prices, and chatting with salespeople when they're looking to purchase shopping goods. These are more one-off purchases and are typically more important and higher economic-impact products compared to toilet paper, soap, and other convenience products.
- **A specialty product** is the *only* one of its kind on the market, which means consumers typically don't feel the need to compare or deliberate as much as they would with other products. Examples include the iPhone or Mercedes car, where consumers have become dedicated to a specific brand.
- **Unsought products** are those that people aren't typically *excited* to buy and, don't buy on impulse. Good examples of unsought goods include fire extinguishers, batteries, and life insurance.
- **Services** normally should not be considered as a separate product classification. Depending on the particular service, they are either consumer or industrial goods. They are activities, benefits, or satisfactions offered for sale or are provided in connection with the sale of goods.

Industrial Products

Industrial products can be further classified into capital, raw materials, component parts, major equipment, accessory equipment, operating supplies, and services.

- **Capital products** are directly used in production. Capital goods consist of installations and accessory equipment. Buildings, plants, and machinery are examples of installations.
- **Raw materials** are used in the making of other products. This category includes natural resources such as forest products, minerals, water, oceanic products, agricultural products, and livestock. In most instances, raw materials lose their individual identities when used in the final product.

- **Component parts**, unlike raw materials, parts usually have been processed before being used in the finished product. Although they may not be visible, parts are left intact and assembled into the total product.
- **Major equipment** comprises industrial products used to make, process, or sell other goods. These include machinery, typewriters, computers, automobiles, tractors, engines, and so on. This equipment includes industrial products used to facilitate the production process or middleman sales. It does not become part of the finished product but aids in the overall production or selling effort.
- **Accessory equipment** includes tools, shelving, and many other products that tend to have a lower cost and shorter life than major equipment.
- **Operating supplies** include office stationery, repair, and maintenance items. Supplies can be treated as convenience products of the industrial market as they are purchased with minimal effort.
- **Industrial services** include maintenance and repair services, factory premise cleaning, office equipment repair, and business consultancy services. These services are generally provided through contracts by small producers and manufacturers of the original equipment.

The Scope of the BoK

The fundamental principles and framework used in this book are applicable to all the above categories and for profit and not-for-profit organizations, with the application of specific techniques and the emphasis with which they are applied.

The definition of a *new product* as applied in the BoK

The PDMA BoK regards a *"new product"* as being a product that is new in any aspect. This includes improvement to an existing product, a line extension, a new to the organization product or a new to the world product.

8 WHAT IS PRODUCT INNOVATION?

Innovation is turning a creative idea into value.
Product innovation is the creation and subsequent introduction of a good or service that is either new or an improved version of previous goods or services.

In BoK3, product innovation is used as an all-embracing term to include all aspects of bringing a product to market from strategy and initial idea through to commercialization, and includes the processes and tools required throughout. It encompasses product improvement, line extensions, cost reductions, and new-to-the-company products.

Although the term product innovation specifically refers to *product*, most of the principles discussed throughout the book are equally applicable to services and not only products (either where the service is a product in its own right or where the service is part of a product offering).

The scope of product innovation

Put simply, successful product innovation is about choosing the right products to develop (doing the right things) and using the right processes, practices, and tools to develop the products (doing things right). Figure 4 extends this to include the essential ingredients of people (culture, organization, and teams), and performance metrics used as a basis for continuous improvement.

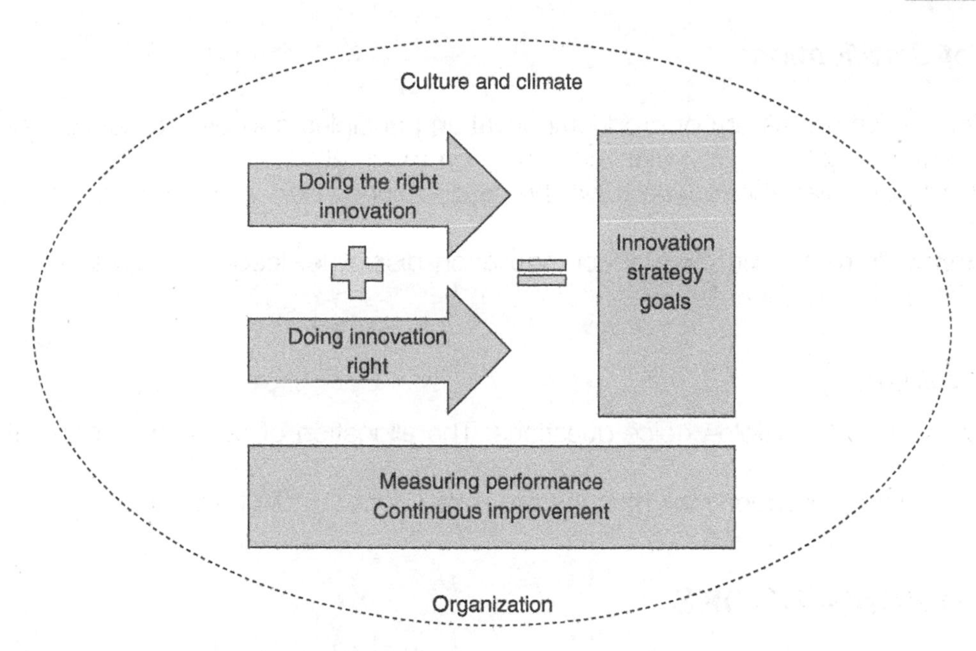

FIGURE 4 A macro view of product innovation

9 THE ROLE OF PRODUCT INNOVATION IN THE ORGANIZATION

Most organizations rely on their products or services for both sustenance and growth. The ongoing review and refreshing of an organization's product offering, through product improvement and new product innovation, is fundamental to its survival. New products and services are frequently referred to as the *lifeblood* of an organization, providing new revenue. Here are some insights from business leaders about the value of product innovation:

- "Investing in new product innovation and expanding the product catalog are the most difficult things to do in hard times, and also among the most important." *Bill Hewlett and David Packard.*
- "I would rather gamble on our vision than make a 'me too' product." *Steve Jobs.*
- "We see our customers as invited guests to a party, and we are the hosts. It's our job to make every important aspect of the customer experience a little better." *Jeff Bezos, founder of Amazon.*
- "I think that too often, companies tend to have engineers working in individual cubes. They are isolated. They often don't see themselves as part of a larger process with a complex web of interdependencies." *Jim Morgan, senior advisor, Lean Enterprise Institute.*

10 APPLICATION OF PDMA'S BODY OF KNOWLEDGE

The BoK has been applied in several ways, including self-learning for career advancement, personal and in-company training, university courses, and its major focus, preparation for PDMA's professional certification, the NPDP.

11 THE NEW PRODUCT INNOVATION PROFESSIONAL CERTIFICATION (NPDP)

The NPDP was first introduced in 2001, mainly in the USA. It is now a recognized qualification for product management and product developers worldwide.

Detailed information on the NPDP can be obtained from the Certification section of the PDMA website: www.pdma.org.

The Benefits of Certification:

For *individuals*: Confirms mastery of product innovation principles and best practices leading to professional advancement, new job opportunities, and greater remuneration.
For *management*: Identifies those who have the product innovation skills and knowledge to move into leadership roles.
For organizations: Promotes better product innovation discipline, leading to greater new product success with associated obvious benefits.

About the Examination

The exam consists of 200 multiple-choice questions. The allocation of questions across the seven chapters of the BoK.
To pass the examination, 150 questions must be answered correctly (75 percent).

12 RELATED PDMA BOOKS

BoK3 is a guide to the product innovation body of knowledge, drawing on research and practice from multiple sources, including other PDMA resources. Over a number of years PDMA has supported the publication of a range of books related to product innovation. For additional information, refer to the PDMA website at www.pdma.org.

PDMA Toolbook 1
The first of the Toolbook series, it provides practical cross-functional coverage of the entire product innovation process from idea generation through delivery of the final assembled product. Includes sections on benchmarking and changing your new product innovation process and managing your product portfolio (Belliveau et al., 2002).

PDMA Toolbook 2
This book covers all aspects of product innovation, from the creation of the concept through development and design to the final production, marketing, and service (Belliveau et al., 2004).

PDMA Toolbook 3
The third volume in the Toolbook series covers the best practices of product innovation, including critical aspects of product innovation from the creation of the concept through development and design, to the final production, marketing, and service (Griffin & Somermeyer, 2007).

PDMA History, Publications, and Developing a Future Research Agenda
This book describes the many publications that PDMA has created and provides influences that may impact the future of PDMA itself (Hustad, 2013).

PDMA Handbook of New Product Development 3rd Edition
The handbook provides a comprehensive picture of what managers need to know today for effective new product innovation (Kahn et al., 2013).

Open Innovation: New Product Innovation Essentials from the PDMA
Many organizations have adopted Open Innovation. This book provides a comprehensive guide to the theory and practice of Open Innovation (Griffin et al., 2014).

Design Thinking: New Product Innovation Essentials from the PDMA
This is a guide to better problem-solving and decision-making in product innovation and beyond. You'll learn how to approach new product innovation from a fresh perspective, with a focus on systematic, targeted thinking that results in a repeatable, human-centered problem-solving process (Luch et al., 2015).

Leveraging Constraints for Innovation
This PDMA Essentials Book, the third in this series, provides a framework of individual, organizational, market, and societal constraints that guides managers in identifying specific constraints related to their innovation activities, and provides them with corresponding tools and practices to overcome and leverage those constraints (Gurtner et al., 2018).

PDMA Handbook of New Product Development and Innovation 4th Edition
The 4th edition continues in providing an updated contribution to product innovation professionals for effective new product innovation, "revisiting familiar topics with fresh approaches and insights and will likely introduce you to entirely new concepts that take you to the leading edge of this exciting world of new product development and innovation" (Bsteiler & Noble, 2023).

13 THE 10 PRINCIPLES OF SUCCESSFUL PRODUCT INNOVATION

There is no singular recipe for product innovation success. The diversity of industry, organizational size and structure, products and services, markets, internal capabilities, geographic spread, and external environment are some considerations that preclude a "one size fits all" approach to applying product innovation processes and practices. However, there are some basic principles that provide a foundation for success, irrespective of organizational differences. Based on concepts in this book, the principles are:

1. **People first**, affirmed by the creation and maintenance of the right culture and climate to enable the successful application of process and practice, not vice versa.
2. **Empower all employees** to identify opportunities for creating new value instead of designating specific individuals or teams as the sole innovators.
3. **Clarity of direction**, provided through a well-developed organizational strategy and vision, communicated to, and understood by the whole organization.
4. **Effective product portfolio management**, ongoing, and founded on clear strategic goals and criteria throughout all stages of the Product Life Cycle. This includes rigor in selecting the right projects, managing resources responsibly, and investing in the future.
5. **A product innovation process or processes** tailored to the specific needs of teams and the whole organization, well communicated, understood, and adhered to.
6. **A strong emphasis on the Front End of Innovation** to ensure that continuation to the more expensive and risky stages of product innovation is well founded.
7. **A strong focus on the customer**, capturing the voice of the customer throughout product innovation with the application of appropriate market research techniques.
8. **Development and maintenance of high-performing teams**, based on cross-functional membership and the practices that create high performance (i.e., trust, accountability, positive conflict, result-focused, clear communications, and appropriate recognition/rewards).
9. **A strong focus on sustainability**, including the triple bottom line or the three Ps of People, Profit, and Planet.
10. **Continuous improvement** through applying performance metrics and benchmarking that underpin a culture of learning, both from successes and failures.

REFERENCES

Belliveau, P., Griffin, A., and Somermeyer, S. (ed.) (2002). *The PDMA toolbook 1 for new product development*. John Wiley & Sons.

Belliveau, P., Griffin, A., and Somermeyer, S. (ed.) (2004). *The PDMA toolbook 2 for new product development*, 2e. John Wiley & Sons.

Bsteiler, L. and Noble, C.H. (2023). *The PDMA handbook of innovation and new product development*, 4e. John Wiley and Sons.

Griffin, A., Noble, C.H., and Durmusoglu, S.S. (2014). *Open innovation: New product development essentials from the PDMA*. John Wiley & Sons.

Griffin, A. and Somermeyer, S. (ed.) (2007). *The PDMA toolbook 3 for new product development*, 3e. John Wiley & Sons.

Gurtner, S., Spanjol, J., and Griffin, A. (2018). *Leveraging constraints for innovation: New product development essentials from the PDMA*. John Wiley & Sons.

Hustad, T.P. (2013). *PDMA history, publications and developing a future research agenda: Appreciating our PDMA accomplishments—Celebrating people, lasting friendships, and our collective accomplishments*. Xlibris Corporation.

Kahn, K. B., Evans Kay, S., Slotegraaf, R. J., & Uban, S. (Eds.). (2013). *The PDMA handbook of new product development* 3 John Wiley & Sons. doi:https://doi.org/10.1002/9781118466421

Luchs, M.G., Swan, S., and Griffin, A. (2015). *Design thinking: New product development essentials from the PDMA*. John Wiley & Sons.

Markham, S.K. (2023). An innovation management framework: A model for managers who want to grow their business. In: *The PDMA handbook of innovation and new product development*, 4e (ed. L. Bsteiler and C.H. Noble), 45–58. John Wiley and Sons.

Project Management Institute (2021. APA). *A guide to the project management body of knowledge (PMBOK® Guide)*, 7e. Project Management Institute.

1

Product Innovation Management

Maximizing the return from product innovation through application of sound management practices throughout the product life cycle

Product Development and Management Body of Knowledge: A Guidebook for Product Innovation Training and Certification, Third Edition.
Allan Anderson, Chad McAllister, and Ernie Harris.

WHAT YOU WILL LEARN IN THIS CHAPTER

This book presents a comprehensive overview of product innovation from the definition of clear strategic goals, through the development of a balanced portfolio to the processes required for successful new product outcomes. Supporting this overall product innovation framework are the tools related to design and development, market research, business analysis, culture, teams, and leadership. This breadth of activities requires input from, and coordination of, wide-ranging disciplines, making management of product innovation both complex and challenging. In this chapter, we present the key factors for product innovation success and the attributes required in Product Innovation Management from strategic planning to initial product concept through to launch and onwards through the product's entire life cycle.

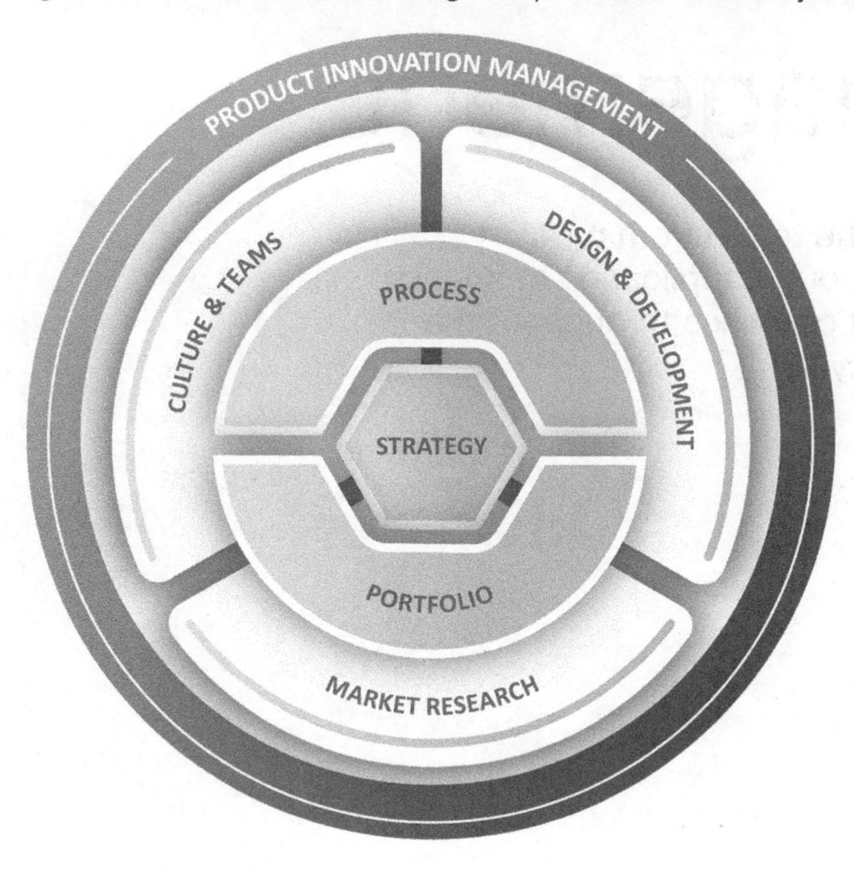

CHAPTER ROADMAP

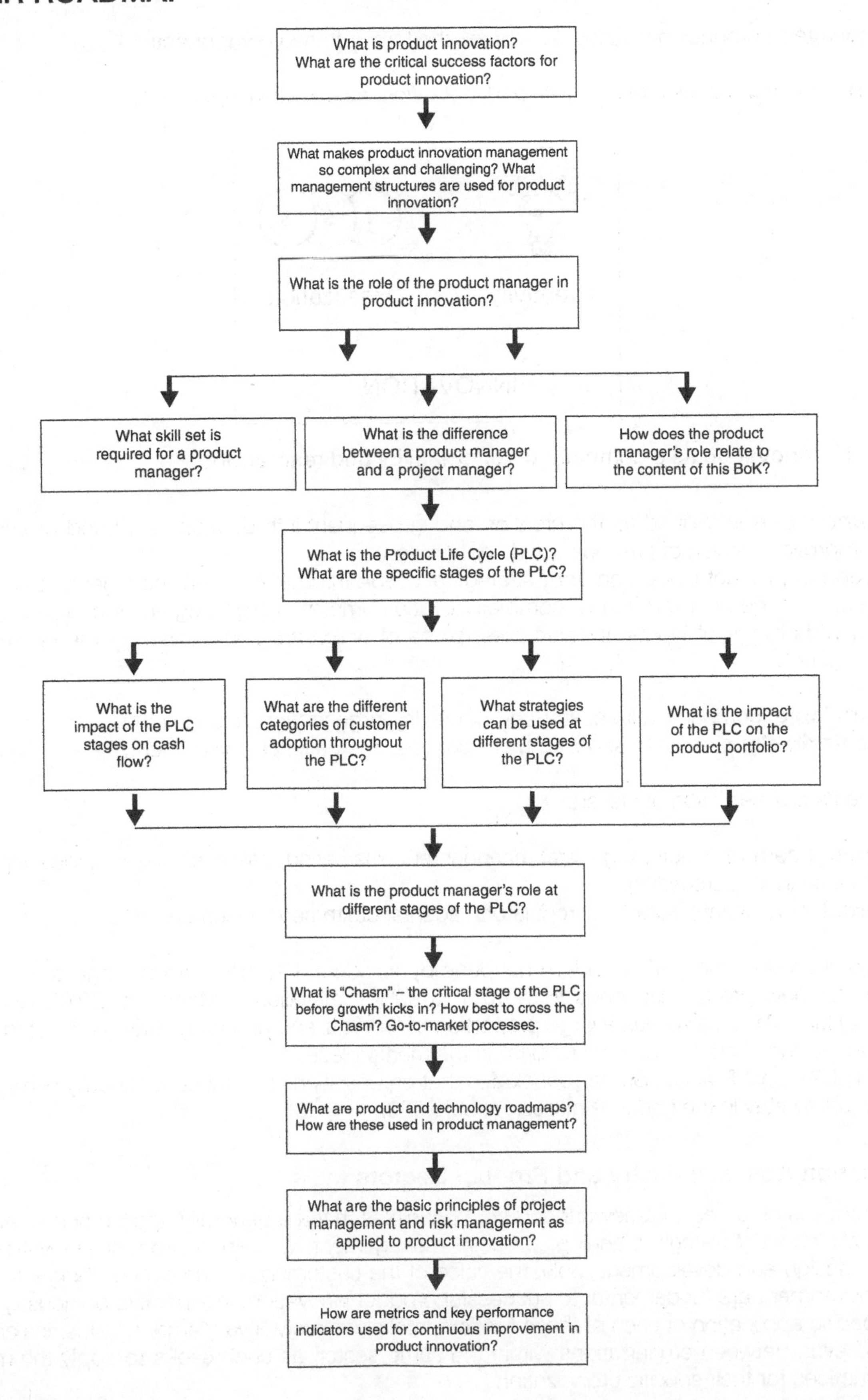

1.1 WHAT IS PRODUCT INNOVATION?

To best understand product innovation, let's break the term into two components.

Innovation is the combination of *creativity* and *realization*, as shown in Figure 1.1.

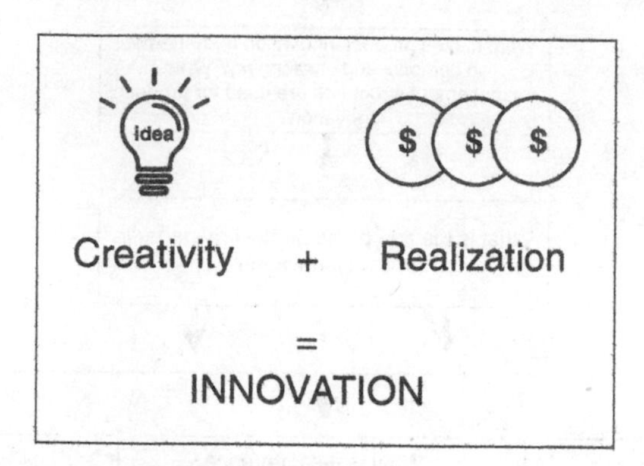

FIGURE 1.1 Innovation as a combination of creativity and realization

Product innovation is defined as the creation and subsequent introduction of a good or service that is new or an improved version of previous goods or services.

The scope of product innovation from strategy direction through portfolio management and from individual project management through to commercialization, implies a high degree of complexity. It is influenced by a wide range of inputs and variables, some of which are controllable by the organization and others that are not:

- **Controllable:** company culture, strategy, capability, organization, finances.
- **Uncontrollable:** competitors, government policies, international environment.

It involves a cross-section of players:

- **Internal:** board of directors, general management, marketing, sales, engineering/development, production, finance, purchasing.
- **External:** consultants, suppliers, regulators, agents, customer/consumers.

Successful product innovation seldom happens by accident. Reports from a range of sources quote failure rates for new product launches from 40% to 90% (Catstellion & Markham, 2013; Kocina, 2017; Schneider & Hall, 2011). Moreover, a large proportion of financial and human resources devoted to product innovation are focused on products that will fail in the marketplace.

PDMA's 2021 Best Practice Survey concludes that organizations ". . .must continually evolve their NPD capabilities just to stay in the game" (Knudsen et al., 2023).

1.1.1 Variation Across Industry and Product Sectors

This book focuses on a basic framework and core principles that are generally applied across all industries and product sectors. Strategic planning, portfolio management, a defined product innovation process, appropriate design and development tools, the voice of the customer, and an over-arching innovation culture and environment are fundamental to successful product innovation, irrespective of industry and product. The specific application of each of these fundamental elements will vary among industries and product sectors and even between organizations within the same sector, as each seeks to apply the most fit-for-purpose practices for their specific organization.

Introducing the Core Principles of Product Innovation Management

Although examples of the application of various practices in different sectors are provided, emphasis throughout is placed on providing an understanding of the core principles and encouraging the reader to explore how these principles might best be applied within their specific context.

1.1.2 Critical Factors for Product Innovation Success

Results from PDMA benchmarking studies (Markham & Lee, 2013; Knudsen et al., 2023) point to the following differentiating factors for organizations that are high performers in product innovation:

Innovation culture in the best companies:

- Failure is an understood aspect of innovation.
- Managers establish objectives.
- Innovation is considered in recruitment.
- External sources are used.
- Innovation and risk-taking are valued.
- Constructive conflict is encouraged.
- Internal communication is effective.

New product strategy in the best companies:

- Use a well-defined new product strategy to direct all new product innovation activities.
- Likely use a "first to market" strategy.
- Are more likely to focus on sustainability.
- Are more likely to apply intellectual property considerations.
- Apply a global business model (operations in multiple countries).

Portfolio management in the best companies:

- Use portfolio management tools to select projects and to ensure ongoing balance across projects:
  - Radical vs. incremental innovation.
  - Low vs. high risk.
  - New vs. existing markets.
- Specific tools used include:
  - Scoring models.
  - Strategic buckets.
  - Financial models.

New product process in the best companies:

- Use formal, cross-functional processes.
- Continuously re-design the process.
- Use specialized structures to drive new product innovation.
- Adopt flexibility according to product category.
- Have senior managers who understand and support the new product's process.

The front end of innovation in the best companies:

- Spend considerably more effort to understand customer needs.
- Use formal processes for idea assessment (recognizing the importance of making the right decisions at this early stage).

- Use open innovation to gather and develop new ideas.
- Use social media – discussion forums, blogs, innovation hubs, wikis, etc. – to gather customer information and opinion.

Development tools in the best companies:

- Use market research tools significantly more.
- Use engineering design tools more frequently.
- Place more emphasis on project planning tools.

Measures and metrics in the best companies:

- Use formal metrics for measuring and reporting on new product performance.
- Use metrics for both outputs and processes:
  - **Outputs:** such as profits from product innovation over the past five years.
  - **Process:** such as milestones on time or time to market.
- Use metrics as a basis for learning and continuous improvement.

Cooper (2023) provides further insight into *what separates winners from losers* in Chapter 1 of PDMA's 4th edition of its Handbook of Innovation and New Product Development (Bsteiler & Noble, 2023). He summarizes these drivers for success under three headings as shown in Tables 1.1–1.3:

1. Success drivers of individual new-product projects.
2. Drivers of success: organizational and strategic factors.
3. The right systems, process, and methodologies.

TABLE 1.1 Success drivers of individual new-product projects

Driver name	Success driver description
1. Product advantage	A unique superior product – a differentiated product that delivers unique benefits & a compelling value proposition to the customer or user; also the role of "smart products"
2. VoC	Building in the voice-of-the-customer – market-driven & customer-focused NPD
3. Up-front homework	Pre-development work – doing the homework & front-end loading the project; doing the due diligence before Development gets underway
4. Product definition	Sharp, early, & fact-based product definition to avoid scope creep & unstable specs; leads to higher success rates & faster to market
5. Iterative development	Building into the project a series of "build & test" iterations; putting something in front of the customer, early & often, to get the product right; early & frequent technical validations of versions of the product
6. Speed & order of entry	The impact of speed and time to market; timing & the order of market entry (first to market)
7. Agility	Responsive to making rapid changes in the portfolio of projects; able to react to changing customer needs; ability to pivot from the original plan of action
8. Proficient launch	A well-conceived, properly executed market launch, driven by a solid, properly resourced marketing plan
9. Global orientation	A global or "glocal" product (global platform, tailored for local markets) and targeted at international markets (as opposed to the product designed to meet home-country needs)
10. Expectations of success	Expect success, get success – success is a self-fulfilling prophecy: when project team members expect success, they are empowered & thus realize success

Source: Cooper, 2023 / John Wiley & Sons.

TABLE 1.2 Drivers of success – organizational and strategic factors

Driver name	Success driver description
1. A Product Innovation Strategy	Having a product innovation strategy for the business: clear NPD objectives; defined strategic arenas (markets, product types) for the business to focus its NPD efforts on; and the product roadmap to define placemarks for future products
2. Sustainability Focus	Social responsibility, sustainability, and a green orientation; launching green new products
3. Leveraging Core Competencies	Leverage, synergy, and familiarity: getting a good fit between the firm's core competencies and the resource needs of the project to gain competitive advantage
4. Targeting Attractive Markets	Targeting large, growing, high-need markets with good long-term potential; avoiding intensively competitive markets with low margins and heavy price competition
5. Resources	Adequately resourcing projects; having dedicated project teams
6. Effective Teams	Cross-functional teams, with the right skills; a strong capable team leader; having clear goals and accountability for the project
7. Integrating the Supply Chain	Building the supply chain into NP projects; good communication with suppliers; seeking ideas and solutions from suppliers
8. Climate and Culture for Innovation	The right environment – a climate that supports innovation activities, risk taking, and freedom to experiment; also: rewards teams and idea-generators; allows time for creative work; and no micro-managing of teams
9. Organizational Design	An organic structure – fluid and flexible in task execution, open channels of communication, decentralized decision-making, and few formal procedures
10. Top Management	Transformational leadership that is committed to NPD; develops the organization's vision, mission, NPD objectives, and innovation strategy; commits the necessary resources to NPD; is engaged in the NP process, makes Go/Kill decisions, and supports committed product champions

Source: Cooper, 2023 / John Wiley & Sons.

TABLE 1.3 The right systems, process, and methodologies

Driver name	Success driver description
1. A NP Idea-to-Launch Gated Process	An effective gating process: a multistage, disciplined idea-to-launch system, such as Stage-Gate, with defined Go/Kill decision points or gates and stages with success drivers and best practices built in
2. Portfolio Management	Focus (doing fewer development projects) and making sharp project selection decisions (picking better projects and getting the right mix and balance of projects in the portfolio); using effective portfolio and project selection methods
3. Accelerated Development	Including: Lean Development; "build-and-test" iterations; and Agile Development. Also success drivers from this chapter, such as an innovative climate; effective cross-functional teams; and resourcing projects adequately (prioritizing projects, focusing resources)
4. Integrating Agile Into the NP Process	Borrowing Agile Development from the software world and applying Agile to physical products (includes sprints, scrums, demo's, retrospectives, new roles, and the Agile values and mindset)
5. Effective Ideation	VoC methods; Open Innovation and co-development; lead user analysis; strategic approaches; and Design Thinking

Source: Cooper, 2023 / John Wiley & Sons.

1.1.3 A Synthesis of the Various Studies into Product Innovation Success Factors

There have been several other research studies into the factors contributing to successful product innovation, including Cooper and Kleinschmidt (2010, 2015), Lester (2016), and Montoya-Weiss and Calantone (1994).

Although there may appear to be differences among the conclusions from these various studies, there is significant commonality. Following is a synthesis of these common success factors, based on an article by Kahn et al. (2013). These success factors are grouped under three headings: project level, people and environment, and organization and strategy.

Critical success factors – project level:

- Unique, superior products.
- Strong market orientation.
- Pre-development homework.
- Sharp, early, and stable product definition.
- Planning and resourcing the launch.
- Quality of execution of key activities from idea to launch.
- Speed – but not at the expense of quality of execution.

Critical success factors – people and environment:

- The way project teams are organized.
- The right environment – climate and culture.
- Top management support.

Critical success factors – organization and strategy

- Using a product innovation and technology strategy.
- Leveraging core competencies.
- Targeting attractive markets.
- Portfolio management.
- The necessary resources.

1.2 MANAGING PRODUCT INNOVATION

Clearly, managing product innovation is diverse and challenging. Its complexity is summarized by Johnson (2017) in the Quartz Open Framework for managing product innovation, shown in Figure 1.2. The six key elements of this framework are connect, discover, commit, describe, create, and deliver. All elements contribute to learning and continuous improvement, which is essential for the ongoing delivery of successful products. This critical element of product innovation management is described further in Section 1.10.

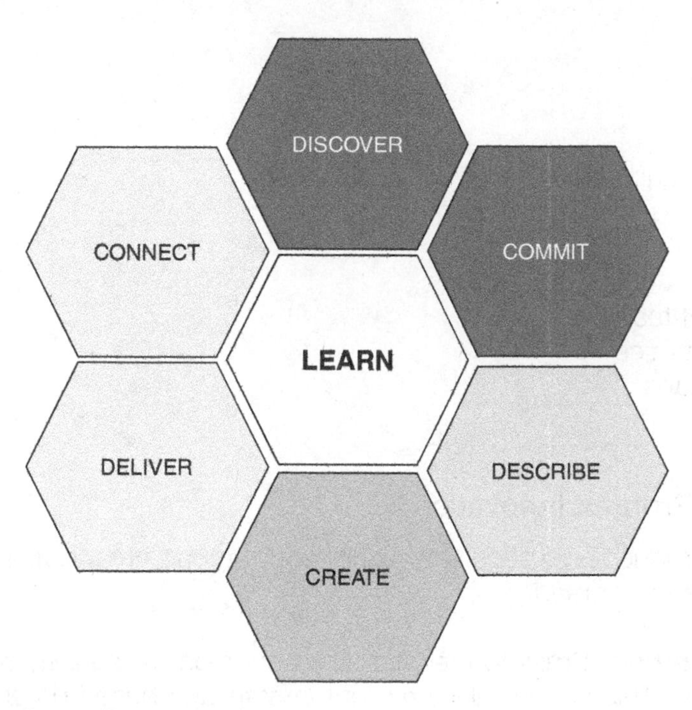

FIGURE 1.2 The Quartz Open Framework.
Source: Johnson (2017).

Connect:

- Monitor success metrics
- Monitor market feedback
- Monitor customer usage
- Perform process retrospectives

Discover:

- Review market feedback
- Conduct problem discovery
- Isolate problems and personas
- Create product vision canvas

Commit:

- Engage leadership team
- Create business deliverables
- Define success criteria
- Get go-ahead for next steps

Describe:

- Engage technical teams
- Roadmap initiatives or epics
- Describe personas and problems
- Define acceptance criteria

Create:

- Prioritize deliverables
- Formalize acceptance
- Communicate status
- Minimize interference and change

Deliver:

- Engage go-to-market teams
- Determine launch activities
- Develop readiness plans
- Prioritize deliverables

1.2.1 How to Manage Product Innovation

Structures for managing product innovation vary significantly among organizations, depending on the size, product range, and geographic spread.

Size of the organization: Small to medium-sized companies regularly rely on the chief executive to manage innovation. This is particularly evident in start-ups based on an innovation by the owner who has taken responsibility for managing the company with a small number of supporting staff. It's interesting to note that the role of product innovation management is often likened to the role of chief executive because of its wide-ranging responsibilities and integration across the organization. However, this quickly changes as organizations scale, with a founding CEO focusing more on optimizing operations and less on innovation.

Product range: Diversity in the product range may necessitate delegation of specific aspects of product innovation management. So, for example, where an organization has extended its current product range into areas of greater technical specialization, specialist knowledge may be required to facilitate communication on aspects of technical development during the product innovation process. Similarly, where an organization is catering to a range of markets, an allocation of product innovation roles may be necessary. For example, a pharmaceutical company marketing directly to the consumer and to medical practitioners will need different roles.

Geographic spread: Organizations with multiple markets will almost certainly need to develop products specifically or modify existing products for individual markets. Knowledge of an individual market and responsibility for product innovation in that market will provide significant benefits.

Regardless of an organization's internal and external environment, focus on the key elements shown in Figure 1.2 in the product innovation management structure remains essential for long-term success.

1.3 THE PRODUCT MANAGER

The responsibility of managing product innovation can lie in a range of job titles – from the CEO of small to medium-sized companies, VP of innovation in large companies, product innovation manager, brand manager, product manager, and more. Currently, the role of the product manager is growing in prominence around the world. Product manager roles are increasingly coveted positions, with high salaries and ample opportunities for growth. In fact, product management ranks tenth on Glassdoor's 2022 list of best jobs in America, with average salaries that rival many professional roles (Glassdoor, 2022).

The product manager is **responsible for both product planning and product marketing**. This includes managing the product throughout the product life cycle, gathering and prioritizing product and customer requirements, defining the product vision, and working closely with engineering to deliver winning products.

Martin Eriksson, curator of Mind the Product, describes product management as the intersection between business, user experience, and technology, as shown in Figure 1.3 (Eriksson, 2011). A good product manager is experienced in at least one of these, deeply interested in all three, and can work with and learn from those who are experts in all three. As Marty Cagan, founding partner of Silicon Valley Product Group and a 30-year veteran of product management, puts it, "The job of a product manager is to discover a product that is valuable, usable, feasible, and viable" (Cagan, 2017).

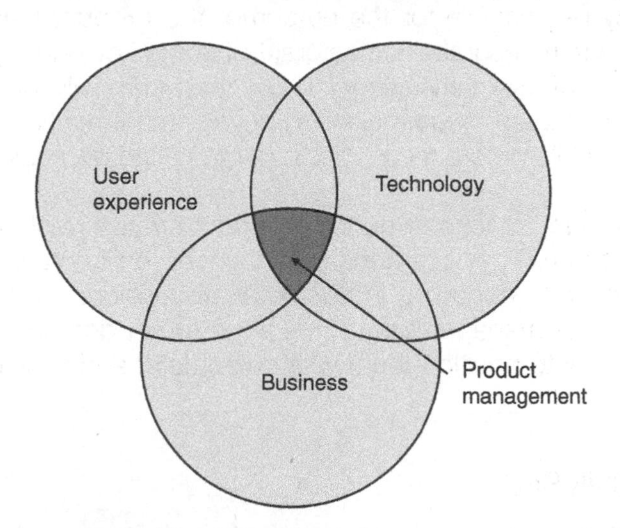

FIGURE 1.3 The product management intersection

Business: First and foremost, product managers are responsible for delivering value to their organization through revenue generated from products sold to customers. Meeting business objectives and goals is paramount, as success in those areas ensures the ability of a company to succeed and thrive, returning value to owners, shareholders, etc.

User experience: In current vernacular, "UX" is often associated with interface design, human factor design, etc., and while those are definitely a part of the "user experience," UX ultimately comes down to understanding the customer; the jobs they are trying to get done and problems they are trying to solve. Product managers don't have to be full-time designers, but they do need to make sure they are spending plenty of time with customers to discover these things.

Technology: Product managers need to have a level of understanding of not just "what" is being built but also "how" it's built. They don't need to be engineers or developers, and they don't need to know how to code or design a PCB, but they do need to have a level of understanding that allows a shared language between themselves and developers and engineers.

The product manager is the person who creates the internal and external product vision and leads product management from scratch. The product manager develops the positioning strategy while working with stakeholders and teams throughout the process. Some of the key responsibilities of the product manager are to:

- understand customer experience.
- develop vision.
- prioritize processes and activities.
- contribute to product pricing and positioning strategies.
- negotiate with stakeholders.
- build and follow a roadmap.
- arrange product testing groups.
- drive product launch.
- participate in the promotion plan development.
- build and maintain product awareness on all levels among product teams.

The product manager is responsible for bringing an idea to life and commercializing it for the organization. That includes providing methods to evaluate ideas and guiding the product innovation process. The product manager is ultimately responsible for the outcome of a product launch. It's hard to measure all these activities with independent, measurable indicators, but some key ones typically include monetization, user engagement, and the level of user satisfaction. These measures will vary depending on the company and industry. Some product managers focus mainly on development, writing specifications, and supervising development progression, while others focus more on marketing and sales by developing a marketing plan and training a sales team.

It may seem as if a product manager performs administrative tasks rather than actually "making" something, but that's not true. They are constantly working on improving the existing product, analyzing data, doing market research, engaging in customer discussions, and observing current trends of the industry. Eventually, a product manager has to make the tough decision of what products to build or improve, how to price them, how to position them with customers, and how to guide them to commercial success.

1.3.1 Product Manager Skill Set

We have already emphasized the scope and complexity of product innovation. Clearly, the product manager must have the skills and experience to cope with this breadth in scope and complexity. The product manager's skill set includes:

- understanding the product and related needs of the customers.
- market knowledge.
- innovation awareness.
- strategic thinking.
- technical knowledge.
- expert communication skills.
- relationship management.
- understanding user behavior.
- empathy.
- ability to explain business and technical requirements to all members of a team.
- ability to measure the success of a product.

1.3.2 Product Manager vs. Project Manager

The difference between a project management role and a product management role is largely prescribed by the definitions of product and project.

> **A product** is a good that satisfies the needs of a particular group, also known as a target market. A product can be anything, such as software, heavy machinery, pharmaceuticals, cosmetics, etc. It can even include services such as consulting or banking. Each product goes through a cycle from initial idea, validation, development, testing, market launch, and eventually, its life in the marketplace from introduction to growth, maturity, and decline.
> **A project,** on the other hand, is a collection of tasks to accomplish a specific goal. Projects should have outcomes and deliverables, which can be anything from a website revamp to a new internal process.
> **Product management** aims to oversee the products being developed within an organization, which includes managing every aspect of the product life cycle.
> **Project management** helps teams organize, track, and execute work within a project. This includes managing stakeholders, tasks, and progress; seeing the project through to completion; and implementing necessary tools for success.

The key differentiating factors between product and project management are summarized in Table 1.4.

TABLE 1.4 Product management vs. project management

	Product Manager	Project Manager
Key role	Oversees all product needs throughout its life cycle from birth to death, consistent with organization strategy.	Oversees all project needs through assigning so as to achieve project completion on time, in budget and in specification.
Major focus	Product strategy.	Task coordination.
Significant tasks	Coordinate all functions to focus on product needs throughout its life cycle. Measure performance and learn for continuous improvement.	Plan and regularly monitor tasks and resources to ensure successful completion of a project, on time, in budget, and in scope.
Measure of success	Meets projected return on investment. Maintaining a good level of new product contribution to overall organization profits, consistent with strategy. Continuous improvement in practices and processes.	Positive team environment and performance. High level of inter- and intra-team communication. A high standard of ongoing reporting of project progress against forecasts. Completion of project, on time, in budget, and in scope.

1.3.3 Relating the Product Management Role to PDMA's BoK

PDMA's Body of Knowledge is designed to provide a framework and specific tools to support those who are charged with managing product innovation, which is often the product manager. The following is a brief discussion of how each chapter of BoK3 is relevant to product managers:

> ***Chapter 2 – Strategy:*** In almost all organizations, new products and product updates provide the foundation for growth. The product portfolio and the contribution from product innovation are critical to this foundation and, as such, should form a key component of the corporate strategy. The product manager's direct or very close involvement with strategy development ensures strong alignment between product innovation initiatives and organizational goals.

Chapter 3 – Product portfolio management: A key function of the product manager is the oversight or direct management of the product portfolio. The product portfolio provides the revenue source that supports the organization now and in the future. Managing this portfolio requires close adherence to organizational goals and strategy, resulting in a balance of product innovation projects, ranging from new to the company, line extension, and product/brand refreshments.

Chapter 4 – Product innovation process: Several processes have been put forward to optimize the effectiveness and efficiency of product innovation. Each of these processes has its merits, but no single process is appropriate for all situations. Selection and implementation of the process or combination of processes is a critical function of product innovation management.

Chapter 5 – Design and development tools: A wide range of tools are available to assist product management in its decision-making. This chapter focuses on tools that improve the efficiency and effectiveness of the design and development aspects of product innovation. Selection, facilitation, and application of the best tools are essential to optimizing design and development outcomes.

Chapter 6 – Market research: Successful products are born out of a clear understanding of customer needs and the delivery of these products in a way that is readily accessible to the customer. A wide range of market research tools is available to provide greater insights into articulated and unarticulated customer needs. Selection and application of the right market research tools at different stages of the product innovation process are essential to maximizing product success.

Chapter 7 – Culture, teams, and leadership: Research over many years has underlined the importance of organizational environment and culture to successful product innovation. Product innovation management is complex and, in particular, requires input from a range of functions and disciplines. Formation, development, and leadership of multifunctional teams, together with clear and effective communication channels, are all essential components of product management.

1.4 THE PRODUCT LIFE CYCLE

What is the product life cycle, and how is product innovation managed through this life cycle?

1.4.1 Introduction to the Product Life Cycle

Most marketing texts define the product life cycle (PLC) as the sequence of stages from introduction to growth, maturity, and decline through which most products progress. In this chapter, we will extend this definition to include development, or product innovation, and retirement. Product innovation is not only important for the *birth* of the product but should continue throughout its life until it is removed from the market. This is important in the context of our Body of Knowledge as it emphasizes both the *birth of the product* and the ongoing contribution of innovation throughout the life of the product. The six stages of the PLC, shown in Figure 1.4, are briefly described as follows:

Development: generation of the product idea, further concept development, design and testing, and business analysis through to market launch.
Introduction: the first stage in the product life cycle is where a company tries to build awareness about the product or service in a market where there is often little or no competition.
Growth: having established a foothold in the market where there is greater customer awareness and limited competition, sales start to grow.
Maturity: competitors enter the market, and although total market sales may be expanding, the share of that market will be declining, resulting in a "flattening" of the life cycle curve.
Decline: increasing competition and possible lack of product novelty and differentiation will lead to a decline in sales.
Retirement: (often called *end of life*) occurs when a company decides to remove the product from the market. Products may be retired for various reasons, such as technology changes that make the product obsolete, competitive pressure that makes the product no longer viable, or the product simply can't meet the required revenue or profitability thresholds.

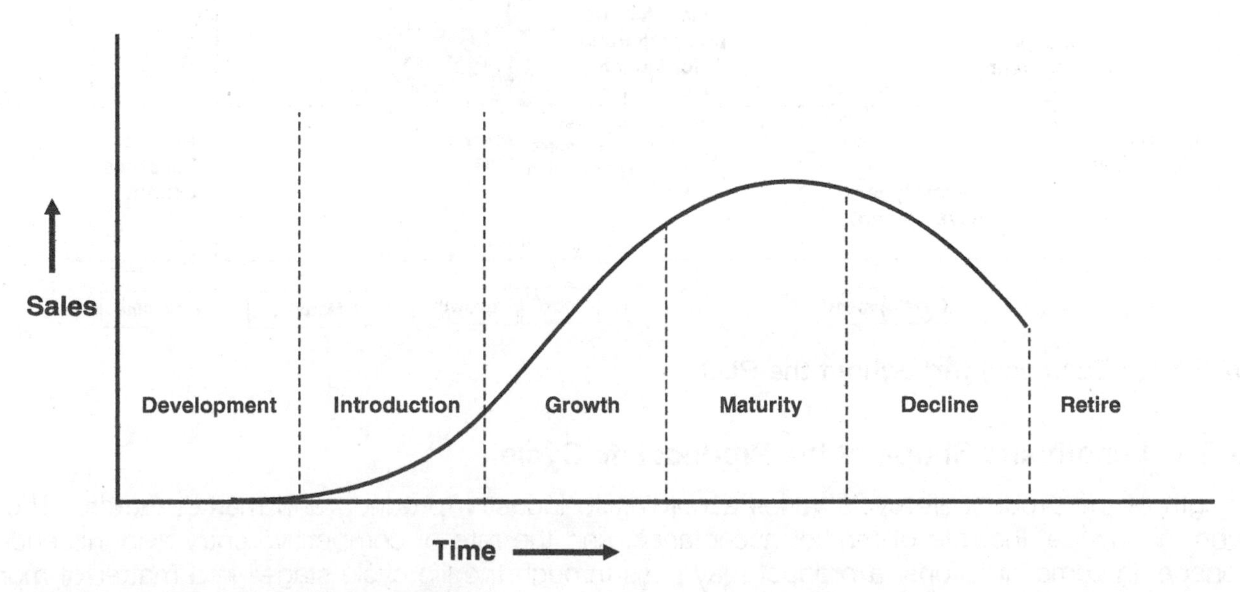

FIGURE 1.4 Product life cycle stages

1.4.2 The Impact of the PLC Stages on Cash Flow

Development: The organization identifies, plans, and resources effort to develop a product intended to be commercialized and generate revenue. Investments are made, and cash flow is negative.

Introduction: During the introduction stage, there is generally little-to-no competition for the product. However, organizations still often experience negative cash flow at this stage as returns from sales still don't exceed costs associated with establishing the product in the market.

Growth: During the growth phase, the product becomes more popular and recognizable. Although the organization may choose to continue investment in promotion and product differentiation, returns from sales will generally well exceed costs resulting in a significantly positive cash flow.

Maturity: As a product moves into maturity, cash flow becomes positive. Economies of scale result in lower unit cost and higher profitability. As the product reaches full maturity, competition increases and demand becomes more price elastic leading to falling price and reduction in profit margins.

Decline: Sales decline and the organization needs to make tough decisions on what to do with the product. Cash flow declines as any additional investments do not drive incremental sales. Investments stop and are focused on other products.

Retirement: Zero cash flow.

Figure 1.5 provides an overview of the expected cash flows throughout the product life cycle.

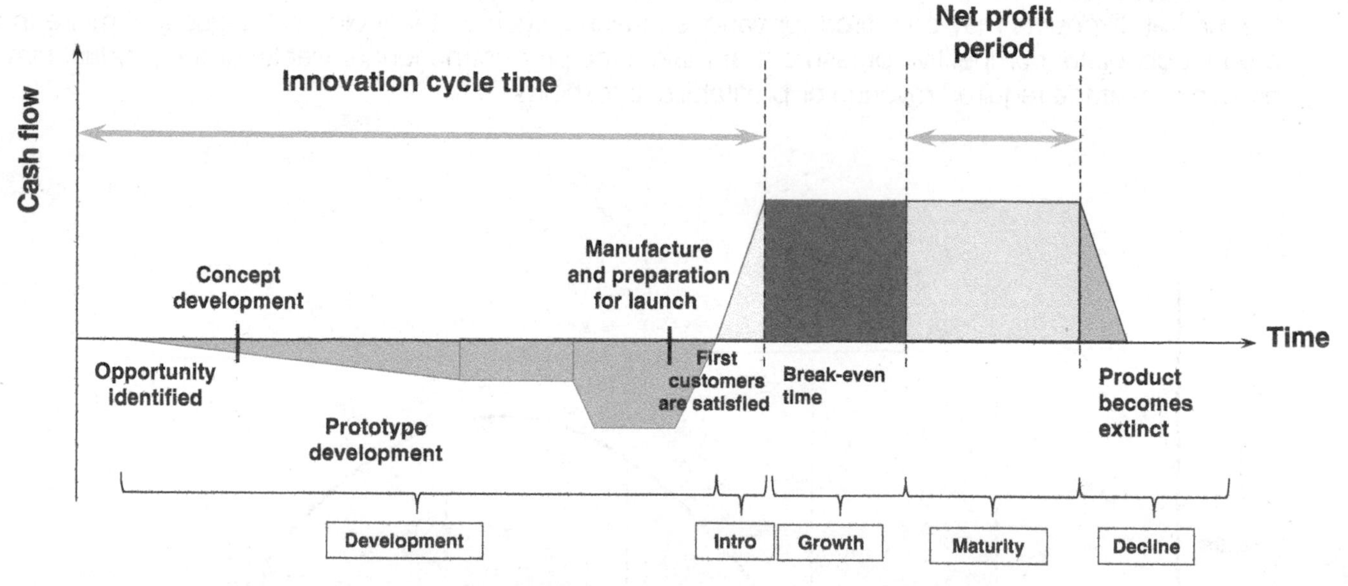

FIGURE 1.5 Cash flow throughout the PLC

1.4.3 The Length and Shape of the Product Life Cycle

The length of the product life cycle varies according to industry, product, and market factors. The rate of technical change, the rate of market acceptance, and the rate of competitive entry also influence the PLC shape. In some situations, a product may pass through the life cycle stages in a matter of months. However, in other cases, a product's life cycle may last for several years. But, in almost all cases, product life cycles are getting shorter, as shown in Figure 1.6. The factors influencing this include:

- Customers become more demanding;
- Competition is increasing;
- Technology is constantly changing and improving;
- Global communication is increasing.

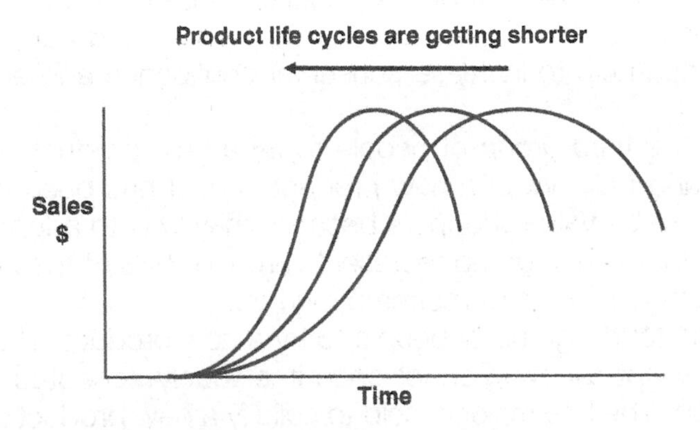

FIGURE 1.6 Shortening of the product life cycle

The shortening of the life cycle for most products has put significant pressure on:

- Constantly refreshing the organization's products – including both new products and modifications and improvements to existing products.
- Strategic management throughout the product's life cycle to extend its life through improvements (new features, improved functionality, reduced cost, enhanced branding, etc.) that will rejuvenate consumer interest and sales, as shown in Figure 1.7.

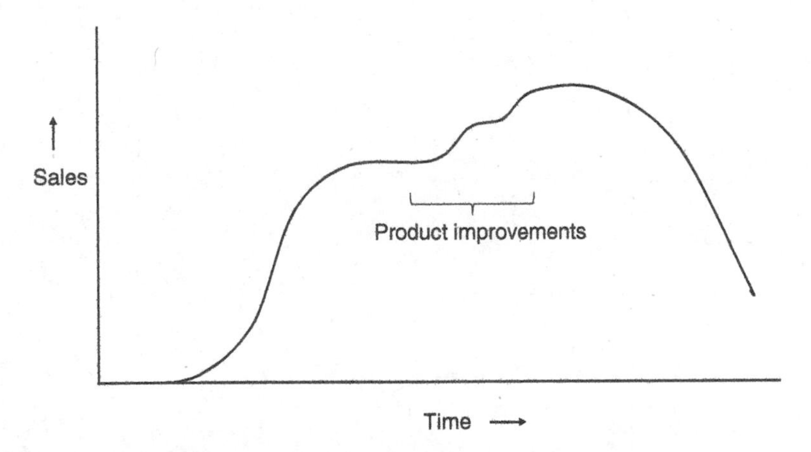

FIGURE 1.7 Product improvements to extend the PLC

1.4.4 Customer Adoption of New Products

Rogers (1962) identified that people adopt new products and ideas in five distinct categories. These categories are based on the degree of risk that adopters are willing to take: innovators, early adopters, early majority, late majority, and laggards. You'll see these categories overlaid on a similar model discussed later in Figure 1.10.

The ***innovators*** are the first group of people to try a new product. They are typically early adopters who are willing to take risks. They are often thought- and opinion-leaders in their field. This group is typically small, representing less than 5% of the population. Innovators are important to the product adoption process because they help to create awareness of the new product.

The ***early adopters*** are the second group of people to try a new product. They are often opinion leaders in their field and are typically more risk-averse than innovators, but they are still willing to take some risks. This group represents 15–20% of the population. Early adopters are especially important in the product life cycle because they help to validate a new product and adopt it before it becomes mainstream. Their endorsement can help to increase consumer confidence and encourage more people to try the product.

The ***early majority*** are the third group of people to try a new product. They are typically more risk-averse than early adopters and only adopt a new product once it has been validated by their peers. The early majority are considered to be value shoppers because they wait to adopt new products until they are sure they will get value from them. This group represents around 35% of the population. The early majority is important in gaining acceptance by the mainstream market.

The ***late majority*** are the fourth group of people to try a new product. They are more risk-averse than the early majority and only adopt a new product once it is widely accepted. This group also represents around 35% of the population. The late majority help to solidify a new product's position in the market.

The ***laggards*** are the last group of people to try a new product. They are typically the most risk-averse and only adopt a new product out of absolute need with no other choices. This group represents around 15% of the population. Although late to the market, laggards add "icing on the cake" in terms of product sales and profitability.

1.5 MANAGING THE PRODUCT LIFE CYCLE

Different strategies are required at each stage of the PLC. The following is a brief description of the situation at each stage of the PLC and the strategies that may be employed for elements of the marketing mix (product, price, distribution, and promotion). Refer to Table 1.5.

TABLE 1.5 Managing the product throughout the PLC

PLC stage	Situation	Strategy
Development	• New product progressing from concept to launch • Investment required: R&D, market testing, capital equipment, launch	• **Concept selection** according to portfolio criteria. • **Concept development and design** specifications consistent with production, marketing, and consumer constraints. • **Business analysis**, return on investment. • **Prototype development** and testing (functional and usability). • **Market testing.** • **Design of marketing mix** consistent with product concept: distribution, promotion, and price. • **Launch plan.**
Introduction	• Low sales • High cost per customer • Financial losses • Innovative customers • Few (if any) competitors	• **Product** branding and quality level are established, and intellectual property protection (such as patents and trademarks) is obtained. • **Pricing** may be low **penetration pricing** to build market share or high **skim pricing** to recover development costs. • **Distribution** is selective until consumers show acceptance of the product. • **Promotion** is aimed at early adopters. Communications seek to build product awareness and to educate potential consumers about the product.
Growth	• Increasing sales • Cost per customer falls • Profits rise • Increasing number of customers • Increase in competitors	• **Product** quality is maintained, and additional features and support services may be added. • **Pricing** is maintained as the organization achieves increasing demand with little competition. • **Distribution** channels are added as demand increases and more customers accept the product. • **Promotion** is aimed at a broader audience.
Maturity	• Peak sales • Cost per customer stabilizes • Profits high • Mass market • Stable number of competitors	• **Product** features may be enhanced to differentiate the product from that of competitors. • **Pricing** may be lower due to new competition. • **Distribution** becomes more intensive and incentives may be offered to broaden the opportunities for customers to purchase. • **Promotion** emphasizes product differentiation and new features as they are added.
Decline	• Cost per customer low • Falling sales • Falling profits • Contracting customer base • Decrease in competitors	• **Maintain the product**, possibly rejuvenating it by adding new features and finding new uses. • **Harvest the product** by reducing costs. Continue to offer the product but to a loyal niche segment. • **Discontinue the product**, liquidate remaining inventory, or sell it to another organization.
Retirement	• Technology changes making product obsolete • Competitive pressure makes product unprofitable	• **Completely remove** the product from the market without replacing. • **Replace the product** with a new version.

1.5.1 Impact of the PLC on the Product Portfolio

The management strategies at different stages of the PLC, outlined above, emphasize the importance of product improvement, addition of new features, line extensions, and cost reductions. All must be accounted for in the new product portfolio (further discussed in Chapter 2). The overall business strategy and the innovation strategy provide the direction and framework for product portfolio management. These strategies should establish the priorities across the various product innovation options, including:

- New-to-the-organization/world products
- Line extensions
- Cost reductions
- Product improvements

New-to-the-organization and **line extensions** have the potential to establish new products for the organization, while **cost reductions** and **improvements** are essential tools in PLC management to rejuvenate and prolong a product's life.

The question, "What defines a product?" in terms of the classic product life cycle description, is an interesting one. In the case of the iPhone, is the iPhone itself the product or are the various model iterations – SE, 13, 14, etc. – the products? The answer is that the iPhone itself has a life cycle, and the individual models also have their own life cycles that contribute to extending the overall life of the iPhone. The extension of a product's life through feature enhancement and new model release has become an important component of most organizations' product innovation portfolio.

The trend toward more frequent product modification and feature enhancements – particularly in the electronics, software, and Internet industries – has placed greater pressure on speed to market and reduced product innovation cycle times. In turn, this strategic focus on product innovation has significantly influenced the new product process with a strong drive toward Lean and Agile.

In addition to ensuring a balance across the types of product innovation, it is also important that portfolio management recognizes the need for a balance of products across the PLC. Obviously, having a high proportion of products in the introduction or launch phase places significant financial pressure on the organization. On the other hand, a high proportion of products in the decline phase does not present a very positive future for an organization. Figure 1.8 shows a distribution of products across the PLC.

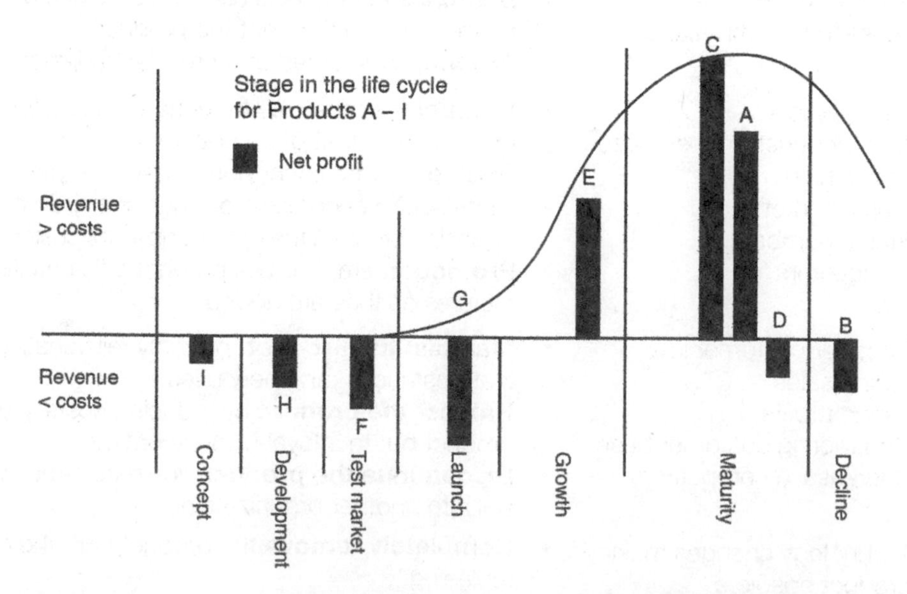

FIGURE 1.8 Balancing the product portfolio across the product life cycle

1.5.2 The Product Manager's Role in the Product Life Cycle

The PLC has a significant impact on marketing strategy, the marketing mix, and product innovation. Product managers play a critical role in guiding a product through its life cycle. This role takes on different responsibilities during the PLC with an overall eye on maximizing the success (revenue, market share, customer satisfaction – pick your KPI) of that product, as shown in Figure 1.9.

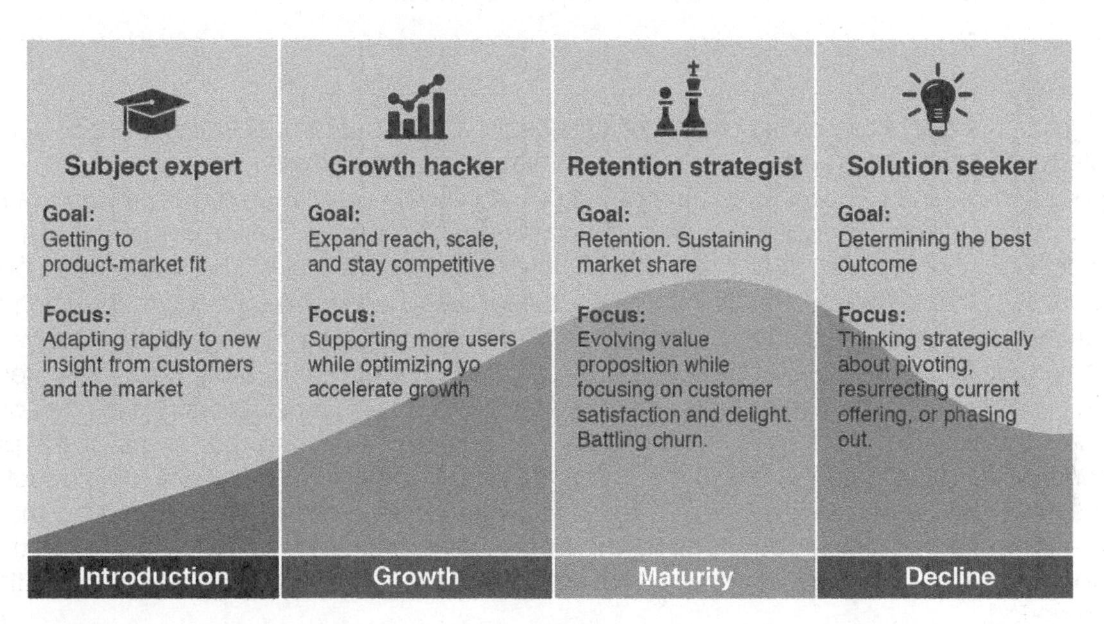

Subject expert
Goal:
Getting to product-market fit
Focus:
Adapting rapidly to new insight from customers and the market
Growth hacker
Goal:
Expand reach, scale, and stay competitive
Focus:
Supporting more users while optimizing yo accelerate growth
Retention strategist
Goal:
Retention. Sustaining market share
Focus:
Evolving value proposition while focusing on customer satisfaction and delight. Battling churn.
Solution seeker
Goal:
Determining the best outcome
Focus:
Thinking strategically about pivoting, resurrecting current offering, or phasing out.
Introduction
Growth
Maturity
Decline

FIGURE 1.9 Product manager roles in the product life cycle. ProductPlan®.
Source: Used with permission.

Product Introduction

Depending on the type of product (line extension, new-to-the-company, or new-to-the-world), a product manager's role during the introduction phase will vary slightly. But in general, their responsibility involves validating the fit for the market, ensuring that the product is priced correctly, and being the expert and champion for the product as it's trying to get a foothold. Validating the fit and being quick to adjust is a key part of what product managers often need to focus on for *new-to-the-company* (NTC) and *new-to-the-world* (NTW) products. There should have been market research and concept validation prior to launching the product, but nothing provides solid, useful feedback as the type that comes from paying customers. Product managers have to be sure they are paying attention to customer feedback during this phase. Just as importantly, they have to take that feedback and build useful information for the development or engineering teams to use. Refining the value proposition is critical at this point as well. Feedback from real users may undermine existing value proposition points, necessitating that they be changed or eliminated. New, previously unexpected values may also be uncovered that have to quickly be folded into the overall value proposition to ensure the product is still addressing its intended customers and solving the problems/doing the job(s) it was intended to do.

Pricing can also be challenging at this phase. For simple line extensions, it may be a relatively straightforward calculation. For products that are NTC or NTW, pricing may present challenges. Depending on the corporate/company focus, new products may be priced aggressively to capture market share early, or absent significant competition, may be priced to maximize operating KPIs like margin or revenue. In both cases, a misstep can have an impact on the potential success of the product. If pricing aggressively for market share, the dangers are that customers become conditioned to expect new features and capabilities in new products at the same (or lower) price as a previous generation product. This may undermine a value proposition or set the market at the low end of the price scale with no room to maneuver as the market matures. The pricing decision will have to be made so that it supports the value proposition and supports the business plan for the product. It is a dynamic decision to balance building momentum during a launch with the need to be sustainable over the long term.

In short, during the introduction phase, product managers have to be sure they are listening to the market and their customers, move the organization to adapt as needed, and provide the encouragement needed through proper pricing to accelerate the adoption of the product.

Growth

During the growth phase, the product manager can begin to focus on accelerating adoption, refining the pricing strategy, and identifying additional product innovation opportunities for the product.

As the product begins to develop some momentum, this is the time when a ramp-up in marketing can be beneficial. As the value proposition is clarified and feedback from the customer has been folded back into the product positioning, efforts to more broadly introduce the product to the market can be effective. During this phase, profit margins start to improve/increase due to economies of scale, reducing the piece-part price of the product (if applicable). During the growth phase, product managers have a few key things to focus on. One is to be aware of the growing set of competitors that will likely begin to emerge, which can have an impact on pricing strategy. Competitors may try to undercut established pricing in order to get a foothold. Another key issue for product managers to focus on is ensuring that some of the profit being captured finds its way back into the product innovation process to help set the stage for product enhancements and extensions. Lastly, this is the time to ensure that feedback from customers and the market is captured to fold back into the product positioning and value proposition, to make sure that the product continues to successfully navigate the chasm that can exist in the transition from introduction through the early growth period.

Maturity

During the maturity phase, most of the uncertainty around positioning and value proposition has been taken care of, which sounds good; however, it's also during this phase that sales growth slows considerably, with most customers who wanted to buy your product having already bought it. Many fewer new buyers are entering the market, so retaining existing customers becomes very important. For product managers, this is often the time to focus on the numbers. Carefully managing product costs, marketing and sales costs, etc., can often extend the life of a mature product. This is also the time that customer feedback from the growth phase can be leveraged into product line improvements/extensions to prolong the effective life of the product.

Decline

All good things come to an end. There is a time for all products when their contribution to the organization begins to decline. This is due to newer products offering more effective competition, a change in the market that has customers pivoting to a new/different solution, or even the decline of an entire market due to disruptive innovation that makes the entire category/market no longer relevant. As an example, Netflix and other video streaming services caused the decline of the DVD player.

Product managers have just as important a role during decline as in the other phases. The most important may be to clearly understand what is driving the decline. If it is due to competitive changes, then additional marketing or promotion may be necessary. If market/customer needs are changing, then extensions or updates to the product may extend its productive life. If there is a disruptive innovation causing the decline, product managers need to look at whether their company can become part of the disruption. Regardless of the diagnosis and the chosen path forward, a final critical responsibility of the product manager is to effectively manage the end-of-life and transition of the product to whatever comes next. An effective end-of-life transition is often overlooked but can make or break the effectiveness and success of future products and product enhancements and extensions.

1.6 THE CHASM IN THE PRODUCT LIFE CYCLE?

For product innovation, a critical part of the product life cycle occurs near the beginning, i.e., after introduction, but before growth has fully kicked in. This section is often called the "Chasm" (see Figure 1.10). This term was introduced by Geoffrey Moore (2014) in his book, *Crossing the Chasm*.

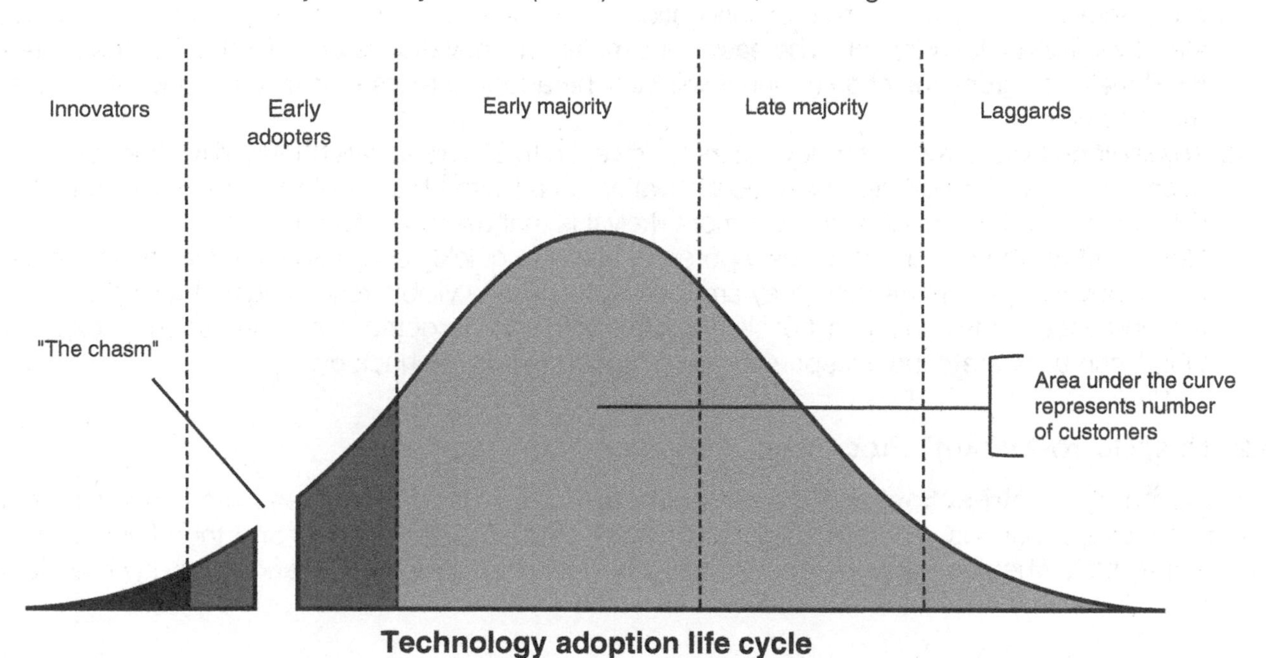

FIGURE 1.10 Crossing the chasm – Technology adoption cycle.
Source: Moore (2014) / Harper Business.

This gap (or chasm) is one that must be navigated to fully realize the growth potential of a product. This entire process by which we describe how and when consumers are willing to accept new products (or innovations) is referred to as Diffusion of Technology, presented previously in Section 1.4.4.

The curve above is a complement to the PLC curve. You can overlay this adoption curve onto the PLC curve as innovators and early adopters typically appear during the introduction and early growth phase of a product.

So, why is there a chasm? In its simplest form, the answer is that innovators and early adopters buy innovation as a change agent. They want to be ahead of the curve and get ahead of the competition, even if it has bugs or is missing some capabilities. The early majority of adopters are buying an innovation as an enhancement to their productivity – to improve what they do. Consequently, they expect a complete (whole) product they perceive to be of good value. But in new product innovation, it's rarely a seamless transition between early innovation and fast-growing successes.

1.6.1 Crossing the Chasm

Understanding why a chasm exists is important to begin to formulate ideas about how to cross that chasm effectively. Some products make the leap, while many don't. Some of the reasons and ways the chasm is successfully negotiated:

1. ***Advantage***: the product innovation has a relative advantage (that can be perceived) over the existing solution it is attempting to replace. This measurable advantage can take many forms – economic, social prestige, convenience, performance, etc. The greater this perceived advantage, the more likely that the product or innovation will make it past the chasm and continue to be adopted. There are no

hard-and-fast rules, though, for what constitutes a relative advantage. It depends on the user group being targeted and its needs and perceptions.

2. ***Consistency***: its perceived consistency with past or existing practices, values, past experiences, and needs of the user group. The degree to which the innovation demonstrates this consistency will influence how quickly it is adopted or if it is adopted at all.
3. ***Simplicity***: the degree to which an innovation is perceived as being either easy or difficult to understand will impact its adoption. The easier or simpler an innovation is perceived to be, the more likely it will make it across the chasm, and it will likely be adopted more quickly than innovations that are more complex.
4. ***Trialability***: how easy is it for new users to "give it a try?" This is called trialability. The easier it is for users to test and experiment with the innovation on a limited basis before committing to it, the less risky the innovation appears and the more likely it is that users will adopt it.
5. ***Referability***: the easier it is for new potential users to quickly and easily see the results of using a new innovation, the more likely they are to adopt it. These visible results again lower the perceived risk and uncertainty and make it easier for peers within the targeted user group to share experiences which can accelerate the acceptance and adoption of new technology.

1.6.2 The "Go-to-Market" Processes

A comparison of the **old-school** and the **new-school** Go-to-Market processes is shown in Figure 1.11. The old-school approach is very much a linear process of making the product and then figuring out how to sell the product, whereas the new-school approach to developing a go-to-market strategy is an iterative process.

FIGURE 1.11 Old-school and new-school go-to-market processes

An Example of the "New-School" Approach

The "Life Bike" was an innovative eBike, designed to be cool with a unique sitting position (refer to Figure 1.12).

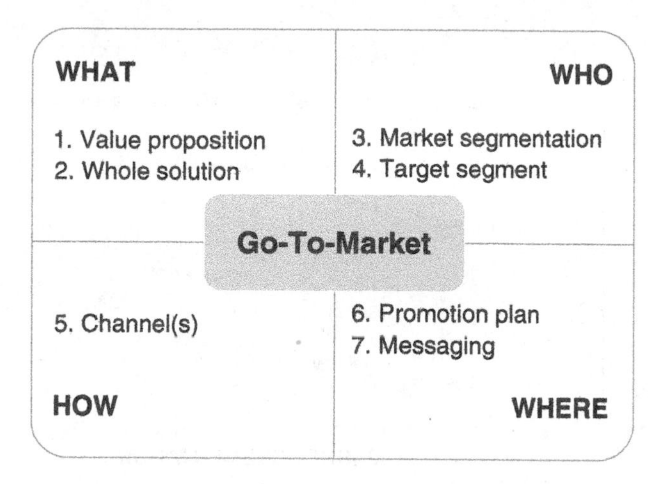

FIGURE 1.12 Go-to-market example

What are you selling?

The first step, shown in Figure 1.13, is to define exactly what it is you are selling. This should be based very much on the concept description and design specifications that evolved throughout the development process. What **unique** benefits does your product provide in comparison to competitors? Typically, competition is based on either cost or differentiation. Remember that you are usually selling in a competitive environment where what you are selling impacts your competitors and vice versa.

The value proposition

This description of what product you are selling can now be developed into a value proposition, defined as "a short, clear, and simple statement of how and on what dimensions a product concept will deliver value to prospective customers. The essence of 'value' is embedded in the trade-off between the benefits a customer receives from a new product and the price the customer pays for it" (Appendix A, PDMA Glossary).

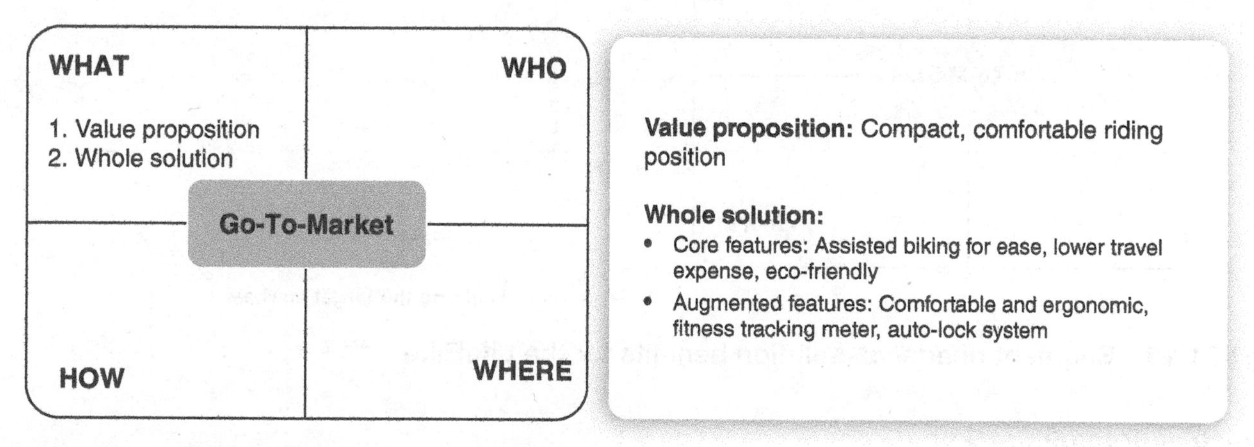

FIGURE 1.13 Defining the "what" in go-to-market

Who are you selling to?

The next key question is, "Who are you selling to?" Think about segments, not market size or share of the market. See Figure 1.14.

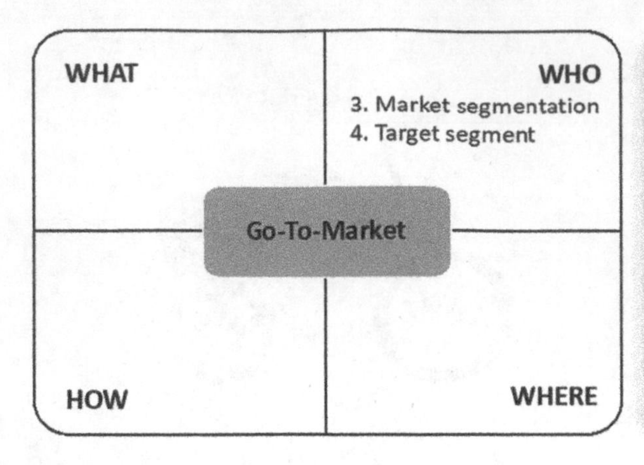

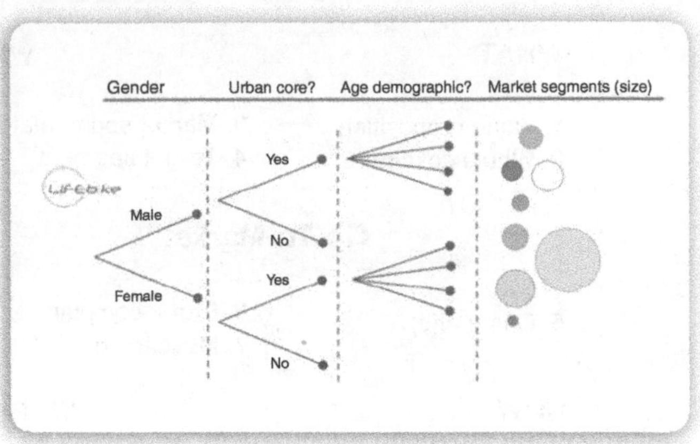

MARKET SEGMENTATION

FIGURE 1.14 Defining the market

Refining the target market

This is the process of comparing the needs of each segment with the primary benefits. Figure 1.15 shows two benefit dimensions for the LifeBike example – ease of storage (compactness) and riding position (comfort).

Various segments of the target market are represented by circles. The two segments whose needs match the LifeBike benefits are university students and female professionals who live and work in a city. The major competition is positioned in a different area of the market and appeals to different market segments.

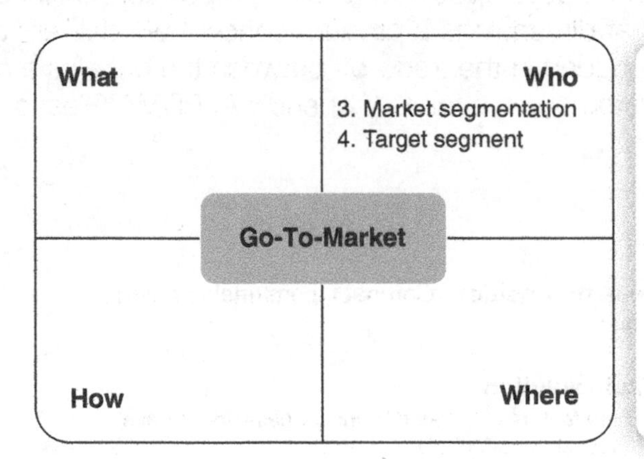

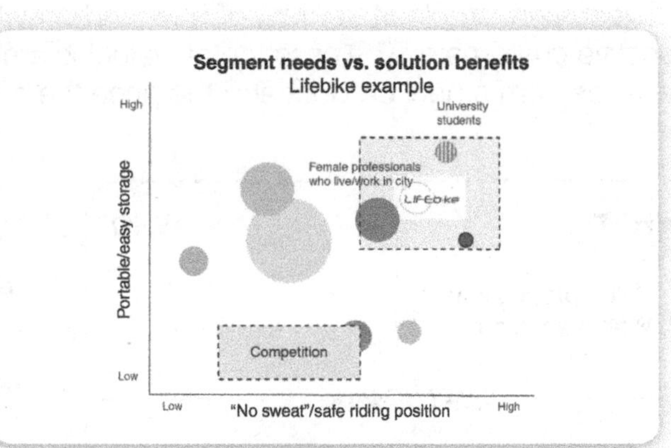

Refining the target market

FIGURE 1.15 Segment needs vs. solution benefits for the LifeBike

Reaching your target market

What channels of distribution will be used? The LifeBike channels are shown in Figure 1.16.

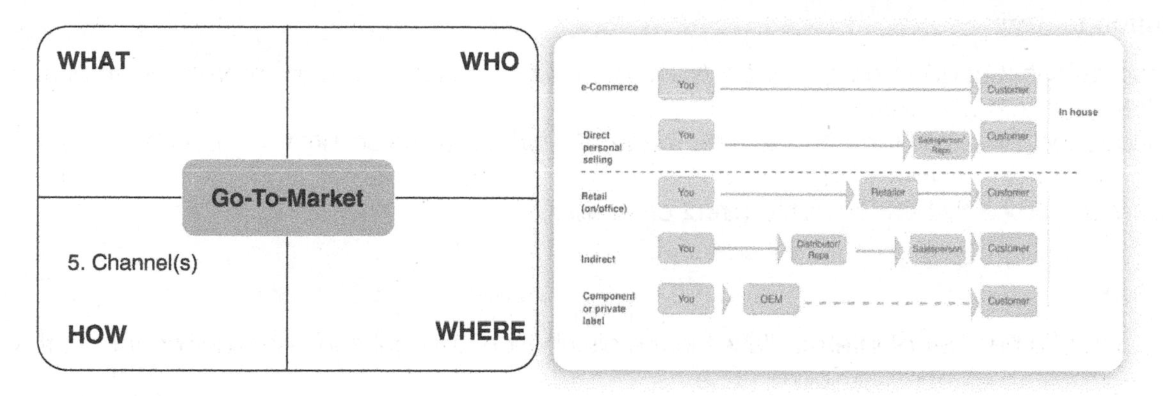

FIGURE 1.16 Getting the new product to the selected market segment

Channel(s) strategy

Many channel options are available for reaching the target market. A selection of these options is shown in Figure 1.17.

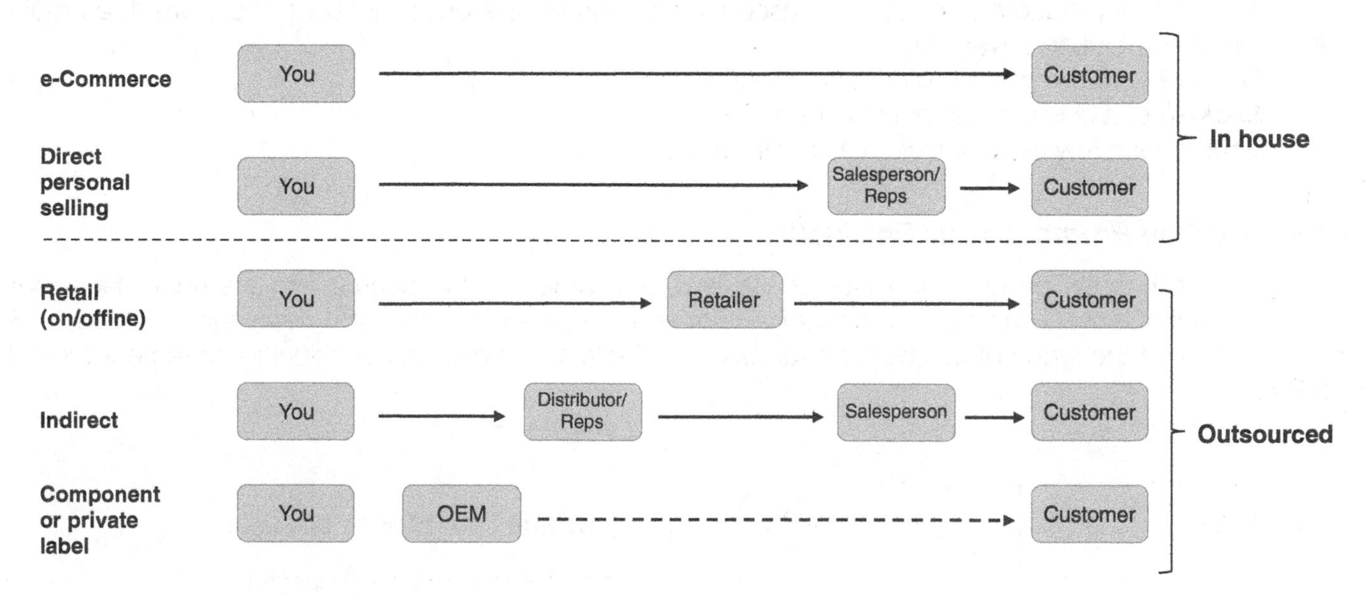

FIGURE 1.17 Channel options

Factors to Consider in Channel Selection

Product factors

- In general, highly sophisticated products are distributed directly to buyers, whereas unsophisticated or standardized (commoditized) products with low value are typically distributed through indirect channels.
- The product may be year-round or seasonal, fragile or sturdy, durable or perishable.
- The stage of the product life cycle is important. At the early stage, gaining market share is important, and it may be an advantage to have a number of distributors. In the maturity phase, distribution efficiency and cost savings are critical.

Organization factors

- Organizations that don't have – or can't afford – their own sales force are more likely to use agents or brokers to reach wholesalers or other buyers.
- Organizations that are targeting a wide range of markets are also likely to use channels external to the organization, whereas organizations that manufacture a range of products for a particular target market are probably best suited to direct channels.

Price factors

- Inclusion of a number of intermediaries in the distribution channel can add significant cost that results in a significantly higher final selling price.

Customer factors

- Ultimately, the key factor to consider is reaching your customer in the most effective way.
- You need to know your potential buyers, where they buy, when they buy, how they buy, and what they buy.
- In the case of products that are purchased by a customer at a physical outlet, there are three main levels of distribution coverage:
  - ***Mass or intensive***: a large number of outlets.
  - ***Exclusive:*** a single outlet or chain of outlets.
  - ***Selective***: a few retail outlets in a specific area.

Where Will You Promote Your Product?

Not only must the new product be made available to the target market segment in the right place, but customers must be made aware of product benefits and features using the right messaging. Figure 1.18 presents some of the types of promotion available, and Table 1.6 shows the positioning statement for the LifeBike.

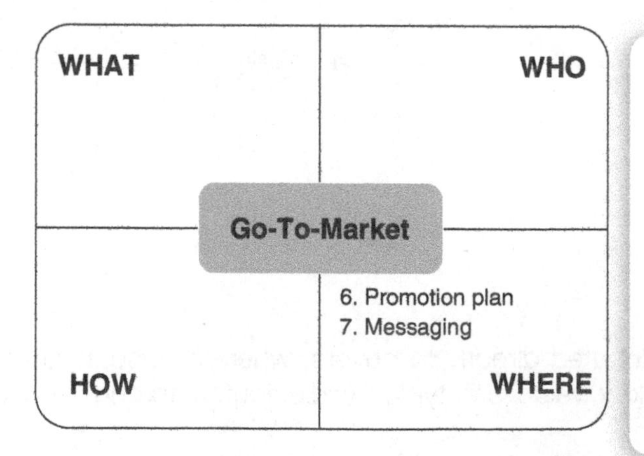

WHERE TO PROMOTE LIFEBIKE

- Align promotion plan with your target market.
- Focus on reaching decision makers in the target market.
- Where do the decision makers get their information?
- Types of promotion include internet/social media (Google AdWords, Facebook), PR (mass market publications vs. specialty publications), word-of-mouth (social media promos), trade shows, product demos, print media, and television.

FIGURE 1.18 Where to promote the product

TABLE 1.6 LifeBike positioning statement

LifeBike . . . [what]	A compact and "cool-looking" bike with a comfortable riding position.
. . . primarily for . . . [target market]	University students and female professionals who live and work in the city.
The compelling reason to buy [benefits] . . .	• Power-assisted biking. • Lower travel costs. • Eco-friendly • Comfortable and ergonomic • "Cool" looking.
. . . unlike [competitors] . . .	• Uncomfortable • Not easy to stow away. • Ugly

1.7 ROADMAPS USED IN PRODUCT INNOVATION

Roadmaps are very important tools to assist in communicating the vision of a product line with internal and external stakeholders. They also help to organize and plan a logical and compelling release of features and capabilities over time. There are a number of dos and don'ts when it comes to roadmaps. There are also different types of roadmaps that can be used alone or in combination with others to properly communicate the product line vision.

There are a number of roadmap types. A roadmap is just a visual way to see the rationale and logic of how/why we're moving from point A to point B. There can be HR roadmaps, sales roadmaps, project roadmaps, etc. When thinking about products or services intended for an end customer, several types of roadmaps can be very helpful.

1.7.1 Product Roadmaps

For product-driven companies (Hardware or Software, B2B or B2C), a product roadmap is essential to organizational alignment. A product roadmap illustrates high-level product strategy and demonstrates how a product will evolve over time. It includes upcoming features and the nitty-gritty of product innovation, like technical considerations and resourcing. A product roadmap is a powerful communication tool that product managers use to align different departments on one vision. It empowers the sales team to lead informed product conversations with prospects and the marketing team to plan campaigns that align with future product releases and product line extensions. Refer to Figure 1.19.

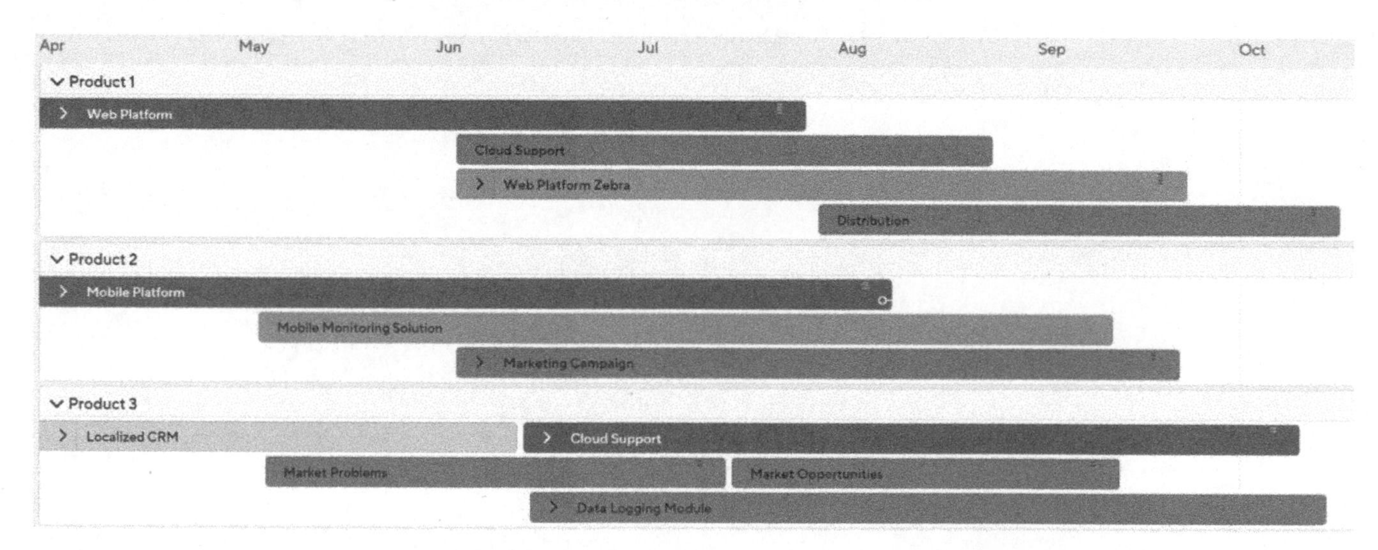

FIGURE 1.19 Multiple product roadmap.
Source: Created using ProductPlan.

Release Plan

A release plan is a tactical document designed to capture and track the features planned for an upcoming release. A release plan usually spans only a few months and is typically an internal working document for product and development teams. Differences between a roadmap and a release plan are shown in Table 1.7.

TABLE 1.7 Product roadmap vs. release plan

Roadmap	Release plan
High-level overview	Detailed view of the specific steps
Covers a longer time frame – potentially multiple releases to customers	Covers a shorter time frame – the immediate release to customers
Mainly used to communicate with internal stakeholders	Often used to create alignment between product and engineering teams and set expectations with customers
Focuses on why you're doing something based on user insights and strategy	Focuses on what you're doing and when you're committing to having it completed

1.7.2 Kanban boards

Kanban (English: signboard) started as a visual scheduling system and is best known as part of the Toyota production system. The Kanban board has become one of the most useful Agile project management tools, used widely by Agile teams and often referred to as Agile task boards. Kanban boards use Cards, Columns, Swimlanes, and WIP Limits to enable teams to visualize and manage their workflows effectively.

- ***Kanban cards*:** This is the visual representation of tasks. Each card contains information about the task and its status, such as deadline, assignee, description, etc.
- ***Kanban columns*:** Each column on the board represents a different stage of your workflow. The cards go through the workflow until their full completion.
- ***Work-in-progress limits*:** They restrict the maximum number of tasks in the different stages of the workflow. Limiting WIP allows you to finish work items faster by helping your team focus only on current tasks.
- ***Kanban swimlanes*:** These are horizontal lanes used to separate different activities, teams, classes of service, and more.
- ***Commitment point*:** A commitment marks a point in the work process where a work item is ready to be pulled into the system.
- ***Delivery point*:** The point in the workflow where work items are considered finished.

A frequently used roadmap in Kanban format consists of three columns for backlog work (available work to be done), work that is in progress, and work that is done. An example is shown in Figure 1.20.

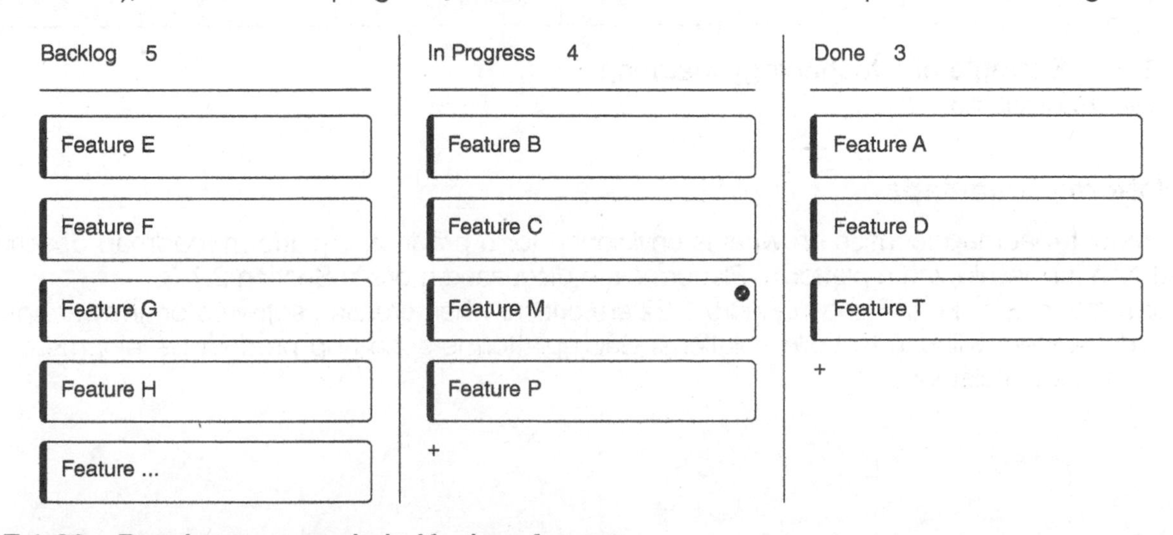

FIGURE 1.20 Roadmap example in Kanban format.
Source: Created using Miro.

1.7.3 Technology Roadmaps

Technology roadmaps are an important complement to the product roadmap in aligning technology planning and development to overall planning for the launch of a single product or a range of products. Technology roadmapping is particularly important in organizations that have a strong strategic focus on technology in underpinning the innovation strategy and new product innovation.

Figure 1.21 shows an example of a technology roadmap that shows work in four areas (operations, infrastructure, compliance, and security) for a new product.

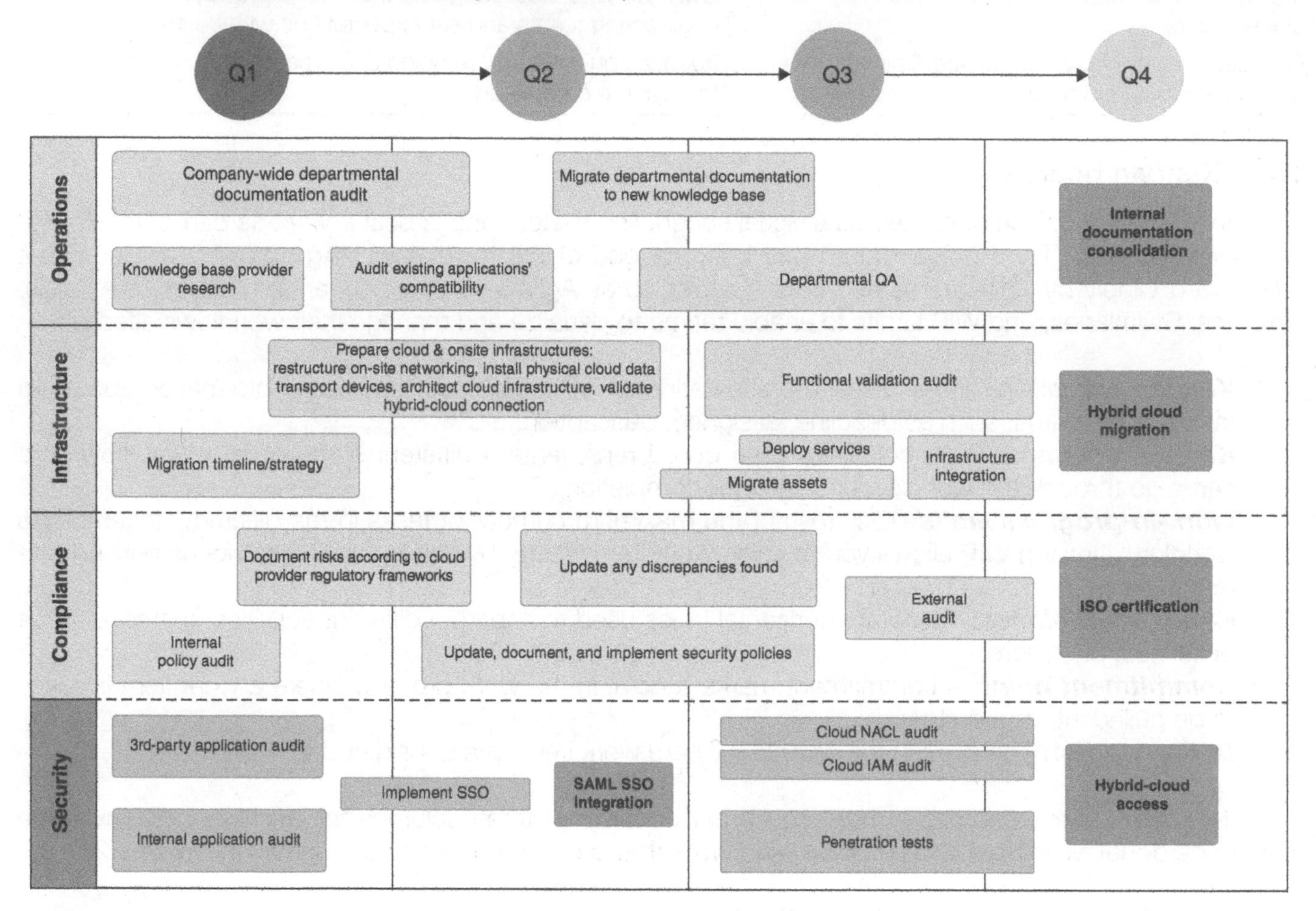

FIGURE 1.21 Example of a technology roadmap.
Source: Created by LucidChart.

1.7.4 Platform Roadmaps

While a product roadmap focuses on what is envisioned for a product, a platform roadmap describes the high-level plan for developing a platform. Platforms are discussed more in Section 2.7.3.

Platform roadmaps like the one in Figure 1.22 are critical when creating software or firmware platforms that other developers will use to build solutions. Such platforms speed up product development, reduce risk, and increase consistency.

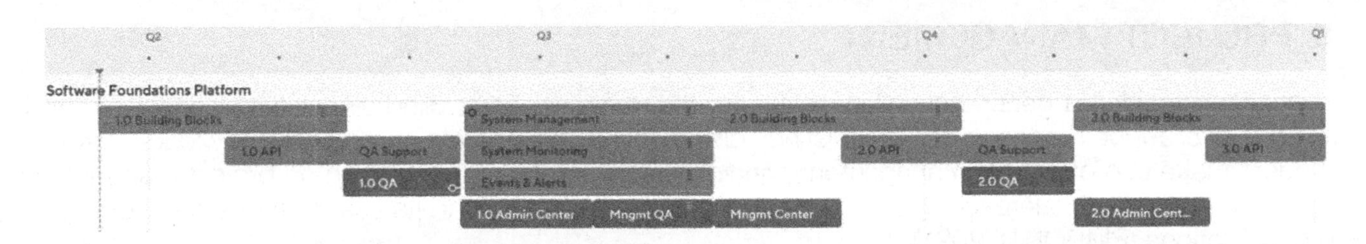

FIGURE 1.22 Example of a platform roadmap.
Source: Created with ProductPlan.

1.7.5 Best Practices for Using Roadmaps

- Roadmaps set expectations with stakeholders. Regardless of how many times *tentative*, *estimate, etc.* are used on a roadmap, stakeholders may interpret the information as a commitment. Consequently, be careful about how timelines and specific features are shown on roadmaps.
- Information on a roadmap must be updated when new information is learned. For example, if new market research shows that a product feature should no longer be implemented, remove it from the roadmap and communicate the reason with stakeholders.
- Use a single repository for roadmaps so that when an update is made, it is automatically pushed to all stakeholders who have access to the roadmap. The common but poor practice of using Excel or PowerPoint to create a roadmap and sending the file to stakeholders makes version control very difficult. You'll never know when a stakeholder is making decisions based on an out-of-date roadmap file.
- Share roadmaps carefully. Roadmaps, by their very nature, contain a lot of information about your company's strategies, plans, etc. Their power can also be their weakness in that they can cause problems if not carefully controlled. We're not necessarily talking about corporate espionage, but even within your own company, an improperly controlled roadmap presentation can cause confusion, concern, and could set back your efforts.
- Be sure to mark your roadmaps as "Confidential – Do Not Distribute" whenever you do share them.
- Consider having versions of your roadmaps with different levels of granularity. This will help you ensure they are consistent, whether used within your company or outside. It also lets you share a high-level version with non-employees and more detailed versions with those directly impacted by the roadmap.
- Be sure your roadmaps are developed collaboratively and get feedback early and often. For a roadmap to be effective, it has to be fully supported by those it affects. Making sure the team fully buys into the roadmap will help ensure its effectiveness.

1.8 PROJECT MANAGEMENT

This section provides a general overview of project management as applied to product innovation. While *project* and *product* management are distinct disciplines with separate bodies of knowledge, product managers make use of project management principles. More in-depth information on project management can be gained through reference to the Project Management Institute and its associated reference material and qualifications (PMBOK®, 2021).

1.8.1 Project Management in the Context of Product Innovation

"The definition of a project includes a specific start and stop date for the work as well as achieving goals with a temporary team" (PMBOK®, 2021). Although the overall process of developing and commercializing a new product may be seen as a single project, it can also be viewed as a composite of a number of small projects, for example:

- Generating a list of potential new opportunities.
- Analyzing the commercial potential.

The five steps in a project are defined in the guide as (PMBOK®, 2021):

1. Initiating
2. Planning
3. Executing
4. Monitoring and controlling
5. Closing

These steps are somewhat different than those defined for a structured product innovation process (such as Stage-Gate®), primarily because the product innovation process defines the overall roadmap from ideation to commercialization, while project management defines the structure and detail required to achieve the individual goals within the product innovation process and, in turn, the overall goal of successful commercialization.

1.8.2 The Triple Constraint

One of the most common challenges with any project is managing the Triple Constraint – scope, schedule, and budget – as shown in Figure 1.23.

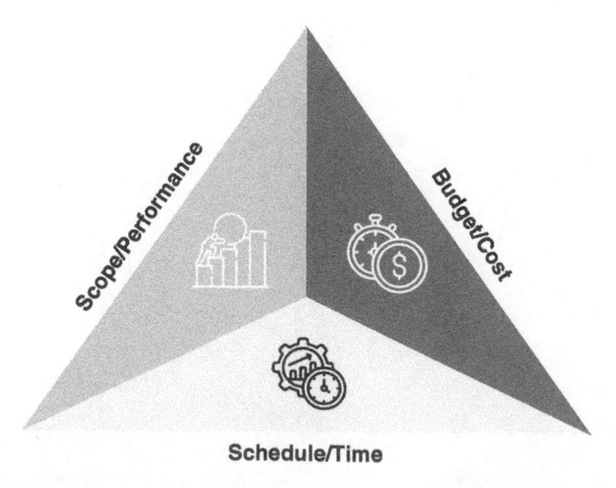

FIGURE 1.23 The Triple Constraint.
Source: Dobson (2004) / Berrett–Koehler.

The Triple Constraint is normally displayed as an equilateral triangle. If one of the core elements of a project changes, the project becomes unbalanced, e.g., if the scope of a project is increased, the schedule and the budget will need to be adjusted. The Triple Constraint emphasizes the very important point that scope, schedule, and budget of a project are highly intertwined.

1.8.3 Project and Product Scope

The Project Management Institute provides two uses for the term scope (PMBOK®, 2021):

1. ***Project scope:*** "The work that needs to be accomplished to deliver a product, service, or result with the specified features and functions."
2. ***Product scope:*** "The features and functions that characterize a product, service, or result."

In product innovation, the scope (both in terms of project and product) is provided in the product innovation charter, a tool presented in the chapter on process.

1.8.4 The Schedule

The schedule includes the activities and key milestones to successfully achieve the project goal. Project schedules are often constructed and illustrated using a bar chart or a Gantt chart, as shown in Figure 1.24. In any project, certain activities cannot begin before other activities have been completed. These activities are defined as critical path items whose delay will directly impact the budgeted completion of the project.

1.8.5 Critical Path Method (CPM)

In a project plan, the critical path is the longest path from start to finish or the path without any slack. Thus, the path corresponding to the shortest time in which the project can be completed. Refer to Figure 1.24.

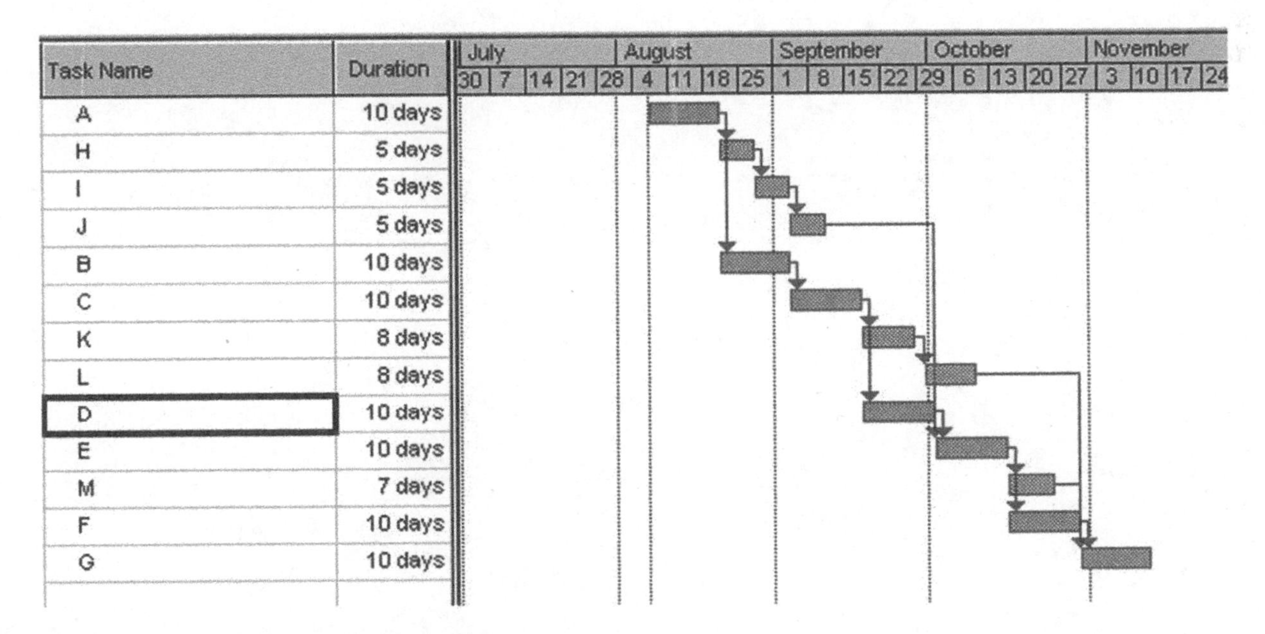

FIGURE 1.24 The Gantt chart and critical path

1.8.6 Schedule Compression

In many cases, a project's end date is fixed, for example, the new product launch date. If a project falls behind, it will be necessary to seek ways to compress the schedule without a significant impact on the scope. Basically, there are two ways to achieve schedule compression:

1. Add resources (and cost) or,
2. Do tasks in parallel that otherwise would have been done in series.

1.8.7 Budget

The project budget is defined as the anticipated cost required to complete the scope of the work on time. There are several ways to prepare a budget:

- ***Bottom-up*:** Identify all individual cost elements and sum these across the project.
- ***Parametric*:** Infer the cost of the project from similar projects or other available data, applying an average, such as worst, best, and expected cost.
- ***Historical data*:** Use specific cost data from past projects, such as prototyping or market research, as a basis for budget estimation.
- ***Company-specific methods*:** larger companies may have specific models and methods that are applied to project budgeting.

1.9 RISK MANAGEMENT

1.9.1 What is Risk Management?

Project risk is "an uncertain event or condition that, if it occurs, has a positive or negative effect on one or more project objectives such as scope, schedule, cost, or quality" (PMBOK®, 2021).

Risk management is the identification, assessment, and prioritization of risks followed by coordinated and economical application of resources to minimize, monitor, and control the probability and/or impact of unfortunate events or to maximize the realization of opportunities.

Project risk management is an important aspect of project management. Risk management is one of the ten knowledge areas defined in PMBOK®. Project risk can be defined as an unforeseen event or activity that can impact the project's progress, result, or outcome in a positive or negative way. A risk can be assessed using two factors: impact and probability.

The causes of risk can come from various sources, including a requirement, such as legal requirements imposed by laws or regulations; an assumption, such as the conditions in the market (which may change); a constraint, such as number of personnel available to work on any given phase of the project; or a condition, such as the maturity of the organization's project management practices.

Known risks are those which can be identified and analyzed beforehand in such a way as to be able to (1) reduce the likelihood of their occurrence or (2) plan a risk response to reduce their impact in the event that they occur. Unknown risks, on the other hand, are those that are not identified beforehand. If they are not identified, they cannot be analyzed, and of course, cannot be managed proactively.

There are four possible responses to a risk, depending on whether there is low or high probability of it occurring and whether the financial impact, if it does occur, is either high or low:

1. ***Avoid*:** for high-probability, high-impact events.
2. ***Transfer (such as purchasing insurance)*:** for low-probability, high-impact events.
3. ***Mitigate*:** for high-probability, low-impact events.
4. ***Accept*:** for low-probability, low-impact events.

1.9.2 Risk Management Steps

According to PMBOK® (2021), risk management includes six main steps, in summary:

1. ***Planning for risk management*:** Start with a risk management plan – plan the work, work the plan; how risks are managed.
2. ***Risk identification*:** Use prior documentation – such as the charter, budgets, schedules, plans, etc. Have the right people involved who know risks; have a risk owner. Use methods such as brainstorming, interviews, Delphi technique, and root cause analysis.
3. ***Qualitative risk analysis*:** Analyze probability and impact. Calculate and rank the risk score.
4. ***Quantitative risk analysis*:** Used for important risks, which can be quantified (limited number of risks). Quantitative models include discount cash flow and internal rate of return with sensitivity analyses.
5. ***Risk response planning*:** Strategies include avoid (don't take risky actions); transfer to somebody else (take insurance, set up contracts); mitigate (make changes to reduce probability); accept (let it happen, establish contingency reserves such as cost, schedule, and performance).
6. ***Risk monitoring and control*:** Reassess new and existing risks. Use audits, variance, and trend analysis.

1.9.3 Risk Management in Product Innovation Projects

In Chapter 4, we introduce the new product process as one of risk and reward. A key element of this process is the recognition of the level of risk associated with the development and commercialization of the new product and taking appropriate steps to manage this risk. The decision-making gates in the

Stage-Gate® process are a critical element of risk management, where informed decisions are made on the basis of sound information and data. The outcomes of the new product's process are impacted by two categories of risk:

Project-based **risks, which include:**

- ***Resource availability***: the right resource at the right place and at the right time.
- ***Finance availability***: sufficient to cover project and capital expenses.
- ***Resource capability:*** the types and numbers of available people with the appropriate knowledge and skills.
- ***Reliability of information***: access to and level of confidence in the information necessary for reliable decision-making.
- ***Scope definition***: clarity and communication of scope to ensure alignment of all people associated with the project.

Product-based **risks, which include:**

- Causes harm to customers.
- Fails to deliver promised benefits.
- Fails to meet regulatory requirements.
- Does not meet customers' expectations with regards to aesthetics, features, functionality, or price.

Risk management in product innovation should be viewed as a combination of the PMBOK six-step process and the application of the decision practices focused on the innovation strategy, the product innovation charter, the specific new product process, and the underpinning tools that enable sound decision-making.

1.9.4 Decision Trees

A decision tree is a decision support tool that uses a tree-like graph or model of decisions and the possible consequences, including event outcomes, resources, and costs. It provides an effective structure to lay out options and investigate the possible outcomes of choosing those options. Decision trees help to form a balanced picture of the risks and rewards associated with each possible course of action. Decision trees can be developed by hand or using one of a number of software tools available. An example is shown in Figure 1.25.

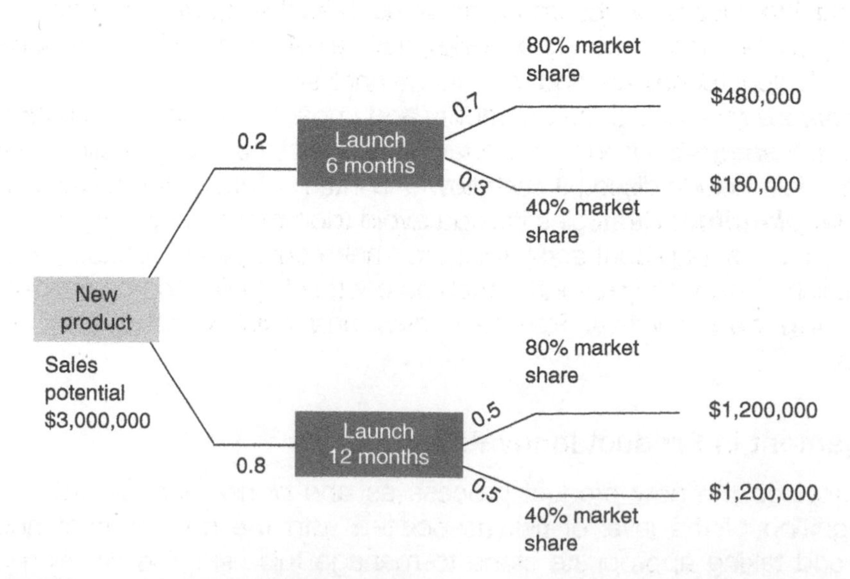

FIGURE 1.25 Example of a decision tree

1.10 METRICS AND KEY PERFORMANCE INDICATORS

Measuring product innovation performance is a key function of product managers, both in reporting on the return on product innovation and for continuous improvement on processes and practices that lead to greater product innovation success.

Key performance indicators (KPIs) are measurable values that show you how effective the organization (or group) is at achieving business objectives. In contrast, metrics are different in that they simply track the status of a specific business process. In short, KPIs track whether you hit business objectives/targets, and metrics track processes. All KPIs are metrics, but not all metrics are KPIs.

1.10.1 The Balanced Scorecard

The principle of the Balanced Scorecard was developed by Kaplan and Norton (1992). The Balanced Scorecard provides managers with a set of measures that give a fast but comprehensive view of the business. The Balanced Scorecard includes financial measures that tell the results of actions already taken. It complements the financial measures with operational measures on customer satisfaction, internal processes, and the organization's innovation and improvement activities – operational measures that are the drivers of future financial performance.

Kaplan and Norton (1992) drew an analogy between organizational management and flying a plane. "Think of the Balanced Scorecard as the dials and indicators in an airplane cockpit. For the complex task of navigating and flying an airplane, pilots need detailed information about many aspects of the flight. They need information on fuel, airspeed, altitude, bearing, destination, and other indicators that summarize the current and predicted environment. Reliance on one instrument can be fatal. Similarly, the complexity of managing an organization today requires that managers be able to view performance in several areas simultaneously."

The aim of the Balanced Scorecard, as designed by Kaplan and Norton, was "to align business activities to the vision and strategy of the business, improve internal and external communications, and monitor business performance against strategic goals." The Balanced Scorecard provides a relevant range of financial and non-financial information that supports effective business management.

The Basis of the Balanced Scorecard

Kaplan and Norton (1992) devised a framework based on four perspectives – financial, customer, internal processes, and organizational capability – each aligned with business vision and strategy. See Figure 1.26. The Balanced Scorecard is based on the principle that no single measure can provide the full picture of the organization's health. A composite of a number of measures provides far more comprehensive and meaningful insight into the organization for learning and continuous improvement. The specific measures that are applied to each perspective will vary according to the organization's vision and strategy and its strengths and weaknesses. Examples of specific measures are presented in Figure 1.26.

1.10.2 Product Innovation Metrics

Before discussing the application of the Balanced Scorecard approach to product innovation, it is important to define further the terms Key Performance Indicator (KPI) and metrics, specifically in the context of product innovation.

As defined earlier, KPIs are measurable values that relate to business goals, whereas metrics are designed to track the status of business processes that contribute to successful KPI outcomes.

The official definition of performance metrics for product innovation is:

> "Performance metrics are a set of measurements to track product innovation and to allow an organization to measure the impact of process improvement over time. These measures generally

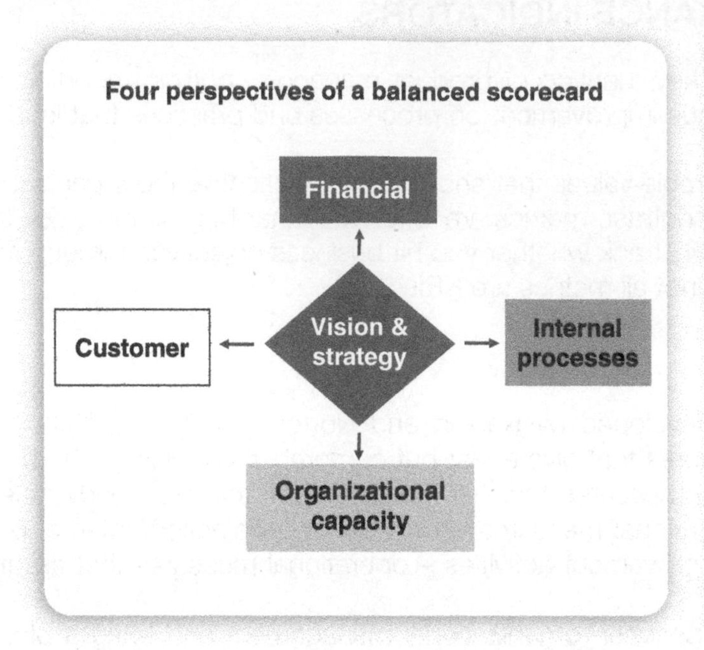

Examples of measures for each of the balanced scorecard perspectives

Financial

- Revenue
- Net profit
- Gross profit margin
- Sales
- EBIT
- ROI

Customer

- Customer retention
- Market share
- Customer complaints
- Brand equity

Internal processes

- Six sigma level
- Unit costs
- Machine downtime
- Energy consumption
- New product time to market

Organizational capacity

- Employee churn rate
- Employee core competency profile
- Employee satisfaction

FIGURE 1.26 The four perspectives of a traditional Balanced Scorecard

vary by organization but may include measures characterizing both aspects of process, such as time to market and duration of particular process stages, as well as outcomes from product innovation such as the number of products commercialized per year and percentage sales due to new products" (Appendix A, PDMA glossary).

Metrics for Reporting vs. Continuous Improvement

In many organizations, product innovation metrics are used as a key tool for management to report the returns on product innovation investment and to justify future investment. Metrics commonly used by senior management for reporting include:

- Vitality index (% of current year's sales from product developed over the past "*n*" years)
- R&D expenses as a percent of revenue
- Break-even time, or time to profitability
- Number of patents filed and awarded
- Number of new products released over a specific time period.

Although these metrics are useful to justify investment in product innovation, they do not necessarily lead to learning and continuous improvement.

Let's take the simple example of weight control. If you set a goal to reduce weight, you can stand on the scales and find your current weight. But the scale measurement, although useful in giving your current weight, doesn't provide you with any direction for losing weight. And so, it is with the reporting metrics for product innovation listed above. You have a measure of the current (or past) situation, but this doesn't provide you with a plan for future improvement.

Looking for Causal Relationships

If we can identify the causes or contributors to the final outcomes we have measured (e.g., break-even time), we can address these causes directly. SaaS product managers examining the retention rate for a product call this the *activation metric*, which is the key metric that indicates higher retention can be expected. For Facebook, their activation metric is the number of friends made during the first two weeks of use.

For our weight-loss analogy, there is ample evidence as to the contributors to weight loss. Addressing these contributing factors will directly lead to weight loss (a better scale measurement):

- Eat less and/or better
- Do more exercise
- Drink less alcohol

By measuring these contributing factors, we have a basis for action. So, we could measure the number of steps we take per day as an indicator of exercise. Setting a target number of steps per day, and achieving this target, should lead us to weight loss. Using target metrics for each of the three contributing factors will probably lead to even greater improvement (Anderson, 2015).

What are the Contributing Factors for Product Development Improvement?

There is a wealth of research carried out over recent years pointing to the key factors for product development success. PDMA literature provides an excellent reference source:

- The Comparative Performance Assessment Studies (CPAS)
- The *Journal of Product Innovation Management*
- Presentations by the Outstanding Corporate Innovation winners

It is the responsibility of all involved in the product, especially the product managers, to research these key success factors and relate them to their own organization.

Following are some of the success factors identified in PDMA's 2012 CPAS – focusing on what differentiates the best companies from the rest (Markham & Lee, 2013). These success factors are categorized under the key headings of "doing the right things," "doing things right," and "culture, climate, and organization" as shown in Figure 1.27 and previously discussed in the Introduction to this book.

Doing the right things

The best companies:
- Spend more time per project, but on fewer projects.
- Adopt a first-to-market strategy.
- Establish global strategies for market and operations.
- Monitor new technology.
- Recognize the importance of intellectual property.
- Have a clear portfolio management strategy.
- Have formal idea generation practices.

Doing things right

The best companies:
- Use a range of engineering, R&D, and design tools (critical path, FMEA, Lean NPD, TRIZ, etc.).
- Use qualitative market research tools to identify customer needs.
- Use social media to gather information.
- Have a customer feedback system.
- Use a formal NPD process—but with flexibility.
- Involve senior management.
- Focus on team development and practices.

Culture, climate, and organization

The best companies:
- Involve senior management.
- Focus on team development and practices.
- Use cross-functional teams.
- Have good recognition and reward systems.
- Support external collaboration and open innovation.

FIGURE 1.27 Success factors for product innovation

From this single research study, we have a range of factors that have been linked to product development success. Just as we suggested measuring exercise as a contributing metric for weight loss, so too can we use a selection of these success factors as a basis for performance measures that are linked to, or contribute to, product development success.

1. Is the organization ***doing the right*** product innovation? Does it have a clear innovation strategy embedded with the overall business strategy? Does it have well-developed and applied portfolio management? Does it have the right KPIs and metrics in place to track strategic decisions and outcomes? Does it learn and seek continuous improvement based on these metrics?
2. Is the organization using the ***right processes***? Does it have a product innovation process that is appropriate to its company and products? Does it have a governance structure to ensure consistent application of the process or processes? Is the right team selection and development structure in place? Is there strong commitment and contribution from senior management?
3. Does the ***organizational structure and climate*** support its product innovation efforts? Does it actively foster the development and maintenance of a creative and innovative culture? Does it support active team development practices? Are there appropriate recognition and reward practices in place?

Table 1.8 presents a summary of specific KPIs and metrics within the context of the simplified product innovation framework.

TABLE 1.8 Examples of innovation KPIs and metrics

	Metrics		
Innovation KPI's	**Doing the right product innovation (PI)**	**Doing product innovation (PI) right**	**Culture, climate, and organization**
Percentage revenue or profits from product innovation over the past three to five years; often described as the vitality index. Targets vary with best practice of 20–25%.	Clearly defined innovation strategy linked to overall business strategy.	A well-developed PI process – appropriate to company and products.	Well-developed team selection and development.
Return on investment: Either individual or new products or the entire development portfolio. Targets will depend on specific "hurdle rates" established for individual companies or specific products or product categories.	Innovation strategy clearly communicated to all staff.	A formal idea generation process.	Cross-functional teams.
Development and/or growth of specific target markets: For example, a target to increase exported products to 20% of sales or to capture a 10% share of the "recent mothers" market.	A pool of new ideas.	Strong voice of customer input throughout development.	A positive climate for innovation.
Develop a new product category based on a recently acquired technology to achieve a 20% contribution to EBIT in three years.	Well-structured and managed portfolio management.	Sound business case analysis.	Evidence of a "learning" culture.
	Sound technology planning and roadmapping.	Fit for practice tools to support all stages of development.	Strong support for ongoing training.
	Sound stakeholder and competitor intelligence.	Gates achieved on time.	Strong support from senior management.
		Break-even cycle time.	Well-managed and appropriate recognition and reward practices.

Success Factor Examples

Examples of success factors and associated contributing metrics for "doing the right things," "doing things right," and "organization, climate, and culture" are shown in Figures 1.28, 1.29, and 1.30.

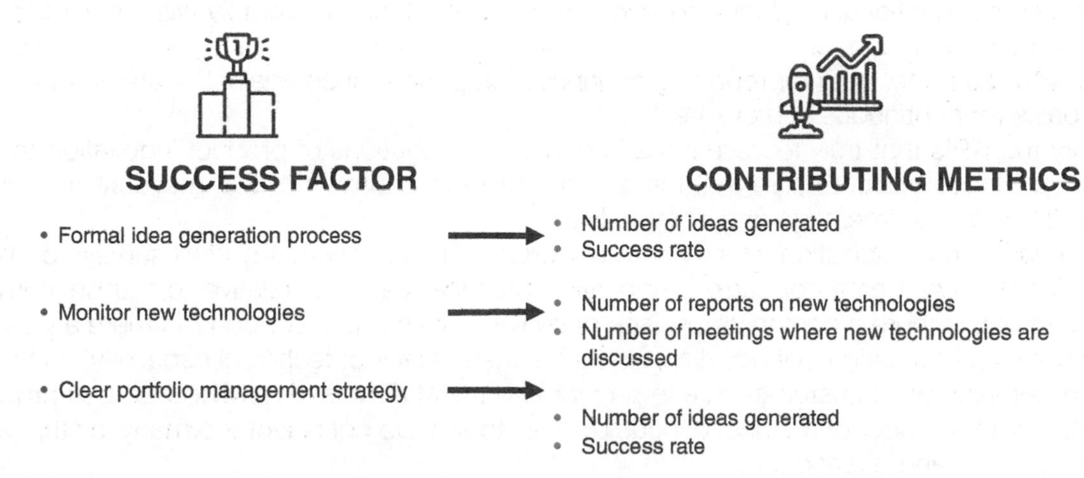

FIGURE 1.28 Example of success factors and contributing metrics for *"doing the right things"*

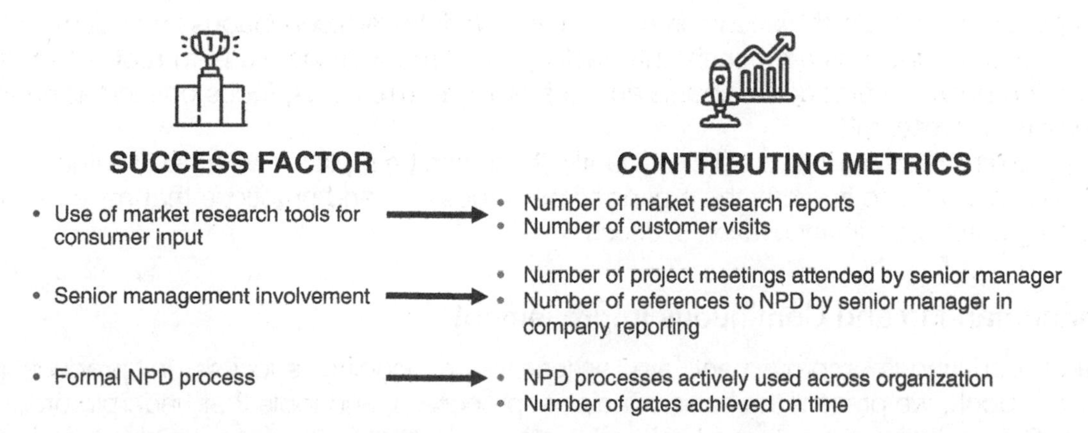

FIGURE 1.29 Example of success factors and contributing metrics for *"doing things right"*

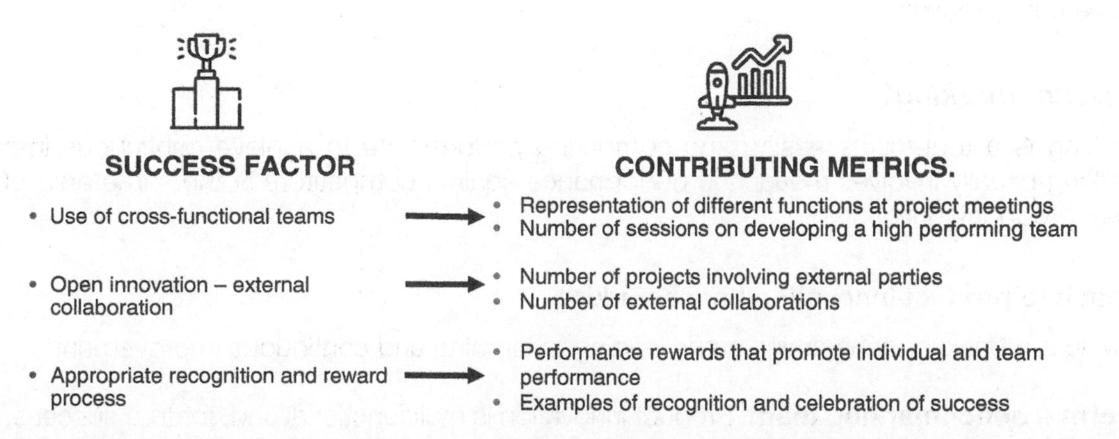

FIGURE 1.30 Example of success factors and contributing metrics for *organization, climate, and culture*

1.10.3 Developing a Balanced Scorecard for Product Innovation

The following process can be used to develop a Balanced Scorecard for product innovation:

1. Form a cross-functional product innovation improvement team, preferably with senior management representation and support. This team should be charged with establishing the balanced scoreboard framework as a tool both for reporting on innovation performance against strategy and for learning as a basis for continuous improvement.
2. Identify the KPIs that truly represent the important contributions of product innovation to achieving overall business goals. Ensure that there is an objective and quantifiable measure associated with each KPI, e.g., a percentage or a dollar value.
3. Benchmark the organization either with other organizations or with reported surveys on best practices. Identify what areas of current innovation practice are weak relative to the benchmark data. For example, is there a demonstrable lack of senior management support? Is there a lack of cross-functional representation in innovation teams? Is there a lack of technical capability? Is there a mismatch between what customers are asking for and the attributes of newly developed products?
4. Select a small number of metrics to focus on (four to six). Do not select too many, as this will lead to a lack of focus and attention.
5. Monitor each metric based on quantifiable measures. For example, which functions are represented at innovation meetings? How often is senior management represented? Other measures may include an innovation climate survey, or the number of new ideas generated monthly.
6. Compare these data with benchmarked data and work toward continuous improvement. For example, if specific functions are consistently lacking in team meetings, seek to rectify this. If there is a mismatch between product attributes and customer requirements, focus on increasing the voice of the customer research.
7. As specific metrics are improved significantly, these can be replaced with new metrics.
8. The overall goal is to improve the organization's processes and practices that have been proven to lead to greater product innovation success.

1.10.4 Benchmarking and Continuous Improvement

Learning and continuous improvement are essential for ongoing success in product innovation. Throughout this book, we present the basic principles, processes, and tools that underpin product innovation success. But, in themselves, these basic principles are just that, "basic principles." All organizations are different and demand their own specific application of these principles and a unified approach to learning and continuous improvement. Benchmarking provides an excellent basis for this learning and continuous improvement process.

What is Benchmarking?

Benchmarking is a tool for assessing and comparing performance to achieve continuous improvement. Benchmarking usually involves measuring performance against competitors in the same area of interest – in our case, product innovation.

An approach to product innovation benchmarking

The following are basic benchmarking steps leading to learning and continuous improvement:

1. **Form a benchmarking team.** Product innovation is multifunctional and, for true success, requires a common understanding and commitment across the organization. A benchmarking team, comprising cross-functional representation, ensures both agreement on the current strengths and weaknesses in product innovation and the commitment to the ways to resolve these strengths and weaknesses.

2. **Identify current strengths and weaknesses.** Appendix 1.A provides a questionnaire relating to specific areas of best practice in product innovation. This questionnaire should be completed by all members of the benchmarking team, and preferably wider if possible. Almost invariably, representatives from different functions will have diverging views on performance. Collation of the results and subsequent discussion leads to stronger cross-functional cooperation in the continuous improvement of the organization's product innovation processes and practices.
3. **Performance metrics:** As previously discussed in Section 1.10.2, performance metrics provide a valuable tool for identifying areas for improvement.
4. **External benchmarking:** A comparison with other organizations will provide insights to how specific practices and processes can be improved. Although comparison with organizations with a similar focus is often recommended, valuable learnings can also be gained from outside the normal industry sector. Clearly, obtaining deep insights into a benchmarked organization requires a high level of agreement and cooperation which can seldom be found with competitors.
5. **External surveys:** A number of surveys have been carried out to determine the success factors for product innovation or to identify what separates the "best" from the "rest" in product innovation performance, including Cooper and Kleinschmidt (1995, 2010), Markham and Lee (2013), and Bsteiler and Noble (2023).
6. **Bringing it all together:** The data and knowledge gathered in (1–5) above will provide the basis for identifying the areas to address for improvement in product innovation. This whole process, facilitated by the cross-functional team, provides the platform for immediate action and for ongoing learning and improvement.

Anderson (2008) provides an example of the application of benchmarking, demonstrating the approach above.

Key findings from PDMA's 2021 Best Practice Survey, Bsteiler and Noble (2023):

- Firms must continually evolve their NPD capabilities just to "stay in the game" as the business and technology environments change.
- No one single practice is required for greater innovation performance. Rather, the Best firms are better at employing and skillfully combining a variety of NPD capabilities and practices.
- The Best firms are much more likely to have a new product strategy that encourages radical innovation, is oriented toward risk-taking and long-term, and strives for growth through entering new markets and new technologies. They also spend more than the Rest on developing all types of innovations.
- The Best are more proactive in dealing with crises such as the global COVID-19 pandemic.

1.11 IN SUMMARY

This chapter has focused on the role of product innovation management.

What is product innovation?

- Product innovation is defined as "the creation and subsequent introduction of a good or service that is either new or an improved version of previous goods or services."
- The scope of product innovation – from strategy direction, through portfolio management, from individual project management to commercialization – implies a high degree of complexity. It is influenced by a wide range of inputs and variables.

Managing product innovation

- The role of managing product innovation can lie in a range of job titles, from the CEO of small to medium-sized companies, VP of innovation in large companies, product innovation manager, brand manager, product manager, and more. The role of the product manager is growing in prominence around the world.
- The job of a product manager is to discover a product that is valuable, usable, and feasible. The role of a product manager lies at the intersection of business decisions, user experiences, and technology.
- A product manager is not a project manager. Project managers manage the process of creating a project or service. Product managers commercialize solutions to solve customer problems and/or meet a market need/demand.
- The principles and fundamentals of product innovation are common across most industries and product types. Differences lie in the specific strategies, processes, and tools applied. The key to successful product innovation management is to understand the basic principles and fundamentals and recognize and implement *fit-for-purpose* strategies, processes, and tools for the specific organization.

Managing the product life cycle

- Most products have a life cycle that follows the stages of development, introduction, growth, maturity, decline, and retirement.
- In general, the product life cycles from introduction to decline have become shorter over recent years, placing greater pressure on organizations to develop new products and regenerate existing ones.
- Product management in terms of all elements of the marketing mix – product, price, promotion, and place – is determined by the stage of a product's life cycle.
- Product managers have a key role in guiding products through the life cycle.
- In the early stages of the product life cycle, the *diffusion of innovation* is very important. This is the process by which new innovations are accepted and begin to achieve market success. Products need to navigate the chasm between the introduction stage and the growth stage.
- Product and technology roadmaps are fundamental tools in product innovation planning.
- Although there is a fundamental difference between the roles of project and product managers, product innovation management, while focused on product, does require the application of project management principles and tools. Complementing the New Product Development Professional (NPDP, from PDMA) certification with the Project Management Professional (PMP, from PMI) certification is ideal.

Measuring product innovation performance with a focus on continuous improvement

- Product innovation performance metrics that measure and report on the outcomes of product innovation are essential to demonstrating the return on investment.

- Although the application of performance metrics for reporting is extremely important, their application to continuous improvement in product innovation processes is even more important to the long-term growth of an organization.
- Sound knowledge of the best practices for product innovation and regular benchmarking, both internal and external, is essential to high-performing product innovation management.
- Benchmarking against the processes and practices of other organizations can provide invaluable input to complement internal performance metrics in support of the process of continuous improvement.

1.A PRODUCT INNOVATION MANAGEMENT PRACTICE QUESTIONS

1. Jane is the product manager for a toy manufacturing and marketing company. The company has just embarked on developing a new product targeted at children between 8 and 12 years of age. Jane is charged with forming a cross-functional team to complete an early-stage feasibility analysis of the product's potential. She has chosen representatives from marketing, technical, R&D, and finance to be part of the team. What important function has Jane failed to include?
   A. Repairs and maintenance.
   B. Manufacturing.
   C. Board of Directors.
   D. Customer.
2. Key difference between old-school "go-to-market" processes and new-school is?
   A. More leadership.
   B. Strong product manager.
   C. Vision-driven.
   D. More iterative.
3. What is a set of measurements to track product development?
   A. TRIZ.
   B. Decision trees.
   C. Metrics.
   D. The critical path.
4. What illustrates high-level product strategy and demonstrates how a product will evolve over time?
   A. Product roadmap.
   B. Product dashboard.
   C. Perceptual map.
   D. Product life cycle.
5. Product management actions taken during the product life cycle are first determined by which of the following?
   A. Life cycle assessment.
   B. Sustainability plans.
   C. Product, price, promotion, and place.
   D. Phase of the product's life cycle.
6. In the introduction stage of the product life cycle, which of the following product pricing strategies is most commonly used?
   A. Penetration pricing.
   B. Skimming pricing.
   C. Either A or B.
   D. Competitive pricing.
7. What type of consumer is most likely to purchase a product in its introduction stage?
   A. Laggards.
   B. Early adopters.
   C. Innovators.
   D. Early majority.

8. Jack is a product manager responsible for a product that has clearly entered the decline stage of its life cycle. What strategy should Jack adopt?
   A. Seek to reduce costs and continue to sell to a loyal market.
   B. Rejuvenate the product by adding new features and finding new users.
   C. Discontinue the product.
   D. Any of A, B, or C.
9. In the management of a product portfolio, it is important to:
   A. Have a good mix of products across the introduction, growth, and maturity stages of the product life cycle.
   B. Focus strongly on products in the introduction and growth stages of the product life cycle.
   C. Place significant emphasis on the product in the maturity stage of the product life cycle.
   D. Place significant emphasis on products in the growth and maturity stages of the product life cycle.
10. In reviewing a product's contribution to the company's product portfolio, the product manager observes that sales growth has slowed significantly, and a significant number of products have entered the market. Price reductions have been necessary resulting in lower profit margins. What stage of the product life cycle has this product reached?
    A. Growth
    B. Maturity
    C. Retirement
    D. Decline

Answers to practice questions: product innovation management

1. B
2. D
3. C
4. A
5. D
6. C
7. B
8. D
9. A
10. B

1.B A QUESTIONNAIRE FOR EVALUATING AN ORGANIZATION'S NEW PRODUCT INNOVATION MANAGEMENT PRACTICES AND PROCESSES

Introduction

This self-evaluation questionnaire is designed to focus on the key areas which define success in product innovation management. Although reasonably comprehensive, the questionnaire is not intended to cover completely all issues that need to be addressed to achieve a high level of product innovation performance (Figure 1.B.1). It is intended that the questions will encourage an organization to critically examine its product innovation management, its strengths, weaknesses, and opportunities for improvement.

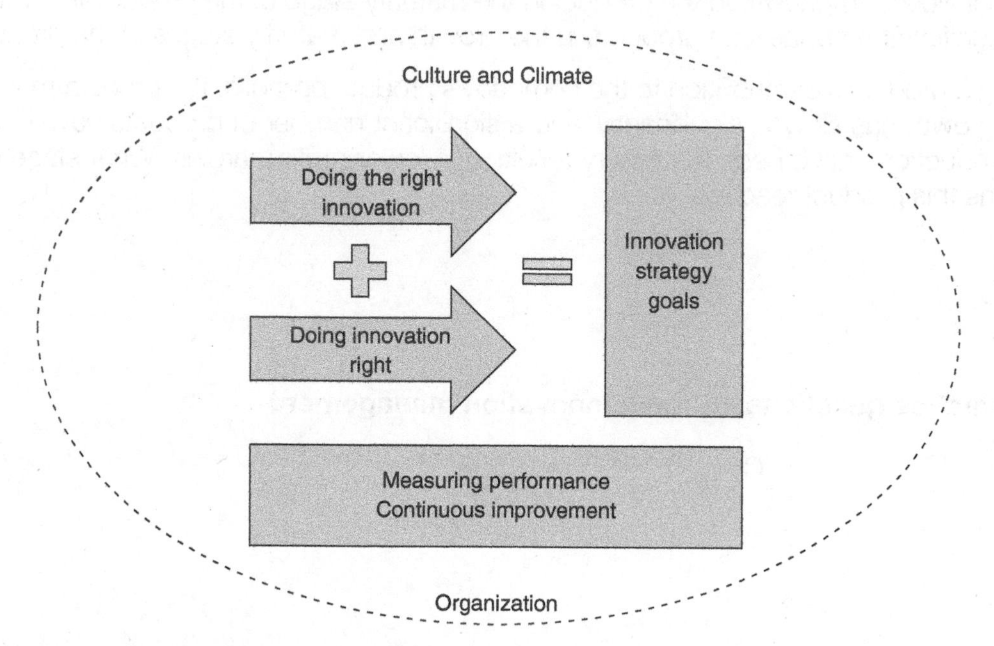

FIGURE 1.B.1 A model for product innovation

Complete the Following Questionnaire

1. Respond to each question by circling a number from 1 to 5 in the associated box.

1	2	3	4	5

Not at all — Always

2. In the summary section at the end of the questionnaire, add the scores for each major section. Briefly describe how the overall performance of **product innovation** could be improved and what specific areas to focus on.
3. There is no optimal or target score for this questionnaire. It is designed to encourage organizations to critically evaluate how they do new product innovation. It also provides an excellent basis for identifying specific metrics that will lead to improved product innovation outcomes.

SECTION 1: DOING THE RIGHT PRODUCT INNOVATION

Statement	Rating
1.1 Our company has a well-developed overall business strategy.	1 2 3 4 5
1.2 This strategy is well communicated and understood by all staff.	1 2 3 4 5
1.3 Product innovation is an integral and clearly identified element of the overall company strategy.	1 2 3 4 5
1.4 The company strategy provides a clear context for an NDP, which defines the approach and direction for all product innovation.	1 2 3 4 5
1.5 Our company has a product innovation strategy which provides a comprehensive direction for all aspects of product innovation including: • Expected contribution to company growth targets; • Product emphasis; • Market emphasis; • Product innovation risk profile; • Intellectual property strategy; • Core internal capability; and • Use of external capability.	1 2 3 4 5
1.6 The product innovation strategy provides clear and meaningful direction for prioritization of all product innovation projects.	1 2 3 4 5
1.7 The product innovation strategy is widely communicated and understood by all company staff.	1 2 3 4 5
1.8 All staff involved in a product innovation project can articulate the strategic relevance of that project.	1 2 3 4 5
1.9 The product innovation strategy is supported by a technology plan.	1 2 3 4 5
1.10 The technology plan is benchmarked, where possible, against competitors.	1 2 3 4 5
1.11 Our company has a well-developed process for technology planning and roadmapping.	1 2 3 4 5
1.12 Our company has a formal set of criteria for project selection which is rigorously applied.	1 2 3 4 5
1.13 Our company has a process for project prioritization to ensure best utilization of finance and resources.	1 2 3 4 5
1.14 Our company has a well-defined framework for defining the optimal product innovation portfolio.	1 2 3 4 5
1.15 Our company has an ongoing process for product portfolio analysis which is used to regularly review the product innovation portfolio.	1 2 3 4 5

SECTION 2: DOING PRODUCT INNOVATION RIGHT

Statement	Rating
2.1 A detailed stakeholder analysis is carried out at the beginning of the NDP project.	1 2 3 4 5
2.2 Regular stakeholder input is sought throughout the product innovation process.	1 2 3 4 5
2.3 All elements of the 4Ps (product, price, promotion, and place) are integrated into the product innovation process from an early stage.	1 2 3 4 5
2.4 Clear criteria are used for the selection of product innovation projects.	1 2 3 4 5
2.5 A well-defined selection process, involving all key parties, is used for all product innovation projects.	1 2 3 4 5
2.6 A well-defined product concept is developed early in the product innovation process. This is communicated to, and agreed upon, by all key parties.	1 2 3 4 5
2.7 A sound project plan is developed, involving all key parties and including key activities, timelines, resourcing, and budgeting.	1 2 3 4 5
2.8 Key project roles are defined before the project begins. These may include the project leader, key team members, and an overall senior management steering committee.	1 2 3 4 5
2.9 A business case is developed early in the product innovation process, including an estimated return on investment based on sound estimates of target market and sales potential, manufacturing and marketing costs, and capital investment.	1 2 3 4 5
2.10 A well-defined process is used for all product innovation projects. Although not necessarily the same for all projects, it is fit for the specific purpose. The process is clearly understood and followed by all parties involved.	1 2 3 4 5
2.11 The product innovation process includes clear go/no-go decision points, or gates, with well-defined deliverables required for each gate.	1 2 3 4 5
2.12 Go/no-go decisions are taken seriously – with projects being approved to proceed, re-directed, or concluded.	1 2 3 4 5
2.13 Review meetings are carried out at the end of all product innovation projects. Learnings from these review meetings are used to bring about continuous improvement.	1 2 3 4 5

SECTION 3: CULTURE, CLIMATE, AND ORGANIZATION

3.1 Senior management is very supportive of product innovation. — 1 2 3 4 5

3.2 Our chief executive regularly demonstrates support for product innovation through: — 1 2 3 4 5

Reference to it in company presentations;

- Recognition of product innovation success;
- Leading initiatives for product innovation improvement;
- Leading committee for portfolio management; and/or
- Active involvement in steering committees for highly significant projects.

3.3 Our company regularly carries out climate surveys. — 1 2 3 4 5

3.4 Dimensions which are important to creativity and innovation are included in the climate survey. — 1 2 3 4 5

3.5 Results from the climate survey are actively used to improve the company's performance. — 1 2 3 4 5

3.6 Our company has a well-organized plan for appointment and development of product innovation capability. — 1 2 3 4 5

3.7 This plan focuses on a balance of core, internal capability, and external sourcing. — 1 2 3 4 5

3.8 This capability plan is based on the company's product innovation strategy and technology plan. — 1 2 3 4 5

3.9 Our company has a clearly defined set of core values, which are used in the recruitment of new staff. — 1 2 3 4 5

3.10 Our company has clear guidelines and processes for recognition and reward of individual contributions to product innovation. — 1 2 3 4 5

3.11 Our company has clear guidelines and processes for recognition and reward of team contributions to product innovation. — 1 2 3 4 5

3.12 Product innovation project teams are made up of a cross-section of company functions. — 1 2 3 4 5

3.13 Involvement in the product innovation team is clearly identified as part of an individual's overall roles and responsibilities. — 1 2 3 4 5

3.14 Members of product innovation project teams are involved in the project from start to finish. — 1 2 3 4 5

SECTION 4: METRICS

Statement	Rating
4.1 Performance measures (metrics) are required by senior management to demonstrate the value of product innovation investment.	1 2 3 4 5
4.2 These metrics are soundly based on well-documented information and provide a reliable valuation of product innovation investment.	1 2 3 4 5
4.3 A comprehensive suite of product innovation metrics is used to address the key areas identified for improvement.	1 2 3 4 5
4.4 Metrics cover all aspects of product innovation and not just those directly related to the product innovation department.	1 2 3 4 5
4.5 Senior management demonstrates a commitment to the implementation of the product innovation metrics and to ensuring their application to product innovation improvement.	1 2 3 4 5
4.6 Product innovation metrics are clearly linked to the performance and development framework within our company.	1 2 3 4 5
4.7 Product innovation metrics are used as a basis for continuous improvement and are taken seriously across all functions and levels of management.	1 2 3 4 5
4.8 Product innovation metrics are changed from time to time, according to specific areas identified for improvement.	1 2 3 4 5

SECTION 5: SUMMARY

Total scores

Doing the right product innovation	/75
Doing product innovation right	/65
Culture, climate, and organization	/70
Product innovation metrics	/40
TOTAL	/250

Areas of strength and weaknesses:

Opportunities for improvement:

Strategies for improvement:

REFERENCES

Anderson, A.M. (2008). A framework for NPD management: Doing the right things, doing them right, and measuring the results. *Trends in Food Science and Technology 11*: 553–561. https://doi.org/10.1016/j.tifs.2008.01.015.

Anderson, A. M. (2015). *Keynote: Performance metrics for continuous improvement* [Conference paper]. The PDMA Annual Conference, Anaheim.

Bsteiler, L. and Noble, C.H. (2023). *The PDMA handbook of innovation and new product development*, 4e. John Wiley and Sons.

Cagan, M. (2017). *Inspired*, (2nd ed.).e. Silicon Valley Product Group.

Catstellion, G. and Markham, S. (2013). Myths about new product failure rates. *Journal of Product Innovation & Management 30*: 976–979. https://doi.org/10.1111/j.1540–5885.2012.01009.x.

Cooper, R. (2023). What separates the winners from the losers and what drives success. In: *The PDMA handbook of innovation and new product development*, 4e (ed. L. Bsteiler and C.H. Noble), 3–44. John Wiley and Sons.

Cooper, R. and Kleinschmidt, E. (1995). Benchmarking the firm's critical success factors in product development. *Journal of Product Innovation Management 12*: 374–391. https://doi.org/10.1111/1540-5885.1250374.

Cooper, R. and Kleinschmidt, E. (2010). Success factors for new product development. In: *Wiley international encyclopedia of marketing* (ed. J.N. Sheth and N.K. Malhotra), 1–9. John Wiley & Sons https://doi.org/10.1002/9781444316568.wiem05021.

Cooper, R. and Kleinschmidt, E. (2015). *Winning businesses in product development: Critical success factors*. Research-Technology Management.

Dobson, M. (2004). *The triple constraints in project management*. Berrett-Koehler Publishers.

Eriksson, M. (2011). *What, exactly, is a product manager?* Mind the Product. https://www.mindtheproduct.com/what-exactly-is-a-product-manager/

Glassdoor. (2022). *Best Places to Work*. https://www.glassdoor.com/index.htm.

Johnson, S. (2017). *Turn ideas into products: A playbook for defining and delivering technology products*. Independently Published.

Kahn, K.B., Evans Kay, S., Slotegraaf, R.J., and Uban, S. (ed.) (2013). *The PDMA handbook of new product development*, 3e. John Wiley & Sons https://doi.org/10.1002/9781118466421.

Kaplan, R.S. and Norton, D.P. (1992). The balanced scorecard-measures that drive performance. *Harvard Business Review 70*: 71–79.

Kocina, L. (2017). *What percentage of new products fail and why?* MarketSmart Newsletters. https://www.publicity.com/marketsmart-newsletters/percentage-new-products-fail/

Knudsen, M.P., Zedtwitz, M., Griffin, A., and Barczak, G. (2023). Best practices in new product development and innovation: Results from PDMA's 2021 global survey. *Journal of Product Innovation Management* 40 (3): 257–275.

Lester, D.H. (2016). *Critical success factors for product development*. Research-Technology Management.

Markham, S.K. and Lee, H. (2013). Product development and management association's 2012 comparative performance assessment study. *Journal of Product Innovation Management 30* (3): 408–429. https://doi.org/10.1111/jpim.12025.

Montoya-Weiss, M.M. and Calantone, R.J. (1994). Determinants of new product performance: A review and meta analysis. *Journal of Product Innovation Management 11* (5): 397–417. https://doi.org/10.1111/1540-5885.1150397.

Moore, G.A. (2014). *Crossing the chasm*, 3e. Collins Business Essentials.

PMBOK®. (2021). *A guide to the project management body of knowledge*, 7e. Project Management Institute.

Rogers, E.M. (1962). *Diffusion of innovations*. Free Press.

Schneider, J., & Hall, J. (2011). *Why most product launches fail*. Harvard Business Review. https://hbr.org/2011/04/why-most-product-launches-fail

2

Product Innovation Strategy

Provides the context, goals, and direction for product innovation and ongoing product management

Product Development and Management Body of Knowledge: A Guidebook for Product Innovation Training and Certification, Third Edition.
Allan Anderson, Chad McAllister, and Ernie Harris.

WHAT YOU WILL LEARN IN THIS CHAPTER

Strategy lies at the very heart of an organization's sustenance and growth. It lays the foundation and provides the framework for all of the organization's functions and activities. In this chapter, strategy is described as it is applied at various levels of the organization. In particular, emphasis is placed on the innovation strategy, providing frameworks for analysis and development of the innovation strategy together with guidance on its implementation throughout the organization.

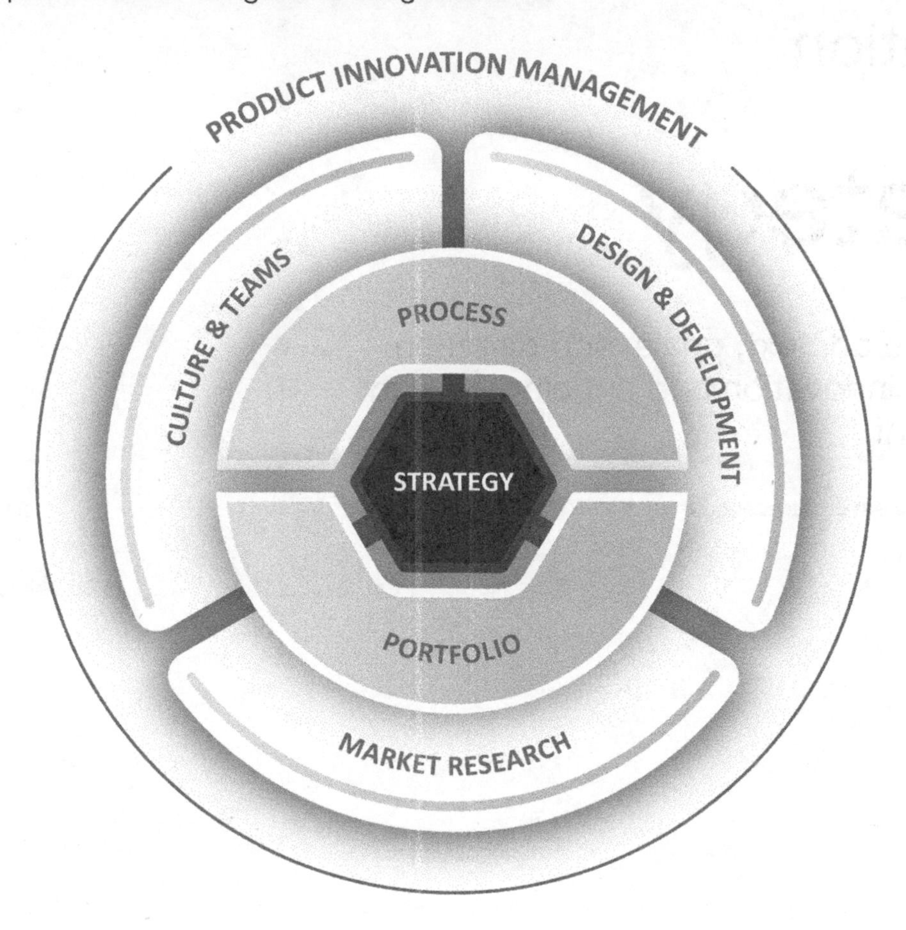

THE CHAPTER ROADMAP

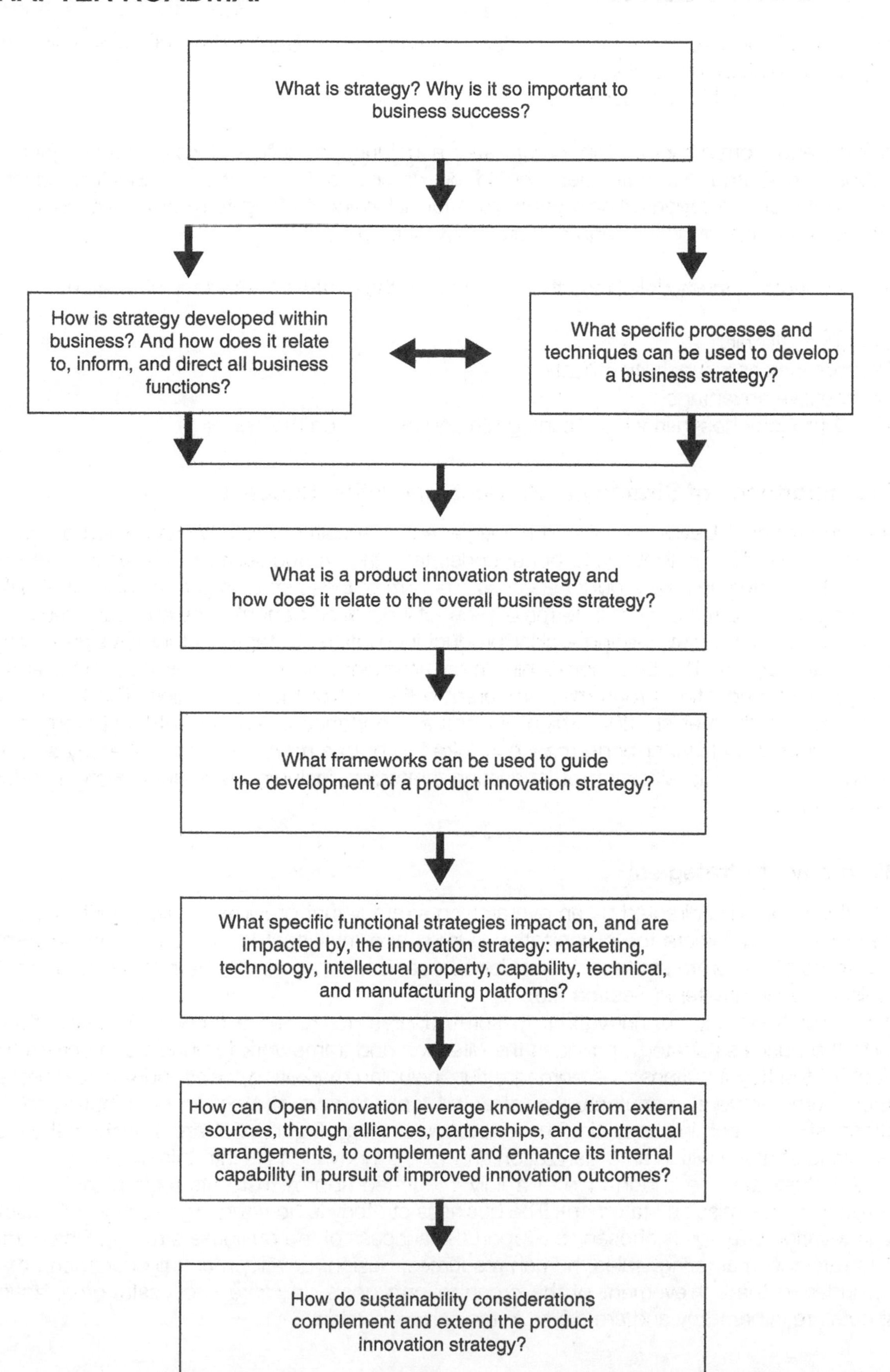

2.1 WHAT IS STRATEGY?

A strategy is broadly defined as a method or plan chosen to bring about a desired future, such as achievement of a goal or solution to a problem.

When applied within the business context:

> "It defines and communicates an organization's unique position and says how organizational resources, skills, and competencies should be combined to achieve competitive advantage" (Porter, 2008), or "An organization's game plan for achieving its long-term objectives in light of its industry position, opportunities, and resources" (Kotler, 2012).

Key components of these definitions that stand out in the context of product innovation:

1. Unique positioning
2. Competency and capability utilization
3. Competitive advantage
4. The approach chosen in which to integrate and execute on the above.

2.1.1 The Importance of Strategy to Product Innovation Success

Since 1990, the Product Development and Management Association (PDMA) has carried out regular studies of a cross-section of organizations to better understand the product innovation practices and processes that lead to improvement in new product success. The aim of these studies is to identify what differentiates the "best" companies from the "rest" in terms of product innovation performance and outcomes.

Each study has found that defining a clear product innovation strategy contributes significantly to overall new product success. The best companies more frequently have strategies that direct and integrate their entire product innovation programs compared to the rest of the companies. The 2021 PDMA Best Practice Survey (Knudsen et al., 2023) highlighted the importance of an innovation strategy: ". . .overall, these results suggest that better performance is linked to both a more innovative strategy and more radical innovation projects." Classification of innovation strategies, including radical innovation, is further discussed in Section 2.6.

2.1.2 Hierarchy of Strategies

An organization should be directed by an overarching strategy that provides goals, priorities, and focus for the whole organization. Where the organization is large and multifaceted, the organizational strategy may be referred to as a corporate strategy. In smaller companies, it may simply be referred to as the business strategy (discussed in greater in Section 2.3).

Most organizations rely on innovation, to some degree, for revenue growth. An innovation strategy, aligned with the business strategy, provides the direction and framework for innovation across the organization. Each of the key functions of an organization including marketing, sales, human resources, design, development, procurement, manufacturing, etc., will form its own strategy in contributing to the overall organizational strategy and in supporting the innovation strategy. Figures 2.1 and 2.2 show the hierarchy of decision-making strategy within an organization and where innovation fits into this hierarchy.

Figure 2.1 illustrates the importance of a fully integrated approach to strategic planning. This begins with the organization's mission statement. The business strategy is how the organization will satisfy its mission. The innovation strategy is chosen to support the success of the business strategy. Finally, the specific functional strategies (marketing, sales, human resources, design, development, procurement, manufacturing, etc.), underpin the achievement of the organization's goals. In more successful organizations, all of these aspects are in harmony and are driven in the same direction.

FIGURE 2.1 The hierarchy of strategies

Figure 2.2 provides an extension to Figure 2.1, emphasizing the importance of the organization's mission in informing project selection, portfolio management, and the overall product innovation strategy. It also emphasizes the importance of alignment of functional strategies through to the development and implementation of the product innovation strategy.

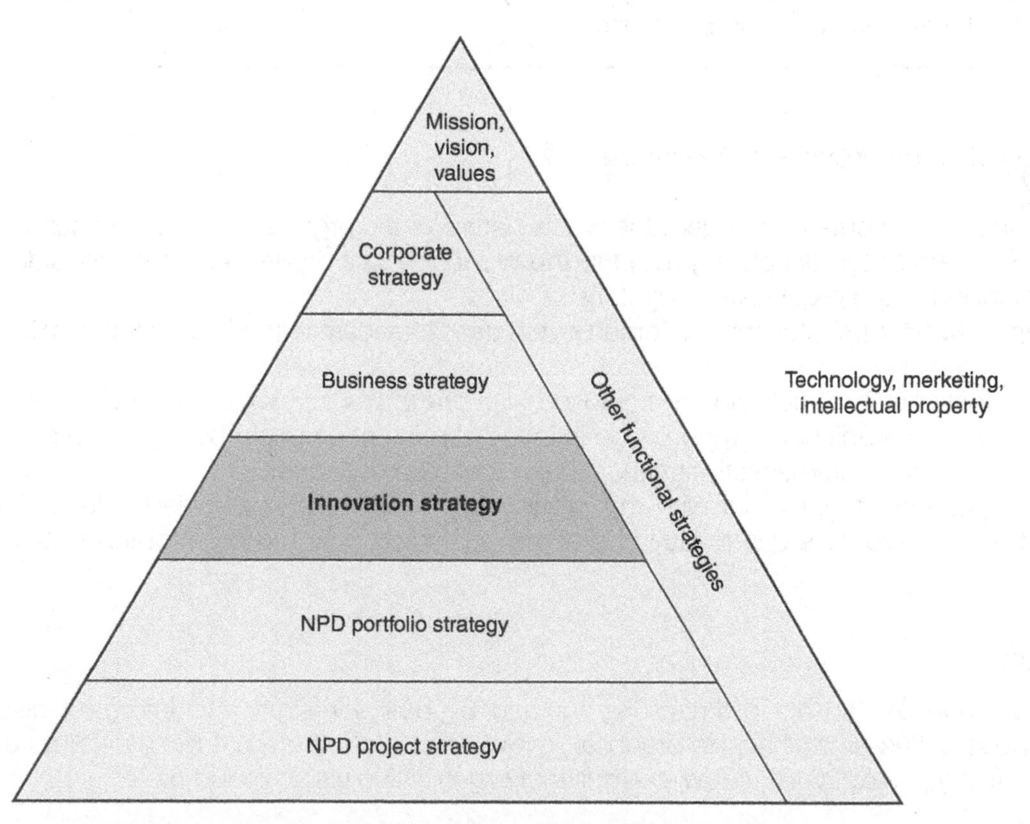

FIGURE 2.2 Innovation and the strategic decision hierarchy

2.2 ESTABLISHING THE ORGANIZATION'S DIRECTION

2.2.1 Organizational Identity

Fundamental to the long-term success of an organization is a clear definition and understanding of what the organization stands for and why it exists. Organizational identity answers the question, "Who are we as an organization?" (Whetten, 2006).

Examples of Organizational Identity

Zappos.com

- Convenient buying with free, no hassle returns
- Customers are king
- Most customer-friendly

Ritz-Carlton Hotel

The hotel was opened by Swiss hotelier César Ritz in May 1906, eight years after he established the Hôtel Ritz Paris with the company's credo specifically stating "authentic care and comfort of guests is Ritz-Carlton hotels' highest mission." For many luxury travelers, the "point of difference" of the Ritz-Carlton hotel brand is its Club Level of enhanced accommodations. Club Level rooms and suites enjoy upgraded amenities such as complimentary Wi-Fi, butler-style services like laundry, and take-home sizes of toiletries, and a personal gift.

Key attributes of an organization's identity are:

- ***Central***: If an attribute is changed, the whole nature of the organization will be changed.
- ***Enduring***: Attributes deeply ingrained in the organization are often explicitly considered sacrosanct or embedded in the organization's history.
- ***Distinguishing***: Attributes that the organization uses to separate itself from other similar organizations.

Organizational identity should not be confused with *brand identity*, which is how an organization wants to be perceived by its audience. Brand image is external perceptions held by consumers, whereas organizational identity is about internal activities that shape these perceptions.

Organizational identity interacts with the organization's vision, mission, and values. These attributes manifest in the organization's day-to-day operations, including how the product innovation strategies are played out.

2.2.2 Vision

A Vision is defined as "an act of imagining, guided by both foresight and informed discernment, that reveals the possibilities as well as the practical limits. . .It depicts the most desirable future state of. . .an organization" (Kahn et al., 2013). A few examples of organization vision include:

- ***Nike***: "Bring inspiration and innovation to every athlete* in the world. (*If you have a body, you are an athlete.)"
- ***Alzheimer's Association***: "A world without Alzheimer's disease."

- ***McDonald's***: "To be the best quick service restaurant experience. Being the best means providing outstanding quality, service, cleanliness and value, so that we make every customer in every restaurant smile."
- ***Huawei***: "To bring digital to every person, home and organization for a fully connected, intelligent world."
- ***Amazon***: "To be earth's most customer centric organization; to build a place where people can come to find and discover anything they might want to buy online."

2.2.3 Mission

A Mission is defined as "the statement of an organization's creed, philosophy, purpose, business principles, and corporate beliefs. The purpose of the mission is to focus the energy and resources of the organization" (Kahn et al., 2013). A few better-known mission statements are listed below:

- ***Starbucks***: "To inspire and nurture the human spirit – one person, one cup and one neighborhood at a time."
- ***Google***: "To organize the world's information and make it universally accessible and useful."
- ***Microsoft***: "To empower every person and every organization on the planet to achieve more."
- ***Kickstarter***: "To help bring creative projects to life."
- ***Tesla***: "To accelerate the world's transition to sustainable energy."
- ***Alibaba***: "To make it easy to do business anywhere. We enable businesses to transform the way they market, sell, and operate and improve their efficiencies."

2.2.4 Values

Values are defined as "principles to which a person or organization adheres with some degree of emotion" (Kahn et al., 2013). Put simply, organizational values are the guiding principles that provide an organization with purpose and direction. They help companies manage their interactions with both customers and employees. Examples of organizations' values are listed in Table 2.1.

TABLE 2.1 Examples of values in organizations

Facebook:	**Starbucks**
• Focus on impact • Move fast • Be bold • Be open • Build social value	• Creating a culture of warmth and belonging, where everyone is welcome. • Acting with courage, challenging the status quo and finding new ways to grow our company and each other. • Being present, connecting with transparency, dignity, and respect. • Delivering our very best in all we do, holding ourselves accountable for results.
Huawei: • Openness • Collaboration • Shared success	
Adobe: • Genuine • Exceptional • Innovative • Involved	

2.2.5 Organizational Identity and Product Innovation

The mission, vision, and values define not only what the organization is seeking to achieve, they also define the *organization's personality* – how it acts and how it feels. The personality:

- Has a significant impact on reinforcing how important product innovation is to the organization.
- Significantly influences the focus for product innovation and how it is implemented. Managers should ensure that the mission, vision, and values provide the appropriate context and direction for product innovation.
- Has responsibility for ensuring the relevance and connection of the mission, vision, and values at all levels of the product innovation process. Communication and regular reinforcement of this relevance and connection across all functions and staff involved in product innovation is vitally important.
- Supports the brand identity and communication of the organizational identity and values to its customers and other stakeholders.

2.3 BUSINESS AND CORPORATE STRATEGY

The corporate or business strategy underpins the goals and direction of an organization's activities. The overall business strategy provides the foundation for functional strategies, including manufacturing, marketing, intellectual property, capability, and, most importantly, in the context of this book, innovation. Business strategy is informed by and informs the product innovation strategy. In turn, the product innovation strategy informs the selection and on-going management of the organization's product portfolio, prioritizing specific innovation projects.

2.3.1 Business Strategy

Tregoe and Zimmerman (1980) define business strategy as "the framework which guides those choices that determine the nature and direction of an organization." Ultimately, this boils down to selecting products (or services) to offer and the markets in which to offer them.

Porter (1996) argues that competitive strategy is "about being different." He adds, "It means deliberately choosing a different set of activities to deliver a unique mix of value." In short, Porter (1996) argues that strategy is about competitive position, about differentiating the organization, and its offerings, in the eyes of the customer, about adding value through a mix of activities different from those used by competitors.

In essence, the amalgam of these two definitions leads to the following:

Business strategy is about choosing a set of activities to deliver a unique mix of value where, for most businesses, the unique mix of value is centered on the product and/or services the business offers to specific markets.

Product innovation and product management contribute to most business strategies. The business strategy provides the context and direction for the innovation strategy and product innovation. The key steps of business strategy leading to product innovation implementation are:

1. Define the business goals, including specific product categories and markets to focus on and respective growth targets.
2. Define the role that product innovation will play in achieving these goals. For example, some goals are best supported by mergers and acquisitions. Other goals are better reached through product innovation. With product innovation, the business has a number of potential options available, including in-house development, external partnering, and licensing for development and/or marketing.
3. Define the *attack plan* for the product innovation strategy, such as focusing efforts on creating first-to-market products or improving on competitor's products. In later sections, we present several innovation strategy frameworks that direct a business into making decisions among major areas of focus for its product innovation. Broadly speaking, these center on decisions related to how much risk the business is willing to take in its product innovation. Does it want to be a market leader or a follower? Does it want its innovation to be technology or market-led? Does it want to differentiate its new products based on cost or specific features? Does it want to target a wide range of product and/or market areas, or does it want to pursue a narrow focus?

The strategic decisions outlined above, discussed in more depth later in this chapter, lay the foundation for further steps in new product planning, including individual project selection, product portfolio management, and resource allocation. It ultimately also provides a basis for selecting the appropriate product innovation methodology for a given product innovation opportunity.

2.3.2 Corporate Strategy

Organizations can vary significantly in size and in the way in which they are structured. Large organizations will often be divided into business units, each centered on specific product categories, brands, services, markets, or regions. Figure 2.3 shows the organization structure for Coca-Cola, demonstrating the complexity of multifaceted global corporations.

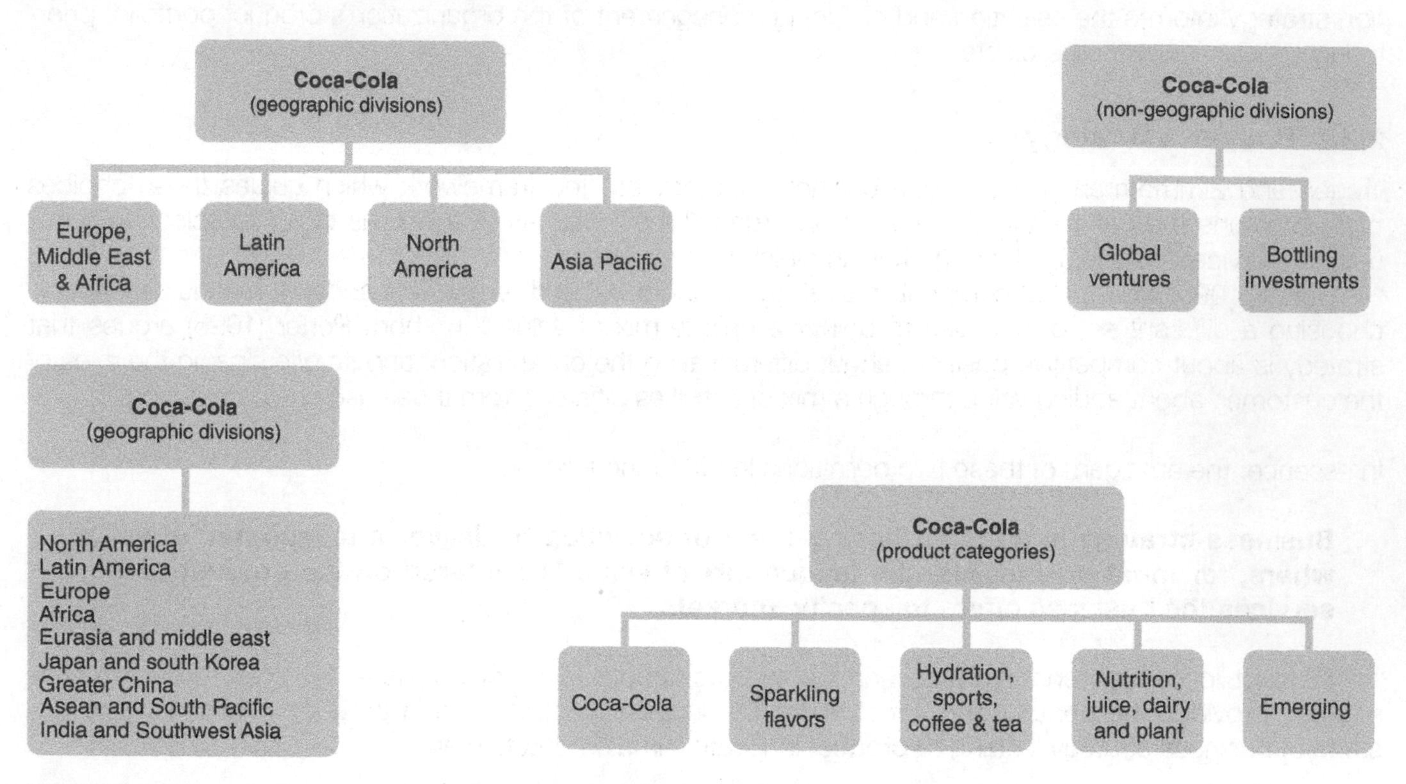

FIGURE 2.3 An example of a corporate business structure, Coca-Cola as of August 2022

In these large and multifaceted organizations, it is generally desirable to have an overarching strategy for the whole organization with separate strategies for the individual business units linked to this overall or corporate strategy. A corporate strategy is, therefore:

- The overarching strategy of a diversified organization.
- It answers the questions of (1) *in which businesses should we compete,* and (2) *how do these individual businesses create synergy and/or add to the competitive advantage of the organization as a whole?*

The role that product innovation plays within a large corporation will depend on the structure of the organization and particularly the level to which the organization seeks to achieve synergies across its business units, as opposed to maintaining a high degree of business unit autonomy. Organizations should have a high-level strategy to optimize synergies across their business units. This will impact the mergers and acquisitions (M&A) strategy as well as the internal innovation strategy. Some hypothetical examples:

- A manufacturer of PCs may define a strategic goal of fully integrating its own operating system into its branded PCs. It could do this by developing its own operating system in-house, partnering externally for the development, or acquiring a company that has either the requisite development skills or an operating system that meets the company's needs.

- A global food manufacturing company with business units in 50 countries may decide to have a core platform of technologies that can be applied to a range of products. But the specific taste preferences in different countries may dictate a requirement for slight variations in product formulations. In turn, the company must decide whether it is better to have a large central R&D facility, focused facilities in selected regions, or a combination of both.
- A large oilfield service company with over 20 business divisions in 80 countries has several R&D facilities worldwide. Each of the facilities has different R&D systems and processes. There is very little knowledge transfer across the R&D facilities. Strategically, the company has an important decision: do the advantages of standardized R&D and knowledge management practices outweigh the benefits of highly focused, individual facilities with their own "fit for purpose" systems and practices? Or, is there a middle ground that creates knowledge sharing between the separate R&D facilities?

2.4 PREPARING A BUSINESS STRATEGY

A comprehensive understanding of the business context is essential to informing the development of business goals and strategy. This includes the organization itself, the industry within which the company operates, and the wider regional and global environment. Refer to Figure 2.4.

A number of tools can be used to provide a structured approach to provide the requisite knowledge on which to develop business goals and strategy. Following are some of the most commonly used tools. Instead of relying on a single approach, a combination of these tools is recommended to provide the most comprehensive coverage of the knowledge required to underpin strategic planning.

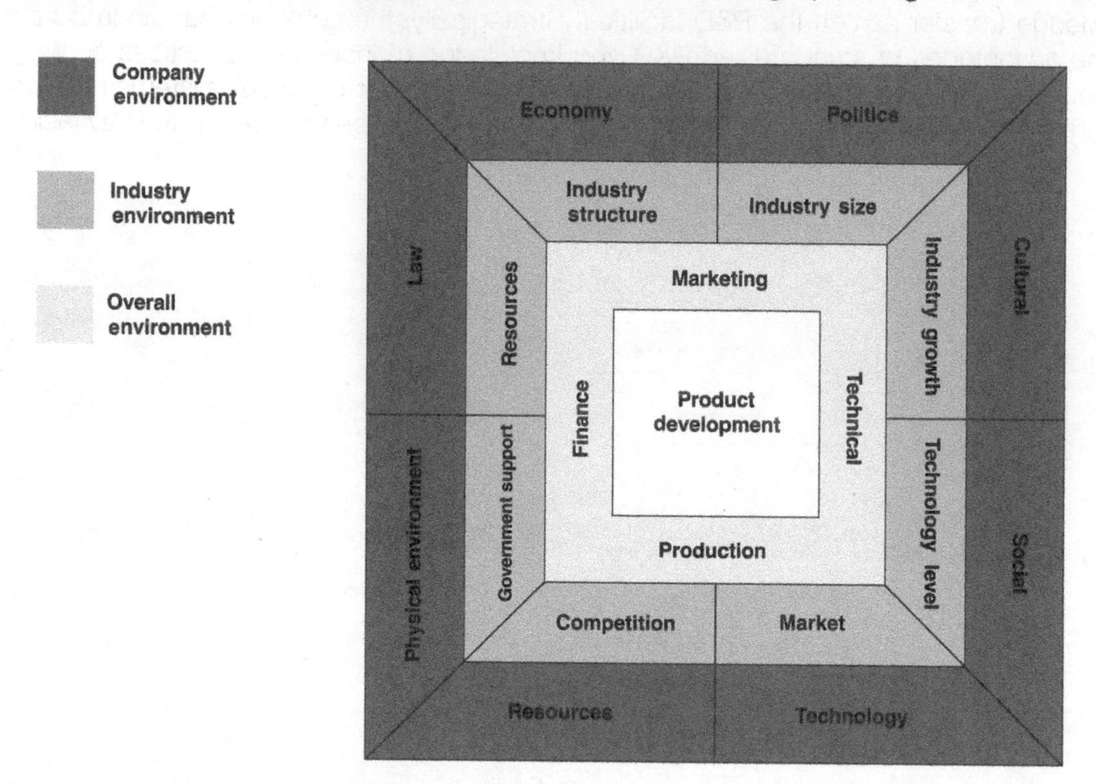

FIGURE 2.4 The context for product innovation

2.4.1 PESTLE Analysis

A PESTLE analysis is a structured tool-based, macro-environmental analysis of **P**olitical, **E**conomic, **S**ocial, **T**echnological, **L**egal, and **E**nvironmental factors. It is particularly useful as a strategic framework for seeking a better understanding of trends in factors that will directly influence the future of an organization – such as demographics, political barriers, disruptive technologies, and competitive pressures. The tool should be used when starting a new business or entering a new foreign market or anytime the external business environment must be better understood (refer to Table 2.2).

TABLE 2.2 Example of PESTLE analysis

Political	Economic	Social	Technological	Legal	Environmental
• Government policy • Political stability • Foreign trade policy • Tax policy • Labor law • Trade restrictions	• Economic growth • Exchange rates • Interest rates • Inflation rates • Disposable income • Unemployed rates	• Population growth rate • Age distribution • Educational levels • Safety emphasis • Lifestyle attitudes • Cultural barriers	• Technology incentives • Level of innovation • Automation • R&D activity • Technological change • Technological awareness	• Discrimination laws • Antitrust laws • Employment laws • Consumer protectionlaws • Patent laws • Health and safety laws	• Weather • Environmental politics • Climate change • Pressure from NGOs

2.4.2 SWOT Analysis

SWOT stands for *Strengths*, *Weaknesses*, *Opportunities*, and *Threats*, as shown in Figure 2.5.

> ***Strengths:*** Characteristics of the business or project that give it an advantage over others.
> ***Weaknesses:*** Characteristics that place the business or project at a disadvantage relative to others.
> ***Opportunities:*** Elements that the business or project could exploit to its advantage.
> ***Threats:*** Elements in the environment that could cause trouble for the business or project.

Opportunities and threats originate from external factors. Advantages can be taken of opportunities and plans made to protect against threats, but as external factors, they cannot directly be changed. Examples include competitors, prices of raw materials, and customer lifestyle changes. Recognizing opportunities and threats allows for plans to be made and appropriate management processes used to respond to external factors.

For a SWOT analysis to be effective, company senior management needs to be deeply involved. This isn't a task that can be delegated to others.

However, organization leadership shouldn't do the work on their own, either. For best results, a group of people who have different perspectives on the company should be involved. People should be selected to represent different aspects of your company, from sales and customer service to marketing and product innovation. Some organizations even look outside their own internal ranks when they perform a SWOT analysis and get customer input to add their unique voice to the mix.

FIGURE 2.5 SWOT analysis

2.4.3 Delphi Technique

The Delphi technique is a forecasting method based on the results of questionnaires sent to a panel of experts. Three or more rounds of questionnaires are sent out, and the anonymous responses are aggregated and shared with the group after each round. It is mainly applied to future forecasting or foresighting, and to longer-term strategic planning. The aim of the Delphi technique is to clarify and expand on issues, identify areas of agreement or disagreement, and then begin to seek consensus.

The Delphi technique consists of following seven steps:

Step 1: Choose and appoint a facilitator
The first step is to choose your facilitator, preferably a neutral person familiar with research and data.

Step 2: Identify the subject-matter experts
The Delphi technique relies on a panel of experts. This panel may be members of the project team, including the customer or other experts from within the organization or industry. An expert is any individual with relevant knowledge and experience of the area being investigated. The choice of subject-matter experts directly influences the quality of the information Delphi can produce, so care is needed when selecting the experts.

Step 3: Define the problem
What is the problem or issue where understanding is being sought? The experts need to know what problem they are commenting on. A precise and comprehensive definition is required.

Step 4: Round one questions
Ask general questions to gain a broad understanding of the experts' views on future events. The questions may go out in the form of a questionnaire or survey. Collate and summarize the responses, removing any irrelevant material and looking for common viewpoints.

Step 5: Round two questions
Based on the answers to the first questions, the next questions should delve deeper into the topic to clarify specific issues. These questions may also go out in the form of a questionnaire or survey. Again, collate and summarize the results, remove any irrelevant material, and look for common ground. The key goal is to achieve consensus among the experts.

Step 6: Round three questions
The final questionnaire aims to focus on supporting decision-making. Focus on the areas of agreement. What is it the experts agree on? In some situations, more than three rounds of questioning are required to reach a closer consensus.

Step 7: Act on your findings
After the third round of questions, the experts should have reached a consensus with a view of future events.

Predicting the future is not an exact science, but the Delphi technique can help in understanding the likelihood of future events and what impact they may have on specific strategies or projects.

2.4.4 Business Model Canvas (BMC)

The Business Model Canvas (BMC), first developed by Osterwalder et al. (2010), is a simple yet effective visual strategy tool that organizations, big and small, use for business model innovation. The BMC provides the basis for the Lean Canvas approach used in Lean startups with its emphasis on entrepreneur-focused business planning, discussed further in Section 4.3.11.

The organization's business model in the context of strategy and innovation is very important. Business models, if incorrectly defined and not supporting the innovation strategy and management, the technology strategy, and the product strategy, will not achieve the goal of value creation. A well-aligned business model produces value for multiple stakeholders, including customers, partners, and the organization.

The BMC addresses nine aspects of a business model: customer segments, value propositions, channels, customer relationships, revenue streams, key activities, resources, partners, and cost structures. A key component of the BMC is the aforementioned aspects of knowledge in a visual form, similar to Figure 2.6. It is designed to depict an entire business model on a single page. The right side of the BMC focuses on the customer, while the left side focuses on the business. The information is derived from asking and answering key questions:

1. ***Customer segments*:** An organization often focuses on a broad range of customers. These can be divided into distinct customer segments, each with specific needs and requirements. This allows for tailoring of specific value propositions, customer relationships, and channels for each segment.
2. ***Value propositions*:** The value proposition defines how an organization distinguishes itself from the competition. This distinction focuses on quantity such as price, service, speed, and delivery conditions on the one hand. On the other hand, it also focuses on quality including design, brand status, and customer experience and satisfaction.
3. ***Channels*:** Which channels will be focused on to reach the desired customer segments? How are those channels integrated? Which are most cost-effective?
4. ***Customer relationships*:** What type of relationship is required for each customer segment? What are the expectations of these customers? How are they established? What would be the associated costs?
5. ***Revenue streams*:** What are the customers willing to pay and for what value? How would they prefer to pay? How are they currently paying? How does each stream add up to the total revenue?
6. ***Key activities*:** What key activities are required to successfully deliver the value proposition? These include R&D, marketing, manufacturing, and distribution channels.
7. ***Key resources*:** Resources can be categorized as physical, intellectual, financial, or human resources. Physical resources may include assets such as business equipment. Intellectual resources include (among other things) knowledge, brands, and patents.
8. ***Key partners*:** Who are the key partners? Key suppliers? Which key resources are they providing? Which key activities do the partners perform? Key partners are the external companies or suppliers the business needs to perform key activities and deliver value to the customers. Buyer–seller relationships are necessary to optimize operations and reduce the risks associated with a business.
9. ***Cost structure*:** What are the most important cost drivers in the organization's business model? Which key resources and activities are most expensive? The business can be either cost-driven or value-driven. A cost-driven organization looks to minimize all costs while a value-driven company is more focused on delivering great customer value in terms of quality or prestige.

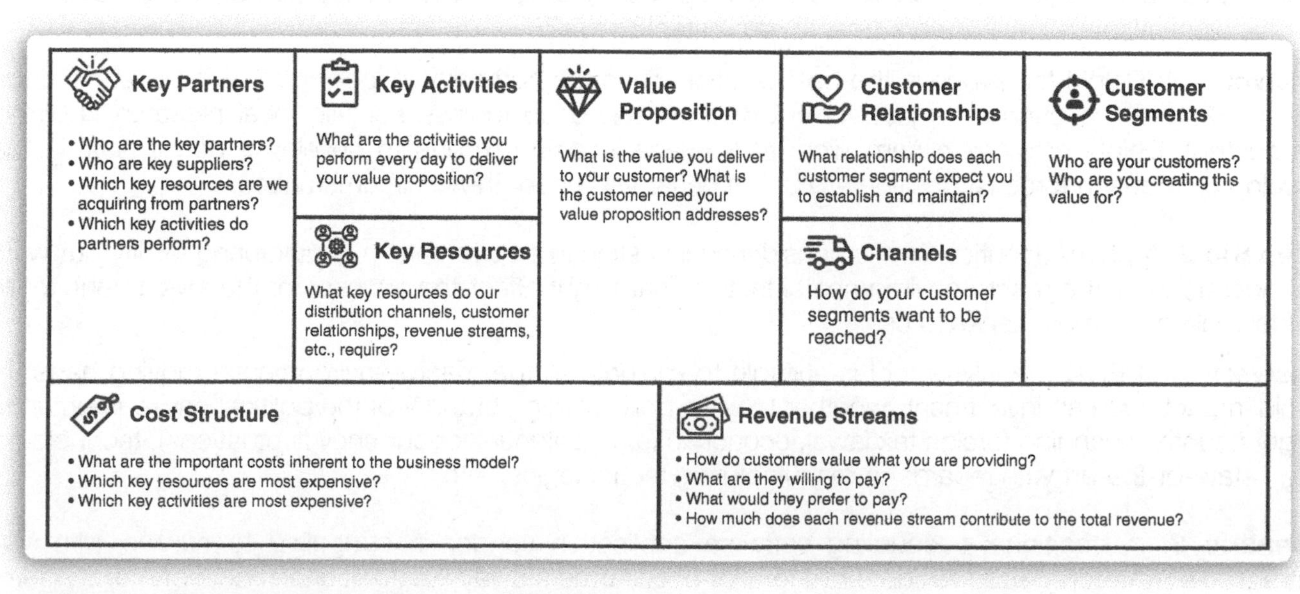

FIGURE 2.6 Business Model Canvas (BMC) framework

2.4.5 Porter's Five Forces

Porter (1979) provides a framework that models an industry as being influenced by five forces. The strategic business manager seeking to develop an edge over rival firms can use this model to better understand the industry context in which the firm operates. Specific issues to address under each of the five forces are shown in Figure 2.7.

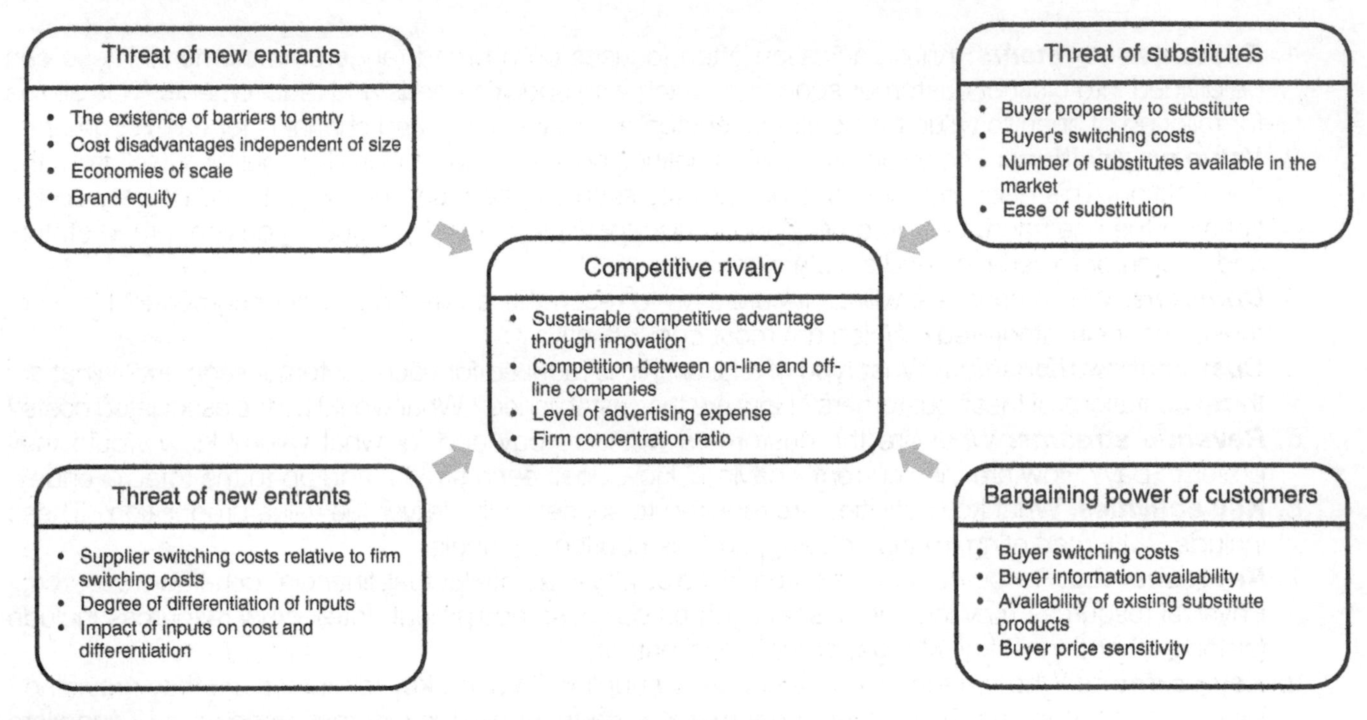

FIGURE 2.7 **Porter's five forces diagram (Porter, 1979)**

2.4.6 Examples of Application of Techniques for Business Analysis

Scenario 1: A company offering telecommunications services to businesses is considering entering the cyber security market, but there is a lot of established competition. It wants to ascertain if the market is growing, can accommodate new players, and will prove lucrative for them. They cannot find good research on the topic. What model will work best to help establish this?

Answer: The Delphi technique is the best solution since the company is looking for a forecast on a specific market, which cannot be just "calculated" or looked up (unless specific local research is already conducted). Delphi provides a framework to leverage industry experts to develop a consensus that will develop as robust a forecast position as can be generated under these circumstances.

Scenario 2: A pharmaceutical firm is considering investing in an overseas manufacturing facility but wants to understand all the macro-environmental factors that might affect the success of the investment. Which model makes the most sense to use?

Answer: The PESTLE analysis tool is suitable to evaluate all the macro-environmental moving parts that could impact such an investment, whether they be political (e.g., stability of the political environment in the target country, changing foreign tax laws), economic (e.g., potential for currency fluctuations), technological (e.g., state-of-the-art with regards to manufacturing technologies), etc.

Scenario 3: A traditional accounting software application vendor is struggling to survive with "new entrant" over-the-top SaaS providers taking a lot of customers away. Its traditional model of major up-front perpetual license fees and modest annual support revenues seems out of favor in the market. Customers are hesitant for large capital expenditures every three years or so to upgrade their software. The vendor now wishes to examine alternative ways to generate net income that align with customer preferences. What might be a good option to consider?

Answer: The Business Model Canvas allows them to map their business model across all its dimensions. Then, they can consider the ways they can potentially alter the business model to regain competitive advantage.

Scenario 4: A company with a chain of fitness clubs across the country has found that its traditional business has stagnated. In some cases, customers are choosing alternative ways of working out (at home, in the park, etc.). However, many customers still love the camaraderie of going to a club, the different classes it offers, and support from the company's expert fitness trainers. How should it evaluate its current capabilities and consider what options it may have for the way forward for its business?

Answer: A SWOT analysis is ideal as it needs to create competitive differentiation and a SWOT analysis can identify where differentiation can be created. It allows the organization to create an inventory of all the good things it has to offer (e.g., premises, skilled fitness trainers, excellent classes), some weaknesses (e.g., lack of technical/online capability), and what are some of the opportunities and threats offered by the changing landscape of its business in the light of these.

2.5 INNOVATION STRATEGY

"An innovation strategy is an essential tool for product innovation and continued growth in difficult times" (Cooper & Edgett, 2010).

"Managers should articulate an innovation strategy that stipulates how their firm's innovation efforts will support the overall business strategy. This will help them make trade-off decisions so that they can choose the most appropriate practices and a set of overarching innovation priorities that align all functions" (Pisano, 2015).

Innovation within an organization should be far more than just a grab bag of good ideas and practices. Instead, it should be an integrated and coordinated effort across the organization aligned to the overall business strategy.

An overarching innovation strategy will provide the goals, direction, and framework for innovation across the organization. Individual business units and functions may have their own strategies to achieve specific innovation goals, but it is imperative that these individual strategies are tightly connected with the overarching organizational innovation strategy.

2.5.1 What Defines a Good Innovation Strategy?

The innovation strategy should be tailored to the specific organization. There is no standard recipe book that defines what an innovation strategy should be. The innovation strategy provides a sound basis for alignment across the organization, for establishing priorities, and for evaluating trade-offs.

A good innovation strategy:

- May be messy and should be embraced as such. While it is a process, it is a combination of art and science. It is most often non-linear in its application as it involves abstract concepts such creativity and perseverance.
- Is about experiential learning and development. It comprises iterations of trying out new things, capturing the lessons, and then reprioritizing efforts. The journey is as important as the end-goal.
- Requires information that is generated and useful during the innovation processes.
- Requires intentional curiosity in order to search for the desired designs, outcomes, processes, systems, or products.
- Ultimately is about creating value, not necessarily new ideas and or inventions.
- Requires stakeholder input as early as possible.
- Requires the ability to integrate seemingly disparate (and unobvious) components into one.
- Is ongoing, so organizations need to design and implement support structures and systems that enable it to be continuous.

Although, in many cases, innovation remains more of art than a science, Reeves et al. (2017) suggest it doesn't need to be. They suggest that "Innovation, much like marketing and human resources, can be made less reliant on artful intuition by using information in new ways. But this requires a change in perspective: We need to view innovation not as the product of luck or extraordinary vision but as the result of a deliberate search process. This process exploits the underlying structure of successful innovation to identify key information signals, which in turn can be harnessed to construct an advantaged innovation strategy." Reeves et al. (2017) use a simple example of three children building new LEGO toys to explain different approaches to innovation.

"Think back to your childhood days. You're in a room with two of your friends, playing with a big box of LEGOs (say, the beloved fire station set). All three of you have the same goal in mind: building as many new toys as possible. As you play, each of you searches through the box and chooses the bricks you believe will help you reach this goal.

Let's now suppose each of you approaches this differently. Your friend Joey uses what we call an impatient strategy, carefully picking LEGO men and their firefighting hats to immediately produce viable toys. You follow your intuition, picking random bricks that look intriguing. Meanwhile, your friend Jill chooses pieces such as axles, wheels, and small base plates that she noticed are common in more complex toys, even though she is not able to use them immediately to produce simpler toys. We call Jill's approach a *patient strategy*.

At the end of the afternoon, who will have developed the most new toys? That is, who will have built the most new products? Our simulations show that this depends on several factors. In the beginning, Joey will lead the way, surging ahead with his impatient strategy. But as the game progresses, fate will appear to shift. Jill's early moves will begin to seem serendipitous when she's able to assemble complex fire trucks from her choice of initially useless axles and wheels. It will appear that she was lucky, but we will soon see that she effectively harnessed serendipity.

What about you? Picking components randomly, you will have built the fewest toys. Your friends had an information-enabled strategy, while you relied only on intuition and chance.

What can we learn from this? If innovation is a search process, then your component choices today matter greatly in terms of the options they will open up to you tomorrow. Do you pick components that quickly form simple products and give you a return now, or do you choose the components that give you a higher future option value?"

Research by Reeves et al. (2017) provide three crucial insights to innovation strategy:

1. Information-enabled strategies outperform strategies that do not use the information generated by the search process.
2. In an earlier phase of the development of the innovation space, an impatient strategy outperforms; in later stages, a patient strategy does.
3. It is possible to have an adaptive strategy, one that changes as a market develops and that outperforms in all phases of the market's development. Developing an adaptive strategy requires you, in effect, to know when to switch from Joey's approach to Jill's (in the above LEGO example). The switching point is knowable and occurs when the complexity of products starts to level off after increasing.

2.5.2 An Information-Advantaged Strategy

Reeves et al. (2017) provide a five-step process for constructing an information-advantaged strategy:

1. Choose your space: *Where to play?*
2. It's not enough to analyze markets and customer needs. To innovate successfully you also need to fully understand everything in your "innovation space," including competitors, regulations, technology, etc.
3. Select your strategy: *How to play?*
4. Look backward, not forward. How complex is the innovation space? Consider the range of product variations, unique components, number of competitors, maturity of technology, etc. If the level of complexity is low and relatively stable, then choose an impatient strategy. If complexity is high, then choose a patient strategy,
5. Apply your strategy: *How to execute?*
6. If you choose an impatient strategy your objective is to bring relatively simple products to market quickly, increasing R&D speed and decreasing time to market, often pointing to development of a minimum viable product.

7. Sense shifts and adapt: *How to extract a switch signal?*
8. Remain adaptable. Continue to source information from your innovation space. Changes in this space may demand a change in your innovation strategy.
9. Brace for disruptions: *How to reset the clock?*
10. Although it is not always possible to pre-empt disruptions in the innovation space, maintaining key channels of information and open mind to changes in technology, regulations, and market demands is critical

2.5.3 The Relationship of Innovation Strategy to Overall Business and Individual Functional Strategies

An Example of Integrating Innovation with Business Strategy

Several years ago, "Bristol-Myers Squibb (BMS), as part of a broad strategic repositioning, decided to emphasize cancer as a key part of its pharmaceutical business. Recognizing that biotechnology-derived drugs such as monoclonal antibodies were likely to be a fruitful approach to combating cancer, BMS decided to shift its repertoire of technological capabilities from its traditional organic-chemistry base toward biotechnology. The new business strategy (emphasizing the cancer market) required a new innovation strategy (shifting technological capabilities toward biologics)" (Pisano, 2015).

Throughout this book, a number of specific strategies and processes are presented, which are guided by the overall innovation strategy, including:

- Across several chapters, technology strategy, marketing strategy, platform strategy, open innovation, intellectual property, sustainability.
- In Chapter 1, the stages of the product life cycle and their strategic importance in defining innovation priorities are discussed.
- In Chapter 3, portfolio management where project selection is aligned with, and directed by, the innovation strategy priorities.
- In Chapter 4, a range of processes that are appropriate to different innovation strategies are discussed.
- In Chapter 7, the various organization and team structures that are appropriate to different innovation strategies are discussed.

2.6 INNOVATION STRATEGY FRAMEWORKS

Several frameworks exist that contribute to creating and assessing innovation strategy. Following is a selection of approaches you should consider. No single one of these frameworks may be sufficient to fully develop an innovation strategy for an organization, but they should serve as foundational starting points.

2.6.1 Porter's Strategic Framework

Michael Porter (2008) argued that an organization's strengths ultimately fall under one of two headings: cost advantage and differentiation. By applying these strengths to a broad or narrow market, he defines three generic strategies: Cost leadership, differentiation, and segmentation. Refer to Figure 2.8.

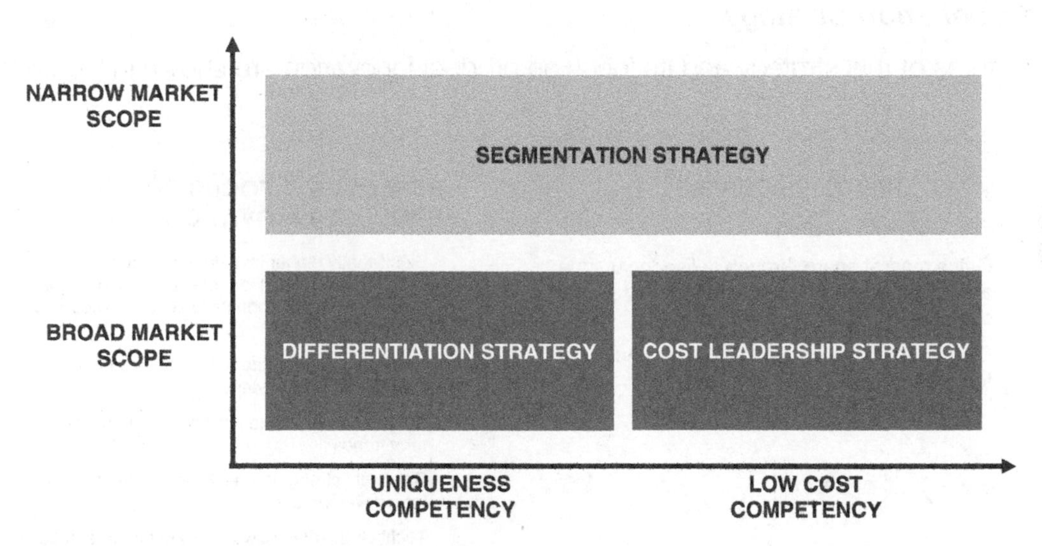

FIGURE 2.8 Porter's three competitive strategies

Porter's Cost Leadership Strategy

The specific features of this strategy and its focus on product innovation are shown in Figure 2.9.

WHAT ARE ITS FEATURES?

- Grows market share by appealing to "cost conscious" customers
- Often adopted by commodity product companies
- Achieved through:
  - Economies of scale
  - Offering "no frills" or "value" products that reduce overall manufacturing cost
  - Optimizing the supply chain

HOW DOES IT FOCUS ON PRODUCT INNOVATION?

- Level of product innovation funding is low—often below 0.3% of sales revenue
- Major emphasis on minor product changes (centered on cost reductions through changes in manufacturing process and raw materials)
- Little/no focus on long-term research or development
- Technology often plays an important role in improving manufacturing systems (automation, robotics)

FIGURE 2.9 Cost leadership strategy: Features and product innovation focus

Advantages of a cost leadership strategy include:

- It is often the only way to break into or maintain a market position in a price-competitive market where there is intense competition, and customers are less concerned about product differentiation.

Disadvantages of a cost leadership strategy include:

- The constant drive for cost reduction can lead to a reduced emphasis on quality, to the point where customers switch to a competitor;
- Generally, lower margins that are constantly being squeezed, resulting in limited R&D investment.

Porter's Differentiation Strategy

The specific features of this strategy and its focus on product innovation are shown in Figure 2.10.

WHAT ARE ITS FEATURES?

- Focus on a broad product base
- Gains market share through delivering unique and superior products and establishing a loyal customer base
- Customers focus on product quality and features

HOW DOES IT FOCUS ON PRODUCT INNOVATION?

- Significantly higher investment in product innovation than cost leadership (around 2% of revenue for food products to 20% for electronic goods)
- Intimacy with customers, to fully understand current and future needs
- Sound foresighting to predict short-to medium-term trends
- Relatively strong emphasis on research and longer-term development
- Technology often plays an important role (focused on product features and functionality)

FIGURE 2.10 Differentiation strategy: Features and product innovation focus

Porter's Segmentation Strategy

The specific features of this strategy and its focus on product innovation are shown in Figure 2.11.

WHAT ARE ITS FEATURES?

- Narrow market focus (unlike cost leadership and differentiation)
- Based on an intimate knowledge of a key niche segment or sub-section of the market—often with specialized needs

HOW DOES IT FOCUS ON PRODUCT INNOVATION?

- Higher product innovation funding than cost leadership or differentiation
- Major emphasis on customer intimacy
- Working with lead user groups in the target segment to identify new opportunities and co-develop new products
- Technology plays an important role in the development of new product features or functionality

FIGURE 2.11 Segmentation strategy: Features and product innovation focus

Advantages of a segmentation strategy include:

- Providing a strong focus for the organization's marketing and product innovation effort.
- Allowing the organization to drive for an in-depth understanding of and relationship with its customers. Therefore, this provides a strong competitive edge against newcomers.
- Providing the opportunity for higher margins and, therefore, product innovation investment.

Disadvantages of a segmentation strategy include:

- The danger of being dependent on a single, narrow market (putting the eggs in one basket). New technologies lead to your product being outdated. Refer to Section 2.6.3 on disruptive technologies, with examples such as Kodak's focus on photography film and the impact of new digital technology.

2.6.2 Miles and Snow Strategic Framework

The timeless Miles and Snow (1978) framework continues to provide a useful method to describe a strategic approach to innovation. It is based on four specific strategy approaches: prospector, analyzer, defender, and reactor. The characteristics of each approach are presented in Figure 2.12.

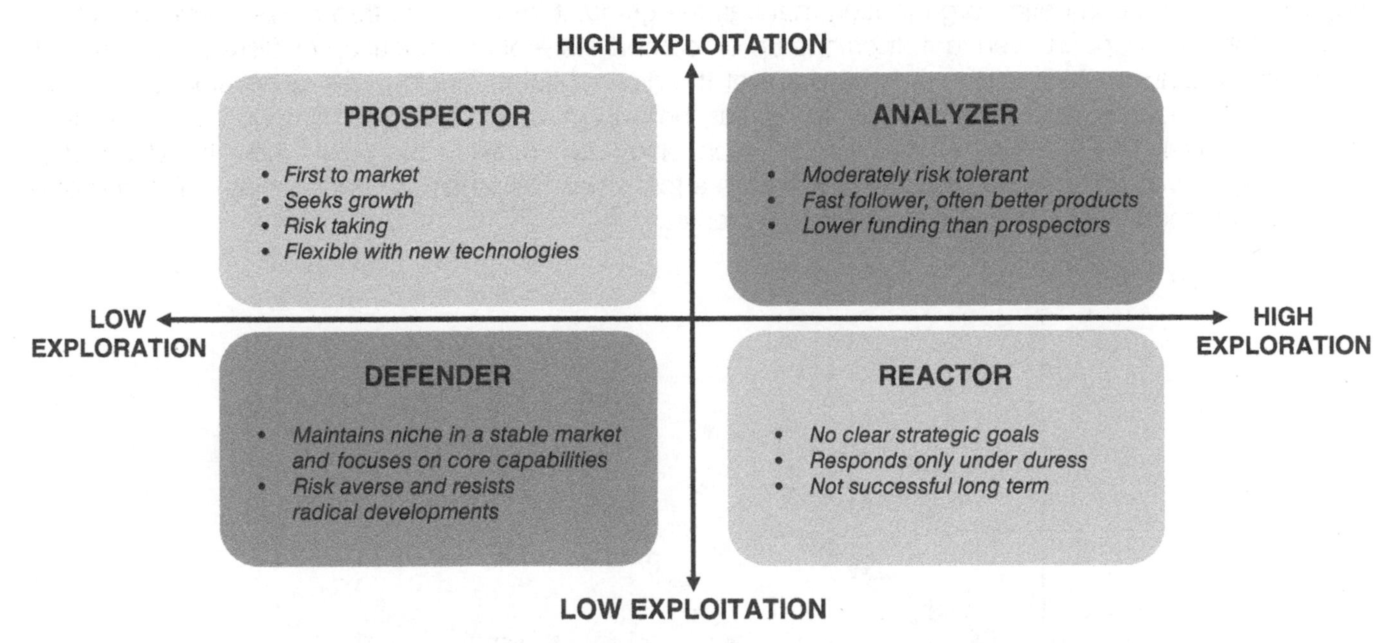

FIGURE 2.12 The characteristics of the Miles and Snow (1978) strategic framework

PDMA's 2021 Best Practice Survey (Knudsen et al., 2023), provide the following insights into the differences between the *Best* companies compared to the *Rest* with regards product innovation results:

- Prospectors are most likely to among to be among the *Best*.
- Reactors are most likely to be among the *Rest*.
- 80% of Prospectors, Analyzers, and Defenders indicate that their innovation efforts are driven by strategy, whereas less than 50% of Reactors claim this.
- A Proactive strategy leads to a higher level of radical innovation projects.
- Reactors are less likely to commercialize radical innovations.

2.6.3 Sustaining vs. Disruptive Product Innovation

A key strategic decision for many organizations is whether to focus their product innovation efforts on sustaining or disruptive innovation. Most organizations choose (or default to) sustaining innovation. Also, considering disruptive innovation opportunities in a product portfolio can create revolutionary opportunities.

The principle of disruptive innovation was first explored by Clayton Christensen (1997). He illustrated the difference between disruptive and sustaining innovation by using the diagram shown in Figure 2.13 and describing it as:

> "A *disruptive technology* or disruptive innovation is an innovation that helps create a new market and value network, and eventually disrupts an existing market and value network. The concept of disruptive technology is widely used but disruptive innovation seems a more useful concept in many contexts since few technologies are intrinsically disruptive. It is rather the business model than the technology that enables and creates the disruptive effect. In contrast to disruptive innovation, a sustaining innovation does not create new markets or value networks but only develops existing ones with better value, allowing the companies to compete against each other's sustaining improvements."

Although an organization can choose a sustaining or disruptive innovation strategy, often, it is a disruptive innovation of a competing organization that has the greatest impact. The trap of disruptive innovation is that it does not happen overnight. It creeps up on companies – often unnoticed or disregarded. A well-documented example is that Kodak was aware of the new digital technology being developed by other companies but disregarded it as a significant threat, believing that the quality of film was far better than digital could offer. This was true at the time until technology developed to the point where the digital deficiencies were overcome. By this stage, it was too late for Kodak, as other companies that embraced digital technology took a significant share of Kodak's market.

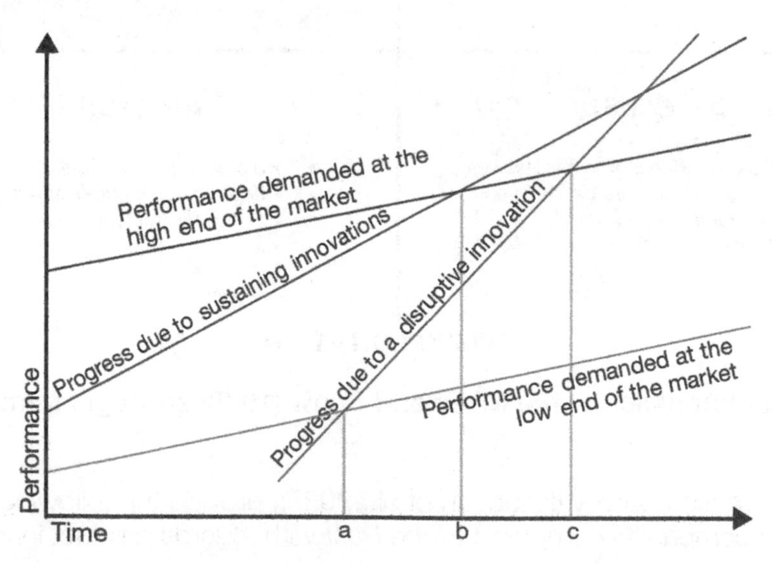

FIGURE 2.13 Sustaining vs. disruptive innovation.

Source: Reprinted with permission from Christensen (2016).

Features of disruptive innovation

- The new product or service targets specific needs of an existing product or market segment.
- Although the new product may be inferior in some ways to the existing product, it offers special features that are truly valued by the small segment of the market with users who are leaders or influential to the market as a whole. This allows the new product to gain a foothold in the market.
- In time, the overall features of the new product are improved to the point where more and more customers are "converted." The value of the new features now significantly outweighs any negative features or under-performance.

Examples of disruptive product innovation:

- ***Digital cameras*:** disrupting traditional film cameras.
- ***Uber*:** disrupting traditional taxi services (some purists may argue that this does not fully fit the true model of disruptive innovation).
- ***TripAdvisor*:** disrupting traditional travel agencies; Mobile phones: disrupting traditional land lines.
- ***3D printing*:** disrupting traditional manufacturing in the future; Artificial intelligence: disrupting human-incumbent jobs.
- ***Cryptocurrency*:** potential to disrupt payment services systems and banking.

2.6.4 The Innovation Landscape Map

Although new technology is often the source of innovation, this is by no means always the case. In fact, over recent years, many leading companies, including Amazon, LinkedIn, and Alibaba, have mastered the art of business model innovation. A key decision in defining an organization's innovation strategy is the relative effort and resources allocated to technological and business model innovation.

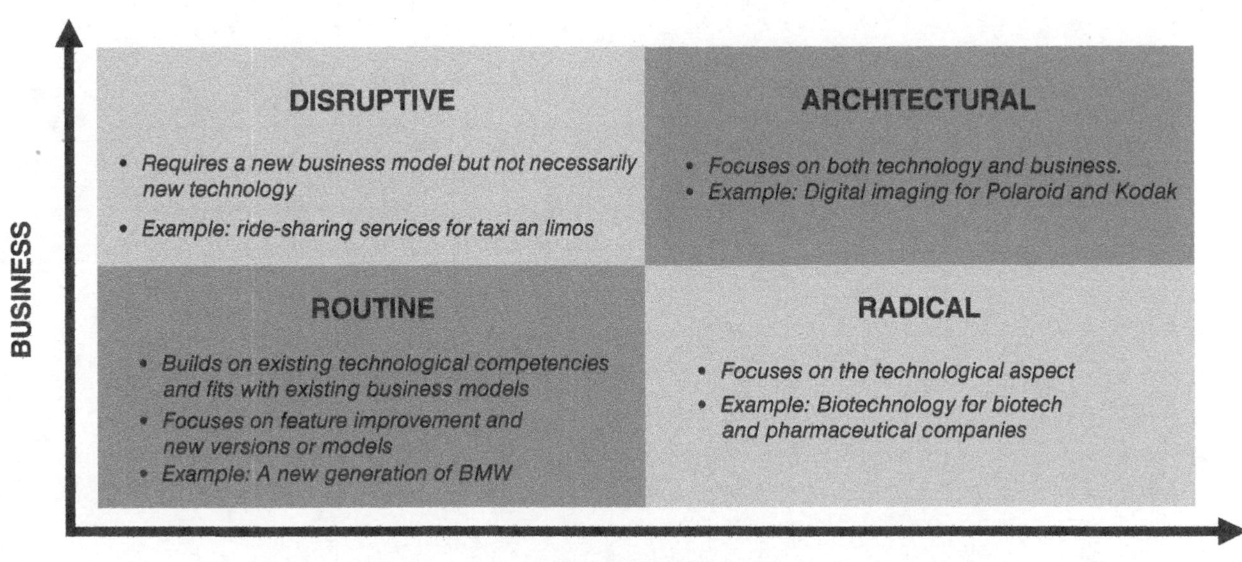

FIGURE 2.14 The innovation landscape map.

Source: Reprinted with permission from Pisano (2015).

The four quadrants of innovation described by Pisano in Figure 2.14 are:

1. ***Routine innovation*:** Builds on an organization's existing technological competencies and fits with its existing business models. Innovation is focused on feature improvement and new versions or models.
2. ***Disruptive innovation*:** Requires a new business model but not necessarily new technology. Note that this is not the identical description of disruptive innovation described by Clayton Christensen previously, but it is similar when new technology is not driving the disruption. An example is the Disruptive innovation introduced by Uber, which used a new business model (ride-sharing) but not significantly new technology (mobile apps).
3. ***Radical innovation*:** Here the focus is mainly on technological advances. As an example, genetic engineering and biotechnology have had a significant impact on the pharmaceutical industry.
4. ***Architectural innovation*:** A focus is placed on both technology and business model change. A well-quoted example is digital photography, which caused significant disruption for traditional film companies.

2.7 STRATEGIES THAT SUPPORT THE INNOVATION STRATEGY

The overarching innovation strategy provides the goals, direction, and framework for innovation across the organization. Contributing to and impacting this strategy are a number of other organizational areas and functions. Each of these may have its own specific and discrete strategy. It is vitally important that these individual strategies, which are critical to successful innovation, be tightly linked to the overall innovation strategy – feeding off and contributing to its development, evolution, and execution. Examples include:

- marketing.
- technology.
- product platforms.
- intellectual property.
- capability.
- digital.

Each of these is discussed below.

2.7.1 Marketing Strategy

A marketing strategy is a process or model to allow an organization to focus limited resources on the best opportunities to increase sales and thereby achieve a unique competitive advantage.

The marketing strategy must be informed by, and be consistent with, the business goals identified in the overall business strategy. It is, therefore, an important contributor to the successful development and implementation of the innovation strategy.

The hierarchy of business goals, marketing strategy, and marketing planning are summarized in Figure 2.15.

1. **Communicate business goals:** start with business goals encapsulated in the vision and mission.
2. ***Develop the marketing strategy*:** high-level direction for the marketing efforts.
3. ***Define the marketing mix*:** product, price, promotion, and place.
4. ***Prepare the marketing plan*:** specific tasks and activities designed to achieve the marketing strategy and business goals.

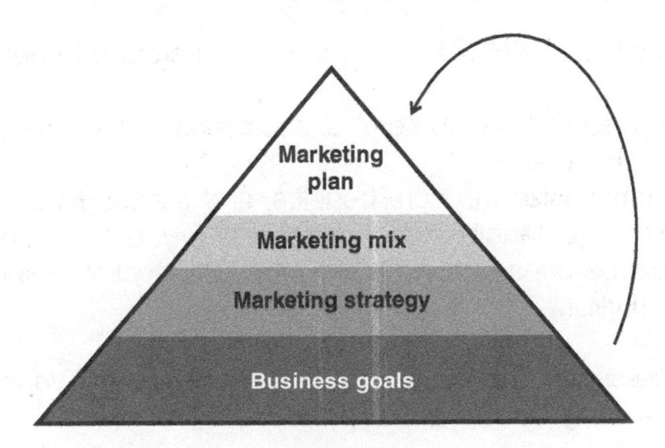

FIGURE 2.15 From business goals to marketing plans

In developing the marketing strategy, the following questions need to be answered:

- What products will be offered? This includes determining the breadth and depth of product lines.
- Who are the target customers? Consider market boundaries and segments to be served.
- How will the customers be informed of the product's availability and of its benefits?
- How will the products reach the customers? This specifies the distribution channels to be used.

The Marketing Mix

The marketing mix comprises the basic tools available to market a product. The market mix is often referred to as the *four Ps – product, price, promotion, and place*. Elements of the marketing mix are shown in Figure 2.16.

FIGURE 2.16 The marketing mix

Developing the marketing mix for a product

In developing the marketing mix for a product, the following should be considered:

- All elements of the marketing mix should be synchronized around target market needs and expectations.
- The price should be aligned with customers' expectations of the product value – evidenced by its functional and aesthetic attributes.
- The promotion should emphasize the core benefits, and the tangible and augmented features (see below for a description of core, tangible, and augmented product features).
- The place of sale should be consistent with the product's quality, brand perceptions, features, and behaviors of the target market.

Chapter 1 further discusses the application of the marketing mix at various phases of a product's life cycle, specifically emphasizing the introductory phase.

Levels of a Product

A product can be described at three levels (refer to Figure 2.17):

1. ***Core product:*** *The benefits that the target market will derive from the product.*
2. ***Tangible product:*** *The physical and aesthetic design features that give the product its appearance and functionality.*
3. ***Augmented product:*** *These are benefits that may be provided extra to the product, either free of charge or for a higher price.*

An example of levels: a luxury automobile

Core benefits: demonstration of wealth, power, and prestige to the owner.
Tangible features: sleek styling, powerful engine, heated seats, etc.
Augmented features: 5-year warranty, special payment terms, two years free service.

CORE: Benefits for target market

TANGIBLE: Physical and aesthetic design features

AUGMENTED: Extra benefit, free of charge or higher in price

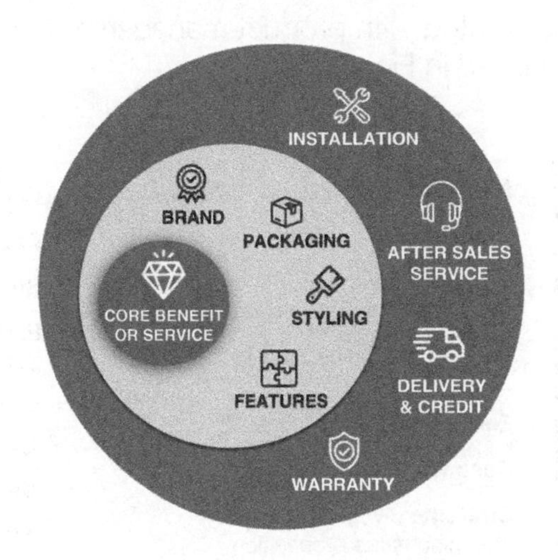

FIGURE 2.17 The three levels of a product

The three levels of a product will be developed further in Chapter 5 when we focus on product design and development.

The Value Proposition

Value proposition is defined as "a short, clear and simple statement of how and on what dimensions of a product concept will deliver value to prospective customers. The essence of 'value' is embedded in the trade-off between the benefits a customer receives from a new product and the price the customer pays for it." (PDMA ToolBook 1).

Clarity of the value proposition associated with a new product under development is central to its ultimate success in the marketplace. Appropriate market research to establish customer needs lays the foundation for the product's value proposition. This, in turn, can be developed into a clear concept description and product design specifications. Throughout the product innovation process, ongoing market research should ensure the product's design continues to align with the value proposition.

Analyzing the Current Product Portfolio

Analysis of an organization's current product portfolio is central to both product management and product innovation planning.

A technique developed by the Boston Consulting Group (BCG) provides a 2×2 matrix framework for analyzing the current product portfolio on the basis of two dimensions, market share and market growth (refer to Figure 1.20). Using the 2×2 matrix, current products are placed into one of four categories – stars, question marks, cash cows, and dogs – defined as follows:

Stars are products that command a significant market share in a growing overall market.

Question marks are products that are in a market of high overall growth but as yet have not captured a significant market share.

Cash cows are products that have a high share of a market that has low overall growth.
Dogs are products that have a low share in a market with low overall growth.

The strategies associated with product management and product innovation for each of these product categories are summarized in Figure 2.18.

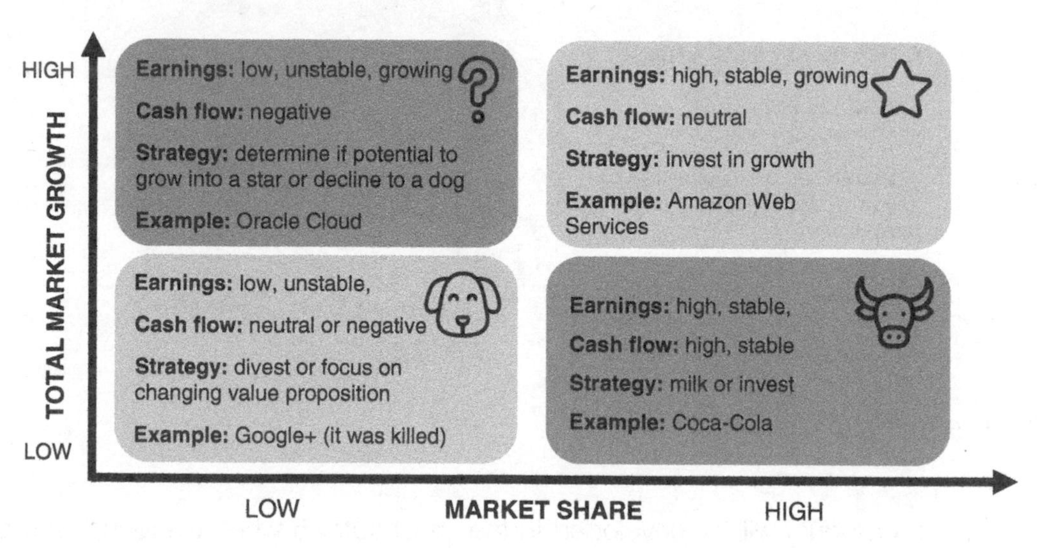

FIGURE 2.18 Adaptation of the BCG growth-share matrix.
Source: Reprinted with permission from NetMBA.com

The Product Roadmap

A product roadmap is a plan that matches short-term and long-term business goals with specific product innovation solutions to help meet those goals.

The purpose of a product roadmap is to communicate direction and progress to internal teams and external stakeholders. It shows the high-level initiatives and the planned steps to get there. Creating a product roadmap should be a continuous process throughout the life cycle of a product. It is also a collaboration tool to keep selected stakeholders informed of plans as well as provide a catalyst for input on plans. Product roadmaps have been further discussed in Section 1.7.

2.7.2 Technology Strategy

A technology strategy is a plan for the maintenance and development of technologies that supports the future growth of the organization and aids the achievement of its strategic goals.

Some questions to consider when developing an organization's technology strategy include:

- Does the organization lead or follow in its adoption and development of new technologies?
  - What are the boundaries that will impact the organization's innovation (the maximum level of risk and uncertainty it has on innovation projects)?
  - If and when it follows, does it acquire or imitate the leaders?
- What level of investment does the organization make in developing and owning new technologies?
- What methods does the organization use to acquire and protect technologies? Patents, trade secrets, standards, speed (refer to Section 1.7.3 on intellectual property).
- What comprises the company's technology platform(s) – the technologies shared across products, services, and processes?
- Does the organization make or buy technologies?
  - To what extent does it open its innovation to the outside world?
  - To what extent and in what ways does it engage partners and suppliers in technology development?
- What core competencies does it need to develop in-house? Core competencies are those absolutely critical to the successful development and implementation of the new technologies.

Technology Foresighting

Technology foresighting is a process of looking into the future to predict technology trends and their potential impact on the company. A wide range of tools is used for foresighting. These include brainstorming, expert panels, Delphi, SWOT analysis, patent analysis, and trend analysis. A basic framework is shown in Figure 2.19.

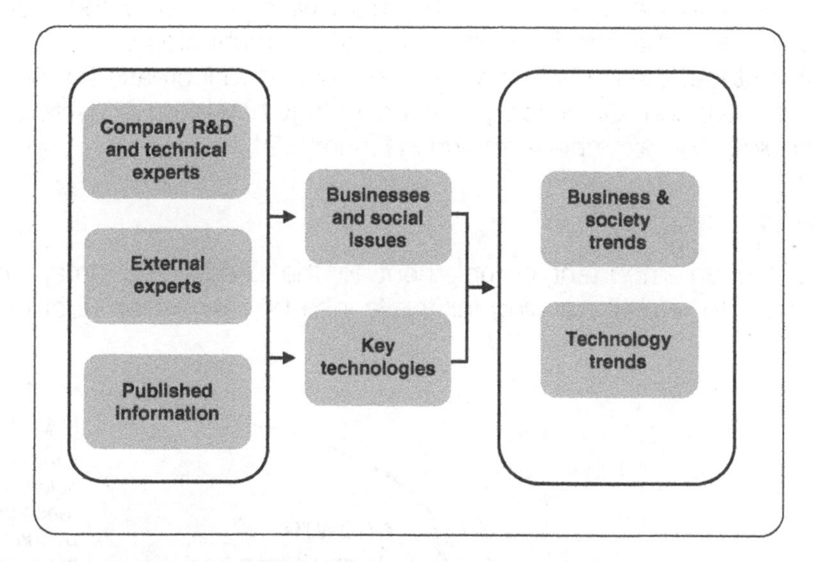

FIGURE 2.19 A basic technology foresighting framework

Technology Strategy: Link to the Business and Innovation Strategies

The role and importance of a technology strategy will depend on the strategic focus for the organization. Consider these guidelines:

- Technology-driven companies are highly dependent on achieving a competitive edge through new and innovative technologies.

- Market-driven companies are heavily focused on meeting customer needs, where the technology may or may not be a significant component.
- Most companies will fall somewhere in the middle ground, with a strong focus on meeting customer needs but where technology is an important consideration in achieving a competitive edge.

In Section 2.6.4, Pisano's innovation landscape map was introduced. This presented four quadrants of innovation strategy based on the relative emphasis on business and technology. Architectural innovation was defined as focusing on both business and technology, while radical innovation was defined as a more singular focus on technology. Organizations that have strategically positioned themselves in either the architectural or radical quadrants of Pisano's innovation landscape will need to place a strong emphasis on the status and trending of relevant technologies (both current and future). Fundamental to technology development and application is the principle of the technology life cycle or the technology S-Curve.

Technology S-Curves

The technology S-Curve (see Figure 2.20) shows the life cycle stages that apply to most technologies as they emerge, grow, and mature. The stages are:

1. ***Embryonic*:** this is where the technology is initially applied, often with limited performance. At this stage, there are significant risks for companies applying the technology. These risks, associated with potential product failure and lack of customer satisfaction, may be a deterrent to organizations that are employing relatively low-risk innovation strategies. On the other hand, organizations that have committed to higher-risk strategies may see technology at the embryonic stage as an opportunity to gain an early foothold in the market and to become a market leader.
2. ***Growth*:** at this stage, there is significant improvement in the technology and its performance. The more risk-adverse organizations will now be more willing to apply the technology. This leads to greater competition for other products also applying the technology.
3. ***Maturity*:** At this stage, the limits of the science required to increase the performance of the technology have been reached. Or possibly, the technology has been superseded with new and more advanced technology. An example is shown in Figure 2.21.

Technology Roadmap

Technology roadmaps are an important complement to the product roadmap in aligning technology planning and development to overall planning for the launch of a single product or a range of products.

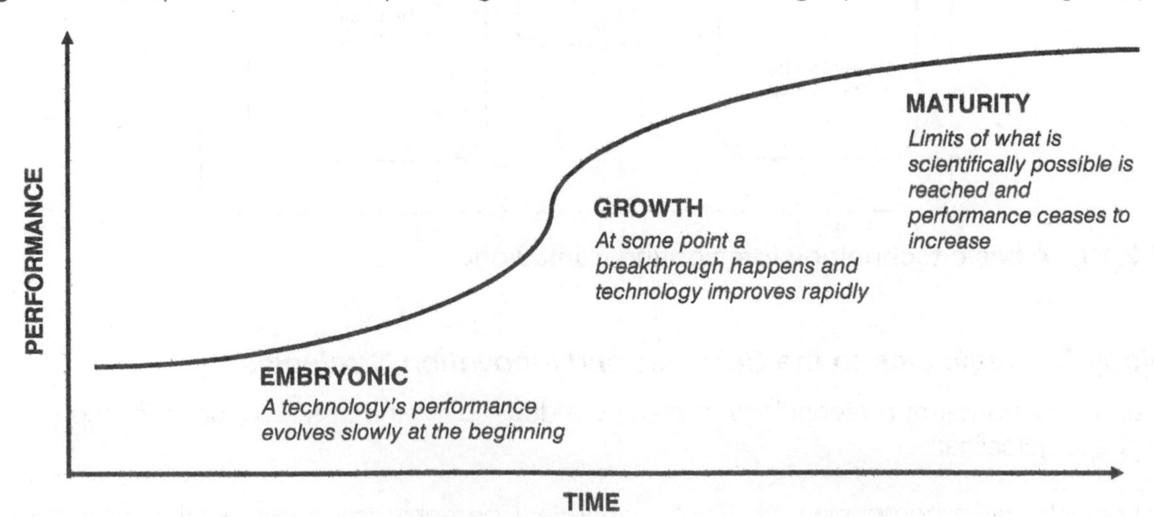

FIGURE 2.20 The technology S-Curve

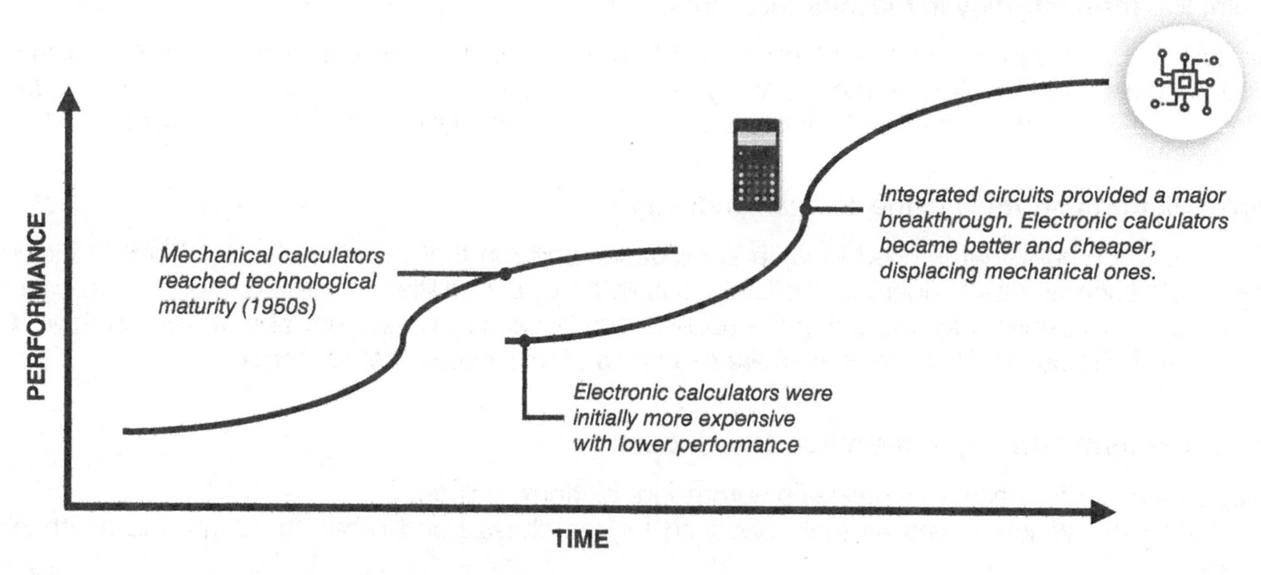

FIGURE 2.21 Technology disruption example: mechanical calculators

Technology road mapping is particularly important in organizations that have a strong strategic focus on technology in underpinning the innovation strategy and new product innovation. Refer to Section 1.7.3 for additional information on technology roadmaps.

2.7.3 Product Platform Strategy

A product platform strategy is fundamental to the development of new products for most companies.

A product platform strategy is defined as a set of subsystems and interfaces that form a common structure from which a stream of derivative products can be efficiently developed and produced.

> "A product platform technology is not a product. It is a collection of common elements, especially the underlying core technology, implemented across a range of products. It is the foundation of product strategy, especially in high-tech companies where many products are built from a core technology" (McGrath, 1995).

Benefits of platform strategy

A product platform strategy offers the following benefits to an organization:

- It enables products to be deployed rapidly, consistently, and with lower technical risk.
- It encourages a longer-term view of product strategy.
- It can leverage significant operational efficiencies.
- It ensures that the underlying elements of the product platform are clearly understood by the organization and the market.
- It can provide a significant point of differentiation from competing products, especially if the platform contains unique IP.

Example: Platform strategy in the internet industry

Google turned its initial product into a platform for future products. In 1998, the Google search engine was released in beta form. Google's search capability has since become a platform for other Google products, including a search hardware appliance for companies, Gmail, Maps, Docs, YouTube, and more.

Example: Platform strategy in the software industry

A software environment that is used to write applications and run them is a product platform. It includes software tools such as GUI builders, compilers, class libraries, and utilities for developing the applications, as well as a runtime engine for executing the applications because they are not able to run on their own. Sun's Java and Microsoft's NET Framework are examples of major software platforms.

Example: Platform strategy in the automobile industry

Key mechanical components that define an automobile platform include:

The floor plan, which serves as the foundation for the chassis and other structural and mechanical components.

- Front and rear axles and the distance between them (the wheelbase).
- Steering mechanism and type of power steering.
- Type of front and rear suspension.
- Placement and choice of engine and power train components.

For example, the Audi TT and the VW Golf share many mechanical components but appear entirely different visually.

2.7.4 Intellectual Property Strategy

Intellectual property (IP) refers to creations of the mind, such as inventions, literary and artistic works, designs, symbols, names, and images used in commerce. Just like other forms of property (land, buildings, etc.), IP can be sold, licensed, exchanged, or given away by its owner.

IP is particularly important in product innovation because it defines the potential for an organization to capture value from new products. This can be done either directly by the organization in its manufacturing and marketing of the new product, by licensing the product to another organization, or by selling the IP rights. Protection of IP ownership is, therefore, an essential component of an organization's business strategy. IP is protected under law in a number of ways that enable the owner to earn recognition or financial reward from what they invent or create.

Types of IP Rights

Patents, copyrights, and trademarks are the most common forms of IP in product innovation. Trade secrets and niche protection for plant varieties also exist.

1. ***Patent:*** a government authority or license conferring a right or title for a set period, especially the sole right to exclude others from making, using, or selling an invention.
2. ***Copyright:*** the exclusive and assignable legal right, given to the originator for a fixed number of years, to print, publish, perform, film, or record literary, artistic, or musical material.
3. ***Trademarks:*** a symbol, word, or words legally registered or established by use as representing an organization or product.
4. ***Trade secrets:*** information related to IP that is retained confidentially within an organization. Unlike the other forms of IP listed here, trade secrets are not registered with a government authority.
5. ***Plant variety rights:*** Based on the USDA Plant Variety Protection Act, this gives exclusive right to produce for sale and sell propagating material of a plant variety.

IP Management Maturity Model

As an organization progresses in its maturity in applying intellectual property as a business strategy, it moves from a reactive state, where the emphasis is simply on tracking IP developments, to a proactive state utilizing IP to drive competitive advantage. The decision on what emphasis is placed on IP as a component of overall business strategy will depend on the organization's goals and the environment in which it operates. Table 2.3 summarizes the management of IP under four specific headings – reactive, proactive, strategic, and optimized.

TABLE 2.3 Approaches to IP management

	Approach			
Activity	**Reactive**	**Proactive**	**Strategic**	**Optimized**
Research and product innovation	Patent as an after-thought	Freedom to operate	Aligned with business strategy	IP drives strategic advantage – R&D investment
IP portfolio and management	Simple portfolio tracking	Relate portfolio to business Build IP awareness	Portfolio management input to R&D and licensing	Portfolio management for competitive advantage
IP acquisition and monetization	Ad hoc response to IP licensing opportunities	Proactively identify licensing partners	IP royalty and revenue goals	Business drives IP monetization & acquisition targets
Competitive intelligence	Ad-hoc or situation-driven intelligence	Competitive intelligence on key industry players	Ongoing analysis of complete IP competitive landscape	Competitive intelligence is key to business strategy
Risk management and litigation	Respond to surprise litigation	Risk profile monitoring Defense patenting	Protective licensing	Non-mitigated risks are insured

Figure 2.22 outlines the approach to managing IP in an optimized manner, where IP is front and center in achieving the organization's strategic goals. Here, IP is used:

- To drive competitive advantage,
- As a key criterion in portfolio management, and
- As a key driver for company profitability and growth.

Competitive intelligence underpins strategic decisions and mitigates risks.

2.7.5 Capability Strategy

With the innovation strategy defined, it is time to move to execution. It is imperative to establish the right set of capabilities to implement that strategy.

Some Key Definitions

Competencies are based on human resources and refer to the combination of skills, knowledge, and ability that lead to superior performance in an organization and that allow it to be competitive in the marketplace. There are two main types, commonly referred to as "soft" or "behavioral" competencies and "hard" or "technical" competencies. Behavioral competencies include cognitive and personality characteristics, while technical competencies include learned expertise such as project management.

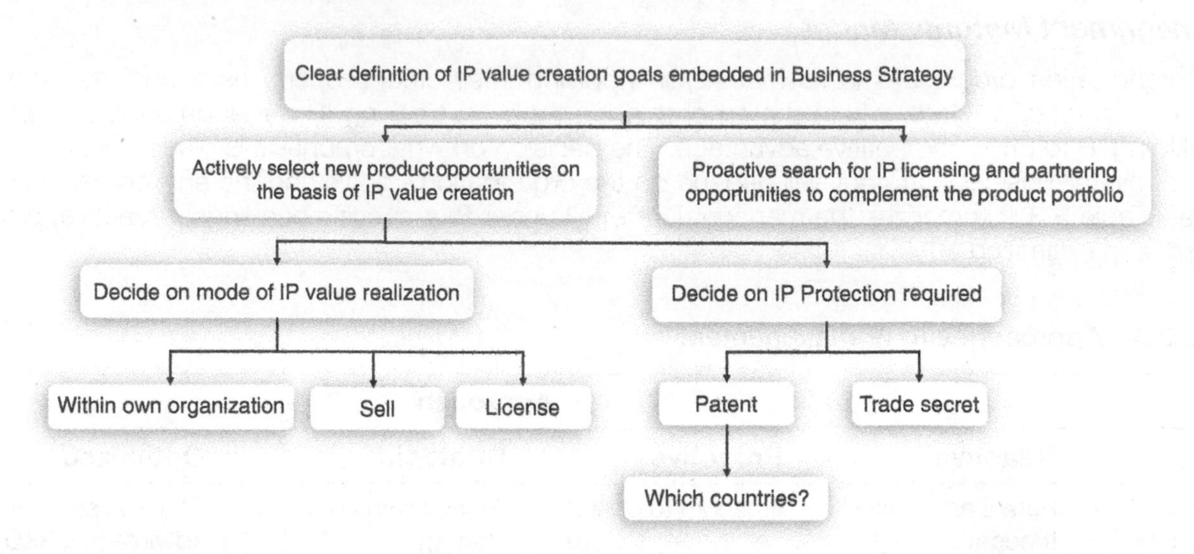

FIGURE 2.22 An optimized approach to IP strategy

Behavioral competencies involve characteristics such as analytical thinking, interpersonal ability, and initiative. Abilities rely on natural or inherent behaviors as opposed to learned ones. Although abilities can be developed to some extent, the majority of what constitutes ability cannot be learned.

Technical competencies involve the knowledge and skills which are learned through study and practice. Skills are the application of knowledge in work (a trade or profession) or leisure.

Core competencies are what give an organization one or more competitive advantages in creating and delivering value to customers. Whereas a standard set of competencies allow an organization to compete in the marketplace, core competencies are those that give the organization a truly competitive edge.

Resources are the operational inputs that allow an organization to perform its business activities. Resources are often divided into three categories:

- Physical assets,
- Human resources,
- Organizational capital.

Capabilities are the activities and functions an organization performs to utilize its resources (physical assets, human resources, and organizational capital) in an integrated way. Capabilities are practiced and developed over time. As they become stronger, the organization enhances its expertise in a particular functional or operational area. This expertise allows the organization to differentiate itself from competitors.

Capability-based strategies, also referred to as the resource-based view of the organization, are determined by those internal resources and capabilities that provide the platform for the organization's strategy and those resources and capabilities that are the primary source of profit. A key management function is to identify what resource gaps need to be filled in order to maintain a competitive edge where these capabilities are required.

A Basic Roadmap for Developing Capability-based Strategies for Product Innovation

1. Ensure a clear definition of organizational mission, goals, business strategy, and innovation strategy. As discussed in previous sections of this chapter, these underpin all organizational decision-making.

A precise alignment is needed between each element. For example, if the business strategy is based on cost leadership, a prospector innovation strategy would be out of alignment. Refer to Figure 2.23.

2. Carry out a SWOT analysis, specifically focusing on the organization's capability to take advantage of opportunities and combat threats. What specific resources (physical assets, human resources, and organizational capital) are required?
3. Carry out a "capability audit" to identify current organizational resources and their strengths.
4. Identify the gaps between the required capability and the current capability. Specifically:
   - What new resources are needed?
   - Where do current resources need to be strengthened?
   - What competencies are required to compete?
   - What core competencies are required to provide a truly competitive edge?
5. How should the organization develop the desired capability and acquire the required resources? Specifically:
   - Develop internally, e.g., train or re-train existing staff.
   - Add roles to supplement or complement existing competencies.
   - Use mergers, acquisitions, joint ventures, and/or open innovation to incorporate external competencies.

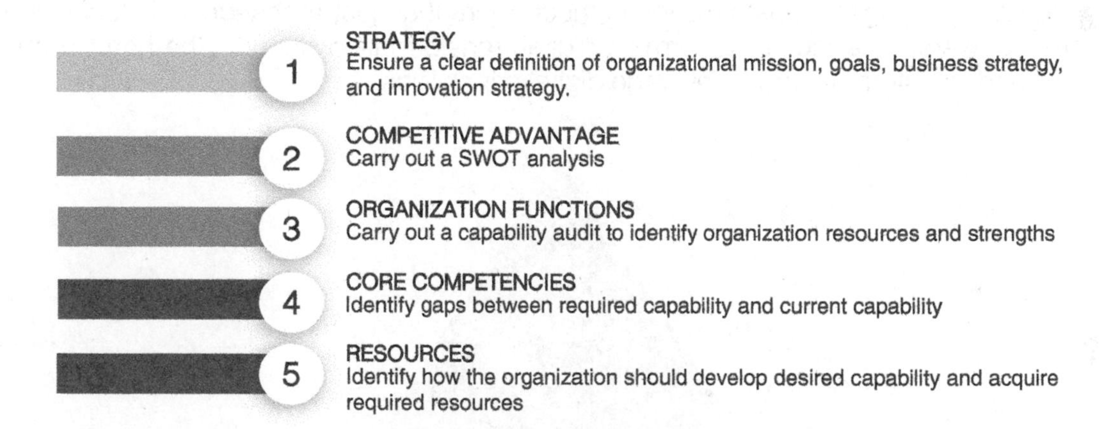

1
STRATEGY
Ensure a clear definition of organizational mission, goals, business strategy, and innovation strategy.
2
COMPETITIVE ADVANTAGE
Carry out a SWOT analysis
3
ORGANIZATION FUNCTIONS
Carry out a capability audit to identify organization resources and strengths
4
CORE COMPETENCIES
Identify gaps between required capability and current capability
5
RESOURCES
Identify how the organization should develop desired capability and acquire required resources

FIGURE 2.23 The hierarchy of capability-based strategy

2.7.6 Digital Strategy

Digital Transformation History

The foundation for digital transformation dates back to the 1940s when digital communications were formalized in Claude Shannon's article, A Mathematical Theory of Communication. The dispersion of computers to organizations and individuals, along with the rapid expansion of the Internet, made digital communications commonplace. By 2010, many organizations had already moved manual systems into a digital world. For example, filling out a paper form became rarer as data was more likely captured on a web form, in a PDF, or other digital formats. This was the first significant step in what most people would recognize as a move from analog to digital, with organizations automating processes that were previously analog and more manually intensive. During this time, the strategy was often phrased as "we are going digital." This was, and continues, to be aligned with the adoption of commercial enterprise-wide information systems provided by SAP, Oracle, Microsoft, and a host of other companies and included Enterprise Resource Planning, Customer Relationship Management, Inventory Management, and many other large IT systems. This paved the way for what is now called the Digital Transformation strategy.

It is generally agreed that the Covid-19 pandemic resulted in a significant acceleration in digital transformation. Walker (2021), in his summary of the findings from the Harvard Business Review Pulse Survey on Accelerating Transformation for a Post-Covid-19 World, reported 95% of respondents said that digital transformation has grown in importance over the first 12 months of the pandemic; 90% said that Covid-19 accelerated the timing or their organization's transformation efforts; and 58% said their organization's transformation strategies since the start of the pandemic had been effective, up for 20% before the outbreak.

Following are some key definitions of terms that, to a large extent, have been used interchangeably over recent years (see Figure 2.24 showing the hierarchy of digital applications):

Digitization refers to creating a digital representation of physical objects or attributes. The goal of digitization is to make information more easily accessible, storable, maintained, and shared.

Digitalization refers to enabling or improving processes by leveraging digital technologies and digitized data. The purpose of digitalization is to enable, improve, and transform business operations through the use of digitized data and technologies in order to transform how organizations conduct business and improve productivity.

Digital transformation: The Institute for Digital Transformation defines digital transformation as, "The integration of digital technologies into a business resulting in the reshaping of an organization that reorients it around the customer experience, business value, and constant change." This definition recognizes that digital transformation is much more than just a change in IT technology such as hardware, software, or a digital platform. A digital transformation goes to the heart of the business processes and transforms them to leverage digital capabilities.

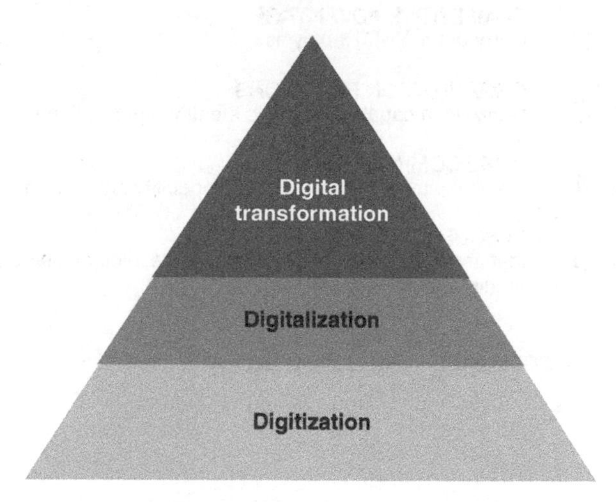

FIGURE 2.24 Hierarchy of digital application

Digital Technology and the Development of New Products and Services

Sheen (2019) provides an illustrative list of categories of digital technologies that impact transformation in the development of new products and services.

- "Smart devices are products or appliances that are characterized by sensors or monitors, a user interface, processing capability, and digital communication. As such, these devices are essentially computers embedded in a product that provides additional functionality. Most smart devices can also be customized with a personal profile or programmable functionality that enables the user to create a unique pattern of interaction with the device.
- The Internet of Things is a digital technology that has proliferated during the past few years. It is characterized by any device or equipment with digital communication capability along with sensors

or monitors. The Internet of Things connects devices digitally across a network and allows either one-way or two-way communication.

- Analytics platforms provide aggregation of data into a dashboard or user interface. This enables a customer to monitor and control a process that has multiple connected devices. The analytics can also be used to troubleshoot and analyze the performance of the different steps in the process being monitored.
  - Digital platforms are software applications that operate over the internet and connect individuals from different user group categories. Through the platform connection, individuals or corporate entities can exchange information, purchase products or services, and promote a cause or product. Digital platforms can be very complex, with many user group categories and extensive sharing capability for presenting and exchanging information. Or the platform can be very simple and focused on a small set of categories or type of information."

Transformation and the Development of New Products or Services

The digital age has impacted product design and development in many ways. Sheen (2019) highlights the following:

- The customer's perspective of the value of a product has extended beyond its aesthetics and performance to the level of connectedness and relationship it provides:
  - Connection with other customers through user groups and product forums.
  - Connection with the manufacturer in providing feedback on the product's performance.
  - Connection with other devices using a common platform.
- Input to the early stages of design and development has been enhanced through the use of social media, crowdsourcing, Big Data, and other on-line tools. Providing rapid feedback on design concepts, often used computer-aided modeling to demonstrate the features and functionality of a product concept.
- Launching of innovative new products – "Crossing the Chasm," discussed in Section 1.6, has been facilitated through digital technologies such as computer-aided design and rapid prototyping that allow customers to experience a product before purchasing. The early adopters can provide feedback through a range of digital media, which can convince mainstream customers to purchase the product.
- The downside of developing products in the digital age is the availability of information that can enable competitors to enter the market more quickly.

Developing a Digital Transformation Strategy

The definition of a digital transformation highlights the integration of digital technologies throughout the entire organization. To be successful, a digital transformation strategy must, therefore, be inextricably linked to the overall vision and goals of the organization and to the supporting strategies of product innovation, marketing, technology, intellectual property, capability, and sustainability. The basic steps in the development of a digital transformation strategy are:

1. Define the overall vision and goals for the organization. Critical steps in this process, discussed in some detail in Sections 2.2–2.4, are the analysis of the internal and external environments and forward projection in terms of consumer trends, competition, and technology. Techniques including Delphi and PESTLE should point to future technology and consumer trends and highlight the potential emergence and growth of new technologies and their potential for product innovation.
2. Identify specific areas of digital technology that are required to underpin the achievement of organizational goals.
3. Define what is required to achieve digitalization.
   - Which specific areas of the business need to be digitized?
   - How will digitization be implemented within and across those areas of the business?

- What capability is required to implement the new technologies?
- What capability is required to operate and maintain these technologies?

4. Define how digital transformation will be achieved.
   - What capability is required for business continuance and improvement based on the new digital technologies?
5. How will the new business model be communicated across the organization?
   - What training programs are required to upskill current staff?
   - What specific KPIs will be implemented to measure and improve the digitalized business model?
   - What structures and systems will be implemented to ensure ongoing improvement?

A Hypothetical Example

Know Your Farm is an agricultural consultancy providing advice and assistance to farmers for over 20 years. Sam, the son of the company founder, has recently taken over the leadership of the company. Sam has been trained in computer science and information technology and has worked for five years in the application of digital technologies in the pharmaceutical industry. *Know Your Farm* is a very traditional company, employing 20 people, mainly with farming backgrounds, working off paper-based systems and basic telephone communication with its farming clients. Recent years have seen significant changes in the farming industry with a growing emphasis on technology. Sam is keen to take advantage of this trend based on a new vision and goal for the company.

The Vision and Goal:

Transforming *Know Your Farm* into a digitally-based company, facilitating easy communication with farmers and across the farming community. Providing rapid feedback and advice to farmers based on the most up-to-date information.

What specific areas of *Know Your Farm* need to be digitized?

The most important area is a digital toolkit for farmers and contractors that will enable them to boost productivity, enhance safety, and reduce paperwork.

What is required to achieve digitalization?

- Input from farmers is required on what they would specifically value in this digital toolkit – its features and functionality. A lead user group of progressive farmers will provide on-going input to the development of the final product.
- Although Sam has experience in digital technology, he recognizes the need for more support in this specific area of product application.

How will digital transformation be achieved?

- Although there is significant support among the company directors and employees for moving into the digital age, the support is not universal, and very few staff have digital technology experience beyond using a mobile phone.
- A training program will be provided to all staff, who will then be encouraged to actively participate in the development of the new digital toolkit through testing its features and functionality and involvement with the farmer lead user group.
- Based on the feedback from staff and farmers, an implementation plan for transitioning to the new digital toolkit will be prepared together with feedback and customer support mechanisms.

2.8 OPEN INNOVATION

2.8.1 Foundations of Open Innovation

Open Innovation (OI) extends innovation resources beyond the boundaries of the organization. It is defined as the strategy adopted by an organization to actively seek knowledge from external sources through alliances, partnerships, and contractual arrangements that complement and enhance its internal capability in pursuit of improved innovation outcomes. These innovation outcomes may be commercialized internally, through new business entities, or through external licensing arrangements.

OI can deliver many benefits, but applying it is challenging. Table 2.4 summarizes some potential benefits as well as challenges of applying OI, organized around the six principles of OI (Tidd & Bessant, 2013).

TABLE 2.4 The benefits and challenges of Open Innovation

Principle of OI	Potential benefit	Challenges in OI application
Tap into external knowledge	• Increase pool of knowledge • Reduce reliance on limited internal knowledge	• How to search for and identify relevant knowledge sources • Breaking down traditional internal views on innovation
Acquire external R&D	• Can reduce the cost and uncertainty associated with internal R&D and increase depth and breadth of R&D	• Less likely to lead to distinctive capabilities and more difficult to differentiate • External R&D may also possibly come from competitors
Do not have to originate research in order to profit from it	• Reduce cost of internal R&D, more resources applied to external search strategies and relationships	• Need sufficient internal R&D capability in order to identify, evaluate, and adapt external R&D
Building a better business model is superior to being first to market	• Greater long-run profitability and success	• First-mover advantages depend on technology and market context and may be difficult to overcome
Best use of internal and external ideas, not generation of ideas	• Better balance of resources to search and identify ideas, rather than generate	• Many companies are better at generating superficial ideas than finding the long-term winners • The cost of evaluation is often high when sifting through potentially thousands of ideas
Profit from others' Intellectual Property (IP) (inbound OI) and others' use of our IP (outbound OI)	• Gets IP in the hands of those companies best equipped to commercialize	• Conflicts of commercial interest or strategic direction • Negotiation of acceptable forms and terms of IP licenses

Source: Adapted from Tidd and Bessant (2013).

2.8.2 Open Innovation Model Types

Searching for and then working with external parties (e.g., customers, suppliers, competitors, or research institutions such as universities) is labeled as "inbound open innovation." Organizations can become different types of "inbound open innovators" depending on the number/types of partners with which they work (aka "partner variety") and the number/types of phases of their innovation process that they "open" to external parties (aka "innovation openness") (Tidd & Bessant, 2013). Figure 2.25 summarizes the four models of OI according to the level of innovation openness and the partner variety. *Closed innovators* have a low level of openness to innovation and a low range of partners. *Integrated collaborators* have a relatively low number of partners with a high level of collaboration. *Specialized collaborators* work with a wide range of partners but with a low level of openness. *Open innovators* have a high level of openness and work with a number of partners.

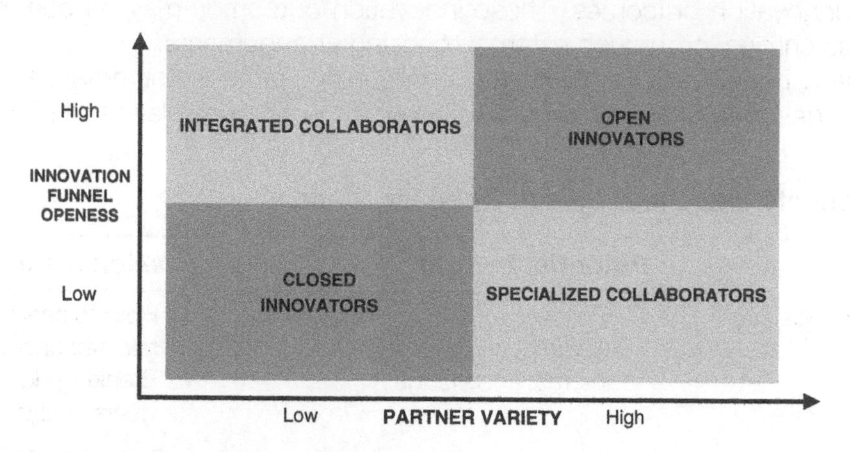

FIGURE 2.25 Open Innovation model types

Organizations must exercise caution when deploying resources for inbound innovation. Research shows that at low levels of openness (being generally closed to outside ideas) and very high levels of openness (considering ideas from a vast array of sources), innovation performance seems to suffer, while organizations with a moderate amount of openness showed superior product and process performance.

The main success factors and appropriate managerial styles differ for these different types of open innovators (Tidd & Bessant, 2013). The main success factors for each of the four OI models and the associated management styles are presented in Figure 2.26.

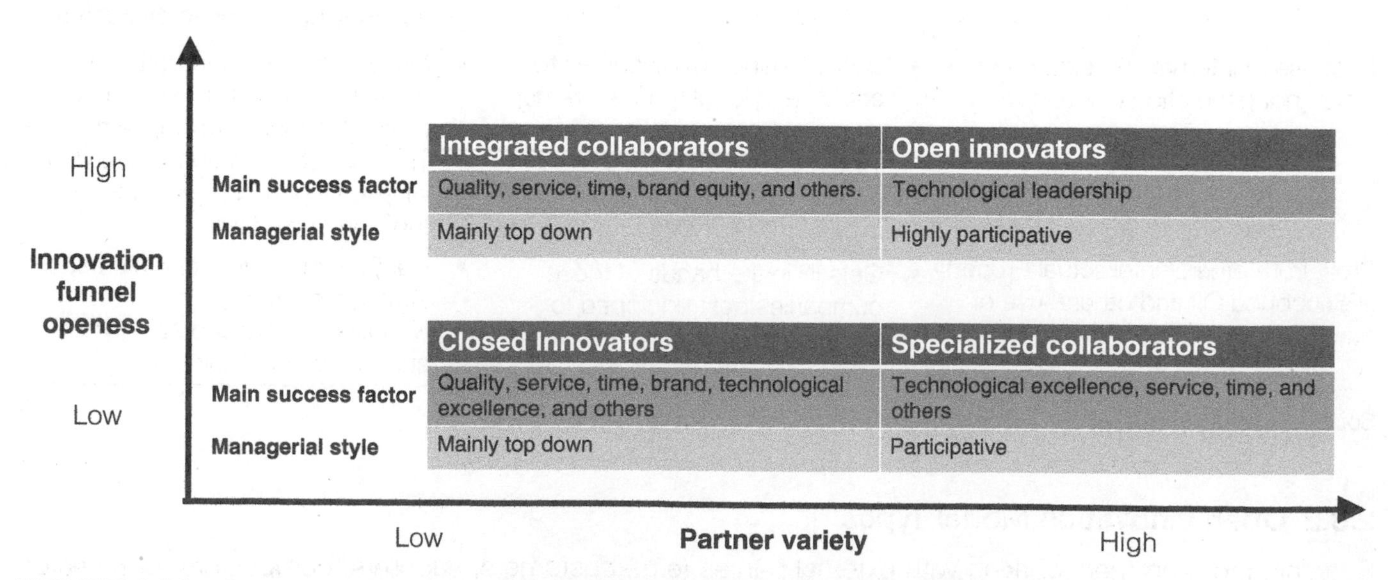

FIGURE 2.26 Success factors and management styles for OI models

Becoming an open innovator improves innovation performance, but this effect is highly contingent on the number of potential partners, the use of pre-screening these partners, and the uncertainty of the industry environment for the organization (Lazarotti & Manzini, 2014).

Organizations can participate in OI in various ways: innovation seeker, innovation provider, intermediary, or open innovator (Ellis et al., 2014). The mechanisms for participation in each are summarized in Table 2.5.

TABLE 2.5 Mechanisms for participation in Open Innovation

OI participation type	Mechanisms to participate
Innovation seeker	Leverage user innovation; outsource and form alliances
Innovation provider	Venture capital, licensing, and alliances
Intermediary	Auctions and partnerships
Open innovator	Outsource, form alliances, mergers and acquisitions, and venture capital licensing

For further reading on Open Innovation, refer to Chesbrough (2003) and Noble et al. (2014).

Examples of Open Innovation

Open Innovation is used by many organizations, including a number of PDMA's Outstanding Corporate Innovation (OCI) award winners.

- ***Novozymes***, a leader in enzymes and microbial technologies, uses OI in various phases of the innovation process. They are particularly active in crowdsourcing in the idea generation stage: they collect ideas online for two weeks, screen these ideas and then mature the selected ones for one month. These matured ideas are then pitched to management as *New Lead Project Proposals*.
- ***Offshore Oil Engineering Co., Ltd (COOEC),*** the OCI winner in 2018, is the world's leading supplier of KIBS marine energy engineering and provides innovative integrated solutions to petroleum and gas field development. COOEC actively uses OI through partnering with university research centers, suppliers, and clients.
- The ***Smart Village*** initiative in impoverished rural Indian states. A number of companies, many of them competing against each other in other markets, collaborated to empower villagers so that the villagers can join in the supply chain and create a win-win environment both for themselves and for the participating companies (Darwin & Chesbrough, 2016).

2.9 SUSTAINABLE INNOVATION

2.9.1 What is a Sustainable Business?

A sustainable business is an enterprise that seeks a positive impact on the global or local environment, community, society, or economy – a business that strives to meet the triple bottom line defined below. Sustainable businesses may also have progressive environmental and human rights objectives.

Sustainability in business generally addresses two main categories in addition to profit:

- The effect business has on the environment;
- The effect business has on society.

The goal of a sustainable business strategy is to make a positive impact on either of those areas. When companies fail to assume responsibility, the opposite can happen, leading to issues like environmental degradation, inequality, and social injustice.

Beyond helping curb those global challenges, sustainability can drive business success. Several investors today use environmental, social, and governance (ESG) metrics to analyze an organization's ethical impact and sustainability practices. Investors look at factors such as a company's carbon footprint, water usage, community development efforts, and board diversity.

According to McKinsey (n.d.) "the strongest motivating factors among company employees to adopting a sustainable mindset are to align with a company's goals, missions, or values; build, maintain, or improve reputation; meet customers' expectations; and develop new growth opportunities."

2.9.2 Sustainability and Strategy

Most organizations now have an overarching framework for incorporating sustainability into standard organization operations. This implies that:

- Organizations develop a formal sustainability plan.
- Sustainability is used to drive competitive advantage.
- Sustainability is used as a driver for innovation and developing new products, following the triple bottom line concept.
- Sustainability is included in the organization's mission statement and values.
- Sustainability metrics are tracked at the executive level.
- A sustainability maturity model is in place, and progress is tracked regularly.
- Sustainable innovation is used to be compliant, market-driven, engaged, or shaping the future, or compliant, reactive, proactive, and purposeful.

The Triple Bottom Line

In traditional accounting, the "bottom line" refers to either the profit or loss at the very bottom of the statement of revenue and expenses. It was the key indicator of business performance. Over recent years, businesses have sought to evaluate their performance from a broader perspective, taking into account social and environmental contributions or impact.

The *triple bottom line* reports performance against three dimensions:

1. ***Financial:*** pursuing profits because that creates a strong organization that can be responsible to the planet and people.
2. ***Social:*** treating people and communities with respect, in part because these are current and future employees and customers.

3. ***Environmental***: interacting with the environment in a way that creates a sustainable (long-term) business model. Some companies have also found ways to reduce costs, such as converting a waste from their manufacturing process into a valued product.

These dimensions are often referred to as the three Ps of the triple bottom line:

- Profit
- People
- Planet

How to Create a More Sustainable Business Strategy

Step 1: Assess the problem and define objectives

The first step to driving change is assessing what sustainability means to your team, company, industry, and client. Consider the big problems each of these groups think is a priority. This information can be used to create a materiality matrix depicting the sustainability topics relative to the impact on your organization and its importance to your stakeholders. A simple example: https://www.greenbiz.com/article/how-make-your-materiality-assessment-worth-effort

To guide this process, consider asking questions such as:

- Which societal and environmental topics are most important for the organization's customer segments?
- Where can societal and environmental topics form a risk for the organization or its customers?
- What is the purpose of the organization?
- How much waste is the organization creating?
- Does the organization have a positive culture?
- Are the hiring practices attracting diverse job candidates?
- Are the organization's products targeted to address the main topics in your materiality matrix?
- What impact does the organization have on the local community?

Answering these questions will help establish the organization's sustainability objectives and improve the quality of input to a PESTLE analysis (Section 2.4.1).

Step 2: Establish the sustainability component of the organization's mission

Decide on specific sustainability objectives and ensure that these are embodied in your company's mission. Ensuring that these objectives are an integral part of the mission is an important part of becoming a more sustainable business. Here are two examples of companies with mission statements that effectively focus on sustainability:

Alignable: "We believe that local businesses are stronger together. Our goal is to help small business owners make the connections that lead to long-term relationships, generate more word-of-mouth referrals, and unlock access to the collective wisdom of the local business community."

Patagonia: "Build the best product, cause no unnecessary harm, use business to inspire, and implement solutions to the environmental crisis."

In each, it is clear what the company values are, and how they are executing against those values.

Step 3: Craft the sustainability strategy

A clearly defined contribution of sustainability to the organization's mission is the starting point for a sustainable business strategy.

In crafting a sustainable business strategy, it is important to ensure that the organization remains profitable. While each component of the triple bottom line is important, profit is the first priority as it enables the other elements of sustainability.

Consider this: does the organization typically leave the electricity and heat on overnight, even while no employees are on site? Imagine how much savings could be realized, in both cost and energy resources, if the last person in the office simply shut them off or sensors were added to turn lights and other devices off when people are not present. Or are consumers willing to pay more for a sustainably produced product? A 2017 Unilever study found that *33 percent of consumers want to buy from brands doing social or environmental good,* creating an untapped opportunity in the market for sustainable goods (Unilever, 2017). Since then, annual surveys show the percentage of consumers choosing to buy from sustainable organizations is increasing. One study shows that 90% of Gen Z is willing to pay more for sustainable products (Petro, 2022).

There are several strategies for specific industries that can increase operational efficiency while driving social and internal value. Putting in the work to build a robust sustainability strategy can help both the organization and the environment in the long term.

2.9.3 Sustainable Product Innovation

This is the process in which new products or services are developed and brought to commercialization. Throughout this process the characteristics of sustainable development are respected from the economic, environmental, and social perspectives, in the sourcing, production, use, and end-of-service stage of the product life cycle. It has a global focus, beyond the primary life cycle of the product/service and takes all stakeholders into account.

Sustainable product innovation is supported by sustainable development, which is defined as "development which meets the needs of current generations without compromising the ability of future generations to meet their own needs" (Brundtland Commission, 1987).

The Importance of Sustainability to Product Innovation

Over recent years, sustainability has become increasingly important in product innovation. A 2011 Sustainability and Innovation Global Executive Study reported that 70 percent of the organizations surveyed included sustainability permanently on the management agenda and invested in it (Haanaes et al., 2012).

STAGE 1: BEGINNING

- Corporate policies do not recognize the triple bottom line (economic, social, environmental)
- Overall strategy is focused on compliance with the minimum legal requirements
- Limited sustainability awareness when setting new product goals and specifications
- Supplier policies do not incorporate sustainability
- Limited sharing and development of sustainability metrics

STAGE 2: IMPROVING

- Centralized sustainability reporting function is in place
- Environmental, health and safety policy publicly states metrics and goals
- Carbon, energy and water footprints are established and measured at the plant level
- Regulatory and policy issues are actively communicated throughout the organization
- Business and product strategy anticipates future customer behavior based on sustainability trends
- Supplier assessment includes a review of supplier's sustainability policies
- Checklists and other tools used to compare sustainability of new products during the NPD process

STAGE 3: SUCCEEDING

- Best practices for improving the triple bottom line are consistently leveraged across the entire enterprise
- Sustainability metrics are established at the enterprise level and linked to the business success of the company
- NPD processes consider and encourage sustainability throughout the phase-gate review process
- Company has shifted focus from meeting government regulations to exceeding requirements (benign by design)
- Supply chain impacts to the corporate sustainability goals are well understood and improvement activities are implemented
- Supplier selection is based upon a supplier's environmental and sustainability policies and efforts

STAGE 4: LEADING

- Company publishes an annual sustainability report that discusses aspects of the triple bottom line
- Corporate sustainability policy is fully integrated into other company policies and viewed as an important lever to drive growth and profitability
- Sustainable ideas and IP are leveraged for broader use by way of the supply chain, licensing, sale of IP, joint ventures, etc. with the specific intent of broader impact
- Organization emphasizes R&D innovation to develop new technologies and design methods that will reduce the overall environmental footprint of new products
- Majority of company's products are "certified" according to industry standards and third party assessments
- Product sustainability metrics shared broadly and seen as competitive edge

FIGURE 2.27 Innovation Maturity Model

In the 2012 PDMA CPAS, a third of the companies responded that sustainability contributed to their profits (Markham & Lee, 2013).

Figure 2.27 presents an Innovation Maturity Model, which shows the characteristics relating to sustainability for organizations moving from what can be described as a "beginner" to a "leader."

Sustainability and Company Strategy Examples

IKEA: Is committed to only using renewable and recycled materials and to reduce the total IKEA climate footprint by an average of 70% per product.

Nike: Aims to use 100% renewable energy across all its owned facilities. In 2020, the company was only using 48% renewable energy in these locations. Plus, the company wants to reduce the use of energy in key operations by 25% (Nike, Inc., 2013).

Apple: Is committed to becoming carbon neutral by 2030and has made sustainability a major selling point in the marketing for new products.

Intel: Is committed to achieving net positive water, 100% renewable electricity, zero waste to landfill, and additional absolute carbon emissions reductions.

Unilever: Is committed to reducing the total waste footprint per consumer use of our products by 32%, and achieving zero waste to landfill across all our factories. Reducing greenhouse gas emissions from our own manufacturing by 65%, and achieving 100% renewable grid electricity across our sites.

2.9.4 Externalities

Externalities are the effects of a product on people or the environment that are not reflected in the market price of the product. Externalities are a consideration for many companies. Government policy or regulation can be used to incorporate externalities into product price. Carbon pricing is an example of a policy that incorporates an externality (i.e., greenhouse gas emissions) into a product (i.e., electricity).

In the absence of regulation or policy, some companies have found advantages in marketing the incorporation of externalities into products. For example, products labeled as having been produced from renewable resources may command a higher price from consumers who value the environment.

2.9.5 The Circular Economy and Innovation

Cradle-to-cradle (the beginning of a product's life to the start of a new product's life) thinking or focusing on the circular economy can become a strategic driver. In the circular economy, the aim is to create closed loops in the product life cycle based on three principles (Ellen Macarthur Foundation, n.d.).

Principle 1: Preserve and enhance natural capital by controlling finite stocks and balancing renewable resource flows.
Principle 2: Optimize resource yields by circulating products, components, and materials at the highest utility at all times in both technical and biological cycles.
Principle 3: Foster system effectiveness by revealing and designing out negative external influences.

Figure 2.28 shows examples of products in the circular economy.

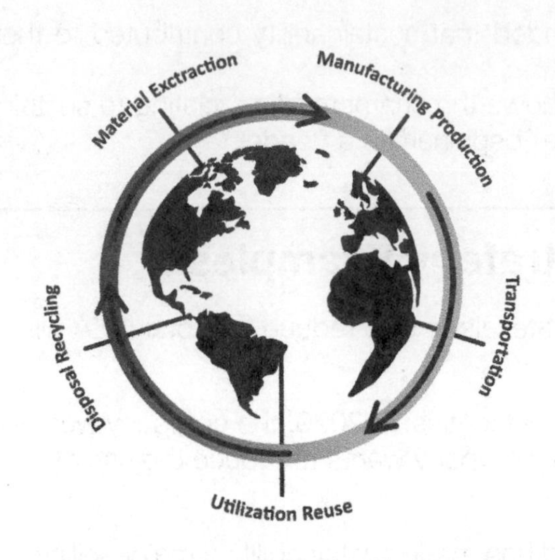

Raw materials

Method detergent
Plant-based formula

I'm green

Braskem, ingeo, corbion
Bio-based plastics

Design

Desso carpet tiles
Designed for disassembly and recycling

Herman miller office chairs
Designed for disassembly and materials

Business model

Caterpillar
Component recovery

Rype office
New, remade, or refurbished furniture

FIGURE 2.28 Examples of products for the circular economy

2.10 IN SUMMARY

- Strategy lies at the very heart of an organization's sustenance and growth. It lays the foundation and provides the framework for all of the organization's functions and activities.
- Strategy starts from the corporate and business levels of the organization and feeds into the various functional strategies of the organization.
- Laying a strong foundation of knowledge related to the organization, the industry, the wider environment, competitors, and customers is essential to good strategy development. A number of tools can assist in this process, including SWOT, Delphi, PESTLE, and Business Model Canvas.
- An innovation strategy is essential to most organizations. It stipulates how the organization's innovation efforts will support the overall business strategy. This provides the basis for trade-off investment decisions across product categories or business units.
- A range of frameworks is available to help with defining an organization's innovation strategy. Examples include those developed by Pisano (2015), Porter (2008), Chesbrough (2003), and Miles and Snow (1978).
- The innovation strategy is complemented by platform, technology, marketing, intellectual property, digital and capability strategies. All strategies must be aligned to achieve common innovation goals.
- Realization of the product strategy (development and launch) can be achieved either using internal or external capabilities – or a combination of both. Increasingly, organizations are adopting an Open Innovation strategy to actively promote innovation with external parties.
- Increasingly, sustainability is becoming an integral part of an organization's overall mission with a specific strategy focused on sustainability. Most organizations now provide triple bottom line reporting based on the three Ps of profit, people, and planet.

2.A STRATEGY PRACTICE QUESTIONS

1. A platform strategy is defined as__________?
   A. A common marketing plan for a series of products.
   B. A common manufacturing system for a range of products.
   C. A set of systems and interfaces that form a common structure from which a stream of derivative products can be effectively developed and manufactured.
   D. A strategy that connects marketing, technology, and manufacturing.

2. A corporate vision is__________.
   A. An NPD project goal.
   B. A philosophy statement about the beliefs of the company.
   C. A set of values.
   D. A picture of some desired future state that the organization hopes to obtain.

3. Mission statements help organizations to__________.
   A. Focus people and resources.
   B. Initiate ideas.
   C. Plan for product launch.
   D. Borrow capital at a low rate.

4. A strategy can be defined as__________.
   A. An organization's philosophy.
   B. A company's game plan for achieving its long-term objectives.
   C. An organization's new product development plan.
   D. The organization's values statement.

5. Which of the following are **not** one of Porter's Five Forces?
   A. Threat of new entrants.
   B. Bargaining power of customers.
   C. Threat of substitutes.
   D. Bargaining power of employees.

6. Who is responsible for setting the corporate and business strategy?
   A. Process management.
   B. Senior management.
   C. The shareholders.
   D. The board of directors.

7. Your competitors have launched a new product. As you evaluate your response to compete with them, you realize that the technology employed in your current product has reached its limits and cannot be improved to increase the performance to effectively compete and the only option is to consider superseding with newer and advanced technology. What stage on the technology S-Curve are you on?
   A. Growth stage.
   B. Embryonic stage.
   C. Maturity stage.
   D. Laggard.

8. A dairy product manufacturer has discovered a new microorganism that may have significant health benefits (probiotic). The company decides to initiate a significant research program to prove these health benefits. It plans to use the probiotic in a range of products including yogurts, health drinks, and infant formula. This is an example of what type of strategy:
   A. Line extension.
   B. Technology.
   C. Platform.
   D. Marketing.
9. Which of the following strategies is most likely to have a strong emphasis on technology?
   A. Routine.
   B. Architectural.
   C. Radical.
   D. Both B and C.
10. In the BCG Growth-Share matrix a "cash cow" is defined as a product with high market share but low growth potential. What strategy should the company adopt for this type of product?
    A. Take as much profit from the product as possible.
    B. Sell the product while it is still making money.
    C. Carry out a detailed analysis of the product, the competitors, and the market, leading to a clear strategy for the future of the product.
    D. Make significant investment in product improvements.

Answers to practice questions: Strategy

1. C	6. B
2. D	7. C
3. A	8. C
4. B	9. D
5. D	10. C

REFERENCES

Brundtland Commission. (1987). *Report of the world commission on environment and development: Our common future*. United Nations. https://digitallibrary.un.org/record/139811

Chesbrough, H.W. (2003). The era of open innovation. *MIT Sloan Management Review 44* (3): 35–41.

Christensen, C. (1997). *The innovator's dilemma: When new technologies cause great firms to fail*. Harvard Business Review Press.

Christensen, C. (2016). *The innovator's dilemma: When new technologies cause great firms to fail*. Harvard Business Press.

Cooper, R. and Edgett, S. (2010). Developing a product innovation and technology strategy for your business. *Research-Technology Management 53* (3): 3–40. https://doi.org/10.1080/08956308.2010.11657629.

Darwin, S., & Chesbrough, H. W. (2016). *Prototyping a scalable smart village to simultaneously create sustainable development and enterprise growth opportunities*. The Berkeley-Haas Case Series. University of California, Berkeley. Haas School of Business. https://doi.org/10.4135/9781526430786

Ellen Macarthur Foundation. (n.d.). *It's time for a circular economy*. https://ellenmacarthurfoundation.org/

Ellis, S.C., Gianiodis, P.T., and Secchi, E. (2014). Advancing a typology of open innovation. In: *Open innovation research, management and practice* (ed. J. Tidd), 39–86. Imperial College Press https://doi.org/10.1142/9781783262816_0003.

Haanaes, K., Balagopal, B., Arthur, D., Kong, M. T., & Velken, I. (2012). *The embracers seize advantage*. MIT Sloan Management Review.

Kahn, K.B., Evans Kay, S., Slotegraaf, R.J., and Uban, S. (ed.) (2013). *The PDMA handbook of new product development*, 3e. John Wiley & Sons https://doi.org/10.1002/9781118466421.

Knudsen, M.P., von Zedtwitz, M., Griffin, A., and Barczak, G. (2023). Best practices in new product development and innovation: Results from PDMA's 2021 global survey. *Journal of Product Innovation Management 40* (3): 257–275. https://doi.org/10.1111/jpim.12663.

Kotler, P. (2012). *Marketing management*. Prentice Hall.

Lazarotti, V. and Manzini, R. (2014). Different modes of open innovation: A theoretical framework and an empirical study. In: *Open innovation research, management and practice* (ed. J. Tidd), 15–38. Imperial College Press https://doi.org/10.1142/9781783262816_0002.

Markham, S.K. and Lee, H. (2013). Product innovation and management association's 2012 comparative performance assessment study. *Journal of Product Innovation Management 30* (3): 408–429. https://doi.org/10.1111/jpim.12025.

McGrath, M. (1995). *Product strategy for high technology companies*. McGraw-Hill.

McKinsey. (n.d.). *Insights on sustainability*. https://www.mckinsey.com/capabilities/sustainability/our-insights

Miles, R.E. and Snow, C.C. (1978). *Organizational strategy, structure and process*. McGraw Hill https://doi.org/10.2307/257544.

Nike, Inc. (2013). *Nike, Inc. FY12/13 sustainable business performance summary*. https://unglobalcompact.org/participation/report/cop/active/75941

Noble, C., Durmusoglu, S., and Griffin, A. (2014). *Open innovation: New product innovation essentials from the PDMA*. Wiley. https://doi.org/10.1002/9781118947166.

Osterwalder, A., Pigneur, Y., and Smith, A. (2010). *Business model generation: A handbook for visionaries, game changers, and challengers*. Wiley.

Petro, G. (2022). *Consumers demand sustainable products and shopping formats*. Forbes. https://www.forbes.com/sites/gregpetro/2022/03/11/consumers-demand-sustainable-products-and-shopping-formats/?sh=55da1b846a06

Pisano, G. (2015). You need an innovation strategy. *Harvard Business Review 93* (6): 44–54.

Porter, M.E. (1979). How competitive forces shape strategy. *Harvard Business Review 57* (2): 137–145.

Porter, M.E. (1996). What is strategy? *Harvard Business Review 74*: 61–78.

Porter, M.E. (2008). *On competition*, Updated and expandede. Harvard Business School Pub.

Reeves, M., Fink, T., Palma, R., & Harnoss, J. (2017). *Harnessing the secret structure of innovation*. MIT Sloan Management Review. https://doi.org/10.7551/mitpress/11858.003.0005

Sheen, R. (2019). *Digital transformation in product development*. Institute for Digital Transformation. https://www.institutefordigitaltransformation.org/digital-transformation-in-product-development/

Tidd, J. and Bessant, J. (2013). *Managing innovation: Integrating technological, market and organizational change*, 5e. John Wiley & Sons http://www.sciencedirect.com/science/article/pii/S002463010900051X.

Tregoe, B. and Zimmerman, J. (1980). *Top management strategy*. Simon and Schuster.

Unilever. (2017). *Report shows a third of consumers prefer sustainable brands*. https://www.unilever.com/news/press-and-media/press-releases/2017/report-shows-a-third-of-consumers-prefer-sustainable-brands/

Walker, M. (2021). *Accelerating transformation for a post-COVID-19 world*. HBR Pulse Survey, Harvard Business School Publishing.

Whetten, D.A. (2006). Albert and Whetten revisited: Strengthening the concept of organizational identity. *Journal of Management Inquiry 15*: 219–234. https://doi.org/10.1177/1056492606291200.

Porter, M.E. (1979). How competitive forces shape strategy. *Harvard Business Review* 57 (2): 137–145.

Tidd, J. and Bessant, J. (2013). *Managing innovation: Integrating technological, market and organizational change*, 5e. John Wiley & Sons.

3

Product Innovation

Portfolio

Establishes and maintains an appropriate balance of new and existing product innovation projects aligned with the business and innovation strategies

Product Development and Management Body of Knowledge: A Guidebook for Product Innovation Training and Certification, Third Edition.
Allan Anderson, Chad McAllister, and Ernie Harris.

WHAT YOU WILL LEARN IN THIS CHAPTER

Portfolio management ensures that the collection of product innovation projects is aligned with strategy and resource availability. In this chapter, we further define the relationship of the product innovation portfolio to the overall business strategy. The types of projects that can be included in the portfolio are discussed, together with criteria and methods that are commonly used to select projects to achieve a desired portfolio balance.

Some of the practical issues in implementing portfolio management are highlighted, together with suggested strategies to address these issues.

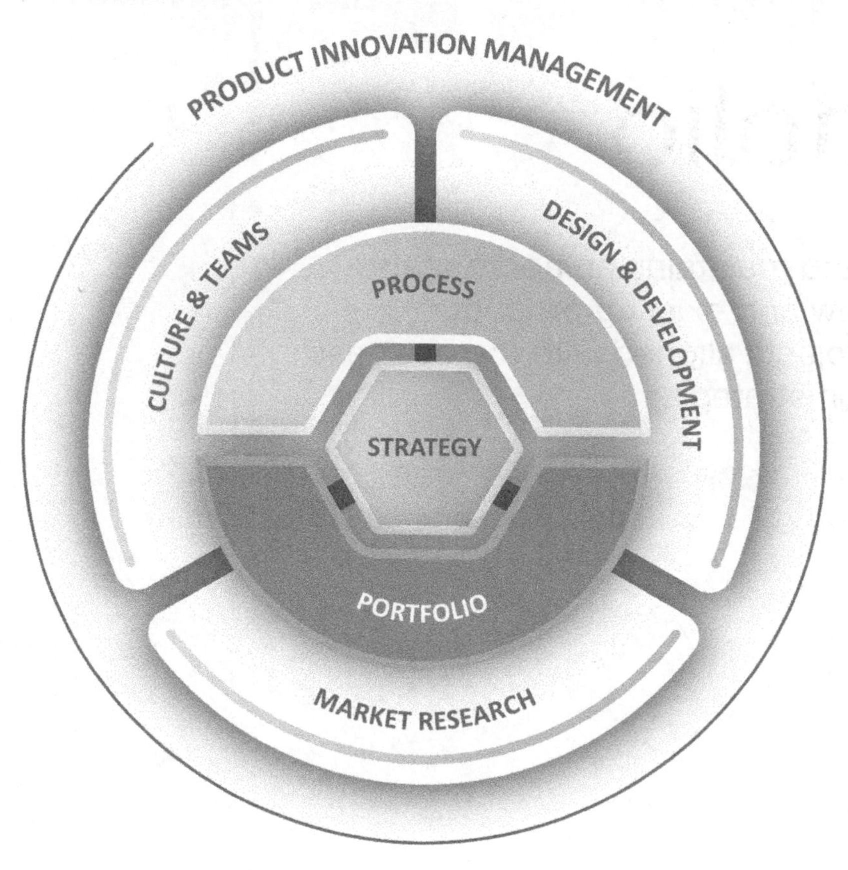

THE CHAPTER ROADMAP

The corporate strategy or business strategy, and, in turn, the innovation strategy, are prerequisites of portfolio management. In both its development and ongoing maintenance, an organization's product portfolio invariably has a set of projects competing for resources and investment. Portfolio management is an essential tool, informed by the innovation strategy, that ensures the right prioritization and balance of product innovation and product management projects.

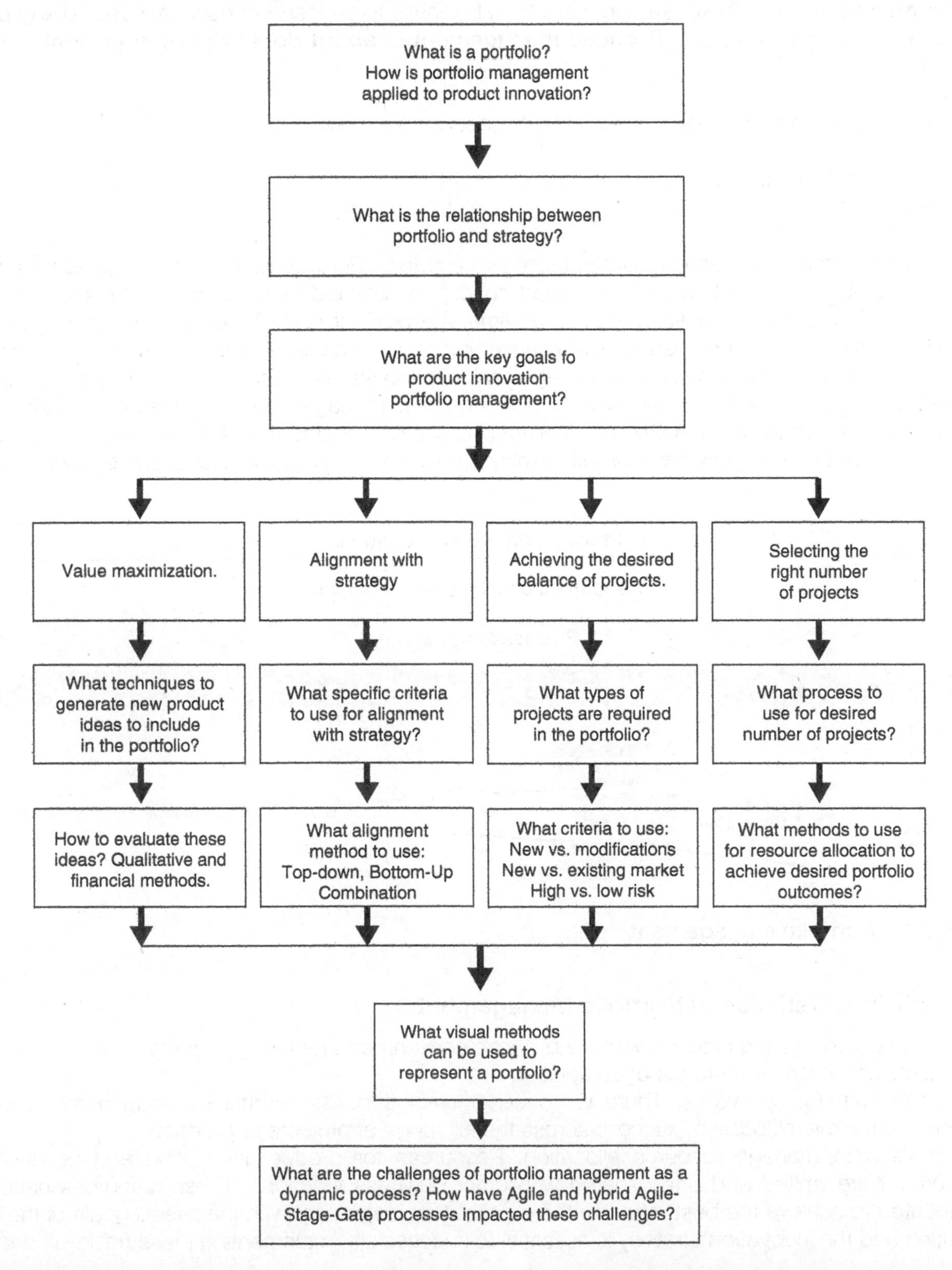

3.1 WHAT IS A PRODUCT INNOVATION PORTFOLIO?

Let's begin our understanding of a product innovation portfolio by considering a couple of terms:

Portfolio: "A portfolio is a collection of programs, projects and/or operations managed as a group. The components of a portfolio may not necessarily be interdependent or even related – but they are managed together as a group to achieve strategic objectives" (Project Management Institute, 2021).

Portfolio management: "There are two ways for a business to succeed at new products: doing projects right and doing the right projects. **Portfolio management is about doing the right projects**" (Cooper et al., 2002).

The activities of portfolio management are often erroneously separated as:

1. portfolio selection, and
2. portfolio review.

There should be no separation between these two activities. Portfolio management should be an ongoing process, applied to projects once the product concept is selected for development through launch and its ultimate decline and final deletion. At any one time, the product portfolio will include existing products, new products, product enhancements, maintenance and support projects, and research and development projects. It is a process that ensures optimal alignment with the strategic goals and maximizes the return on investment. Figure 3.1 shows the overarching role that portfolio management plays; the projects that need to be considered; the scope, from ideation to "end-of-life"; the relationship to individual project execution; the reliance on performance metrics; and the critical linkage with financial planning and resource management.

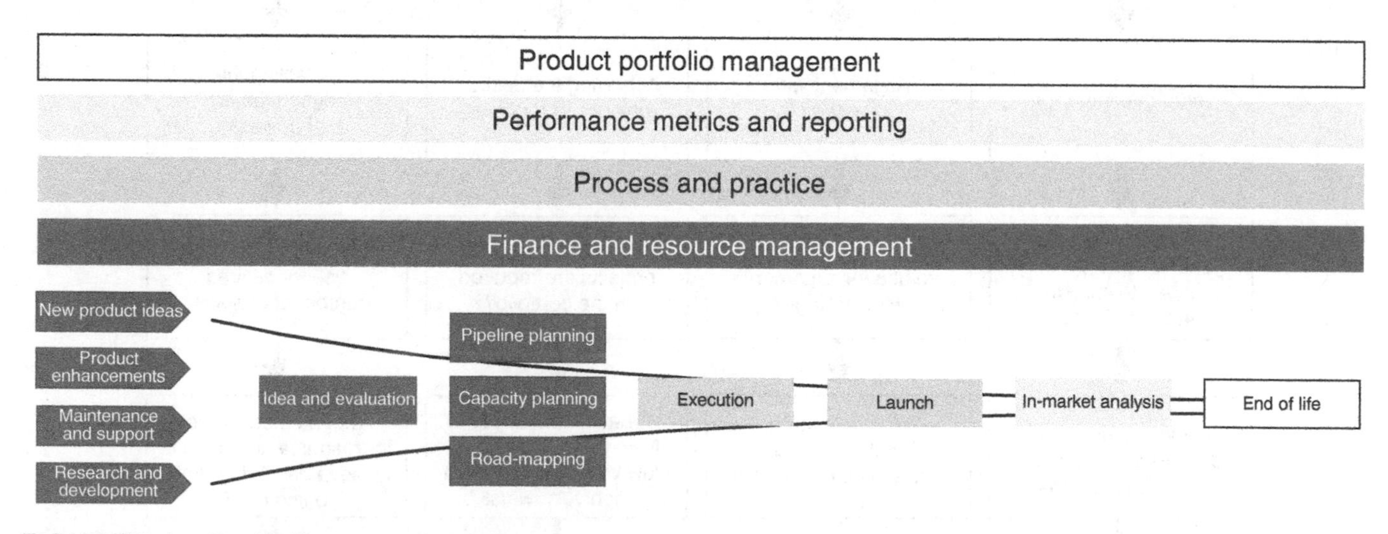

FIGURE 3.1 Portfolio management

3.1.1 Key Characteristics of Portfolio Management

- It is a decision-making process within a dynamic environment, requiring ongoing review.
- Projects are at different stages of completion.
- It deals with future events. There is no certainty of success. Portfolio management is used to increase the overall odds of success across the full range of projects or products.
- It is used to manage resource allocation. Resources for product innovation and product management are limited and often shared with other business functions. These resources need to be allocated to achieve the best return to the organization. Alignment with the overall goals of the organization and the innovation strategy is essential to successfully implementing these trade-off decisions.

3.1.2 Types of Projects Included in a Portfolio

Various project or product classifications are used to derive an appropriate balance of projects in the portfolio. The application of some form of classification system helps with strategic alignment and prioritization. Commonly used projects or product types are:

- **Breakthrough projects** (sometimes referred to as radical or disruptive). These projects strive to bring a new product to the market with new technologies, depart significantly from existing organizational practices, and have a high level of risk.
- **Platform projects,** introduced in Section 2.7, produce a set of subsystems and interfaces that form a common structure from which a stream of derivative products can be efficiently developed and produced. They provide the platform for founding derivative products/projects (see below). They generally involve more risk than product enhancements or incremental improvements, but generally not as much as breakthrough.
- **Derivative projects** are spin-offs from other existing products or platforms. They may fill a gap in an existing product line, offer more cost-competitive manufacturing, or offer enhancements and features based on core organization technology. Generally, they are relatively low risk.
- **Support projects** can be incremental improvements in existing products or improvements in manufacturing efficiency of an existing product. Generally, they are low risk.
- **Joint ventures or M&A projects**. New product opportunities will often arise outside of the organization. These may include joint ventures or mergers and acquisitions. Just like the internal projects, these opportunities will consume finances and resources and should be considered as part of overall portfolio management.

3.1.3 Goals of Portfolio Management

Cooper et al. (2015) defined the goals of portfolio management as:

1. ***Value maximization:*** Allocation of resources to maximize the value of the portfolio (the sum of the commercial worth of the individual projects).
2. ***Business strategic alignment:*** Ensure that the overall portfolio continues to reflect the business strategy and the organization's innovation strategy. Ensure that the investment across the portfolio aligns with the organization's strategic priorities.
3. ***Balance:*** Achieve the right balance of the right projects in terms of some pre-determined criteria such as long vs. short term; high vs. low risk; growth vs. sustaining; specific product, or market categories.
4. ***Right number of projects:*** It is common for companies to have too many projects in their portfolio for the limited resources available. Portfolio management allows for resource allocation to be monitored and controlled.

PDMA's 2021 Best Practice Survey (Knudsen et al., 2023) supports these goals as having a greater focus for the *Best* performing companies compared to the *Rest*.

We now focus our discussion on each of these four goals.

3.2 PORTFOLIO VALUE MAXIMIZATION

A good portfolio starts with a good pool of opportunities or ideas that have met defined criteria for inclusion in the portfolio. Organizations should be continually looking for new product opportunities, either new to the organization, line extensions, or product improvements. Although new opportunities arise randomly, the best organizations have proactive systems in place using a range of techniques to source these opportunities. These techniques can be applied both internally within the organization and externally among customers and other stakeholders.

Having generated this pool of opportunities, the challenge for the organization is then to evaluate and prioritize these opportunities. In the following sections, we present a range of techniques that can be used to generate the pool ideas, followed by techniques for evaluating, prioritizing, and selecting opportunities.

When Do Ideas Enter Portfolio Management?

Every product innovation idea discussed in an organization won't become part of the portfolio. Many ideas fizzle out and don't go anywhere. Others may have value, but lose to ideas that are in some way better for the organization's objectives. This raises the question of when do ideas enter portfolio management. The answer varies by organization, but at a minimum, projects that need wider visibility in the organization, resources committed, and funding allocated should be in portfolio management. Ideas that are still being formulated and are characterized by more unknowns than knowns should remain outside the portfolio. Many organizations have a process for incubating ideas and the ideas that are the most promising may be selected for further investigation and development as part of portfolio management.

3.2.1 Generation of Opportunities

Ideation

Ideation is the creative process of generating, developing, and communicating new ideas. It is an essential part of the design process. Specifically applied to product innovation, "Ideation includes all those activities and processes that lead to creating broad sets of solutions to consumer problems. These techniques may be used in the early stages of product development to generate initial product concepts, in the intermediate stages for overcoming implementation issues, in the later stages for planning launch, and in the post-mortem stage to better understand success and failure in the marketplace" (Kahn et al., 2013, pp. 453).

There are two types of thinking processes involved in ideation:

1. **Divergent thinking** is the process of coming up with new ideas and possibilities with little to no judgment or analysis. It is the type of thinking that allows for free-association, stretching the boundaries, thinking outside the box, and thinking of new ways to solve difficult challenges that have no single, right, or known answer.
2. **Convergent thinking** is associated with analysis, judgment, and decision-making. It is the process of taking a lot of ideas and sorting, evaluating, analyzing the pros and cons, and making decisions.

Ideation Tools

Following is a selection of ideation tools available to managers. A brief description of each technique is provided. Further information and references to these and other techniques are provided by Dam and Siang (2021) and can be found at https://www.interaction-design.org/literature/article/introduction-to-the-essential-ideation-techniques-which-are-the-heart-of-design-thinking.

Scamper

Utilizes action verbs as stimuli for idea generation. It is particularly useful in helping to come up with ideas to modify existing products or for making a new product. SCAMPER is an acronym for the following action verbs:

S – Substitute
C – Combine
A – Adapt
M – Modify
P – Put to another use
E – Eliminate
R – Reverse

Brainstorming

A commonly used technique involving a group of people (typically 6–10) encouraged to generate many ideas where people can speak freely without fear of criticism. Frequently, ideas are blended or built on to create another good idea (1 + 1 = 3). The design firm IDEO has specific rules for its brainstorming sessions. These are: defer judgment, encourage wild ideas, build on the ideas of others, stay focused on one topic, have one conversation at a time, be visual, and provoke quantity of ideas vs. quality.

Although brainstorming continues to be widely used, decades of research also indicate it is a poor generator of ideas compared to other approaches. It is dramatically improved when participants are first given the topic and allowed to individually and silently generate ideas before working as a group.

Brainwriting

Instead of asking participants to verbalize ideas (traditional brainstorming), they are asked to write down ideas pertaining to a specific problem or question. Each participant then passes their ideas over to someone else. This person then adds to the list, and so the process continues. After about 15 minutes, the lists are collected for group-focused discussion. An extension of this technique involves using graphic designs instead of writing ideas and then allowing all group members to enrich the designs of others.

Mind-mapping

A graphical technique for imagining connections between various pieces of information, ideas, or concepts. The participant starts with a key phrase or word in the middle of a page, then works out from this point (branches) to connect to new ideas in multiple directions – building a web of relationships.

Storyboarding

Focuses on the development of a pictorial story about the consumer's use and experience with a product, with the purpose of understanding the problems or issues that might lead to specific product design attributes or new requirements.

Six thinking hats

A tool developed by Edward de Bono, which encourages team members to separate thinking into six clear functions and roles. Each role is identified with six color-symbolic "thinking hats":

1. ***White***: Focus on facts.
2. ***Yellow***: Look for positive values and benefits.
3. ***Black***: The devil's advocate. Look for problems or pitfalls.

4. ***Red***: Expresses emotions, communicating likes, dislikes, fears, etc.
5. ***Green***: The creative that looks for new ideas, possibilities, and alternatives.
6. ***Blue***: Controlling, to ensure that an appropriate process occurs.

Delphi technique

A forecasting method based on the results of questionnaires/inquiries sent to a panel of experts. Several rounds of questionnaires are sent out and the anonymous responses are aggregated and shared with the group after each round. It is applicable to problems that benefit from expert insight, such as future technical and consumer trends, forecasting, and foresighting (further described in the Section 2.4.3 as strategic planning tool).

Ethnographic approaches

This is a research approach based on observing users in their natural environment, such as a kitchen, office, construction site, etc. This method allows for a "deep dive" to understand the complexities, dimensions, and factors that determine behaviors, beliefs, attitudes, and preferences. Further described in Section 6.5.3.

A day in the life

This method uncovers the routines, behaviors, and circumstances faced by users when experiencing products and services. It allows for the identification of the individuals' activities, challenges, and emotions in a day.

Personas

These are fictional characters built based on objective and direct observations of groups of users. These characters become "typical" users or archetypes, enabling developers to envision specific attitudes and behaviors toward product features. These personas are described using demographic, behavioral, attitudinal, lifestyle, and preference data. Later, these characteristics are used to identify possible user segments or possible targets, and the interaction persona-product is analyzed with the purpose of adjusting design.

Journey maps

A journey map is a representation as a flowchart of all the actions and behaviors consumers take when interacting with the product or service. The interaction moments are referred to as "touch points." The map includes the emotions triggered as the experience evolves and attempts to identify gaps that provide value-creating opportunities.

Empathy analysis

Involves the capacity to connect with and understand customers deeply and have a direct emotional connection with them. The designer needs to be immersed in the user's world, understand his/her problems and propose solutions from their perspective. This is often an aspect of other ideation tools, such as ethnography, day in the life, personas, and journey maps.

Big data

Organizations of all sizes are increasingly using data they obtain about their customers, such as grocery chains' loyalty programs, Tesla's driving data from all their vehicles, and hotel brands' data on the preferences of frequent guests. Big data is described as "a collection of large and complex data from different instruments at all stages of the process which go from acquisition, storage, and sharing, to analysis and visualization" (Pisano et al., 2015, p. 191).

Big data and the big data analytics industry have matured significantly in the last few decades. The act of gathering and storing large amounts of information for analysis is widely used. Zikopoulos and Tagaras (2015) provide a useful analogy with gold mining to explain the meaning and potential of big data: "In the old days miners could readily spot nuggets or veins (high value per byte data) because they were visible to the naked eye. But there was more gold out there that wasn't readily visible. Trying to find it would have required the mobilization of millions of people. Today miners work differently. Gold mining leverages new generation equipment that can process millions of tons of dirt (low-value per byte data) to find 'nearly-invisible to the naked eye' strands of gold. And with modern equipment those small strands can be extracted from the mass of dirt and processed into gold bars (high-value data). Today there is a mass of data, residing in different forms in a range of different locations. The challenge for the future is to locate this data and process it into a form that is useful for a specific purpose."

So, Big Data provides enormous potential in many aspects of product innovation, including the identification of new opportunities. The key challenge remains: how to distill the mass of data into those truly valuable new product opportunities.

The applications of Big Data are further discussed in Section 6.5.12.

Crowdsourcing

Crowdsourcing is described as the practice and use of a collection of tools for obtaining information, goods, services, ideas, funding, or other input into a specific task or project from a large and relatively open group of people, either paid or unpaid, most commonly via technology platforms, social media channels, or the Internet. Many companies and organizations also use their own websites as a means of crowdsourcing new product ideas.

As an example, LEGO has a dedicated website (LEGO Ideas) and online community for fans and customers to contribute their own product ideas. The platform motivates and incentivizes participation by allowing users to vote for their favorite idea, state how much they would pay for it, and explain why they like it so much. If more than 10,000 people support the idea, then it goes to the official LEGO review board, which decides whether to put it into production. The creator even receives name recognition on the product package and a one percent royalty on worldwide sales if the idea is commercialized. See the platform in action at https://ideas.lego.com.

As another example, in 2012, Anheuser-Busch ran a crowdsourcing project to create a new beer. It varied slightly from the typical consumer-led crowdsourcing projects as the initial recipes were created during a competition involving the brewmasters at Budweiser's 12 breweries. However, more than 25,000 consumers were involved in the subsequent taste tests to decide the winning brew, so the wisdom of the crowd was involved at some points during the development process. The Black Crown variety came from the recipe created by a Los Angeles brewery and went on sale in 2013.

Additional applications of crowdsourcing are further discussed in Section 6.5.13.

Ideas from within the organization and its stakeholders

Some of the best new product ideas arise from within the organization. Sources of new product ideas include company employees, customers, competitors, outside inventors, acquisitions, and channel members. Both solicited and spontaneous ideas may emerge from the sources, and some even occur by accident. Procter & Gamble's Ivory soap was a result of an accident. Manufacturing over-mixed the ingredients in the manufacturing process, creating air bubbles in the soap, causing it to float. Perhaps the most common cited example is 3M's Post-it Note which was developed through a combination of failed product innovation and an astute connection to the unintended invention. In 1968, a 3M adhesives engineer had invented a pressure-sensitive adhesive compound but could not think of an application for it. The glue could barely hold two pieces of paper together but could be applied to surfaces and removed without leaving a residue. Some months after the "failed development" another 3M engineer was singing in a choir. He usually would place small pieces of paper in songbooks to mark specific sections of songs. In a flash of inspiration, the engineer remembered "weak glue" and recognized its potential in a reusable bookmark. And so the sticky note was born. Initially named Press' n Peel and later, the Post-it Note (Belludi, 2021). Although examples of this type are not uncommon, more formal and structured "in-company" idea generation will lead to more sustained success. Specific suggestions include:

- Create and foster an innovative work environment and culture. A culture of innovation reinforces the value of innovation, encourages employees across the organization to be individually and collectively creative. More on this topic in Chapter 7.
- Idea management software. Provides a platform for employees to ideate and collaborate in a structured and transparent way to share, refine, and promote ideas.
- Invest work time in creative thought. Increasingly, organizations are allowing and encouraging employees to spend a proportion of their paid work time on personal projects that they believe would add value to the organization. Google has allowed employees 20% time for personal projects and 3M has long used 15% time as the foundation of their innovation management.
- Recognition and reward. A clear connection between organization goals and individual employee goals is critical. Individual performance reviews then have real significance in terms of how the individual has contributed to achieving the organization's goals. Identifying new opportunities that have the potential to create real value can be recognized in a number of ways – financial bonus, a box of chocolates or bottle of wine, business class upgrade for next overseas flight, attendance at a conference, and so on. It is important to select an item of recognition that has real value and meaning to the recipient. Recognizing and rewarding in staff meetings, the organization's newsletter, social platform, or other corporate communications can be used. Often, a simple thank you is best – in person rather than through email or phone.
- Use external stakeholders. External stakeholders, including customers, secondary manufacturers, distribution channel parties, etc., can be an extremely valuable source of new ideas; either totally new products or improvements to existing products. Examples include the manufacturer of medical devices using primary health providers as well as dairy ingredient manufacturers seeking regular feedback from secondary manufacturers of their products. The 2016 winner of PDMA's Outstanding Corporate Innovator Award, Sherwin Williams Paint Group Stores, was recognized for its innovative approach to utilizing in-store customer contact for pro-actively seeking new product ideas from home owners and trades people.
- Feedback. Above all else, it is critical that feedback is provided on all ideas submitted by employees and other stakeholders – not just the ideas that become commercial successes. A lack of feedback will quickly translate into a drying up of ideas.

3.2.2 Evaluation of Opportunities

The PDMA 2021 Best Practices Study (Knudsen et al., 2023) identified that the best companies used portfolio management techniques significantly more than others, placing a great emphasis on portfolio management, and are more likely to have a single integrated product innovation and portfolio management process. This necessitates methods for evaluating potential innovation product projects. A summary of the relative usage of various project evaluation tools was collected in the 2012 PDMA research titled Product Innovation and Management Comparative Performance Assessment and is shown in Figure 3.2 (Markham & Lee, 2012).

Broadly speaking, techniques for evaluating and screening new product opportunities can be defined under two headings: qualitative and quantitative.

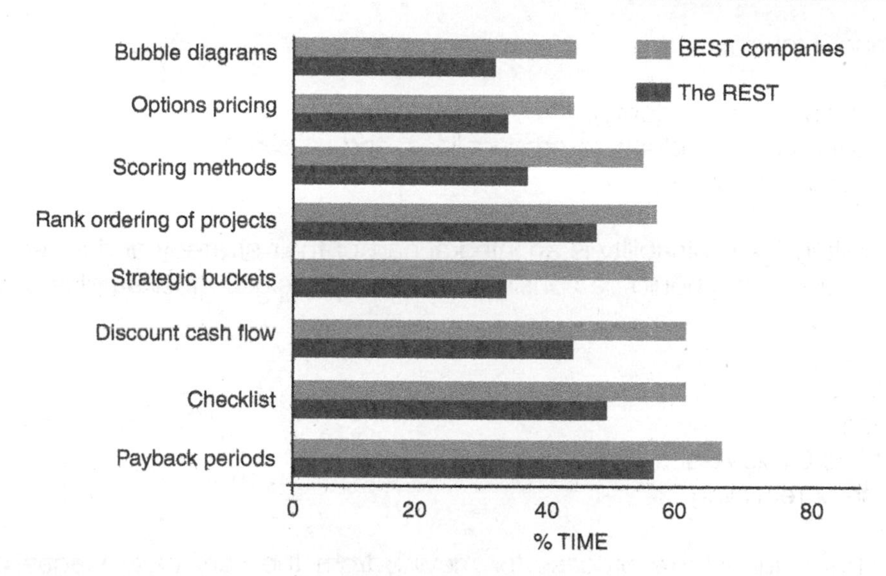

FIGURE 3.2 Project evaluation tools used by companies

Qualitative Evaluation

Qualitative or scoring methods are, by definition, subjective. Nevertheless, there is a sound body of evidence that organizations using scoring methods in portfolio management achieve significant portfolio success (Markham & Lee, 2012; Cooper et al., 2001). These methods are usually based on evaluating opportunities against a specific set of criteria.

Evaluation criteria

These criteria can be a mix of known factors that lead to new product success and requirements specific to the organization.

Over the years, there have been several studies into the factors leading to successful new products. These are further discussed in Section 1.1.2. Some of the more important success factors come from the foundations of creating competitive advantage (Cooper et al., 2001).

- Having a unique, superior product. One that is differentiated from those of competitors offers unique benefits and provides superior value to the customer.
- Targeting an attractive market that is growing, large, has good margins, weak competition, and low competitive resistance.
- Leveraging internal organizational strengths – products and projects that build from organization strengths, competencies, and experience in both marketing and technology.

Most of these success factors are relatively well known and understood by organizations, especially in traditional areas of business. They can be readily applied as criteria to evaluate new product opportunities with a reasonably high level of confidence.

Evaluation criteria specific to the organization include:

- Strategic alignment,
- Technical feasibility,
- Level of risk,
- Regulatory implications,
- Near-term financial return,

- Long-term financial return,
- R&D expense,
- Break-even time (time to profitability),
- Breadth of benefit among products or product lines, and
- Availability of investment funding.

For many organizations, sustainability is an integral part of their strategy and an essential factor in the products created. Sustainability should be considered in project selection. Examples of sustainability criteria are:

- Triple bottom line,
- Carbon emissions,
- ISO-compliant Life Cycle Assessment, and
- Resource re-use or recycling.

Figure 3.1 presents a funnel-like process for moving from the very early stages of idea generation through to final product launch. This funnel process is commonly used to show the importance of evaluating many ideas or opportunities to arrive at those product concepts that are truly worthy of significant investment and inclusion in the product innovation portfolio. Where the organization is presented with many new product opportunities, it may be worthwhile to do a simple pass/fail analysis to quickly reduce the list of opportunities to a more manageable size.

Pass/fail

The pass/fail technique is designed to quickly reduce many product ideas or new opportunities. It assesses whether an opportunity or idea meets some basic criteria. The process requires assigning a pass (P) or a fail (F) to each opportunity against specific criteria. Opportunities must pass all criteria to be carried forward. Basic guidelines for the application of the pass/fail approach include:

- Include a cross-section of functional representatives in this evaluation (marketing, technical, manufacturing) so as to represent a broad range of knowledge and experience.
- Use the individual evaluations from participants as a basis for discussion, leading to consensus evaluations.

Advantages of the pass/fail technique are:

- It does not require extensive research and analysis.
- It is quick and easy to do.
- It offers a relatively simple and quick way to involve a cross-section of people in the early stages of product innovation.
- It serves to highlight areas of uncertainty that need further research.

Disadvantages of the pass/fail technique are:

- It can result in early rejection of good ideas due to a lack of adequate research.

A simple example with three pass/fail criteria is shown in Table 3.1.

TABLE 3.1 Example of pass/fail evaluation

Product idea	Existing distribution channels	Manufacturing capability	Consistency with range	Overall
One	P	F	P	F
Two	P	P	P	P
Three	P	P	P	P
Four	F	P	P	F
Five	P	F	F	F
Six	P	P	P	P
Seven	F	F	F	F

Scoring

The scoring approach usually implies a more detailed analysis and would normally follow a pass/fail screening. More information is required to ensure greater reliability and confidence in the assessments. The process is as follows:

1. Assessment criteria are selected. Usually, 5–7 criteria is sufficient. The criteria applied will depend on the specific circumstances. So, for example, where there is a strong emphasis on the application of new technology, then "technical capability" will be an important criterion. Where a new product is being targeted in a highly competitive market, the level of differentiation of the new product idea will be important.
2. Weightings are assigned to each criterion to reflect its relative importance.
3. Each product idea is scored on a 10-point scale for each criterion and a weighted sum calculated across all criteria. This leads to a ranking of the product ideas.
4. As with the Pass/Fail approach, it is wise to employ a cross-section of people from different functions within the organization to do the scoring.
5. To ensure an objective approach to scoring, especially where a number of people are involved, it is important to provide descriptive reference points on a defined scoring scale.
6. Figure 3.3 shows evaluation of seven new product ideas against three criteria (or screening factors) each with a weighting reflecting its importance – sales potential (10), fit with strategy (7.5), and level of competition (5). Each product idea has been evaluated on a 10-point scale. So, for example, using the scale defined for sales potential, idea 1 is scored 8/10 reflecting an estimated sales potential of around 550,000 units. Idea 5 is estimated at around 80,000 units.
7. The total weighted score for each idea is calculated as follows, thus providing a basis for ranking the seven ideas.

   Total weighted score for idea $1 = (8 \times 10) + (6 \times 7.5) + (8 \times 5) = 165$

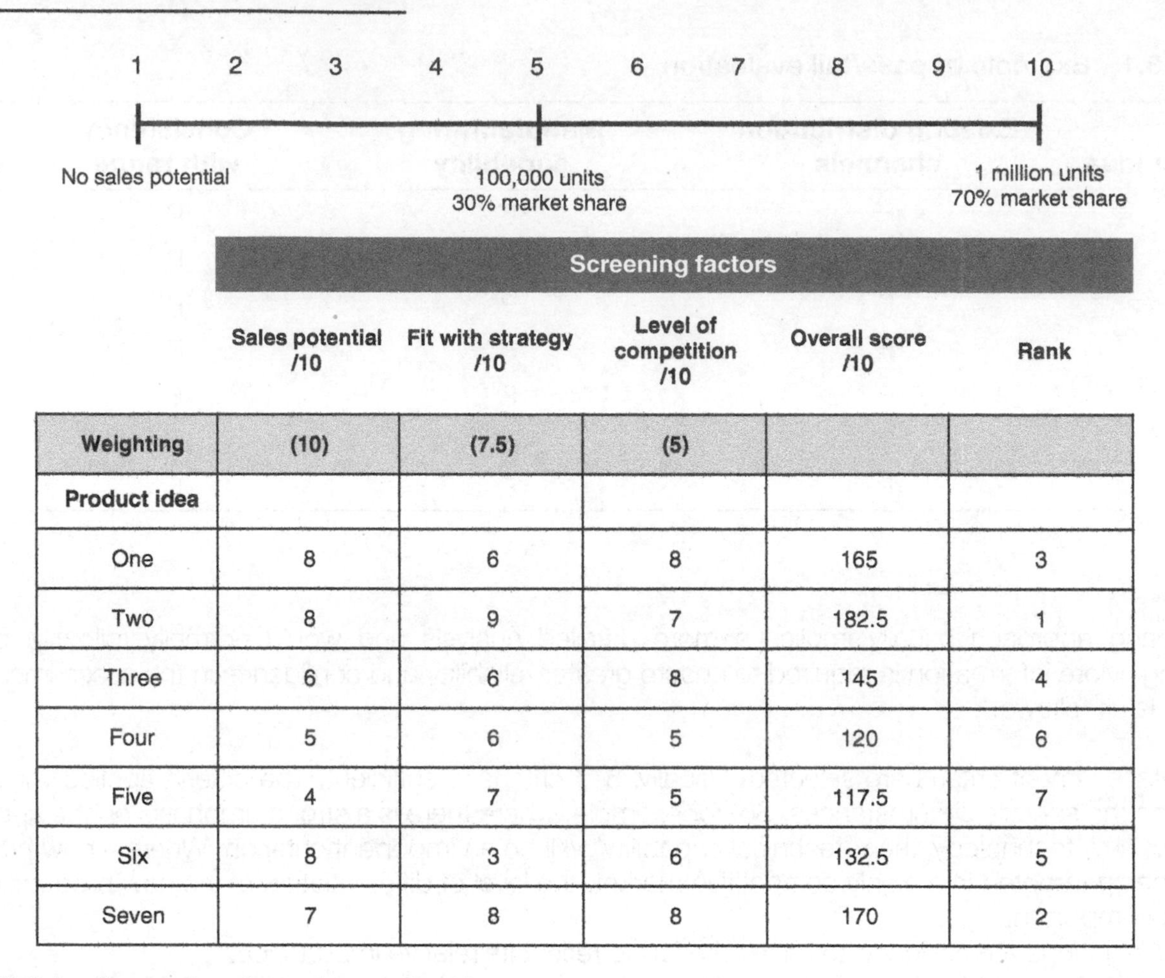

	Sales potential /10	Fit with strategy /10	Level of competition /10	Overall score /10	Rank
Weighting	**(10)**	**(7.5)**	**(5)**		
Product idea					
One	8	6	8	165	3
Two	8	9	7	182.5	1
Three	6	6	8	145	4
Four	5	6	5	120	6
Five	4	7	5	117.5	7
Six	8	3	6	132.5	5
Seven	7	8	8	170	2

FIGURE 3.3 Example of scoring evaluation

Quantitative Evaluation

In the context of product portfolio selection and ongoing management, quantitative evaluation is most financially based and can be used either:

- To decide if the new product is financially viable – does it have the potential to deliver a satisfactory return on investment, or,
- To decide on the prioritization of projects as part of the portfolio selection or ongoing portfolio management process.

Figure 3.4 presents a framework for the financial analysis of a new product opportunity. This framework clearly demonstrates the information required to carry out a quantified financial analysis, namely:

- ***Returns (benefits)***: based on sales volumes and prices,
- ***Costs***: based on the cost of manufacturing and marketing, and
- ***Capital costs***: costs associated with the purchase of buildings, plant, and equipment.

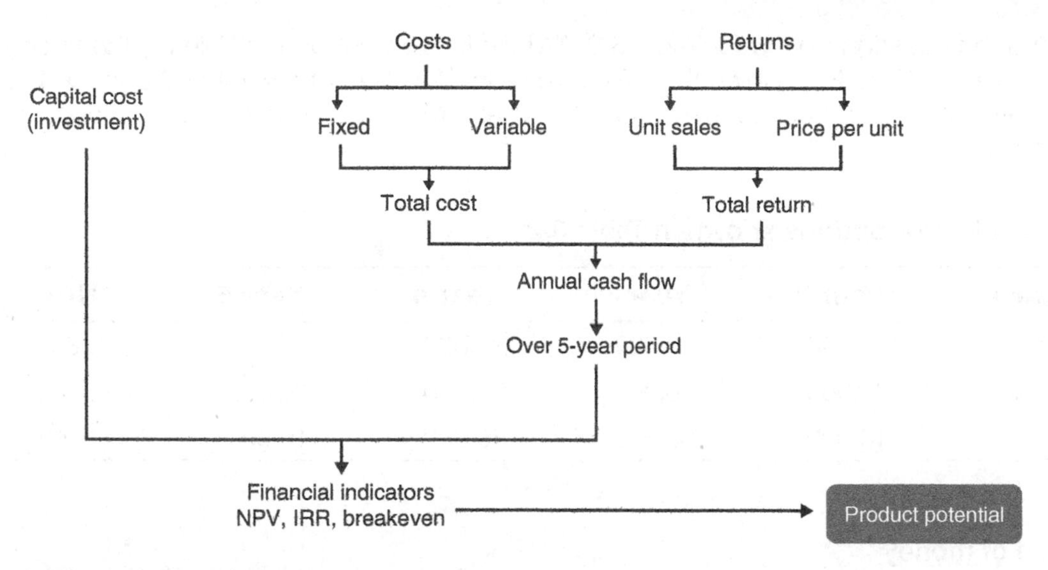

FIGURE 3.4 A framework for financial analysis

Specific financial indicators are used to determine the projected value of the new product opportunity relative to the investment. These include:

- Payback period,
- Net Present Value (NPV),
- Internal Rate of Return (IRR),
- Bang-for-buck ratio,
- Expected commercial value,
- Options pricing.

Payback period

Payback period is defined as the length of time it takes to repay the capital investment. So, for example, if capital costs (for new plant and buildings) are $1 million, how many years does it take for cumulative profits to exceed $1 million?

Although payback period is a commonly used and relatively simple measure of return on investment, it fails to capture the important element of the time value of money. It fails to account for the timing of the receipts of returns on investment. However, the payback period is useful because it is simple and intuitive.

Table 3.2 shows three potential investment options (A, B, and C) with projected returns over a 5-year period. Each option requires the same capital outlay of $100,000 and has the same payback period of five years. Which would you choose? Why? Read on to the next section on Net Present Value for the answer to this question.

TABLE 3.2 Different schedules of returns for the same investment

Option	Year 1	Year 2	Year 3	Year 4	Year 5
A	20,000	30,000	40,000	10,000	
B	20,000	20,000	20,000	20,000	20,000
C		10,000	30,000	50,000	10,000

The key to answering the question posed for Table 3.2, "Which is the best option?" lies in the timing of the receipts over the five years. Table 3.3 shows the NPV calculation for the data in Table 3.2, based on a 10% interest rate. The earlier timing of receipts makes Option A best. More detail on the calculation of NPV is covered later in this section.

TABLE 3.3 NPV calculation for options shown in Table 3.2

Option	Year 1	Year 2	Year 3	Year 4	Year 5	NPV
A	20,000	30,000	40,000	10,000		79,858
B	20,000	20,000	20,000	20,000	20,000	75,616
C		10,000	30,000	50,000	10,000	77,597

More on the time value of money

A dollar that you invest today will bring you more than a dollar next year: having the dollar now provides you with an investment opportunity to earn more in the future.

$$\$1.00 \text{ today at } 10\% \text{ interest} = \$1.10 \text{ in a year's time}$$

Present value (PV) provides a means of putting future money in today's terms (or value). This is achieved by modifying the future value by a factor that represents the change in value of money from today's value.

$$\text{Future value} = \text{original amount} \times (1 + \text{interest rate})^{\text{number of periods}}$$

$$FV = PV \times (1+i)^n$$

Rearrangement of this equation gives the following for present value:

$$PV = FV/(1+i)^n$$

Net Present Value (NPV)

Net present value is the cumulative *present value of returns* (or benefits) minus the cumulative *present value of costs*.

Present value discount factors

Table 3.4 shows the factors that need to be applied at different time periods and interest rates. So, for example, a revenue of $1000 in five years' time, where the interest rate is 10%, equates to a present value of $602.90.

TABLE 3.4 Discount factors for NPV calculation

	Rate			
Year	**10%**	**20%**	**30%**	**40%**
1	0.9091	0.8333	0.7692	0.7142
2	0.8264	0.6944	0.5917	0.5102
3	0.7513	0.5787	0.4552	0.3644
4	0.6830	0.4823	0.501	0.2603
5	0.6029	0.4019	0.2693	0.1859
10	0.3855	0.1615	0.0725	0.0346
20	0.1486	0.0261	0.0053	0.0012

Calculating the cumulative NPV of a new product

Table 3.5 shows a relatively simple calculation of NPV for a new product. Clearly, a lot of work is required to derive the data included in this simple table:

- Decide on the potential life of the product or the time period for the NPV calculation (in this case, five years).
- Project the benefits (returns) for each year of the product's life.
- Project the costs for each year of the product's life.
- Calculate the annual cash flows (the difference between returns and costs).
- Calculate the present value (PV) for each year's net cash flow.
- Add together the individual PVs over the life of the product. This is the cumulative net present value (NPV). In this case, $25,673.

TABLE 3.5 A simple example of NPV calculation (the discount rate applied is 10%)

	Year 1	**Year 2**	**Year 3**	**Year 4**	**Year 5**
Benefits	3500	10,800	15,000	20,000	21,000
Costs	8370	3500	4800	7800	8000
Cash flow	−4780	7300	10,200	12,200	13,000
Present value factor	0.9091	0.8264	0.7513	0.6830	0.6209
PV	−4427	6033	7663	8333	8072
NPV	25,673				

Internal Rate of Return (IRR)

Internal rate of return (IRR) is defined as the discount rate at which an investment has a zero net present value. It is used to evaluate the attractiveness of an investment in a project or product. Most organizations have a required level of return required for investments – this is often referred to as the ***hurdle rate***, e.g., 10% or 15%. The hurdle rate is determined by:

- Rates of return on alternative avenues of investment. An obvious example is, "Can I get a better return from putting the money in the bank rather than investing in a plant to manufacture a new product?"

- The level of risk. A higher hurdle rate is normally assigned to higher-risk investments – just as high-risk shares and bonds may promise a potentially greater return.

Calculation of the IRR provides a comparison with both the company's hurdle rate and with alternate forms of investment, either internally or externally. So, for a simple example, if the IRR for the project is less than the current bank interest rate, all other things being equal, it would be more profitable to put the money in the bank than execute the project.

Financial Sensitivity Analysis

Often there is significant uncertainty regarding the data used in new product investment analysis. This is especially true in the early stages of development – the "the front end of innovation or fuzzy front end." Forecasting sales at different price points and estimating manufacturing costs from product specifications that are yet to be well defined are among the many variables impacting the financial analysis. The most common way to address these uncertainties is to apply what is referred to as "sensitivity analysis." That is, trialing the impact of different scenarios on the financial outcomes. Most commonly used spreadsheets have embedded financial functions, including NPV and IRR, that provide the facility for "what-if" or "sensitivity" analysis. For example, what is the effect on the IRR if the market share is half of what is projected? Or what is the impact if the cost of a major component doubles due to supply chain shortages?

Table 3.6 shows a simple example of a financial analysis spreadsheet for a new product opportunity. Most elements of this spreadsheet are linked to allow for quick and easy "what-if" analyses – an excellent tool to provide a basis for a team discussion or senior management presentation, demonstrating the potential risk of certain outcomes.

The use of a simple spreadsheet can also provide an excellent tool for on-going updating of financial analysis as more reliable data becomes available. Highlighting specific variables within the spreadsheet where there is a high degree of uncertainty over the quality of the data, especially where sensitivity analysis points to the importance of theses variables, reinforces the need for more reliable data.

TABLE 3.6 Example of financial analysis spreadsheet

	Year				
MARKET DATA	**1**	**2**	**3**	**4**	**5**
Market growth (%)		14.00	1.20	1.20	1.20
Market (units)	100	1400	1680	2016	2419
Market share (%)	100%	100%	60%	54%	49%
Market share (units)	100	1400	1008	1089	1176
COSTS ($)					
Variable cost/unit ($)					
Direct materials ($)	120				
Packaging ($)	20				
Sales and marketing ($)	40				
Direct labor ($)	240				
Variable overhead ($)	40				
Total	460				

(Continued)

TABLE 3.6 (Continued)

Fixed costs ($)	
Manufacturing ($)	40,000
Marketing ($)	30,000
Salaries ($)	80,000
Admin ($)	20,000
Total ($)	170,000
PRICING	
Retail price, GST-incl ($)	3300
Retail price, GST-excl ($)	2933
Retailer gross margin (%)	0
Retailer gross margin ($)	1173
Manufacturing selling price ($)	1760
Cost of capital (%)	20%
Equipment cost ($)	2,600,000

SUMMARY OF OPERATING COSTS

	Unit	**Year**					
	Prices	**0**	**1**	**2**	**3**	**4**	**5**
Sales volume (units)			100	1400	1008	1089	1176
Sales revenue ($)	1760		175,980	2,463,720	1,773,878	1,915,789	2,069,052
Variable cost ($)	460		46,000	644,000	463,680	500,774	540,836
Contribution margin ($)	1300		129,980	1,819,720	1,310,198	1,415,014	1,528,215
Fixed costs ($)			170,000	170,000	170,000	170,000	170,000
Operating cash flow ($)			−40,020	1,649,720	1,140,198	1,245,014	1,358,215
Investment ($)		−2,600,000	=	=	=	=	=
Total cash flows (4)		−2,600,000	−40,020	1,649,720	1,140,198	1,245,014	1,358,215
Cost of capital, *r*	20%						
NPV	$318,374						
IRR	24.4%						

Bang for Buck Index

Bang for the buck is an idiom meaning the *worth of one's money or exertion*. The phrase originated from the slang usage of the words "bang" which means "excitement" and "buck" which means "money."

NPV does not consider the resources required for a project. A project may have a very large and attractive NPV but requires significant resourcing, whereas another project with a slightly less attractive NPV may require significantly lower resourcing. A way to address this issue is to determine the "bang for buck" by calculating the ratio of the NPV to projected resource utilization:

Bang for Buck ratio = NPV of the project/Resources utilization

Expected Commercial Value

One of the weaknesses of the NPV and bang for buck methods is that they fail to consider risk. The probability of technical and commercial success is not factored into these methods. The Expected Commercial Value (ECV) method seeks to maximize the expected commercial value, or worth, of a project using the following formula:

$$ECV = \left(\left(PV * P_{cs} - C\right) * P_{ts}\right) - D$$

ECV = Expected Commercial Value
PV = Present Value
P_{cs} = Probability of commercial success
C = Commercialization (launch) costs
P_{ts} = Probability of technical success
D = Development costs

Table 3.7 shows an example of the ECV calculation for a range of potential projects. It demonstrates the impact of the application of the probability of commercial and technical success to the project rankings based on PV and ECV.

TABLE 3.7 ECV calculation example

Project	PV	PV rank	Probability of technical success	Development cost	Probability of commercial success	Commercialization cost	ECV	ECV rank
A	20	4	0.8	3	0.6	5	2.6	5
B	30	3	0.9	5	0.8	3	13.9	1
C	5	5	1.0	1	1.0	1	3.0	4
D	60	1	0.5	4	0.5	2	10.0	3
E	45	2	0.7	4	0.6	3	12.8	2

ECV is particularly suitable for highly uncertain, iterative Agile or Agile-Stage-Gate projects. ECV provides a model for handling incremental or stepwise investments in a product innovation project. This is done using a decision tree approach that evaluates the investment a step at a time, making ECV appropriate where there are a number sprints, each with a defined level of investment. Further detail on the derivation and application of ECV to Agile-Stage-Gate projects can be found in Cooper (2019) and Sommer (2018).

Option Pricing Theory (OPT)

In a discount cash flow or net present value analysis, the assumption is that a project is an "all or nothing" investment decision. This is seldom the case in product innovation, where investments are made incrementally, leading to Go/Kill decisions as new and improved information comes to hand. The detailed theory and computation of options-based pricing is complex and, in its purest form, may be beyond the capability of most organizations. At least the ECV method incorporates the principles of a stage-wise process with options together with probabilities of outcomes. The OPT approach is of greatest value for organizations with a strong research emphasis in their R&D. In these situations the research often has a long lead time to a commercial outcome. Directions may change throughout the research journey. The journey may end with no commercial value generated or, hopefully, a range of possible opportunities not conceived at the start.

Following is an example of how OPT could be applied in pharmaceutical company's drug development. Pharmaceutical drug development is a long process, often several years, requiring several phases to final commercialization, including R&D, pre-clinical research, clinical research, and regulatory review.

In this case, the company is embarking on an R&D project with a $20m investment over a 3-year period. Management estimates a 5% chance of successful product outcome.

Each of the phases required for final commercialization has a high cost associated with it. For our example, we will assume a total cost of $200m. But if development is successful, the prize is a period of market exclusivity. In this case, let's say 10 years. In this situation, the R&D investment should not be viewed as an end but as generating an option to invest or the right to spend more money.

Potential returns for each of the 10 years is estimated at $100m per year. The Net Present Value (NPV) for these returns over the 10-year period at a discount rate of 6% is $225m.

Now, let's just calculate the NPV based on the R&D investment period of three years.

$$\begin{aligned} NPV &= -20 + 5\% \times 225 / (1+6\%)^3 \\ &= -\$11m \end{aligned}$$

Based on R&D alone and the very low probability of development success, the project is unlikely to proceed. Using the options approach presents a different view of the project's potential. Refer to Figure 3.5.

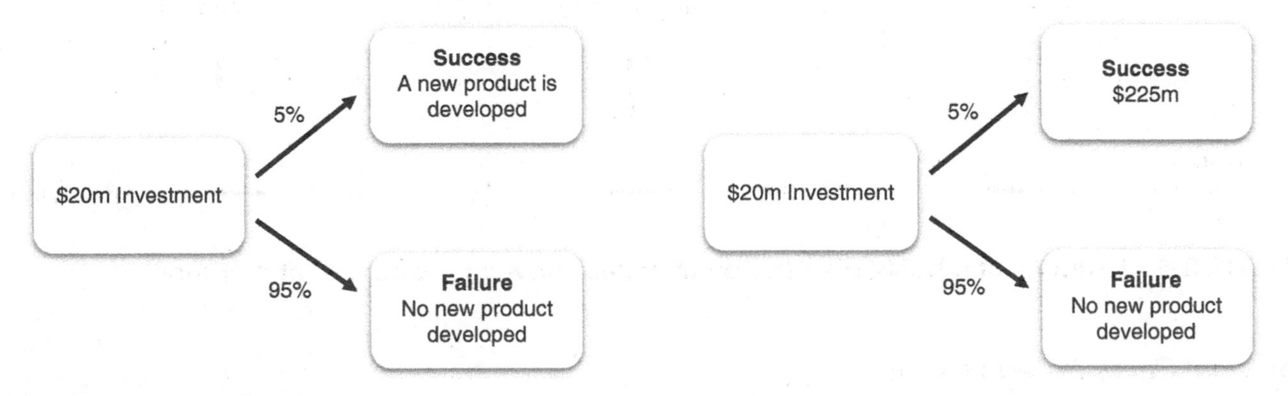

FIGURE 3.5 Example of options pricing analysis

Deriving Data Input for Financial Analyses

Three key elements of data required for new product investment analysis are costs, sales, and selling price.

Cost determination

The basic elements of cost are fixed costs, variable costs, and capital costs:

> **Fixed costs** are expenses whose total does not change in proportion to the activity of a business within the relevant time period or scale of production. Examples include administration, rent, rates, and general management.
> **Variable costs** are expenses that change in proportion to the activity of a business. Examples include production labor, power, cleaning materials, and manufacturing materials.

Total cost = Fixed costs + Variable costs

Capital costs are costs incurred on the purchase of land, buildings, construction, and equipment to be used in the production of goods or the rendering of services.

Working capital is the money spent on direct and variable costs associated with a product or service while awaiting sales. This would include all costs of manufacturing and marketing, together with capital costs of new equipment, etc.

Selling price determination

The ex-factory price is the sum of all costs associated with the product plus any margins accruing to the company. It is the price at which the organization makes the product available to a buyer. The buyer is responsible for transportation and other costs to send the product where it is needed or on-sold to other customers.

When the product is sold beyond the "factory door," the selling price to the final customer will be determined by the sum of the ex-factory price and all costs associated with getting it to the final customer, including the margins charged by distribution channel members. An example of a basic distribution channel is shown in Figure 3.6. The complexity of the distribution channel varies greatly from product to product and organization to organization. It will significantly impact the markup from ex-factory costs to the final customer selling price.

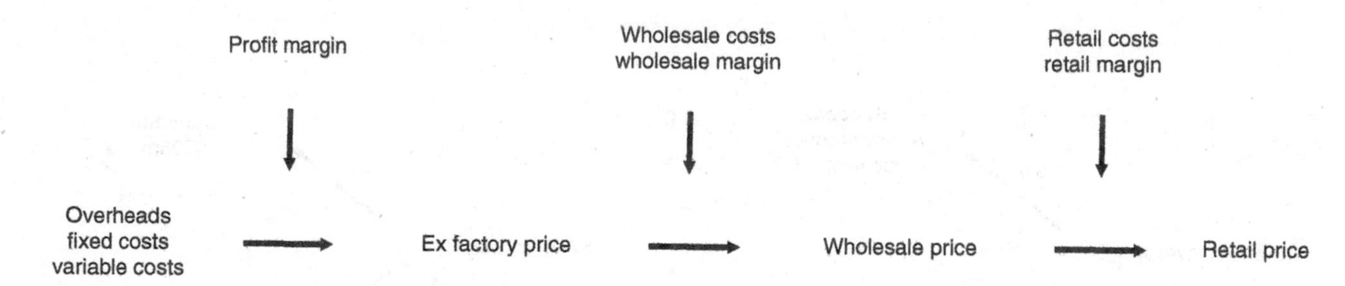

FIGURE 3.6 Example of a basic distribution channel, showing product cost markups

Potential Sales Determination

There are several approaches used to assess demand for new products. Given a particular demand, the organization can forecast the sales potential, which is critical for the overall new product feasibility analysis, financial projections, and production planning.

Several demand models include the Bass model, ATAR, purchase intention, chain ratio method, exponential methods, and time series analyses. A description of the most important methods follows.

Bass model

The model is appropriate to forecast sales of an innovation, new technology, or a durable good. It assumes that the diffusion process is binary (either the consumer adopts or waits to adopt). The key application of this model is when you need to quantify the number of customers that will adopt the product over time and when they will adopt it. When customers adopt a product is based on the rate of adoption, known as the diffusion speed. This model is used in industry because of its effectiveness and simplicity.

The Bass model takes two forms: a basic and a generalized form. The basic model requires the assessment of three parameters: p, q, and N. The parameter p is the coefficient of innovation (initial trial of the product or penetration) and q is the coefficient of imitation (diffusion rate parameter). Both coefficients p and q take a value between 0 and 1 and are affected by either product-related (low complexity, high compatibility, observable benefits) or market-related (links among potential users, communication, etc.) factors. *N* represents the total number of customers in the adopting target who will eventually adopt the product. No repeat or replacement purchase is considered in the estimation. The model either requires some historical sales data or the input of the values for *p* and *q* to estimate demand over time. The demand generally shows an S-shaped curve.

ATAR model

Crawford and DiBenedetto (2011) presented a model for forecasting sales potential when historical sales data does not exist. The ATAR (Awareness-Trial-Availability-Repeat) model is a forecasting tool that tries to mathematically model the diffusion of an innovation or new product. For a person/organization to become a regular buyer/user of a new product or service, there must first be awareness that it exists. There must be a decision to try it. They must find that the item is available to them. Finally, the person or organization must be satisfied enough to adopt the product or service or become a repeat buyer; refer to Table 3.8 for an ATAR example.

TABLE 3.8 Example of ATAR analysis

Number of buying units	3,000,000
Percent aware	40%
Percent trial	20%
Percent availability	40%
Percent repeat	50%
Annual units bought	1.5
Units sold (product of above	72,000
Revenue per unit	$25.00
Cost per unit	$12.50
Profit per unit	$12.50
Profit = Units sold × Profit per unit	**$900,000**

Purchase intention methods

These methods use the results of concept testing as initial input and then adjust these with probability estimates from historical results or past experiences. For example, the results of a concept test for a hand cleanser (using an intent-to-buy scale from "definitely would buy" to "definitely would not buy") showed that:

5% of potential consumers will definitely buy the product.
36% will probably buy the product.
80% of those who answer definitely buy, actually buy.
33% of those who answered probably buy, actually buy.

The final calculation of an estimated market share will be:

$$\text{Market share}\left(\text{forecasted}\right) = (0.8) \times (5\%) + (0.33) \times (36\%) = 16\%$$

Chain ratio method

This method estimates market demand by multiplying a base number by a chain of adjusting percentages. For example, let's consider a clothing manufacturer planning to develop a new line of specialty sports training shirts with built-in capability for key physiological measurements – smart shirts. A simple explanation of the chain method to estimate sales potential for the new product is:

Total population of the country	=	50 million
Percentage involved in sports activities	=	30%
Percentage who value physiological monitoring	=	5% of those involved in sports activities
Estimate of those who would buy	=	10% of those who value physiological monitoring

$$\text{Estimated sales potential} = 50 \times 0.3 \times 0.05 \times 0.1 = 75{,}000$$

While the chain ratio approach and ATAR have similarities, note that only the ATAR model distinguishes the reachable market from the total market. Consequently, the chain ratio approach can result in higher and unrealistic estimated sales.

3.3 PORTFOLIO ALIGNMENT WITH STRATEGY

"It all boils down to strategy. From your business's new product innovation and technology strategy all else flows. The goal to maximize value of the portfolio is meaningless unless value is measured in terms of a company goal" (Cooper et al., 2015).

An organization's vision, goals, and strategy should provide the context and direction to which product innovation projects it undertakes. So, for example, if a company has the vision to become the world leader in agribusiness pharmaceuticals, then the need for breakthrough products requiring a strong emphasis on R&D should be reflected in the emphasis on these projects in the portfolio.

3.3.1 Approaches for Linking Strategy to Portfolio

Cooper et al. (2001) suggest three broad objectives for achieving strategic alignment in portfolio management:

Strategic fit: Are the projects consistent with the articulated strategy? For example, if certain technologies or markets are specified as areas of strategic focus, do the projects fit into these areas?
Strategic contribution: What specific goals are defined in the business strategy? For example, "capture a larger share of an existing market" or "break into a new product category." How well will the projects contribute to achieving these goals?
Strategic priorities: Does the breakdown of the investment across the portfolio reflect the strategic priorities? For example, if the organization seeks technology leadership, the portfolio's balance of projects should reflect this focus.

Cooper et al. (2001) suggest three methods of project selection and ongoing review to ensure a clear link between strategy and product portfolio:

- Top-down,
- Bottom-up, and
- A combination of top-down and bottom-up.

All three methods are designed to ensure that the combination of projects in the portfolio optimally achieves the strategic goals subject to resource constraints.

The Top-down Method

The top-down approach is firmly based on the business vision, goals, and strategy. This provides the direction for portfolio composition and resource allocation. There are two general approaches to the top-down method:

1. ***Product roadmap***: this approach is based on a strategy-led definition of key projects that need to be undertaken and the timing required for each project's completion. This leads to a product roadmap and, in turn, in helping to shape the product portfolio.
2. ***Strategic buckets***: this approach focuses more on resource allocation. Firstly, the business strategy defines the amount of product innovation funding available. This funding is then allocated across various categories of projects. These could include product, market, or technology-based categories or a combination of these. So, for example, a biotechnology company may define its strategic buckets research, development, and improvements/modifications. These may be further divided into markets e.g., industrial, consumer, and technologies or genetic modification and microbial.

This method involves the following steps:

1. Clear definition of the organization and business strategies, and the strategic goals and priorities relating to innovation.

2. Definition of the level of resourcing available for the entire project portfolio.
3. Prioritization of business units or product categories in strategic importance to the organization as a whole.
4. Identification of "strategic buckets" and allocation of the ideal spend across these various business units or product categories. Two examples are:
   a. 10% breakthrough, 20% platform, 20% derivative, 50% support
   b. 60% Business A, 30% Business B, 10% Business C
5. Allocation of projects into the strategic buckets according to their prioritization.
6. Prioritization of projects within strategic buckets according to potential added value.

The Bottom-up Method

As the name implies, the bottom-up method starts with a list of opportunities for projects that may arise from various sources throughout the organization. This requires a process of strict project evaluation and screening to ensure that the best opportunities rise to the surface and are funded. The bottom-up approach, therefore, relies on a sound project evaluation process that is clearly linked to the business's strategy and goals. The key difference to the top-down approach is that the bottom-up approach starts with a list of potential projects and ends up, by a process of strict screening criteria, with projects that align with the business strategy:

- Potential projects are identified.
- Strategy criteria for evaluating projects are defined.
- Each potential project is evaluated against the selection criteria.
- Projects are selected primarily as a result of meeting the selection criteria with no specific consideration of business unit or product category priorities, or for achieving any sense of balance across the projects in the portfolio.

The Combination Method

Both the top-down and bottom-up approaches have their deficiencies. The combination of the two approaches seeks to overcome these deficiencies. It starts at the top with a clear definition of goals and strategy, a product roadmap, and strategic buckets that define the allocation of funding. It then proceeds through generating opportunities per the bottom-up approach with a decision process that reconciles the project selection. Again, as the name implies, this approach combines the benefits of both the bottom-up and top-down methods:

- Strategic priorities with regard to the business unit or product category spend are established.
- All potential projects are ranked against strategic criteria and the spend for each project is estimated.
- The combination of individual project priority and budgeted spend is used to allocate the projects into strategic buckets in line with the priority of the business unit or product category.

3.4 BALANCING THE PORTFOLIO

Most organizations should seek to have a range of new product opportunities in the portfolio. This helps balance risk with reward. Also, a more diverse portfolio is resilient to the effects of market change. The range and proportional representation of these new product opportunities should be directed by the overall corporate or business strategy, aligned with the innovation strategy. Categorization of the new product opportunities can be by business unit, product category, target market, or based on the characteristics of the project, for example:

- Breakthrough, derivative, platform, or support,
- Cost of R&D, or of commercialization,
- Potential returns or benefits,
- Level of risk – in development or in commercialization,
- Technical difficulty, to develop or to maintain,
- Time to market from the start of development to commercial returns,
- Capital investment in plant and equipment, or
- Potential for intellectual property value creation.

As described earlier, the representation and balance of new product opportunities in the portfolio should be directed by the business and innovation strategies. These strategies should provide management with the basis for:

- Deciding on the key dimensions and criteria of the portfolio. For example, the proportion of the portfolio that can be allocated to high-risk product ideas, or to new product opportunities that target new markets for the organization, or the emphasis on product improvements relative to "new to the organization" opportunities.
- Allocating product innovation opportunities to the portfolio to achieve the optimal balance by applying the key dimensions and criteria, thus ensuring alignment with strategy.
- Ongoing management of the portfolio to ensure appropriate selection and balance throughout the development pipeline and the products' life cycles.

Figure 3.7 shows a simple table representing a proportional allocation of projects across product and market newness. This is an example of a way to represent the state of balance in a portfolio with regard to product and market newness.

	Low market newness	High market newness
Low product newness	Improvements to existing products 30%	Additions to existing product lines 15%
Medium product Newness	Cost Reductions 25%	New Product Lines 10%
High product newness	Repositioning 15%	New to the world products 5%

FIGURE 3.7 An example of a product portfolio model

3.4.1 Visual Representation of the Product Portfolio

Visual representations of the portfolio are useful in both the development and presentation of the portfolio. Bubble diagrams are a commonly used form of visual display.

Typically, a bubble diagram shows projects on a two-dimensional *X–Y* plot. The *X* and *Y* dimensions relate to specific criteria of interest (for example, risk and reward). Bubbles are plotted according to the projects' ratings on the *X* and *Y* dimensions. The size of the bubbles represents a third criterion of interest, for example, the level of capital investment or the resource requirements.

Figure 3.8 shows bubble diagrams with different dimensions for representing balance in the portfolio.

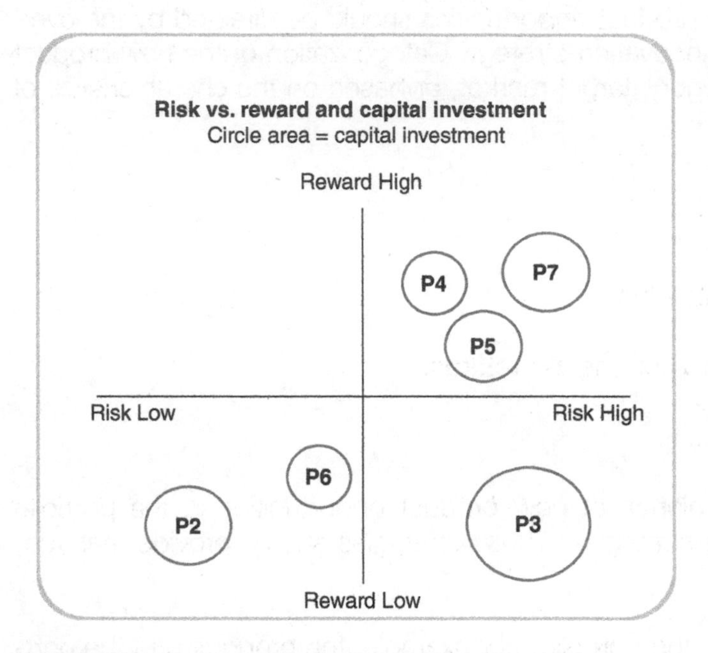

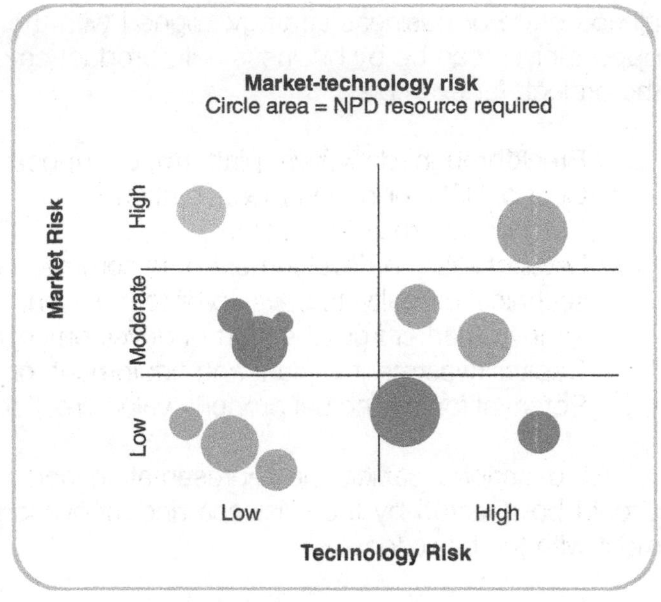

FIGURE 3.8 Bubble diagram portfolios

Bubble diagrams and similar charts are particularly useful in senior management presentations to give a relatively simple at-a-glance overview of the state of a portfolio. It is easy to identify gaps in the portfolio or inconsistencies with regard to business strategy and goals. For example, the right-hand portfolio in Figure 3.8 emphasizes low to moderate-risk projects with respect to both market and technology. Only one project is categorized as high risk for both market and technology, but this project demands a higher resource commitment. This quick overview provides management with insight as to the alignment of the portfolio with the organization's overall strategy with respect to risk.

3.5 THE RIGHT NUMBER OF PROJECTS

3.5.1 Resource Allocation

A critical aspect of portfolio management is resource allocation. It is imperative that projects are adequately resourced. Most organizations discover their product innovation success and ongoing product management are impacted by:

- Too many projects at the same time,
- Poor project planning and execution,
- Product innovation projects competing with other business priorities,
- Project delays that require heroic efforts to complete on time,
- Last-minute efforts to complete a task, only for it to sit in a queue downstream,
- Constantly changing priorities that force pulling of resources from one task to another,
- Lack of support (material, vendor, engineering), and
- Managers' time consumed solving routine, urgent problems.

Invariably this results in delayed launches, lost opportunities, and poor product acceptance due to deficiencies in features and functionality.

Resource allocation is a complex process. It is not simply a question of the numbers of individuals. It is having the right individuals with the right skills available at the right time.

The benefits of proper resource allocation are:

- Better project flow (fewer delays),
- Higher output (more launches),
- Higher employee satisfaction, and
- More effective portfolio management.

A key element of the innovation strategy is capability planning. That is, aligning internal and external capabilities to match the innovation goals and strategy. Application of portfolio selection criteria, based on the innovation strategy, provides a sound basis for resource allocation.

3.5.2 Methods for Resource Allocation

Cooper et al. (2015) recommend two fundamental bases for resource allocation: project resource demand and new product goals. These are best used in association with the overall portfolio management process.

Based on project resource demand:

- Prioritize the current project list from best to worst (using a scoring model or some form of financial analysis).
- Develop a detailed plan for each project. Use a project management software package.
- Allocate resources in the project plan according to specific categories (engineers, designers, manufacturing, etc.).
- Determine the cumulative resource requirements for each category by time.
- Match the resource demand for each category and time requirements against resource availability.
- Identify areas of resource overload. Re-prioritize projects in timing or delete projects from the portfolio, or source additional resources through additional hiring or outsourcing.

Based on new business goals:

- Begin with new product goals. Ask, "What returns or profits are desired from new products?"
- Calculate the potential returns or profits from each potential new product in the portfolio using financial analysis, e.g., EVA or DCF.

- Prioritize the projects and their potential returns against the business goals and select the projects that deliver the cumulative financial returns required to meet the business goals.
- Carry out the same exercise in individual project planning outlined in the "project resource demand" method above to determine the resource requirements by time against resource availability.

3.5.3 Resource Allocation as a Business Process

Without understanding the supply (resources) and the demand (projects), deploying a pull-based resource planning process is impossible.

Figure 3.9 shows four significant roles when establishing a resource allocation process. Depending on the size and structure of the organization, one or more people may fulfill these roles; not all organizations use separate employees for each role.

Figure 3.10 illustrates a typical resource allocation process.

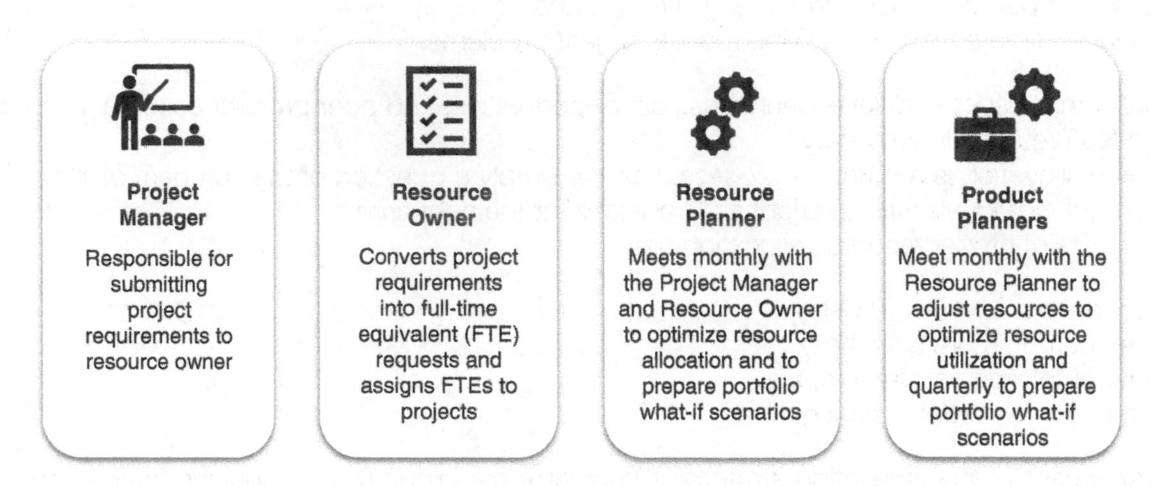

Project Manager
Responsible for submitting project requirements to resource owner
Resource Owner
Converts project requirements into full-time equivalent (FTE) requests and assigns FTEs to projects
Resource Planner
Meets monthly with the Project Manager and Resource Owner to optimize resource allocation and to prepare portfolio what-if scenarios
Product Planners
Meet monthly with the Resource Planner to adjust resources to optimize resource utilization and quarterly to prepare portfolio what-if scenarios

FIGURE 3.9 Roles and responsibilities for resource planning

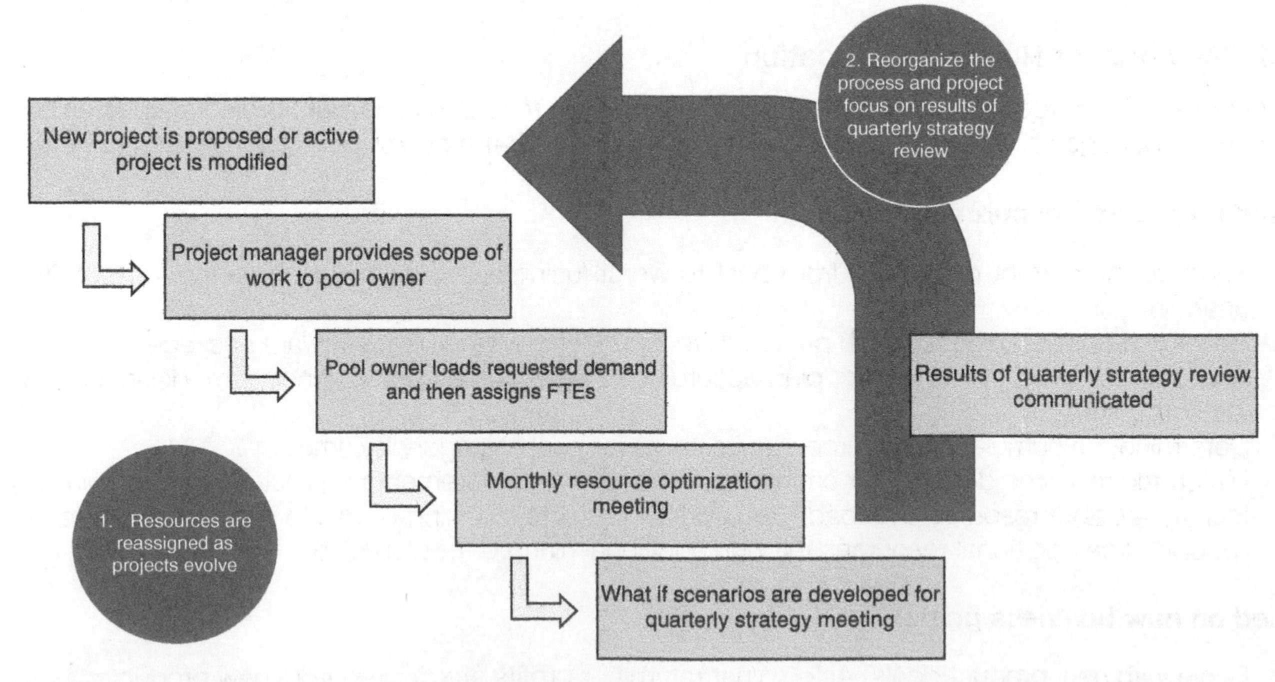

New project is proposed or active project is modified
Project manager provides scope of work to pool owner
Pool owner loads requested demand and then assigns FTEs
Monthly resource optimization meeting
What if scenarios are developed for quarterly strategy meeting

2. Reorganize the process and project focus on results of quarterly strategy review
Results of quarterly strategy review communicated

FIGURE 3.10 Typical resource allocation process

3.5.4 Tools to Support Resource Allocation

In large organizations, collecting information about available resources can be difficult. Optimizing the allocation of potentially competing resources across a range of projects can be extremely challenging. Figure 3.11 presents a process for collecting resource information based on resources required by individual projects and resources available. Resources required are identified by project managers and resources available are identified by the resource owners – functional or departmental managers.

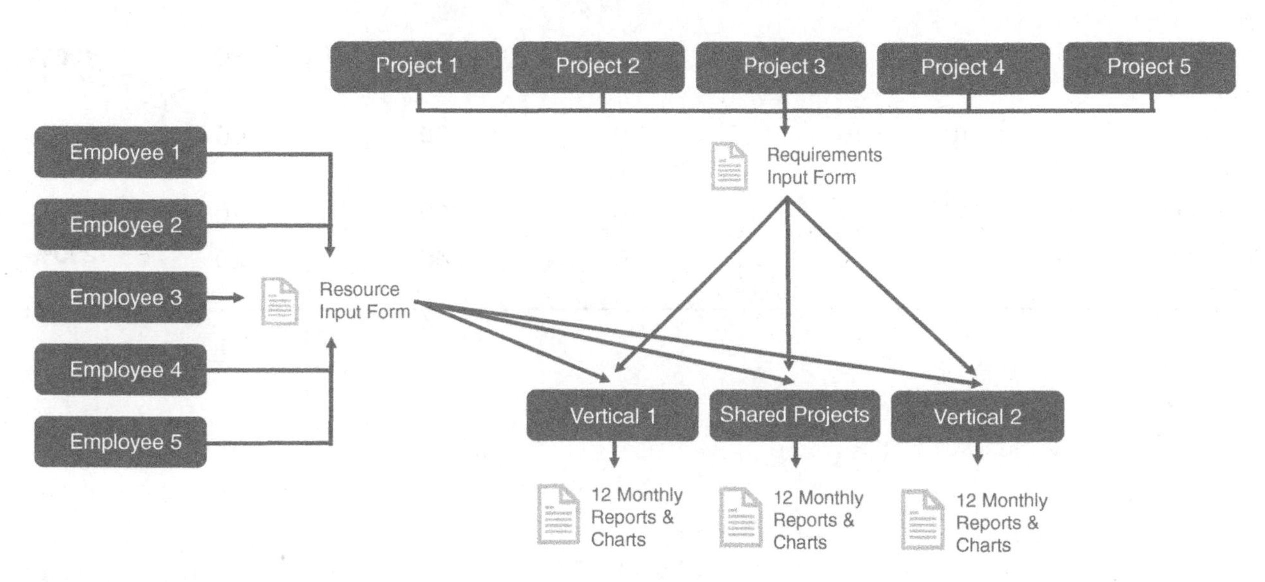

FIGURE 3.11 Process of collecting resource requirements

A resource requirements input form provides a standard and consistent template to capture the resource demand from all project teams. This form contains pre-defined fields requesting specific types and formats of information. Consistency in defining the type of resource required is essential to the resource owner. For example, simply stating a requirement for three engineers does not necessarily provide the level of detail to the resource owner on the type of engineers required. Are they three process engineers or one structural engineer and two process engineers? The answer impacts the resources actually needed.

The resource owners use the requirements input forms to consolidate the resources required by the project teams. Inevitably, as projects start and end, this will lead to an imbalance of resources available relative to the resources required.

In Table 3.9, the data from the forms have been input into a table to show which types of resources are scarce, or possibly where resources are in over-supply. After assessing the resource gaps, where "percentage utilization" exceeds 100%, action can be taken to address them, such as adding resources, shrinking projects, or choosing smaller projects. A clear innovation strategy and goals are critical in providing the basis for cross-functional discussions on the allocation of competing resources.

TABLE 3.9 Output showing where and when the resource gaps will occur

All Roles	Required Hours from Roles on Projects					Total Hours Req'd.	Capacity	Utilization %
	Project 1	**Project 2**	**Project 3**	**Project 4**	**Project 5**			
Project Manager	120	80	40	80	40	**360**	500	**72%**
Mechanical Engineer	0	100	60	20	0	**180**	180	**100%**
Precision Engineer	10	10	10	0	0	**30**	60	**%0%**
Tool Designer	30	40	0	20	10	**100**	120	**83%**
Service Part Technician	20	10	5	0	5	**40**	20	**200%**

3.6 NEW PRODUCT PORTFOLIO MANAGEMENT: A DYNAMIC PROCESS

As mentioned earlier in this chapter, the new product portfolio comprises a range of products at different stages of the product life cycle and with different requirements:

- ***Initial ideas and concepts*:** Are these worthy of inclusion in the portfolio? Do they meet the defined screening criteria?
- ***Products in development*:** Once an idea has passed the screening criteria, it can enter the development process, based on one of the processes to be described in the Process Chapter 4.

Most well-developed processes will have review points where the project can be assessed and a decision made on whether to proceed, change direction, or "kill" the project. Each of these decisions can potentially impact the product portfolio and the resources required. Proceeding may have no immediate impact on the portfolio. Changing direction may require a change in resources and funding requirements. Killing the project may provide an opportunity for the addition of another project to be included in the portfolio.

As described in Section 1.5, products invariably have a life cycle, basically defined by six stages: development, introduction, growth, maturity, decline, and retirement. Strategies employed at each of these stages will have an impact on the portfolio. In particular, there may be a demand for product variants or improvements during the growth and maturity stages. In the decline stage, decisions need to be made regarding retention, which may require product modifications or retiring the product, which may free up resources to be put to better use.

3.6.1 New-Product Portfolio Management Challenges

The overall aim of new-product portfolio management is a higher proportion of high-value projects and a better balance between projects and resources (Cooper et al., 2004). Cooper and Sommer (2020) suggest that only a small proposition of businesses achieve these outcomes as they struggle with portfolio management challenges. The two main challenges are overloaded pipelines compared to the resources available and the lack of reliable data on which to base portfolio decisions.

1. Overloaded pipeline. A common problem is too many projects for the resources available, resulting in resources being spread too thin and most projects being under-resourced. This typically results in lower revenue as the most strategically important projects are completed later than they would have been with appropriate research allocation. One of the reasons for too many projects is the failure to kill bad projects either due to adequate information, poor decision processes, or an overcommitment to *pet projects*.
2. Lack of reliable data. Sound go/kill decision-making through the product innovation project relies on rigorous analytical methods underpinned with reliable inputs – sales estimates, pricing, manufacturing costs, etc. The front end of innovation, which is the project's early phase from ideation to serious commitment to design and development, has long been recognized as a vital decision-making period. It is characterized as requiring low levels of investment. Sound go/kill decisions during this period can significantly reduce significant downstream investment in failed projects and loss of opportunity to add new projects and resources to the portfolio (Markham, 2013). The front end of innovation is discussed further in Section 4.1.4.

3.6.2 New-Product Portfolio Management: Stage-gate and Agile

Managing the new product portfolio is clearly a complex and dynamic process. Until recent years, much of this has been founded on the Stage-Gate process. The advent of Agile and Agile-Stage-Gate has led to the potential for even greater challenges in new product portfolio management. More in-depth descriptions of each of these processes are presented in Chapter 4. If you are unfamiliar with these processes, you may wish to read the relevant sections of Chapter 4 before reading the following section on the impact of Agile on portfolio management.

Since its inception in the early 1980s, Stage-Gate® has become one of, if not the most popular methods for managing the product innovation process. The structure of the Stage-gate process is designed to break the product innovation process down into a series of stages with defined decision points, or gates, between stages. These gates allow decisions to be made on whether to continue as planned, change direction, or stop a project. The results of these decisions can be fed into the portfolio management process to ensure the ongoing management of projects in the portfolio and consequent resource allocation.

The Agile process was developed, initially for the software industry, to promote rapid responses to change, including constant customer collaboration and feedback.

Over recent years, many organizations have recognized the value of combining the advantages of the Stage-Gate® and Agile processes into a hybrid system, Agile-Stage-Gate. In such hybrid processes, each stage in the Stage-Gate® process comprises a series of time-specific iterations called sprints. Each successive sprint is planned in real-time in response to the outcomes of the previous sprint. At the end of each sprint the project team produces a tangible outcome such as a concept, design drawings, or a prototype that can be shown to stakeholders or management. Retrospective team meetings at the end of each sprint assess the outcomes and team performance.

Whereas the Stage-Gate® process is typically linear, with the project team moving from task to task and stage to stage according to a strict timeline with limited customer interaction or iteration within stages, the Agile-Stage-Gate model encourages experimentation and iteration throughout. This process of constant iteration and refreshing of information through inputs from a range of stakeholders helps address the portfolio management challenges discussed above.

The importance of the front-end analysis remains – market, competition, customer needs leading to the business case and product concept definition. But the application of short sprints with defined outcomes provides the opportunity for more regular reviews than offered by the traditional stage-gate process. These regular reviews demand increasingly more reliable estimates of customer acceptance, projected sales, and product pricing. As development proceeds through successive iterations of the product, from proof of concept through early-stage design and prototyping, increasingly more reliable information is gathered on the overall costs and feasibility of manufacturing and marketing the final product. This whole process is designed to gather more and increasingly reliable data on which to base the viability of the new product and the regular Go/Kill decisions.

3.6.3 Addressing the Challenges of Agile Portfolio Management

The Agile-Stage-Gate hybrid helps address some of the challenges of portfolio management, but it also creates new challenges (Cooper & Sommer, 2016). The flexibility and dynamic nature of Agile, with ever-changing plans and product designs that are evolving over time, makes deriving accurate estimates of manufacturing costs, selling price, and sales, extremely challenging.

Another key challenge highlighted by Cooper and Sommer (2016) is in the area of portfolio metrics. In the traditional Stage-Gate® model, many organizations assess a project's progress and continued value by whether it meets defined milestones and is on budget. If these two criteria are achieved, the default decision is to continue the project. But meeting these criteria, although important, does not necessarily lead to a successful product. Also, a project that goes off budget may do so for a very good reason, for example, additional research on a component part that could add significant value to the product.

Different tools and metrics are needed to support decision-making within the fluidity and uncertainty of a purely Agile or hybrid Agile-Stage-Gate model. Three useful metrics proposed by Cooper and Sommer (2016) are the burndown chart, the productivity index (PI), and the expected commercial value (ECV).

A Burndown Chart

A standard Agile tool, *a burndown chart,* is a graphical representation of work left to do versus time. The outstanding work (or backlog) is often on the vertical axis, with time along the horizontal. Burndown charts are a run chart of outstanding work. It is useful for predicting when all of the work will be completed. Refer to Figure 3.12.

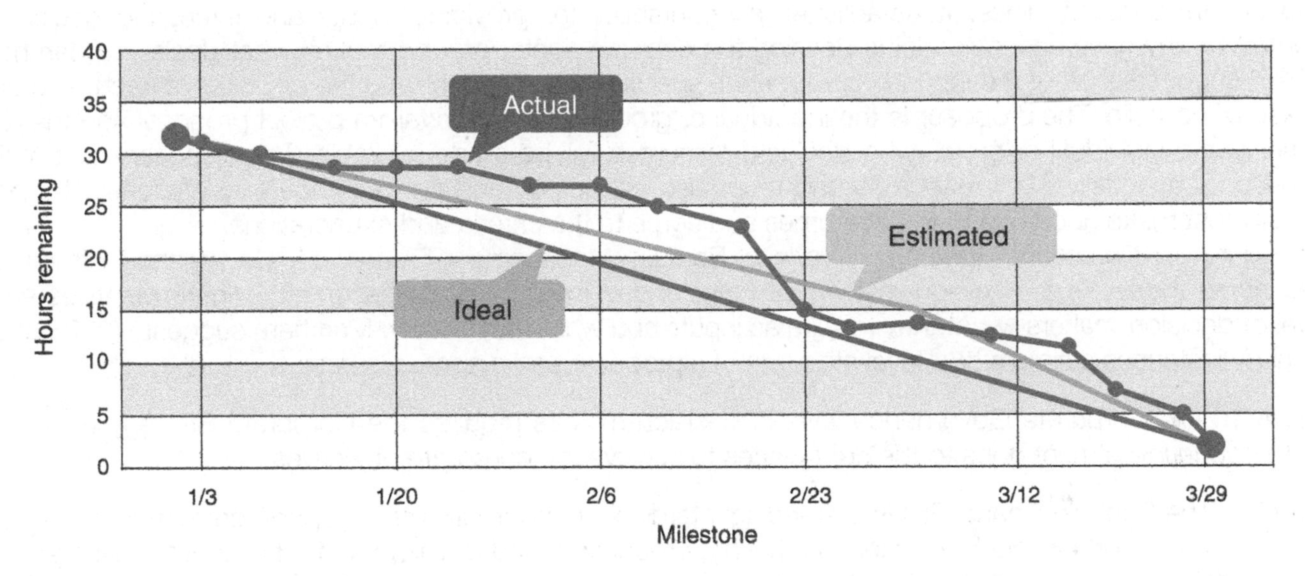

FIGURE 3.12 A burndown chart

The Productivity Index (PI)

The PI gauges the economic well-being of a project in real-time (Matheson et al., 1994). An example of the formula for the productivity index is:

$$PI = NPV/\text{Money remaining to be spent}$$

or

$$PI = NPV/\text{Number of work days left to be done}$$

The PI can be plotted over time. If the project is proceeding well, the PI should increase and its curve should asymptotically approach infinity just before product launch.

3.7 PORTFOLIO GOVERNANCE

Governing the portfolio is a complex political process. It is not a simple decision made by the portfolio manager about one project at a time that everyone else supports.

Markham (2023) provides levels of decision-making authority in portfolio management. It involves all levels in the organization agreeing to what decisions they make and what decisions other levels make – and not crossing the lines. It establishes accountability for providing inputs and executing decisions. Central to any governance model is defining the decision rights, or who makes what decision. Markham (2023) suggests that, at a minimum, governance specifies the *proposer* and the *disposer* at each level and phase of portfolio. The proposer is the individual or group who put forward a project proposal and the disposer is the individual or group who decide if resources will be made available. The disposer must make good on timely delivery of resources and must also set criteria and performance expectations. The proposer must make good use of the resources and agree to the criteria and expectations.

Figure 3.13 presents Markham's five-level Portfolio Management Framework Governance Structured describing the levels of decision-making in a typical development organization. The figure identifies each level of decision-makers, what they require as inputs and what they deliver. Markham suggests (2023) that, in many instances a simple two-level structure of a proposer and disposer is best.

Level 1. The Corporate Governance level of decision-makers requires the Corporate Strategy to ensure that enough investment goes to the right places to achieve the corporate objectives.

Level 2. The Portfolio Board (PB) is a standing interdisciplinary group, often department heads with direct interests in the portfolio such as sales, marketing, production, and development. The Board makes portfolio investment decisions based on optimization of revenue, resources, competition, new opportunities, and internal capacity.

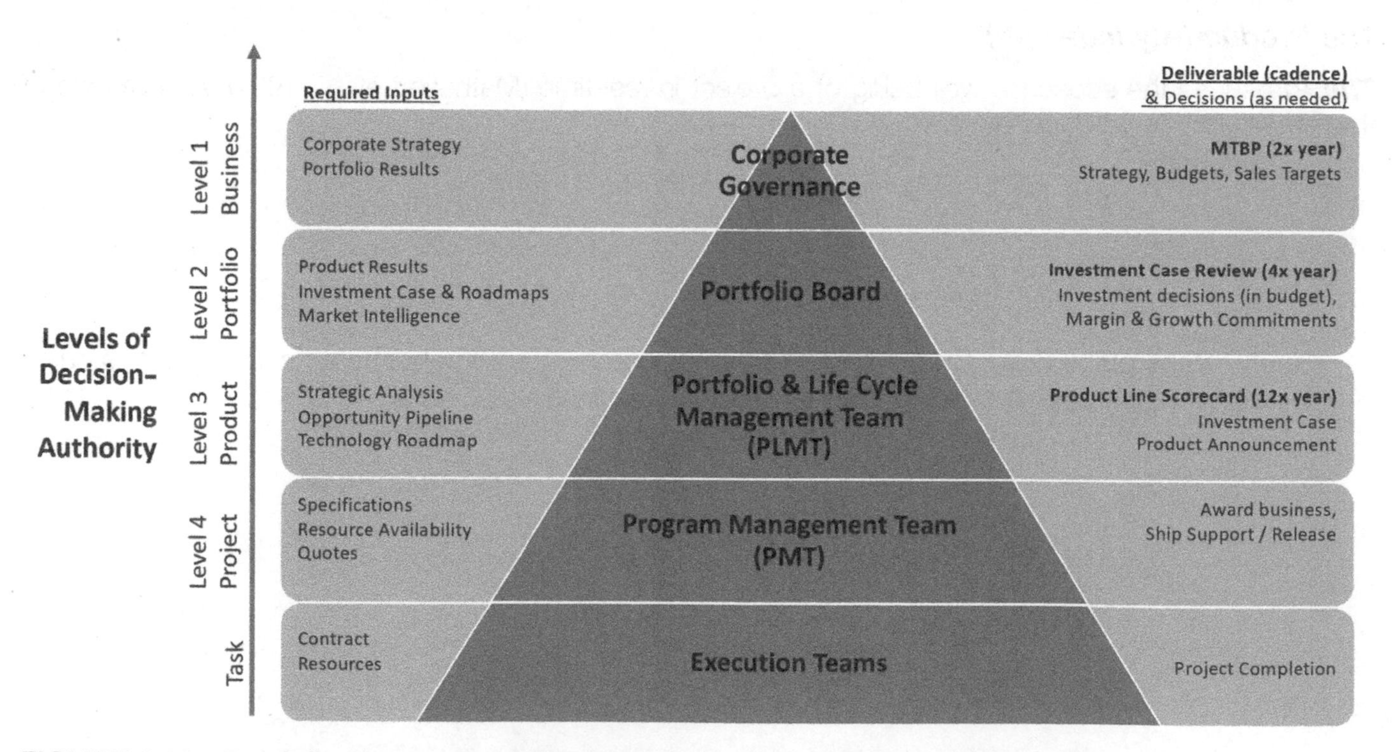

FIGURE 3.13 Portfolio Management Framework Governance Structure.
Source: Markham, 2023 / John Wiley & Sons.

Level 3. The Portfolio and Lifecycle Management Team (PLMT) is also a standing interdisciplinary group, usually representatives of the department heads. To fulfil the portfolio strategy put forward by the Portfolio Board and approved by the Governance body, the PLMT proposes individual projects to the Board. The PLMT needs strategic analyses, a transparent opportunity pipeline, and technology roadmaps.

Level 4. The Program Management Team (PMT) is another standing interdisciplinary group created to meet the development needs of the portfolio. The PMT creates and manages the actual work done by the Execution Teams. The PMT ensures resources are assigned to projects that deliver results to the PLMT, the Board and the Governance body.

Level 5. The Execution Team does the actual work of development. In this chapter it is portrayed as a technical development team, but it could just as easily be a services development team or a maintenance program team, an internal process team or another type of team.

Markham (2023) further suggests that "in smaller programs or smaller organizations there may only be two levels – the proposer and the disposer. Their level in the organization does not matter. The important point is there is a person that makes decisions separate from the person that carries out the decision. For example, there may only be a single director of product development and one product development team."

3.8 PORTFOLIO MANAGEMENT BENEFITS

The PDMA 2021 Best Practices Study (Bsteiler & Noble, 2023) found the best innovating companies rank the use of portfolio management higher than other companies. Further, incorporating a formal portfolio management process was consistently aligned with the best innovators.

As seen in Figure 3.14, the best innovating companies are more likely to use portfolio management to:

- Align projects with business strategy
- Make kill decisions to focus on the most important projects
- Prioritize projects based on the portfolio structure
- Only commit to projects they can properly resource
- Apply a repeatable decision analyses approach
- Consider all projects and how they relate to each other
- Evaluate portfolio investments along with other types of investments

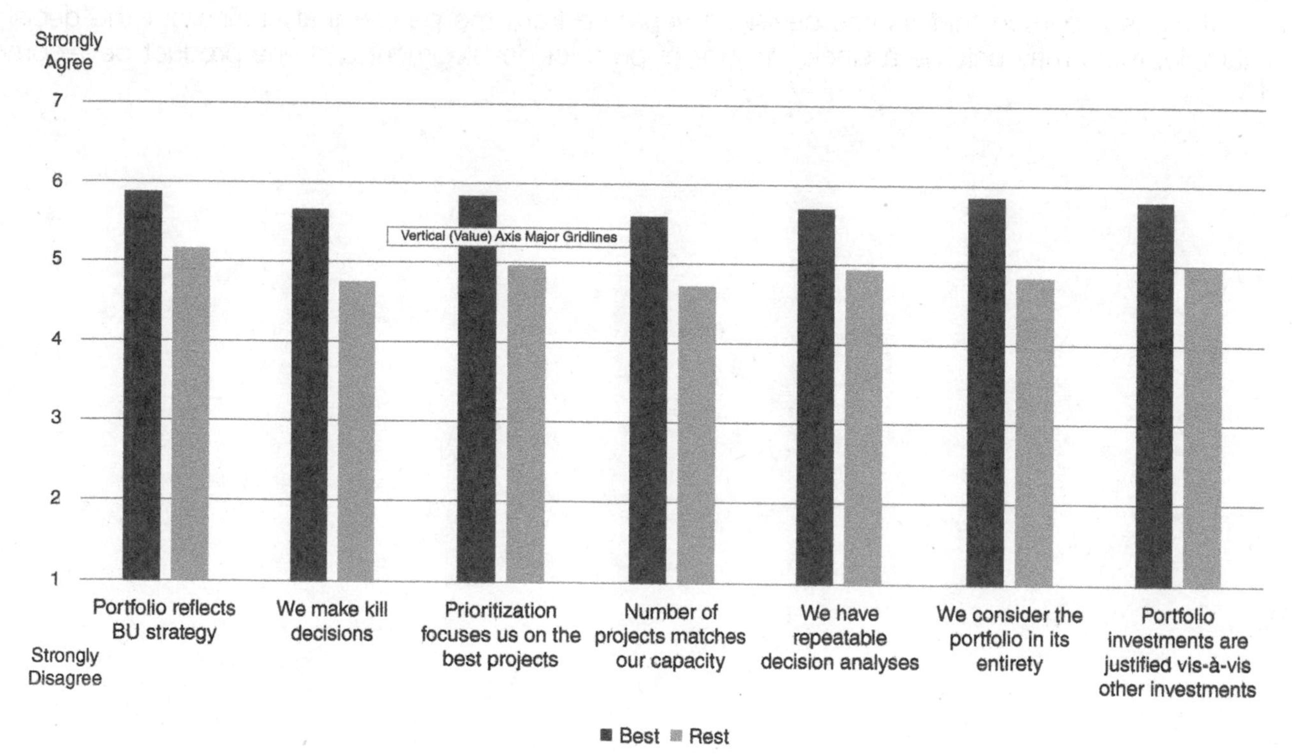

All differences $p < .01$

FIGURE 3.14 Portfolio Management findings from the PDMA Best Practices Study (2021)

3.9 IN SUMMARY

- "A portfolio is a collection of programs, projects, and/or operations managed as a group. The components of a portfolio may not necessarily be interdependent or even related – but they are managed together as a group to achieve strategic objectives" (Project Management Institute, 2021).
- A product portfolio is defined as *a set of projects or products that an organization is investing in and making strategic trade-offs against*.
- The four key goals of portfolio management are:
  1. ***Value maximization*:** Allocation of resources to maximize the value of the portfolio (the sum of the commercial worth of the individual projects).
  2. ***Business strategic alignment*:** Ensure that the overall portfolio continues to reflect the business strategy and the organization's innovation strategy. Ensure that the investment across the portfolio aligns with the organization's strategic priorities.
  3. ***Balance*:** Achieve the right balance of the right projects in terms of some pre-determined criteria such as long vs. short term, high vs. low risk, specific product, or market categories.
  4. ***Right number of projects*:** Most companies have too many projects underway for the limited resources available.
- The overall organization strategy and, in turn, the innovation strategy provides the criteria for product portfolio selection.
- Projects included in the portfolio can be from different business units, including new to the organization products, product improvements, and cost reductions. They can be at different stages of the new product pipeline and product life cycle.
- Typically, organizations will work toward a balanced product innovation portfolio based on criteria including risk vs. reward; new to the organization vs. line extensions; new vs. existing markets, etc. It is critical that these criteria are clearly aligned with strategy.
- Alignment of strategy with individual project selection and a balanced portfolio can be achieved through a "top-down" or "bottom-up" approach (or a combination of both).
- Resource allocation is a critical element of portfolio management. Many organizations are plagued by too many projects – resulting in poor execution, delayed launches, and commercial failures. Capability planning, embedded within the innovation strategy, provides a sound basis for resource planning as part of the portfolio management process.
- To avoid becoming overwhelmed when setting up a portfolio management system, start with a simple process and build upon it over time.
- Portfolio performance metrics can be used to help measure and maintain portfolio balance over time and to assess how well the portfolio changes in response to changes in strategy. Portfolio performance metrics are often created by combining the portfolio assessment criteria.
- The application of Agile and hybrid Agile-Stage-Gate product innovation processes over recent years has helped to address some of the traditional challenges of portfolio management, including overloaded pipelines and lack of reliable data on which to base portfolio decisions. But, at the same time, the new Agile-based processes have presented new challenges mainly related to the dynamic and iterative nature of these processes and the need to employ different approaches to project metrics and financial evaluation for decision-making.

3.A PORTFOLIO PRACTICE QUESTIONS

1. A weakness of the net present value (NPV) method for assessing projects in portfolio management is that:
   A. Projects cannot be rank ordered.
   B. NPV doesn't allow for resource allocation.
   C. Tools for calculating NPV are not readily available.
   D. It is difficult to determine accurate cash flow data for projects, especially early in the development process.
2. Allocating resources across a set of projects to optimize performance is known as:
   A. Value maximization.
   B. Profit-seeking.
   C. Rationalization.
   D. Pruning.
3. "Top-down" and "bottom-up" are ways to think about:
   A. Forecasting.
   B. Leadership.
   C. Connecting customers with the organization.
   D. Linking strategy and the product portfolio.
4. You are the CEO of company A. Your company has grown organically through lots of acquisitions and you now have a wide range of products that have been launched into a variety of markets, and your product teams have ideas to develop new products. You need to optimize the investment across all these existing and new products. What would you do?
   A. Tell the teams not to develop any new products.
   B. Give each team the same amount of money and ask them to proceed with development.
   C. Establish a portfolio management process.
   D. Approve projects that only cost less than $500K to develop.
5. In evaluating product opportunities, which of the following is a financial method for evaluation?
   A. Strategic alignment.
   B. Return on investment.
   C. Technical feasibility.
   D. Time to market.
6. In evaluating product opportunities, which of the following is a non-financial method for evaluation?
   A. Level of risk.
   B. Net present value.
   C. Payback period.
   D. Internal rate of return.
7. Company A has recently implemented a range of new product development practices. As a first stage, it has used a range of ideation tools to generate several potential new product ideas. It is now seeking to evaluate and prioritize the 150 product ideas for further evaluation and development. What technique would you recommend for the first stage of evaluating the 150 new product ideas?
   A. Pass/fail evaluation.
   B. Financial analysis.
   C. Ask the boss to do it.
   D. Detailed scoring of each idea against strategic criteria.

8. Mary is a product manager for ACE Electronics. She has been asked by the senior management executive to prepare a list of criteria as a basis for the evaluation of new opportunities to be included in the new products portfolio. She presents the following list: (1) Potential market share, (2) Potential contribution to company profitability, and (3) Availability of product development resources. Mary has omitted the most important criterion from her list. What is it?
   A. Support from the chief executive.
   B. Alignment with the company's strategy.
   C. Sufficient marketing budget.
   D. Sufficient manufacturing capability.

9. You need to estimate the sales potential of a new product to determine its overall financial feasibility. Your sales manager estimates there is a 40% chance of achieving sales of 500,000 units and a 60% chance of selling 1,000,000 units. What sales potential would you use in your feasibility analysis?
   A. 600,000 units.
   B. 700,000 units.
   C. 800,000 units.
   D. 900,000 units.

10. Jane has been evaluating the state of the innovation portfolio at ACME Bricks and Construction. Several simultaneous projects use the same set of product development experts. Moreover, many of these projects are being delayed and compete with other business priorities. What critical aspect of portfolio management should Jane look at to address these issues?
    A. Resource allocation.
    B. Project flow.
    C. Agile project development.
    D. Strategy execution.

Answers to practice questions: Portfolio management

1. D	6. A
2. A	7. A
3. D	8. B
4. C	9. C
5. B	10. A

REFERENCES

Belludi, N. (2021). *Creativity-it takes a village: A case study of the 3m post-it note*. Right Attitudes. https://www.rightattitudes.com/2021/04/15/creativity-case-study-post-it-note/

Bsteiler, L. and Noble, C.H. (2023). *PDMA handbook of innovation and new product development*, 4e. John Wiley and Sons.

Cooper, R. G. (2019). *Determining the value of ambiguous agile projects with multiple iterations using ECV*. Bob Cooper. http://www.bobcooper.ca/ECV-Article.pdf

Cooper, R.G. and Sommer, A.F. (2016). The agile-stage-gate hybrid model: A promising new approach and a new research opportunity. *Journal of Product Innovation Management 33* (5): 513–526. https://doi.org/10.1111/jpim.12314.

Cooper, R.G. and Sommer, A.F. (2020). New-product portfolio management with agile. *Research-Technology Management 63* (1): 29–38. https://doi.org/10.1080/08956308.2020.1686291.

Cooper, R.G., Edgett, S.J., and Kleinschmidt, E.J. (2001). *Portfolio management of new product*, 2e. Perseus Book Groups.

Cooper, R.G., Edgett, S.J., and Kleinschmidt, E.J. (2002). Portfolio management: Fundamental to new product success. In: *The PDMA toolbook for new product innovation* (ed. P. Belliveau, A. Griffin, and S. Somermeyer), 331–347. John Wiley and Sons.

Cooper, R.G., Edgett, S.J., and Kleinschmidt, E.J. (2004). Benchmarking best NPD practices 2: Strategy, resources and portfolio management practices. *Research-Technology Management 47* (3): 50–60. https://doi.org/10.1080/08956308.2004.11671630.

Cooper, R. G., Edgett, S. J., & Kleinschmidt, E. J. (2015). *Portfolio management: Fundamentals for new product success*. Stage-Gate. http://www.stage-gate.net/downloads/wp/wp_12.pdf

Crawford, M. and DiBenedetto, A. (2011). *New products management*, 11e. McGraw Hill.

Dam, R. F., & Siang, T. Y. (2021). *Introduction to the essential ideation techniques which are the heart of design thinking*. Interaction Design Foundation. https://www.interaction-design.org/literature/article/introduction-to-the-essential-ideation-techniques-which-are-the-heart-of-design-thinking

Kahn, K.B., Evans Kay, S., Slotegraaf, R.J., and Uban, S. (ed.) (2013). *The PDMA handbook of new product development*, 3e. John Wiley & Sons https://doi.org/10.1002/9781118466421.

Knudsen, M.P., von Zedtwitz, M., Griffin, A., and Barczak, G. (2023). Best practices in new product development and innovation: Results from PDMA's 2021 global survey. *Journal of Product Innovation Management 40* (3): 257–275. https://doi.org/10.1111/jpim.12663.

Markham, S.K. (2013). The impact of front-end of innovation activities on product performance. *Journal of Product Innovation Management 30* (S1): 7–92. https://doi.org/10.1111/jpim.12065.

Markham, S.K. (2023). An innovation management framework: A model for managers who want to grow their business. In: *The PDMA handbook of innovation and new product development*, 4e (ed. L. Bsteiler and C.H. Noble), 45–58. John Wiley and Sons.

Markham, S.K. and Lee, H. (2012). Product development and management association's 2012 comparative performance assessment study. *Journal of Product Innovation Management 30* (3): 408–429. https://doi.org/10.1111/jpim.12025.

Matheson, D., Matheson, J.E., and Menke, M.M. (1994). Making excellent R&D decisions. *Research-Technology Management 37* (6): 251–269. https://doi.org/10.1080/08956308.1994.11671006.

Pisano, P., Pironti, M., and Rieple, A. (2015). Identify innovative business models: can innovative business models enable players to react to ongoing or unpredictable trends? *Entrepreneurship Research Journal 5* (3): 181–199. https://doi.org/10.1515/erj-2014-0032.

Project Management Institute (2021). *A guide to the project management body of knowledge (pmbok guide)*, 7e. Project Management Institute.

Sommer, A. (2018). *Determining the value of ambiguous projects with multiple iterations using expected commercial value*. Sommer Systems. https://www.sommersystems.com/ECV

Zikopoulos, C. and Tagaras, G. (2015). Reverse supply chains: Effects of collection network and returns classification on profitability. *European Journal of Operational Research 246* (2): 435–449. https://doi.org/10.1016/j.ejor.2015.04.051.

4

Product Innovation

Process

Provides an approach, which is commonly understood and accepted by the whole organization, for successfully and sustainably developing new products or making improvements to existing products

Product Development and Management Body of Knowledge: A Guidebook for Product Innovation Training and Certification, Third Edition.
Allan Anderson, Chad McAllister, and Ernie Harris.

WHAT YOU WILL LEARN IN THIS CHAPTER

A structured and disciplined product innovation process is recognized as a critical factor in achieving repeatable, successful outcomes. This chapter introduces several processes and ideologies for sustainable product innovation. A summary of each process/ideology is presented, together with discussions of their benefits and limitations.

You are encouraged to select the best process, or combination of processes, that align with your specific organizational and product strategies.

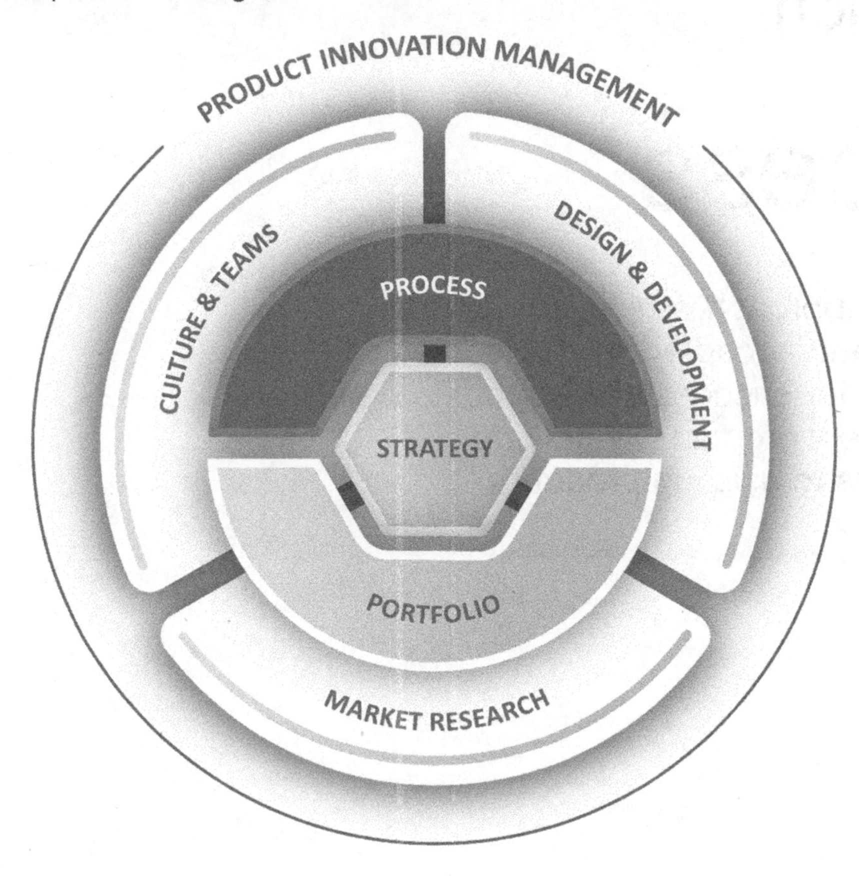

THE CHAPTER ROADMAP

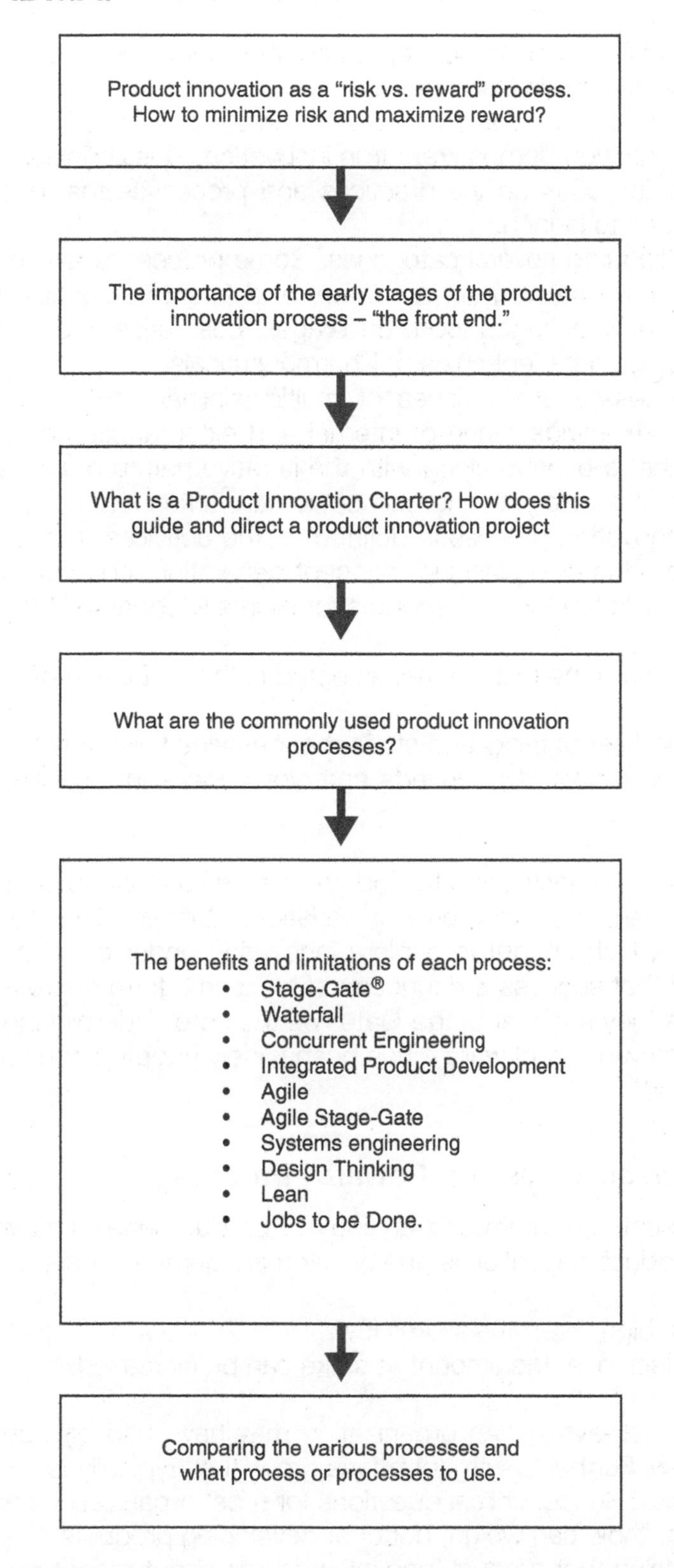

Product innovation as a "risk vs. reward" process. How to minimize risk and maximize reward?
The importance of the early stages of the product innovation process – "the front end."
What is a Product Innovation Charter? How does this guide and direct a product innovation project
What are the commonly used product innovation processes?
The benefits and limitations of each process:
Stage-Gate®
Waterfall
Concurrent Engineering
Integrated Product Development
Agile
Agile Stage-Gate
Systems engineering
Design Thinking
Lean
Jobs to be Done.
Comparing the various processes and what process or processes to use.

4.1 INTRODUCTION TO PRODUCT INNOVATION

This section presents several foundational concepts about product innovation to help set the stage for the innovation processes later explained in Section 4.3.

By combining the input of the ideation process with the structured activities of realizing social or commercial value from those ideas, you demonstrate true innovation. This definition of innovation encapsulates the basis of this book and its focus on the practices and processes that underpin end-to-end product innovation – from product idea to launch.

Products can be classified into several categories. Some processes and practices work best for one category but not for another. Some work for a number of different categories. For example, Fast Moving Consumer Goods (FMCG), like packaged food, beverages, cosmetics, etc., are defined, built, and delivered very differently than Consumer Electronics or Pharmaceuticals.

Product innovation processes are a series of multidisciplinary activities requiring sound decision-making based on inputs from a wide range of internal and external sources to the organization. As the variety of products continues to expand along with the iterative nature of process improvement, an ever-growing number of processes, ideologies, and/or models have taken shape.

Therefore, a product innovation process is defined as the activities, tools, and techniques consisting of strategy development, product-line planning, concept generation, concept screening, and research, to achieve successful outcomes in the form of products for clients (Cooper, 2017).

This definition continues to stand the test of time, as noted in the 3rd Edition of the PDMA Handbook:

> "A disciplined and defined set of tasks and steps that describe the normal [and appropriate] means by which an organization repetitively converts embryonic ideas into salable products or services" (Kahn et al., 2013).

Much like physics, we constantly seek to find the single, unifying process for a standard product innovation process. Even then, there is room for variation. Gurtner et al. (2018) published *Leveraging Constraints for Innovation,* which sought to explore innovation under constraint – in other words, doing more with less. They found that success is a function of ". . .some form of cross-functional, phased development process. . .". While they suggest Stage-Gate® as the core underpinning framework, it was not the first product innovation framework, and while it has been widely adopted and expanded into agile forms, it is also not the only framework.

4.1.1 Product Innovation as a "Risk vs. Reward" Process

Robert Cooper (2001) presents an interesting analogy of product innovation with the gambling process. Basically, the process of product innovation is one of risk management where the rules are:

- If the uncertainties are high, keep the stakes low.
- As the uncertainties decrease, the amount at stake can be increased.

Most organizations are risk-averse. An organization may have had previous products fail, creating a fear of trying something new. Further, as organizations grow, they typically focus on what made them successful and resist taking risks. Hence, critical questions for most organizations are: "is it possible to reduce product failures?" and, if so, "how can we get better at developing products?"

This creates the impression that product innovation is just about managing risk vs. reward – the main objective should be developing a suitable product/solution for the market, thereby balancing risk/reward while delivering on (and possibly exceeding) customer expectations and requirements. Modern product innovation aims to enlarge the "space of possibilities" within the risk/reward balance to achieve the optimal "solution space."

The 2021 PDMA Best Practices Survey (Knudsen et al., 2023), which includes responses from participants in 38 countries representing small and large businesses offering products of all categories showed, among other things, that product success rate was highly dependent on the quality and consistency of execution of product innovation practices and processes adopted by the companies:

- The best companies had success rates of 76%.
- The rest of the companies had success rates of 51%.

To put that in perspective, for every four projects, the best companies have one more product success than the rest of the companies. Given the size of product innovation projects, that difference is a substantial portion of the resources dedicated to innovation projects.

4.1.2 Managing the Risk of New Product Failure

The cumulative costs associated with product innovation increase throughout the product innovation process. The challenge for the product developer is to ensure that the risks related to product failure (the level of uncertainty) are reduced as costs increase (refer to Figures 4.1 and 4.2).

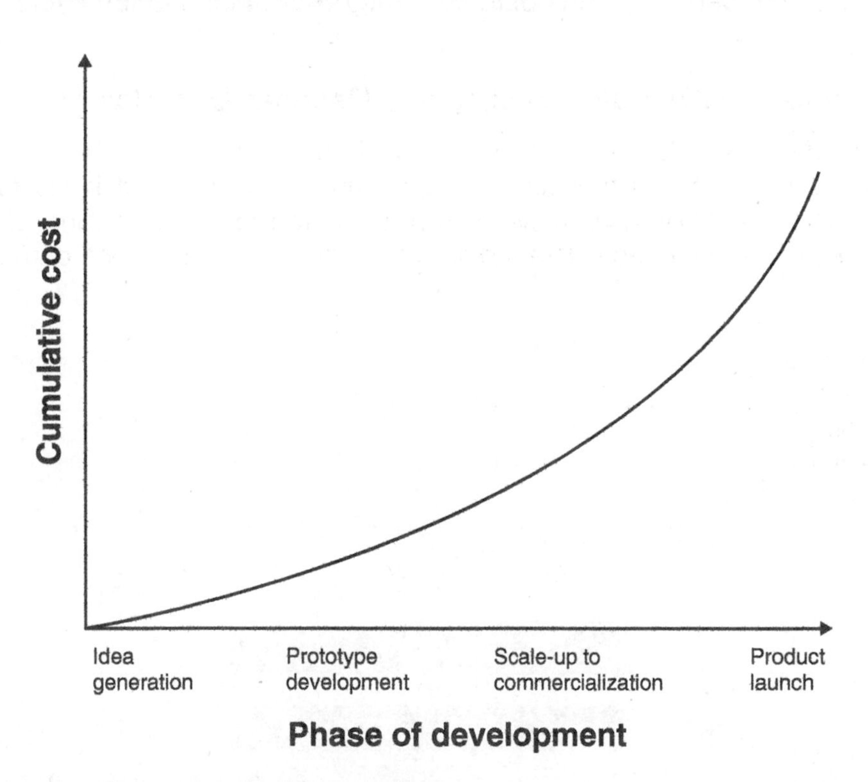

FIGURE 4.1 The costs of new product innovation

Figure 4.1 illustrates how product innovation costs increase as a product innovation project progresses through the development phases, e.g., due to increased resourcing, effort, and materials.

While the cumulative costs increase over time, the uncertainty element decreases (see Figure 4.2). This decrease is a function of better-defined product parameters and project deliverables. In other words, in well-run projects, the uncertainty about the product, the market, and other key factors are reduced by increased understanding or knowledge, which translates directly to an improved probability for success.

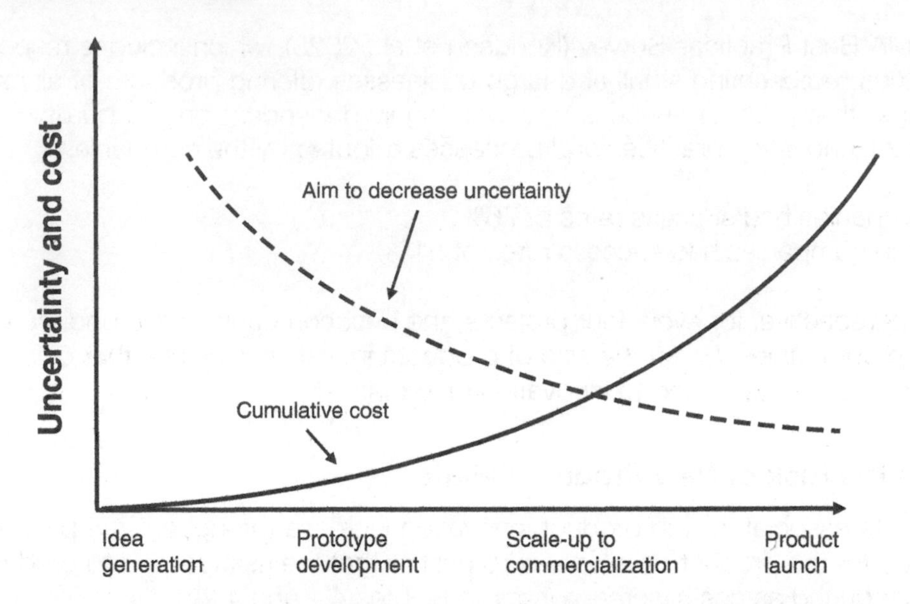

FIGURE 4.2 Managing uncertainty and costs over the product innovation cycle

4.1.3 Knowledge Improves Decision-making and Reduces Uncertainty

Figure 4.3 presents a standard decision framework that is used in a wide range of professions. As the different product innovation process models are discussed, notice that the basic fundamentals of this decision framework underpin all of these models. The aim is to reduce uncertainty and increase predictability. The knowledge, information, and data required to make sound decisions come from a wide range of sources:

- Organization records
- Organization staff
- External advisors
- Published literature
- Patents
- Competitors
- Customers

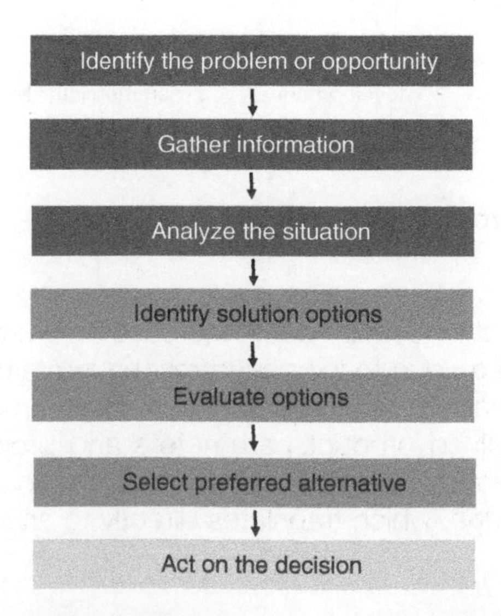

FIGURE 4.3 A standard decision framework

4.1.4 The Early Stages of Product Innovation: The Front End

Sections 4.1.2 and 4.1.3 describe how costs increase significantly as the innovation process proceeds. Making key decisions early in the process is critical to providing confidence in the on-going expenditure and minimizing the likelihood of failure. This early phase in the process is generally called the *front end of innovation,* or the *fuzzy front end*, because of high uncertainty and lack of clarity.

The product innovation process is frequently displayed in the form of a funnel. Figure 4.4 shows three basic stages of product innovation, *front end, development,* and *commercialization*. The transition through the funnel from many ideas or opportunities in the discovery stage diminishes as the process proceeds from development to commercialization.

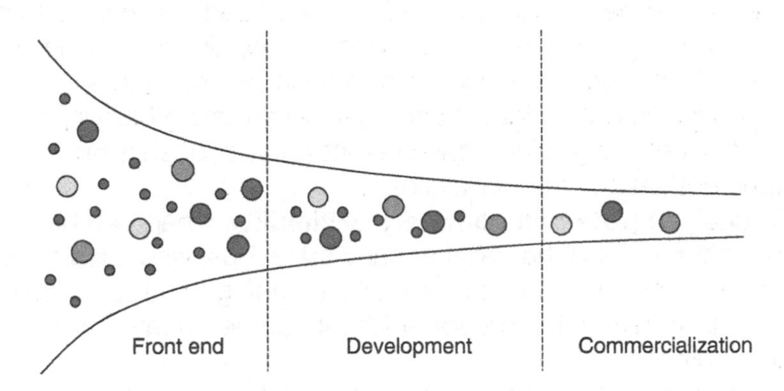

FIGURE 4.4 The product innovation funnel

The large number of possibilities in the early stages, or the front end of innovation, often leads to a lack of clarity, confusion, and even conflict. The greatest challenge in product innovation is to logically work through the fuzziness, to clarify future project direction, and minimize wasted expenditure and risk of failure, as shown in Figure 4.5.

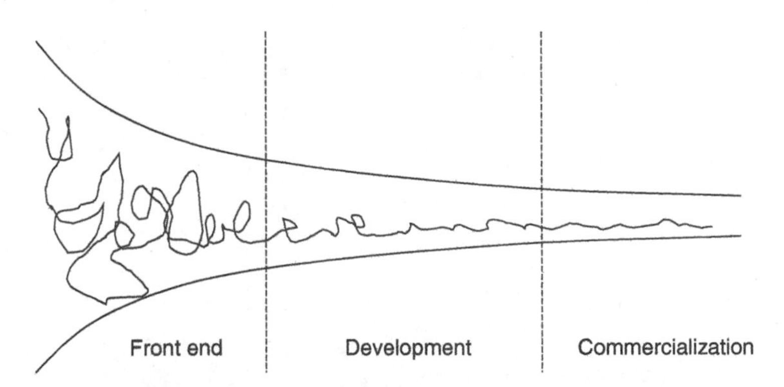

FIGURE 4.5 Addressing the front end of innovation, from fuzziness to clarity

Activities in the Front End of Innovation (FEI)

The basic activities usually included in descriptions of the front end of innovation are:

- **Pre-work to discover opportunities.** These opportunities may include line extensions, modifications to an existing product, or a totally new product. They may also include potential solutions to problems, for example, utilization of waste material or addressing under-capacity manufacturing equipment. Ideation techniques, extensively covered in Section 3.2.1.

- **Confirming the innovation problem or goal.** Clarity of the goal and what is being sought to achieve is critical to the success of any project. This preliminary analysis allows you to figure out what problem you need to solve for your customers before you develop a product. For example, one big failure in the annals of product history is Google Glass. Google made a device without considering what problems they were solving for their customers. Therefore, their product was a monumental (and costly) failure.
- **Scoping feasibility.** Quick and inexpensive assessments of the feasibility of the project are made during scoping. This involves learning what your customers think about your goal and whether you have the technical and human resources to achieve it. For example, in 1985, Coca-Cola Company released New Coke, a revamp of their classic Coca-Cola beverage formula to win over Pepsi drinkers. This reformulation changed a 100-year-old recipe based on market taste research. However, once New Coke launched, consumer outcry was overwhelming. Within 79 days, the company replaced New Coke with the original Coca-Cola formula repackaged as "Classic Coca-Cola." New Coke is now widely considered the biggest commercial marketing blunder of all time. Chapter 6 presents a range of techniques for gaining consumer insights, and Chapter 3 presents screening techniques for early-stage evaluation of opportunities.
- **Business case to analyze potential commercial viability.** Chapter 3 presents a range of financial evaluation methods that can form the basis of early-stage business case development. As shown in Table 3.6, a simple spreadsheet provides an excellent starting point for early-stage financial analysis. Based on relatively rudimentary data, this spreadsheet can be updated with more reliable data as the innovation process progresses.
- **Planning for development.** A plan is needed to funnel these potential products into your product development process. As described in Section 4.2, the *product innovation charter* provides the framework for this early-stage planning. It includes the estimated development schedule, project staffing, budget estimates, and much of the information described above.

4.2 THE PRODUCT INNOVATION CHARTER (PIC)

4.2.1 Providing Guidance and Direction for the Project

Regardless of how an organization manages the front end of innovation, key information is required for each product that moves forward to development. This includes:

- The market trends.
- The customer benefits and acceptance metrics.
- Product definition and specifications.
- Economic analysis of the product.
- The estimated development schedules.
- Project staffing and the budget estimates.

Generation of this information is best accomplished by a cross-functional product team and not solely a product and/or project manager. Cross-functional teams are described in Chapter 7.

Beyond this key information, every new product idea must meet some common success criteria including:

- Fit with the company skill sets and available resources.
- Fit the strategy of your company.
- Solve a problem for someone.
- Be something that someone will buy.
- Be scalable.

A helpful framework encompassing these criteria, by Marty Cagan, is the Four Risks of Product Discovery (2017):

- Value risk (whether customers will buy it, or users will choose to use it)
- Usability risk (whether users can figure out how to use it)
- Feasibility risk (whether our engineers can build what we need with the time, skills, and technology we have)
- Business viability risk (whether this solution also works for the various aspects of our business)

A successful product adequately addresses each of these areas. The challenge is keeping the product team on the same page as the product project evolves. The team needs to understand how each risk is addressed while sharing a clear product vision. Some product innovation projects drastically evolve from the original idea or insight into something that looks very different. A classic example is the Sgt. York *tank*, a massive project that lost sight of its original vision (New York Times, 1985).

So, how do you ensure you capture and convey the information the product team needs to have a clear vision for their work? One way is to create the Product Innovation Charter (PIC). The PIC is typically started and built out during the Front End of Innovation (FEI) phase and will contain assumptions about market preferences, customer needs, and sales and profit potential. As the project enters the execution phase, these assumptions are challenged through prototype development and in-market testing. While business needs and market conditions can and will change as the project progresses, scope creep is responsible for increasing project risk, which in turn, decreases the likelihood of success.

In the 2021 PDMA Best Practices Survey (Knudsen et al., 2023), we see this reflected in the cycle time, or time to complete projects, shown in Table 4.1. The numbers represent the average number of weeks spent on each type of innovation project. You will see that there has not been much change in the Incremental Innovation project cycle time over the past 20 years. That said, better application of product innovation and project management processes have led to a decrease of more than 70% in the time it takes to complete a radical innovation project.

TABLE 4.1 Reduction in weeks to complete product innovation projects

	2020	2012	2004	1995
Radical innovation	59	85	104	181
More innovative	50	59	62	78
Incremental innovation	35	34	29	33

Although many factors have led to the decrease in project time, creating an effective *Product Innovation Charter (PIC)* and then managing against the parameters of that PIC will aid significantly. A PIC helps keep the product team members on the same page and innovation projects on track, thereby reducing risks and time to market.

4.2.2 Content of the PIC

The PIC is normally a relatively short (1–3 pages) document. It contains four specific sections:

1. Background
2. Focus arena
3. Goals and objectives
4. Special guidelines

The content of these sections is detailed as follows:

1. **The background**
   - *Validation of the project*: its purpose and relationship to the business and innovation strategy. Why is the organization pursuing the project?
   - *Scope of the project*: how wide or narrow is the project's focus?
   - The role of the project team in achieving the project's goals.
   - *Project constraints*: resources, funding, manufacturing, marketing, etc.; anything that is likely to impact a successful project outcome.
   - The current and future state of any key technologies.
   - Environment, industry, and market analysis that shows the context for the new product, its customers, competitors, regulations, etc.
   - The desired benefits of the product innovation project.
   - Aspects relating to sustainability in the current and future environment.
2. **The focus arena**
   The term arena is mainly applied in sporting or theatrical terms – denoting where the team plays or the venue where the play takes place. This definition has been extended to business to denote the *playing field* for a business activity. The PIC should contain:
   - The target market (where the game is being played).
   - Key technology and market dimensions that underpin project success.
   - The strengths and weaknesses of competitors (the other players in the arena). Their technology, marketing, brand, market share, manufacturing, etc.
   - The approach taken to identify and incorporate all (other) project/product stakeholders impacted by the product, e.g., society, supply chain.
   - Specific aspects of sustainability that underpin project success.
3. **The goals and objectives**
   - Specific goals related to contribution to the business strategy key stakeholders expect, such as a percentage share of a new market or an increase in the current market share.

- Additional business operation goals include profit, sales volume, cost reduction, increased throughput, and sustainability goals. Sustainability goals extend beyond the market customer segment and consider greater socio-economic impacts. Examples include decreasing negative impacts on planet and people, such as a plan for the use/reuse of materials.
- Project-related objectives including financial, budget, time to launch, benefits tracking and realization, and effective stakeholder expectations management.
- Note that each goal or objective should have specific and measurable success criteria. These are referred to as performance metrics (refer to Chapter 1 for a more detailed discussion).

4. **The special guidelines**
   - Working relationships of the project team – how and when meetings are held, collaboration technology used, etc.
   - Project reporting – frequency, format, specific stakeholders.
   - Budget expenditure responsibilities.
   - Involvement of external agencies, e.g., regulatory bodies.
   - Specific aspects relating to time of launch or product quality.
   - Project governance and leadership.

4.3 SPECIFIC PRODUCT INNOVATION PROCESSES

4.3.1 Evolution of the Product Innovation Process

Over the past 50 years, there has been a dramatic increase in the research into, and application of, product innovation processes. This has resulted in a wide range of processes – many designed for specific industry, product, or market contexts.

It must be emphasized that the new product process is not a uniformly defined process that can be applied in the same way to all organizations or products. It should be designed to meet the specific needs of the organization and its customers, products, or services.

Further, there appears to be a preference for different product innovation models according to the geographical region; for example, Integrated Product Development is popular in the Asia-Pacific region, while agile-stage-gate is common in North America. Some regions utilize systems engineering.

Historically, defined processes for product innovation can be traced to an eight-phase process for developing chemical products in the 1940s. In the 1960s, NASA practiced the concept of phased development, the phase review process, which divided the development project into phases with reviews following each phase.

In the mid-1960s, Booz, Allen, and Hamilton (1982) designed a process that was divided into six basic stages, which laid the foundation for many of the processes developed more recently. These stages are:

1. Exploration
2. Screening
3. Business evaluation
4. Development
5. Testing
6. Commercialization

Perhaps the most defining step in the formalization of the product innovation process and its application across a wide range of industries came in the mid-1980s with the Stage-Gate® process (Cooper & Kleinschmidt 1986).

Over the past three decades, the evolution of a number of processes, designed to meet the needs of specific organizations with different product and market contexts, is evidenced. The remainder of this chapter describes a selection of these modern product processes and outlines their advantages and disadvantages in applications within specific contexts. These processes include but are not limited to:

- Stage-Gate®
- Agile
- Integrated Product Development (IPD)
- Lean Product Innovation
- Lean startup
- Systems engineering
- Design thinking
- Jobs to be done

We have chosen to use the term "process" in this book, although it may well be argued that Agile, Lean, Design Thinking, and IPD might best be described as ideologies, philosophies, or a set of principles.

4.3.2 The Stage-Gate® Process

The Stage-Gate® was first developed by Robert Cooper in the mid-1980s and has been continuously updated over the years in response to changing industry needs. A generalized view of Stage-Gate® process is shown in Figure 4.6, consisting of stages and gates.

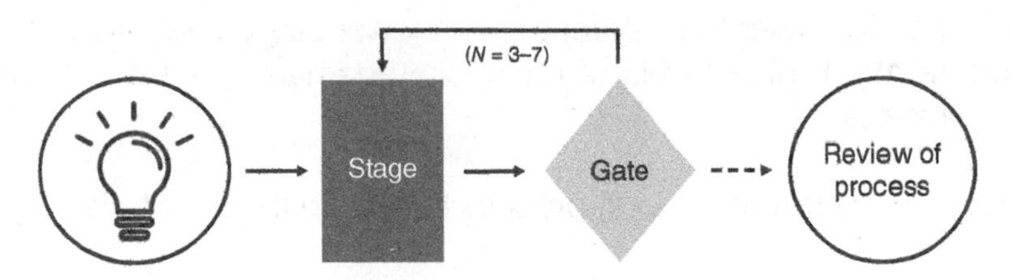

FIGURE 4.6 Generalized view of the stage gate process

What is a stage?

Basically, a stage is a defined work section of the overall product development process, including:

Activities: The work the product team must do according to the project plan.
Integrated analysis: The project leader's and team's integrated analysis of the results of all the functional activities, derived through cross-functional interaction.
Deliverables: The presentation of the results of the integrated analysis, which must be completed by the team for submission at the gate.

What is a gate?

Basically, gates are where key decisions are made on the future of the project, including:

Deliverables: Inputs into the gate review. These are defined in advance and are the results of actions from the preceding stage. A standard menu of deliverables is specified for each gate.
Criteria: What the project is judged against to make the go/kill and prioritization decisions. These criteria are usually organized into a scoreboard and include both financial and qualitative criteria.
Outputs: Results of the gate review. Gates must have clearly articulated outputs, including a decision (go/kill/hold/recycle) and a path forward (approved project plan, date, and deliverables for the next gate agreed on).

Following is a brief description of the main stages for Stage-Gate®.

- ***Discovery:*** looking for new opportunities and new product ideas.
- ***Scoping:*** examining the market opportunity, technical requirements, and capabilities available.
- **Business case:** a critical stage that builds on the scoping stage with a more in-depth technical, marketing, and business feasibility analysis.
- ***Development:*** the product design, prototyping, design for manufacture (if applicable), preparation of manufacturing and launch plans.
- ***Testing and validation:*** testing all aspects of the product and its commercialization plans to validate all assumptions and conclusions.
- ***Launch:*** full commercialization of the product, including full-scale manufacturing and commercial launch into the market.

Figure 4.7 shows what Cooper describes as a *full 5-stage process*. The number of stages is not fixed and should be adapted according to the specific context and will depend on:

- **The urgency for new product launch**. Greater urgency leading to a more compressed process with fewer stages.
- **The current knowledge** about the technology and market related to the new product. The greater the basis of knowledge, the lower the risk resulting in fewer stages required.
- **The level of uncertainty.** Higher levels of uncertainty demand greater information to mitigate risk, leading to a longer process.

Figure 4.8 shows adaptations to the 5-stage process based on these criteria.

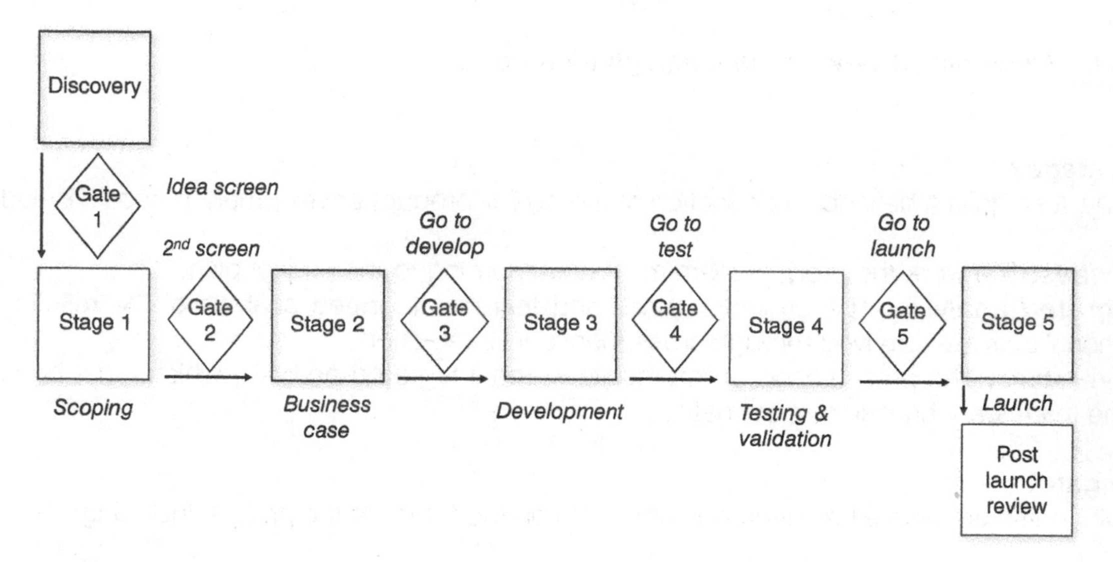

FIGURE 4.7 The 5-stage process

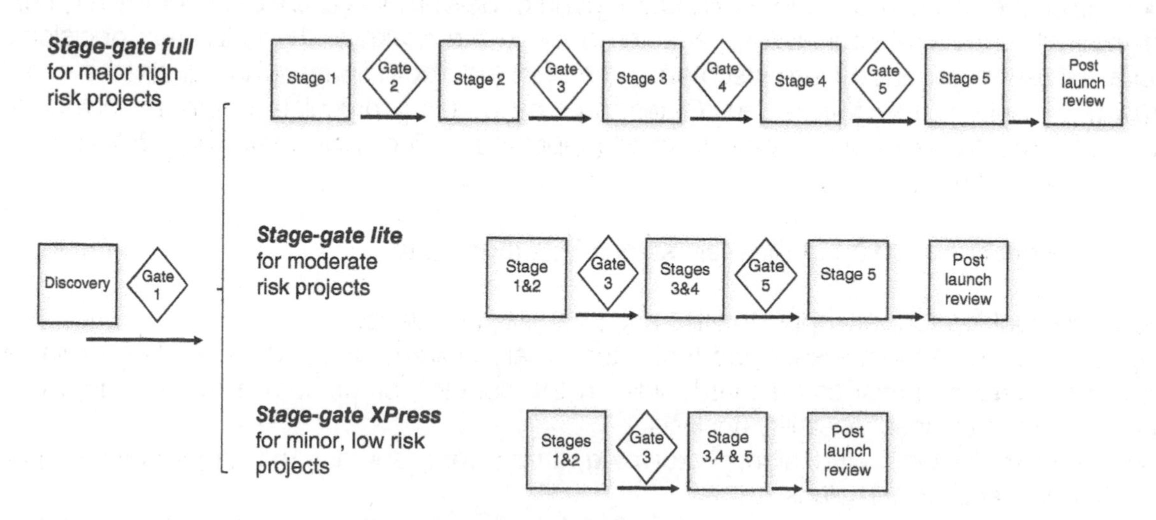

FIGURE 4.8 Three levels of the Stage-Gate® process

Evolution and Application of the Stage-Gate® Process

The Stage-Gate® process, at least in its early iterations, appeared quite linear and sequential. This has never been the intention of its originators (Cooper, 2014). Further information on the history of Stage-Gate® and its current status is provided by Stage-Gate® International (stage-gate.com). Over recent years, Cooper (2014) has written a number of articles demonstrating new evolutions of the Stage-Gate® process and its adaptation to different product development contexts and organizational requirements. Cooper (2014) emphasizes that, although the basic principles of the process may remain the same, its application should be modified to be fit for specific situations. Specific adaptations of the Stage-Gate® process include:

- ***Parallel processing:*** The traditional Stage-Gate® activities are "compressed" into parallel (concurrent) streams of activities (as opposed to linear and sequential), with the aim of expediting the activities within each stage. They then converge at strategic decision gates to allow for progression. The parallel processing requires the organization to allocate more resources to activities in the same time period.
- ***Circular design and product management:*** Here, the focus is on sustainability. This involves incorporating the customer and product users early in the product innovation design stage, with the aim of reusing materials, designs, reducing waste in the product's development and life, and the ability to recycle the product at the end of its use.
- ***Fast Stage-Gate:*** Also known as Agile-Stage-Gate (Cooper & Sommer, 2016), the aim is to incorporate principles of Agile approaches (e.g., sprints, standups, backlogs) into the product manufacturing environments. It required modifications to Scrum, the most common Agile approach, to cater to the differences in the manufacturing environments. This hybrid approach aims to foster innovation, minimize risk, speed up decision-making, and speed up product innovation and launch to market. It is further described in Section 4.3.7.

Benefits and Limitations of the Stage-Gate® Process

Table 4.2 presents the benefits and limitations of the standard Stage-Gate® process.

TABLE 4.2 Benefits and limitations of Stage-Gate®

Benefits	Limitations
• Adds discipline and constraints into product innovation. • Contributes to limiting and managing risk. • Places an emphasis on quality decision-making. • Is transparent to all involved. • Can be adapted to a wide range of organizations.	• Has the potential to become overly bureaucratic. • Where it is not fully understood, it can be seen as too rigid and costly. • The discipline and constraints can be perceived as stifling creativity. • While the process is intended to decrease time-to-market, difficulty scheduling gate reviews results has resulted in increasing time-to-market.

Hypothetical Examples of How the Stage-Gate® Process can be Adapted to Specific Situations

The development of a simple line extension for a novelty ice cream where the company has a long history in the market, sound technical expertise, and existing manufacturing capability. In this case, the development complexity is very low, with a relatively high certainty of success and low risks associated with failure. A short process focusing on an initial business case followed by iterative formulation (or design) stages, informed by product testing, leading to final product launch, is probably all that is required. The time-to-market could be expected to be from one to three months.

The development of a model car that can be driven off-road by 10- to 12-year-old children, where the company has some experience with the target market but has limited expertise with the technology involved. There is also significant uncertainty surrounding the product features desired by the target market and the potential demand. In this case, there is a high level of uncertainty associated with not only the product concept and design specifications but also with the potential complexity in design and manufacture. Here, one would expect more stages in the process with a greater emphasis on gates to ensure informed decision-making and reduce uncertainty. Specific focus should be placed on the FEI stages of concept development and initial business analysis. More activity at the end of the process (focusing on design for manufacture, market testing, final feasibility analysis, and scale-up for launch) could be expected. The time-to-market could be expected to be from six months to two years.

An example of how a company adapted the Stage-Gate® process.

Figure 4.9 shows an example of how General Electric (GE) adapted the Stage-Gate® process to its unique requirements. The GE process is based on 4 stages comprising 10 activities across these stages.

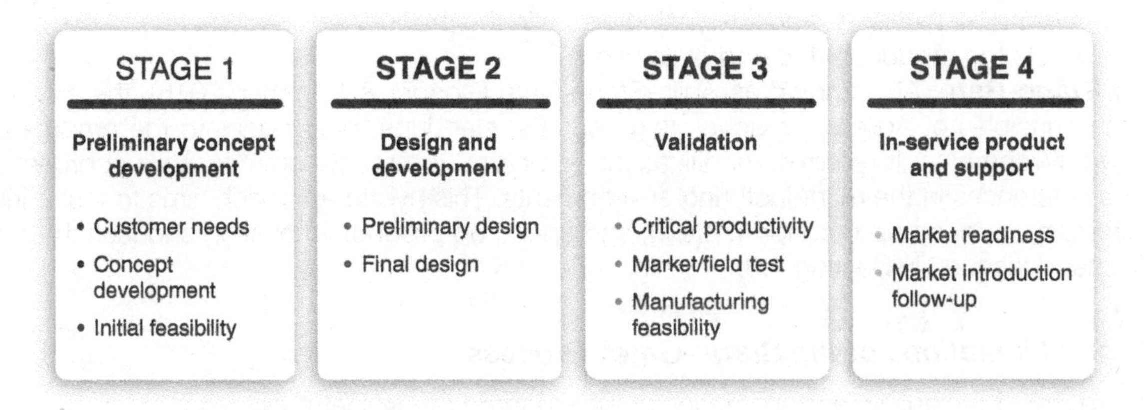

FIGURE 4.9 Example of Stage-Gate® process adaptation.
Source: Phillips et al. (1999).

GE's choice of this Stage-Gate® process and its 10 activities drove a cross-functional focus on early market needs prior to full market production. Additionally, it instilled discipline in the company's project management techniques (Phillips et al., 1999, p. 292).

4.3.3 The Waterfall Method

The first development of the waterfall process is generally attributed to Winston Royce (1970). During the early 2000s, it was widely applied within the software industry. The five phases of the classical waterfall process are shown in Figure 4.10 and include:

1. ***Requirements gathering*:** Understanding what is required to design the product: function, purpose, user needs, etc.
2. ***Design*:** The attributes and features of the product are designed to meet customer needs. It also includes planning and feasibility assessments.
3. ***Implementation:*** The product's design plan is enacted and mainly entails development or production activities.
4. ***Verification:*** Ensuring that the product is meeting customer expectations.
5. ***Maintenance:*** Use of the customer to identify shortfalls or mistakes in the product design that require change.

In recent times, the waterfall method has been frowned upon for being "too slow" or inflexible for a product/project's completion. Some organizations have applied a hybridized approach to the overall framework. This provides the basis of an overall structure while providing an iterative approach with greater flexibility, which speeds up development. The teams performing these projects would typically be smaller and are normally co-located.

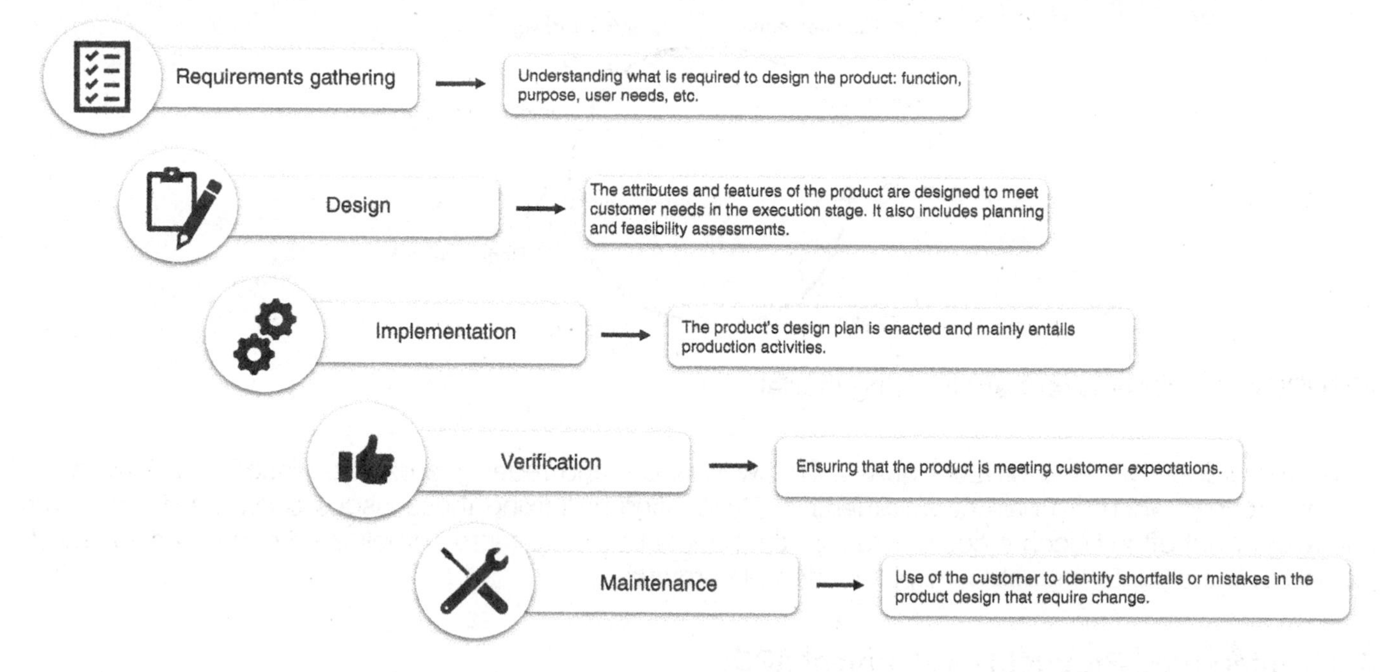

FIGURE 4.10 The waterfall method

Regardless, the waterfall method continues to be used, especially when the requirements are well understood and not expected to change. The benefits and limitations of the waterfall method are shown in Table 4.3.

TABLE 4.3 Benefits and limitations of the waterfall method

Benefits	Limitations
• Needs each phase to be completed before moving to the next. • Is suited well for smaller, well-defined projects with well understood and complete requirements. • Needs minimum client intervention.	• Is not suitable for large, complex projects that have frequent requirement changes. • Has testing of the product in later phases, increasing risk. • Requires detailed documentation which may be time consuming.

4.3.4 Concurrent Engineering

Concurrent engineering was seen as a replacement for the more traditional sequential *waterfall model*. "Concurrent engineering is a systematic approach to the integrated, concurrent design of products and their related processes, including manufacturing and support. This approach is intended to cause the developers from the very outset to consider all elements of the product life cycle, from conception to disposal, including quality, cost, schedule, and user requirements" (Winner et al., 1991). The concurrent engineering model is shown in Figure 4.11.

The basic premise for concurrent engineering is founded on two concepts. The first is the idea that all elements of a product's life cycle, from functionality, producibility, assembly, testability, maintenance issues, environmental impact, and finally, disposal and recycling, should be taken into careful consideration in the early design phases. The second concept is that the preceding design activities should all be occurring at the same time, i.e., concurrently. The idea is that the concurrent nature of these processes significantly

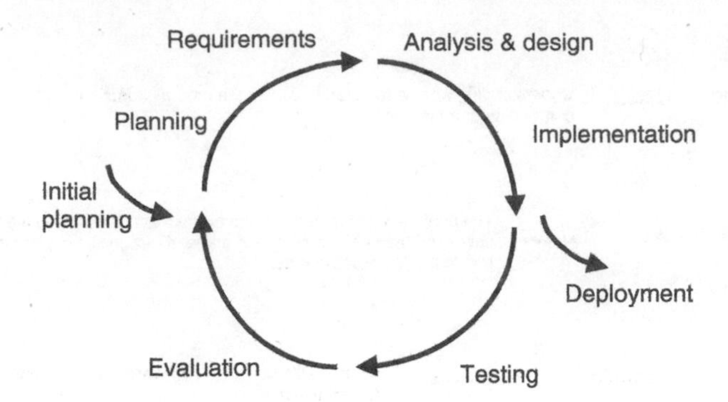

FIGURE 4.11 Concurrent engineering model

increases productivity and product quality. This way, errors and redesigns can be discovered early in the design process when the project is still flexible. By locating and fixing these issues early, the design team can avoid what often become costly errors as the project moves to more complicated computational models and eventually into the actual manufacturing of hardware.

4.3.5 Integrated Product Development (IPD)

Integrated product development has evolved from concurrent engineering. It is defined as "a philosophy that systematically employs an integrated team effort from multifunctional disciplines to [effectively and efficiently] develop. . . new products that satisfy customer needs" (Kahn et al., 2013).

Over recent years, some organizations have focused on a stepwise approach to improving the overall product innovation system, centered on IPD principles. The aim was to progress from a focus on the application of basic tools for product innovation through to the application of project management, voice of the customer, linkage to strategy, and, at the highest level, a learning culture founded on knowledge capture and management, as illustrated in Figure 4.12. This shows IPD more as a framework for evolutionary and capability maturity improvement in product innovation rather than simply a specific model or process. This

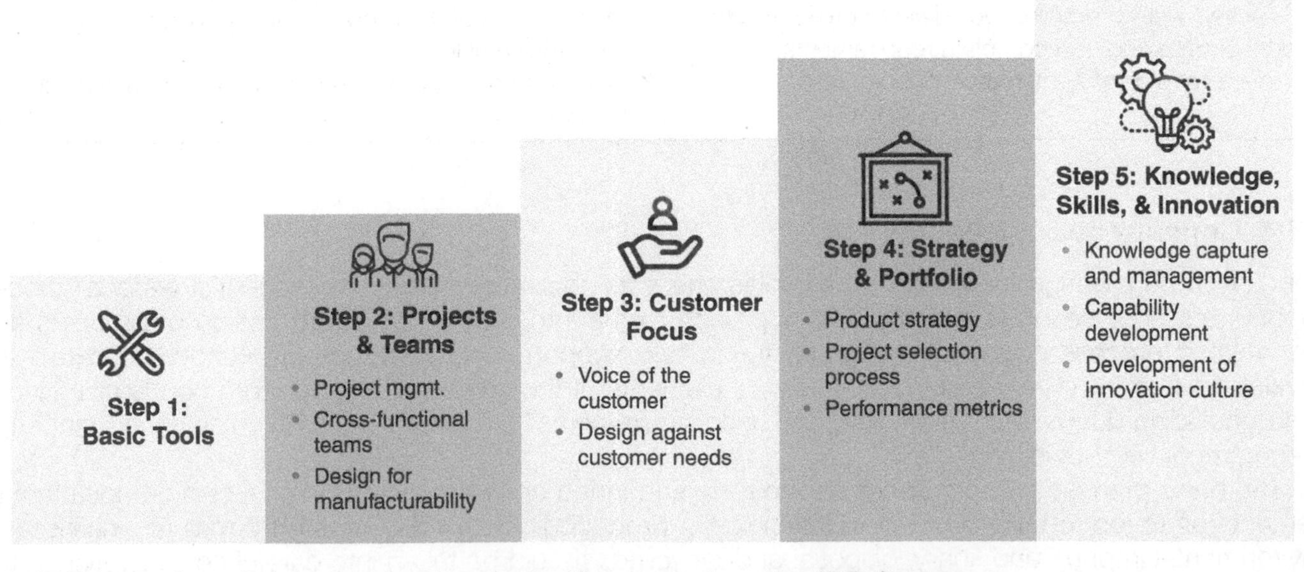

FIGURE 4.12 Levels of organizational practice based on the IPD System

framework captures many of the fundamental principles underpinning the most commonly used product innovation processes, emphasizing understanding the customer, learning, and continuous improvement.

The benefits and limitations of IPD are summarized in Table 4.4.

TABLE 4.4 The benefits and limitations of integrated product development

Benefits	Limitations
• As the organization develops its capabilities and matures through the IPD steps, product innovation and delivery should become more efficient. • Increased efficiencies translate into improved cost management and increased profitability. • Due to the strong emphasis on design principles in the FEI stages (in relation to defining the product/project goals), and thus clearer product definition, the risks are managed sooner in the product innovation life cycle. • Quality, proactive risk management, and a focus on customer requirements ensure precise delivery of value. • Multifunctional teams (internal and external) that effectively collaborate in a product innovation project can pool their skills and capabilities toward common goals. • In the context of sustainability and the circular economy, IPD offers t multi-stakeholder engagement to encourage sustainability objectives.	• Customer requirements must be clear early on in the development cycle so as to meaningfully contribute to the downstream creation and achievement of value. This requires active stakeholder engagement. • The latest tools and methods are needed. • IPD requires the right competencies (people and skills) to be deployed in the process. • Multi-stakeholder collaboration can be problematic and complex if the teams are unable to effectively collaborate and co-create. This is especially true of organizations that have used linear work phases instead of a cross-functional product team approach. • The appropriate balance of FEI and design controls is paramount to project success, otherwise inefficiencies will ultimately delay product delivery (Naveh, 2004).

IPD and sustainability integration – an example

An example of a company that successfully implemented IPD with a focus on integrating sustainability is Nestlé. Espinoza-Orias et al. (2018) describe how Nestlé, "Guided by values rooted in respect, [it] works alongside partners and stakeholders to create shared value (CSV) across all the activities of the company, which contribute to society while ensuring the long-term success of our business." Its commitments go so far as to continually align with the SDGs (sustainable development goals) in terms of scope and timelines.

4.3.6 Agile Product Innovation

Agile is neither a methodology nor a framework. Instead, it is a set of principles and values. Agile methods don't attempt to prescribe a specific magic solution that works for every team. Instead, they focus on helping a team to think clearly, creatively, and collaboratively to achieve the best solution for that team.

The philosophy was first constructed from experiences with software product development and has since been applied across industries and product categories.

Manifesto for Agile Software Development

In February 2001, 17 software developers met in Utah to discuss lightweight development methods. They published the *Manifesto for Agile Software Development*. The manifesto states (Agile Manifesto, 2001):

> "[They] are uncovering better ways of developing software by doing it and helping others do it. Through this work, [they] have come to value:

- ***Individuals and interactions*** over processes and tools,
- ***Working software*** [or product] over comprehensive documentation,
- ***Customer collaboration*** over contract negotiation,
- ***Responding to change*** over following a plan.

That is, while there is value in the items on the right, [they] value the items on the left more."

This means that while processes, tools, documentation, contracts, and plans are important, Agile practitioners prioritize the interactions of individuals, a working product, customer collaboration and active inputs, and the ability to adjust to moving plans (change) as the guiding principles by which product design, development, and its launch are realized.

Agile methodologies are iterative approaches that can be incorporated into product innovation that self-organizing teams perform in a collaborative environment. It helps teams respond to unpredictability through incremental, iterative work cadences. The most common agile methodology is Scrum. Scrum has overwhelmingly been accepted in the software industry but has also been included in a hybridized Agile-Stage-Gate approach (discussed further in Section 4.3.7) for the manufacturing of tangible products. Unlike hardware, software is continuously and infinitely changeable. Recommended further reading on Agile in product innovation includes Pichler (2013) and Cooper and Sommer (2016).

Key Principles of Agile Product Innovation

Following is a summary of the key principles of Agile product innovation (from http://agilemanifesto.org/principles.html)

1. [Their] highest priority is to satisfy the customer through early and continuous delivery of valuable software.
2. Welcome changing requirements, even late in development. Agile processes harness change for the customer's competitive advantage.
3. Deliver working software frequently, from a couple of weeks to a couple of months, with a preference for the shorter timescale.
4. Businesspeople and developers must work together daily throughout the project.
5. Build projects around motivated individuals. Give them the environment and support they need and trust them to get the job done.
6. The most efficient and effective method of conveying information to and within a development team is face-to-face conversation.
7. Working software is the primary measure of progress.
8. Agile processes promote sustainable development. The sponsors, developers, and users should be able to maintain a constant pace indefinitely.
9. Continuous attention to technical excellence and good design enhances agility.
10. Simplicity – the art of maximizing the amount of work not done – is essential.
11. The best architectures, requirements, and designs emerge from self-organizing teams.
12. At regular intervals, the team reflects on how to become more effective, then tunes and adjusts its behavior accordingly.

Key Elements of Most Agile Product Innovation Processes

Although the application of Agile product innovation may vary across organizations and contexts beyond just software development, the basic elements usually remain the same, incorporating Scrum elements into the product innovation process, as shown in Figure 4.13.

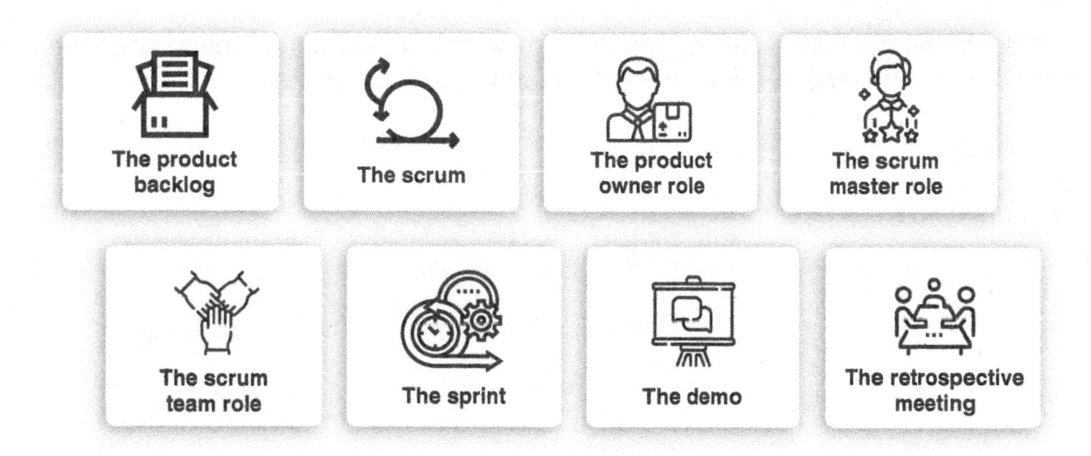

FIGURE 4.13 Basic elements of agile product innovation

Scrum

The scrum is a process created by Jeff Sutherland in 1993 based on an analogy with the "scrum" formation used by rugby teams (Sutherland, 2014). With Scrum, the product is built in a series of fixed-length iterations (sprints), giving teams a framework for shipping product on a regular cadence.

Product backlog

The product backlog contains the requirements for a system, expressed as a prioritized list of product backlog items. These include both functional and non-functional customer requirements, as well as technical team-generated requirements. While there are multiple inputs to the product backlog, it is the sole responsibility of the *product owner* (arguably, the role of the product manager as it is the product manager's responsibility to understand the prioritized needs of customers) to prioritize the product backlog. A product backlog item is a unit of work small enough to be completed by the team in one sprint iteration.

Sprint

The sprint is a set period of time, typically two to four weeks, during which specific work has to be completed and made ready for review.

Demo

At the completion of each sprint, the product (in incremental state) and/or new features that were developed in the previous sprint(s) are reviewed by project stakeholders, such as the customer or the company's management. This entails demonstrating the "product" up to this point in time. The aim is to achieve validation and to continue with the next sprint.

Retrospective meeting

The retrospective meeting aims to identify and record "lessons learned" from the previous sprint(s) and takes place at the end of every sprint. The team evaluates its performance to continuously improve.

Typically, they would discuss aspects such as:

- What went well and why (to keep doing what worked and/or identify new items to incorporate into the working practices).
- What did not go well and why.
- How to address the items that did not go well, which provides a basis for sprint planning.

Figure 4.14 illustrates the events, roles, and outcomes that are expected during and after a sprint planning session. Inputs into the meeting and hence the planning outcomes include:

- Product backlog,
- Team capabilities,
- Business conditions,
- Technology,
- Current product.

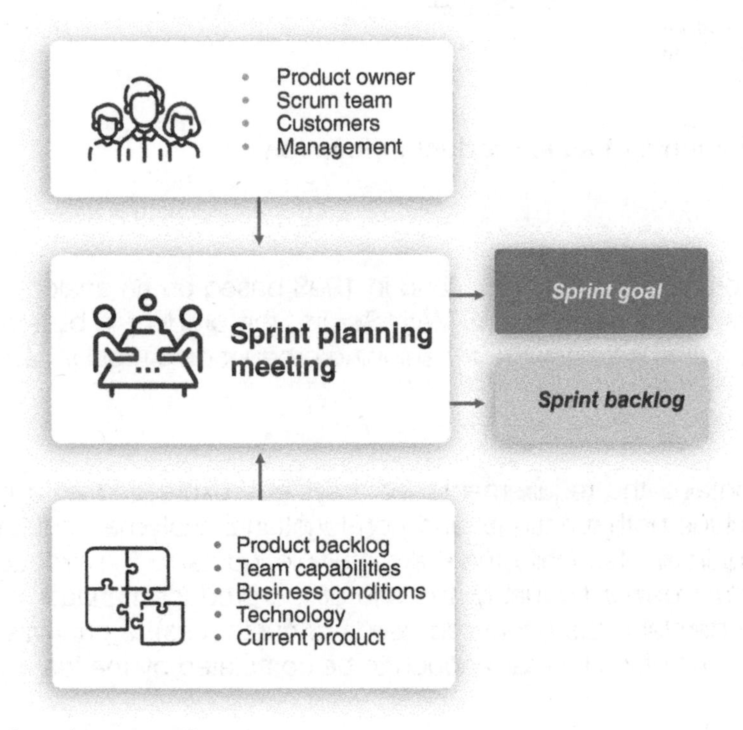

FIGURE 4.14 Sprint planning meeting

Ultimately, the team will want to establish and agree on the next sprint's goals and the sprint backlog is addressed and/or updated.

Each sprint starts with a planning meeting, during which the ***product owner*** (the person requesting the work) and the development team agree on what will be accomplished during the sprint. The duration of the sprint is determined by the ***scrum master***. After the sprint begins, the ***product owner*** steps back to allow the team to do its work. At the end of the sprint, the team presents its completed work to the ***product owner,*** who uses the criteria established in the sprint meeting to accept or reject the work.

Product owner

The product owner is the single person who must have final authority representing the customer's interests in backlog prioritization and requirements questions. This person must be available to the team at any time, especially during the sprint planning meeting. The product owner should not manage the team and should resist the temptation to change the work required after the sprint has started. A key responsibility of the product owner is to balance the interests of competing stakeholders.

Scrum master

The scrum master is the facilitator for the team and product owner. Rather than manage the team, the scrum master works to assist both the team and the product owner through:

- Removing barriers between the team and the product owner.
- Facilitating creativity and empowerment in the team.
- Improving team productivity.
- Improving engineering tools and practices.
- Ensuring that information on team progress is up to date and visible to all parties.

Scrum team

The scrum team usually comprises seven, plus or minus two, members. The team usually includes a mix of functions or disciplines required to successfully complete the sprint goals (cross-functional team). In software development projects, the team could comprise software engineers, architects, programmers, analysts, quality experts, testers, UI designers, etc. During the sprint, the team self-organizes to meet the sprint goals. The team has the autonomy to choose how best to meet these goals and is held responsible for them. Figure 4.15 illustrates the cycle that the scrum team applies in order to meet product and sprint backlogs. Scrum teams meet as often as daily (or as appropriate but often starting the day with a physical or virtual 15-minute standup meeting), with the aim of consolidating information and agreeing on:

- What was achieved,
- What was not achieved,
- What is needed to progress.

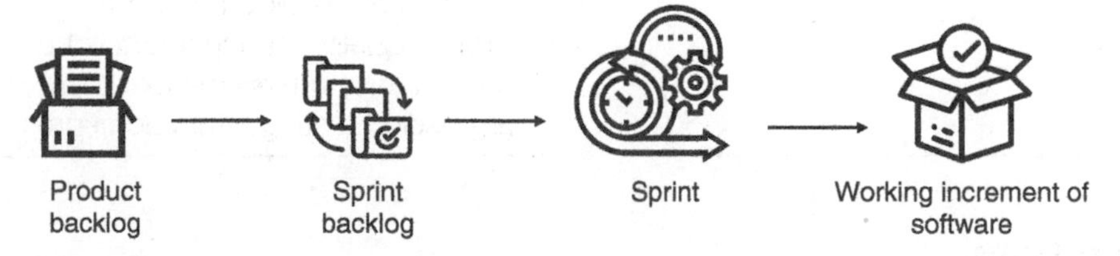

FIGURE 4.15 How scrum works

Sprint events typically have durations of no more than 30 days, and in the process, the product and the sprint backlogs are continually addressed to the point whereby incremental development, or enhancements, are achieved to the product. Typically, a working version of the product is achieved, often referred to as the minimum viable product (MVP). It is a functional product but may not have all the product features included yet; hence the product backlog serves as the record for future development to be performed.

The benefits and limitations of agile product innovation are summarized in Table 4.5.

TABLE 4.5 The benefits and limitations of Agile product innovation

Benefits	Limitations
• Scrum methodology enables projects where the business requirements are initially less clear or uncertain but will emerge during the project. • Fast-moving, cutting-edge developments can be quickly coded and tested using this method, as a mistake can be easily rectified. • It is a lightly controlled method that insists on frequent updating of the progress through time-boxed work. Thus, there is clear visibility of the project's development and progress. Also, it is iterative in nature, continuously involving feedback from the user. • Due to short sprints and constant feedback, it becomes easier to cope with the changes. • Daily meetings make it possible to measure individual productivity. This leads to improvement in the productivity of each of the team members. • Issues are before becoming larger problems through the daily meetings and hence can be resolved speedily. • Scrum has been successfully applied to software, hardware, and mixed products. The overhead cost in terms of process and management is minimal, thus leading to a quicker, less expensive results.	• Scrum is prone to scope creep if not managed well, unless there is a definite end date, the customer stakeholders will be tempted to keep demanding new functionality. • Items in the backlog must be well defined before they are worked on during a sprint and the customer must be available to respond to questions and provide input. • If the team members are not committed, the project will either never be completed or will fail. • Scrum works best with small teams. Consequently, it is best suited for small, fast-moving projects. • This methodology expects the Scrum team consists of experienced team members who have worked together and remain intact for the duration of the project. If the team consists of people who are novices, the project may not be completed in time. • Scrum works well for project management when the scrum master trusts the team being managed. If they practice too strict control over the team members, it can be extremely frustrating, leading to lack of motivation. • If any of the team members leave during a development, it can have a huge adverse effect on the project development. • Project quality management is hard to implement and quantify unless the test team is able to conduct regression testing after each sprint.

4.3.7 Agile-Stage-Gate

Agile-Stage-Gate combines the structure (stages and gates) of the classic Stage-Gate with the self-organized teams and short cycle iterations of Agile methodologies, as shown in Figure 4.16.

In practice, the project's stages, for example, the Development stage, are broken into short time-boxed increments called *sprints*, each about one to four weeks long (the small circles in the diagram below).

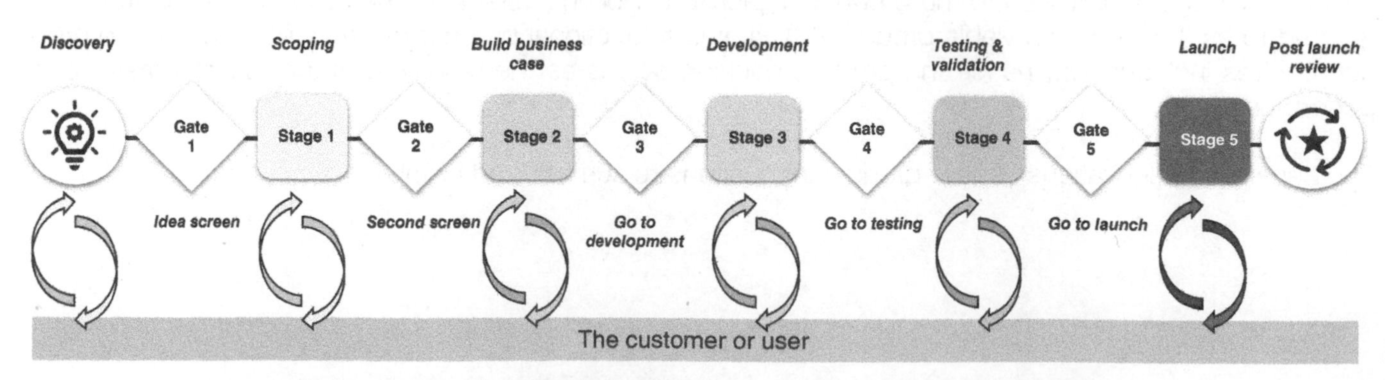

FIGURE 4.16 Model of Agile-Stage-Gate process

These short time-boxed iterations focus on delivering a working product (though a working product does not necessarily mean a product ready to be released to the public). This allows the team to seek feedback from customers and users much more frequently and quickly than with a traditional Stage-Gate process, which in turn allows the organization to focus on what really matters to them and adapt quickly to changes in the market.

Throughout this book, we have emphasized the need for organizations to apply processes and techniques to fit their own specific context and needs. All processes have their specific benefits and limitations. Recognizing these benefits and limitations and adapting to your organization's environment and needs are all part of being a *learning organization*.

Subsequently, Cooper and Sommer (2018) proposed that the company that wants to implement Agile-Stage-Gate processes do so cautiously and with much thought. Every organizational environment and project will be different, so a "one-size-fits-all" expectation will not work. They suggest that initially, a small task force is constituted to streamline current processes and identify and manage the challenges along the journey. Recognizing that this concept is still new will require all practitioners to assess, experiment, and continually improve on its possibilities.

The Benefits and Limitations of the Agile-Stage-Gate Model:

The benefits and limitations relate to how manufacturing companies have adopted Agile principles (that are typically applied in software development companies) and the outcomes this has delivered. Hence, the Agile-Stage-Gate approach leverages the strengths of both methods by "[linking] focus, structure, and control [of Stage-Gate], with the benefits of an Agile approach and *mindset*, namely *speed, agility,* and *productivity*" (Cooper & Sommer, 2016, p. 17). The benefits and limitations were extracted from research published by Cooper and Sommer (2018), shown in Table 4.6.

TABLE 4.6 The benefits and limitations of the Agile-Stage-Gate

Benefits	Limitations
• Flexibility of product design. • Quicker time to market of the manufactured product. • Increased productivity. • Improved ability to respond to market changes. • Improved ability to proactively respond to customer needs. • Improved team morale due to better communication and coordination. • Increased project outcomes due to intentional and intensified focus.	• Management's skepticism because they do not understand Agile and/or the required mindsets that it demands. • Resource assignments and participation intensity increase (note: this can be positive and negative). • Managing fluid product definitions and or designs ("ambiguity") can be difficult. • Managing ever-changing development plans. • Agile teams tend to be become insular, thus disconnected from the rest of the organization. • Clashes with bureaucratic organizational processes and/or performance reward systems, e.g., the degree to which the organization fosters and allows for experimentation and failure, or the drive for absolute quality perfection. Another example relates to short-term vs. long-term planning.

4.3.8 Systems Engineering

Systems engineering (SE) principles combine the concepts of systems thinking and SE process models to take a *problem* through a systematic and integrated process of design and project management tools and methods into a *solution*. All systems comprise *parts/elements* (building blocks) that have *properties/functions,* and the elements are linked to each other through relations. Multiple *systems* (that is, multiple elements in a boundary of one system that relate to another system) can also be called a system or environment. Systems complexity is determined by structure, dynamics in the system/sub-systems(s), changeability, variety, multiplicity, and size.

Hence, systems engineering overlaps and integrates numerous technical and human-centered disciplines. The aim is to combine and integrate elements of industrial, mechanical, manufacturing, control, software, electrical, and civil engineering, as well as cybernetics, organizational studies, and project management into a systems life cycle. The system's user requirements are encapsulated and delivered by means of technical and management processes.

Systems engineering focuses on all the pieces of a product to enhance the entire hole. It can be applied to apparently simple and complex products, such as a toaster, refrigerator, automobile, and the Falcon Heavy spacecraft. It also applies to services, such as examining and improving how a coffee shop delights customers.

Figure 4.17 shows the steps in the systems engineering design framework (extracted from Pahl et al., 2007):

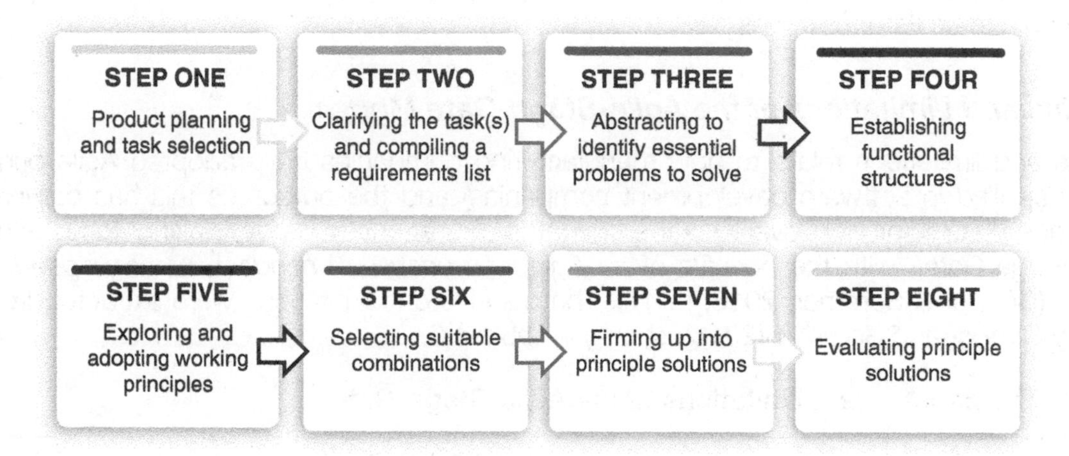

FIGURE 4.17 Systems engineering steps

Various methods are applied to each step, such as selection methods, trend studies, tests and measurements, brainstorming, design catalogs, quality assurance, and costing methods.

Features of systems engineering include:

- Upfront, intentional, and intensive design thinking
- Framing a "problem" from a broad general perspective to detail through analyses
- Being interdisciplinary
- Complexity management
- Cost reductions
- Risk reduction
- Shorter project scheduling
- Optimizations
- Product quality improvement

Systems engineering ensures that all likely aspects of a project or system are considered and integrated into a whole.

The systems engineering process is a discovery process that is quite unlike a manufacturing process. A manufacturing process is focused on repetitive activities that achieve high-quality outputs with minimum cost and time. The systems engineering process begins by discovering the real problems that need to be resolved and identifying the most probable or highest impact failures that can occur – systems engineering involves finding solutions to these problems.

Benefits and limitations of systems engineering are shown in Table 4.7.

TABLE 4.7 Benefits and limitations of systems engineering

Benefits	Limitations
• The benefit of multiple systems' strengths become evident and useable. • Design-related decisions are fairly detailed and made upfront due to intense customer involvement. • Learning opportunities (increased know-how as the project progresses) start early on in the project and are conveyable to all parties. • Impending changes are easily identified and managed.	• Over-analyses of the problem and too much detail upfront may cause the risk of delay and a focus on the incorrect elements for solutioning. • As the project progresses over time, the initial requirements may become outdated because they were based on knowledge that was generated at a historical point in time, and may no longer be appropriate or relevant as more knowledge is gained • The same applies to the planning elements, as prolonged planning and or development risk makes the solution irrelevant. • Exerting influence and the ability to effect change become increasingly difficult as the product innovation progresses toward the final phases due to "operational and resource commitments" to the project.

Haberfellner et al. (1986) offer the following useful advice to behaviorally manage some of the limitations:

1. **Beware of super-integrated solutions** that are too large and take several years to carry out, especially when they are located within a dynamic environment and are subject to uncertain conditions. In case of doubt, it is preferable to lean toward smaller solutions that facilitate a quicker benefit.
2. **Implement flexibility** in overall concepts, pursue Agile systems. Keep options open regarding later adaptations, expansions, dismantling, etc. Plan for modular building blocks that can later be replaced with better or more efficient ones. Be open to or plan for opportunities for expansion or reduction.
3. **Consciously plan for flexibility** or possibilities of multiple use, even though this may require somewhat increased investment.
4. **Consciously make partial introductions** to provide interim benefits.
5. **Dispense with optimizing unproductive details**.
6. **Postpone decisions based on uncertain premises** for as long as possible – as long as this is compatible with the logical process.

An important aspect relating to items 2 and 3 above is designing for changeability (DFC), a set of principles that enable changes in systems throughout their life cycle. This has especially become relevant in the product platforming environment related to the automobile manufacturing industry (Fricke & Schulz, 2005).

4.3.9 Design Thinking

Design thinking is a human-centered problem-solving methodology that borrows from the process that designers follow when designing solutions for people, whether that be graphical user interfaces, physical products, services, buildings, or any other form of product offering.

The term "design thinking" was popularized, initially in North America, by the design firm IDEO and the Stanford School of Design, or d.school, in the early 2000s. Around the same time, a discipline called 'service design' was growing in popularity in the business world in Europe.

While service design is intended to focus solely on the design of services, in practice, it takes the same holistic approach to problem-solving as design thinking. These two terms – design thinking and service design – are often used interchangeably. Both are human-centered methodologies that share many characteristics. This chapter will use the term design thinking to refer to both.

The Evolution of Design Thinking

The term design thinking is derived from research conducted in the 1950s and 60s to identify the common approaches that different types of designers use as they go about their trade. Its adoption into the business world started with the question, "What can design bring to the world of business?" As principles of design were applied in the business environment, it proved to have enormous value, particularly for guiding team-based collaboration and as an innovation method to get to the root of problems using empathy.

In the mid-1960s, design thinking came into the world of business when Rittel and Webber (1973) coined the term "Wicked Problems" to denote issues with multiple layers of complexity. His work focused on how businesses could easily adopt design thinking to work through these "Wicked Problems" in a systematic way.

In the 1970s, Herbert A. Simon, a cognitive scientist and Nobel Prize laureate for economics, introduced core concepts that are now regarded as principles of design thinking. His book, The Sciences of the Artificial) focused on the power of the visual in problem-solving (McKim, 1972).

In the 1980s, Cross (1982) compared designers' ways of problem-solving to the problem-solving methods of other disciplines, such as architecture. He demonstrated that scientists were problem-focused solvers: they systematically explored every possible combination to formulate a hypothesis. The designers tended to create multiple arrangements of potential solutions and then test each one against the problem to determine if it solved its requirements. They would discard any "solutions" that did not work.

From the 1990s, organizations including IDEO and d.school at Stanford University brought design thinking into mainstream use with the development of user-friendly terminology and application tool kits.

The Human-Centered Mindset

The term design thinking is useful because it refers to how designers solve problems – or, rather, how they seek solutions to complex challenges. It is also useful because it refers to a specific way of thinking and looking at a problem space. It is not that designers have some special talent or biological wiring in their heads. It is more about the methodology they use to approach problem-solving and how various design disciplines are taught.

Tim Brown (2008), in his seminal introduction to design thinking, *Change by Design*, introduced the idea that truly innovative solutions are found at the intersection of *feasibility* (is it technically possible), *viability* (is it financially sustainable), and *desirability* (is it something that people want or need). The relationships are shown in Figure 4.18.

START WITH DESIRABILITY

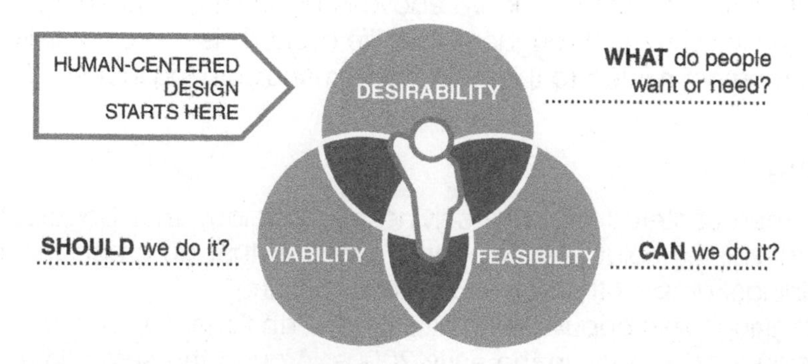

FIGURE 4.18 Design thinking starts by looking for human needs

Business thinkers tend to start by looking at a potential solution's viability: whether the solution can make money. Technical thinkers tend to think about whether the solution is feasible: can we make this work technically? Design thinkers, instead, start the problem-solving process by exploring the human aspects of the problem to determine what the needs and desires are of the key stakeholders (humans) at the center of the design problem space.

Designers seek to understand people's needs based on methodical research, probing the context, environment, actors and other factors to gain a deep understanding of the problem space, or design challenge. One of the key skills required in this research is the ability to empathize with the people at the center of the *design challenge*. Although generally not taught in technical or business fields, empathy is at the core of design thinking.

Empathy is the secret sauce of human-centered design.

So, the easiest way to understand design thinking is to think of it as a specific mindset – that of putting people at the center of the process of problem-solving. The key hypothesis is that if people do not need or want the solution you are proposing, it really doesn't matter whether it is feasible or viable. The mindset of being human-centered throughout the design process is the most important part of design thinking.

Methodology and Frameworks

The methodology of design thinking consists of a few core principles. Two of the most popular frameworks representing the different phases of the design thinking process are the Stanford d.school's Five Hexagon model, shown in Figure 4.19, and the UK Design Council's Double Diamond model, shown in Figure 4.20.

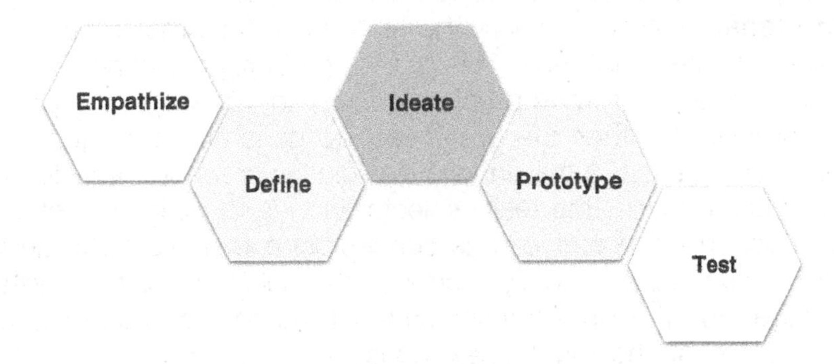

FIGURE 4.19 The Stanford d.school's Five Hexagon Framework

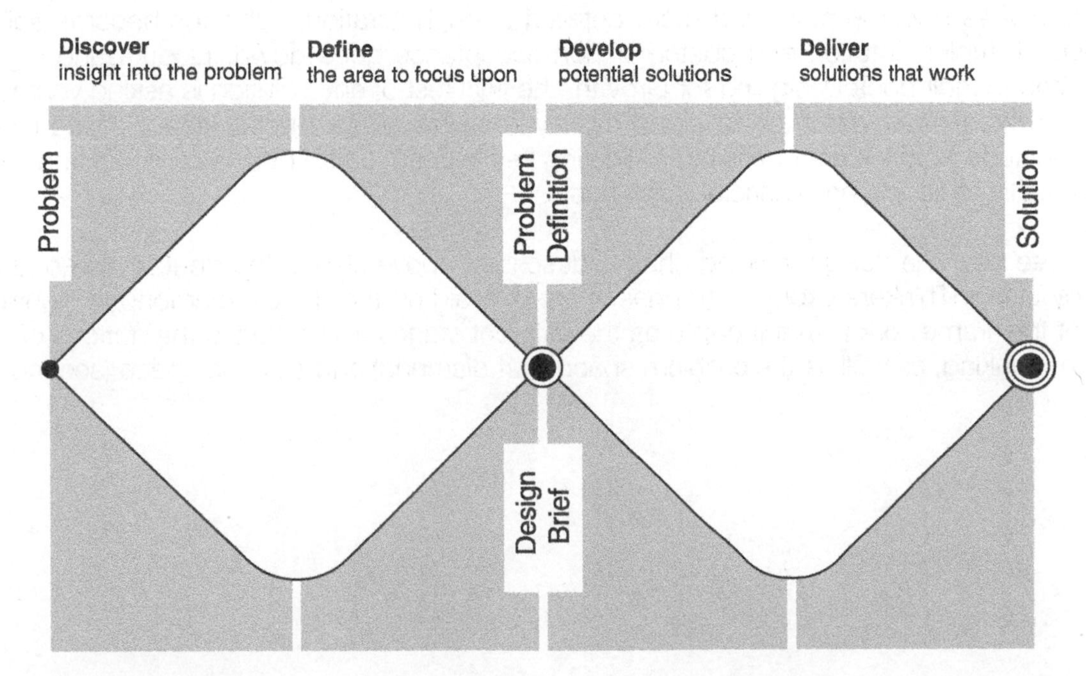

FIGURE 4.20 The UK Design Council's Double Diamond Framework

Fortunately, most frameworks used to describe design thinking share the same core phases or stages of work, just depicted visually in different ways. Those phases are as follows:

- **Start with research:** To find innovative approaches to problems, you need to generate insights (new knowledge based on deep understanding) that can serve as the starting point for problem-solving. Research often comprises surveys, interviews, small focus groups, and other specialized design research techniques such as cultural probes, contextual inquiry, and observational research.
- **Make sense of the data:** The more research you do, the more data you generate and the more that data requires analysis. Instead of spreadsheets and algorithms, however, design thinkers use "tools," templates, and collaborative activities to "make sense" of the (mostly qualitative) data, looking for patterns, stories, insights, touchpoints, stakeholders and deep understanding of people at the center of the problem. This is where true empathy is established between problem-solvers (designers) and those they are serving.
- **Reframing the problem:** A critical point in the process of design thinking is when the design team restates or reframes the problem based on the research and sensemaking activities. This step requires creativity and skill to ensure that the newly stated problem is neither too broad ("trying to boil the ocean") nor too narrow (for example, where the solution is embedded in the new problem statement) and allows ample room for creative solutions to emerge.
- **Generate lots of ideas:** Ideation is where the team generates many varied ideas, often only partially formed, to solve the new challenge. This is a form of brainstorming but is more structured and is actually based on theories of creative problem-solving. In this stage, the possible solutions are not being judged and eliminated; rather, the objective is to generate large quantities of ideas, knowing that with more ideas, the seeds of a truly innovative solution are more likely to emerge.
- **Build prototypes:** After ideation, the team selects an idea or collection of ideas to move into the prototyping stage. This type of prototyping is conceptual and nothing like an engineering or manufacturing prototype. It is all about rapidly modeling the solution concept with the least amount of effort and detail needed to (1) align the team on what the solution must include and (2) be able to present the solution(s) to customers in the next phase.
- **Test and iterate:** In an interactive cycle, prototypes are tested by presenting them to target users or customers (and also other key stakeholders like funders, decision-makers, and implementers). As the prototypes become more and more detailed through iteration, solutions become solidified, and the risk of implementation and customer non-acceptance goes down. Liedtka and Ogilvie (2011) articulated in their book Designing for Growth, the final test of any solution is asking your target users or customers to part with their time and money in exchange for the solution. So, the testing phase should include (to the extent possible) validating key aspects of the business model and value proposition (ideally, including any financial exchanges).

You can see how the design thinking phases described above also fit the double diamond framework developed by DesignThinkers Group. The phases are overlaid on the double diamond in Figure 4.21. The advantage of this framework is that it captures the different stages and includes the notions of convergent and divergent thinking, as well as the problem space (first diamond) and solution space (second diamond).

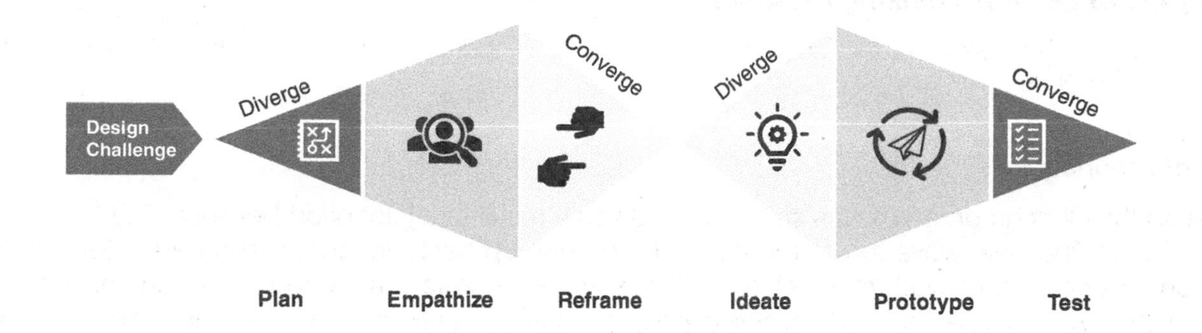

FIGURE 4.21 The Six-Stage Double Diamond by Design Thinkers Group

Design Thinking as a Social Technology

Liedtka (2020) refers to design thinking as a "social technology." This is a very powerful way of referring to design thinking to people who are not directly involved in applying the method but have some stake in its successful application in an organization.

Practitioners are better able to build support for design thinking by communicating that it is built around collaboration and teamwork (the "social" part) and that there is a proven structure and process to how it works in practice (the "technology" part). Indeed, like any other new initiative, one of the biggest challenges in applying design thinking is gaining organizational support for the different way of working that it represents.

Following are some ways that design thinking works as a social technology:

Based on empathy: As previously mentioned, tapping into the power of empathy helps the design team move past their own biases and see the world through the eyes of the people they are designing for. Developed through qualitative research, empathy helps to set the stage for more creative solutions by providing a foundation of data from which the team can draw insights and base decisions.

Highly collaborative: Design thinking is not something you can do in isolation. The true power of the process and structure comes from going through that process with a team that has diverse perspectives and expertise. Funders, decision-makers, and implementers are key stakeholders involved in the design process. The collaboration gets to its highest when the design team collaborates directly with the end users or customers, they are designing for in a process called co-creation.

Iterative in nature: A large dose of humility helps form a mindset that initial ideas are only partially formed and need to be tested. The potential solutions have hypotheses embedded in them, and as those hypotheses are tested, the ideas are modified or added to in an iterative process. Prototypes are ways to ask questions and test hypotheses so the team can adjust, learn, and redesign until a refined solution is found.

A structured process: Finding and developing innovative solutions to problems in a human-centric way requires most organizations to change certain deeply embedded ways of working. The structure and process in design thinking provide guidance and boundaries for the design process so that design teams don't get stuck in "analysis paralysis" or skip important steps in a rush to solve the wrong problem.

These aspects of design thinking help teams to collaborate more effectively. They are also qualities that become integral parts of the way each team member starts to think and approach problems. That is, they are key characteristics of the all-important mindset of the human-centered designer.

Examples of Design Thinking in Action

Design Thinking has proven its effectiveness in a variety of industries over the years. A few select cases below illustrate this.

Bank of America

The Keep the Change program was started in 2004 as a result of a joint effort between IDEO and the Bank of America to find new ways to get more people to open up bank accounts. The IDEO team conducted research to understand customer behavior. The key insight was that "savers were intentionally rounding up when writing checks." By observing people's natural relationship with money, they were able to home in on existing customer behavior to achieve their goal. Learn more about this case study at https://www.invisionapp.com/inside-design/applied-design-thinking/.

Slack

Slack co-created their product updates with their customers by gathering their feedback using their very own software. By working with a variety of stakeholders and observing their behavior, the team at Slack was able to understand a few key insights: people will find the things they really want, communication is critical, and to always learn from beginners. Learn more about this case study at https://slack.com/blog/collaboration/designing-the-future-of-slack-with-customers.

GE

The Pediatric Adventure Series from GE Healthcare was an exercise in using reframing to make a difference in the user's experience. Doug Dietz was involved in designing MRI machines. At the time of installation of one of the devices at a children's hospital, Doug noticed children were scared to undergo the test and learned they were often sedated. Through the use of design thinking and co-creation came the idea to create a story with the child as the star. By creating storylines such as fighting off a dragon or going on a submarine expedition, children were more relaxed. This resulted in a dramatic decrease of children needing to be sedated, and more patients could get scanned each day. To learn more about this case study, see https://www.ideou.com/blogs/inspiration/from-design-thinking-to-creative-confidence.

Quick Start Guide: Implementing Design Thinking

It takes time to learn and think the design thinking way. Although it takes time, it is best to learn by doing and start practicing wherever possible. This is actually a key tenet of design thinking, learning by doing. This helps to solidify the design principles and build a culture of innovation within your team and organization. Some quick ways you can start implementing design thinking include:

- Hold a design thinking foundations workshop. Lasting typically for one to two days online or in person, you can introduce your team to the concepts and work through a challenge together.
- Run a design thinking sprint, developed by Google Ventures (Knapp et al., 2016). In a week or less, this allows your team to rapidly develop solutions to a customer's problem and learn by doing.
- Always ask, "What is the problem" and "Who is the user experiencing this problem?" This helps keep the two most important questions front and center in the work you do.

When to Use Design Thinking

- When facing a human-centered challenge. This can mean any problem that has to do with people and requires a deep understanding of the groups of people involved. Examples include improving employee experience, designing for a particular use case of a product understanding why a new feature is not working in a product, and many more:

- When facing complex challenges. Complex problems are challenges where only parts of the problem are definable; the problem area is not fully understood, and there is unpredictability in the problem landscape. Design thinking is great at exploring complex problems as it allows the information to flow and has the natural iterative process built in to adjust as the problem unfolds.

When Not to Use Design Thinking

- When the problem is not open-ended. If you already know the outcome or the specific solution needed to solve the problem, then it is best not to use design thinking
- When the problem is actually a puzzle with a known solution that an expert can determine.
- When another problem-solving method is proven to work. For problems where there is a known and effective approach that works, design thinking may not add any value. For example, a manufacturing quality problem can be effectively solved using lean six sigma and is not appropriate for design thinking.

4.3.10 Lean Product Innovation

Lean product innovation is founded on the fundamental Lean methodology initially developed by Toyota (Toyota Production System, or TPS). TPS is based on the Japanese term Muda, meaning "futility" – uselessness, idleness, waste. Fundamentally, TPS was designed to remove Muda or waste from processes, including manufacturing. These principles have been incorporated into various product innovation processes.

The core concepts/principles of Lean product innovation consist of removing waste, gathering as much information and knowledge upfront, but importantly, requiring constant (and relentless) learning, and seeking opportunities for improvements along the entire life cycle of the product's development (see Figure 4.22). Key components of lean development thus relate to gathering and increasing product knowledge and having the team fully engaged early in the product innovation process.

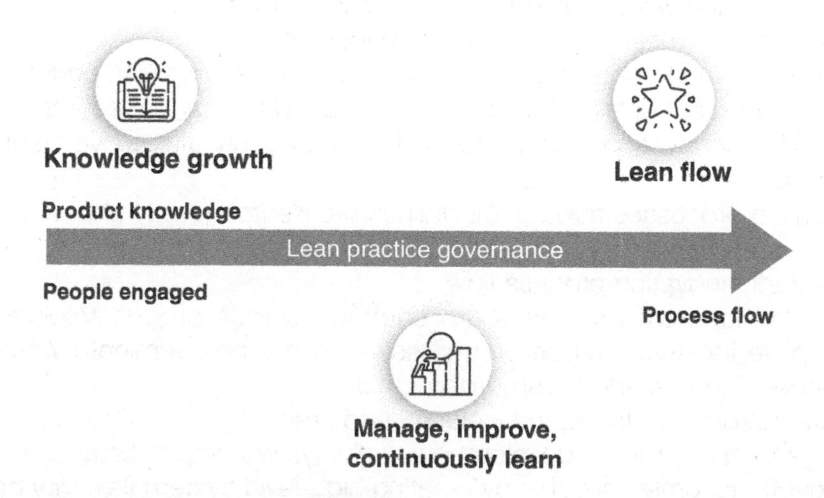

FIGURE 4.22 The concepts of Lean development

Lean product innovation as a contributor to organizational "productivity" endeavors (Mascitelli, 2011), is measured as:

- Profit generated per hour (or unit) of time.
- Efficient utilization of designers/developers.
- Faster time to market.
- More projects completed per unit of time.
- More customers satisfied more of the time.
- Fewer wasteful activities.

Potential sources of waste include:

- Chaotic work environment.
- Lack of available resources.
- Lack of clear prioritization.
- Poor communication across functional barriers.
- Poorly defined product requirements.
- Lack of early consideration of manufacturability.
- Over-designing.
- Too many unproductive meetings.
- Email overload.

An Example of the Application of Lean Product Innovation

Toyota is well-recognized for its application of Lean principles. James M. Morgan and Jeffrey K. Liker (2006) provide a summary of Toyota's lean innovation principles in their book, *The Toyota Product Innovation System, Integrating People, Process and Technology*, summarized below.

Toyota Product Innovation Guidelines

Process Sub-System

1. Establish customer-defined value to separate value-added from waste.
   Separate value-add from waste. Two key wastes to focus on:
   - Poor engineering, leading to underperforming products and/or process performance.
   - Poor product development processes, resulting in wasted time and resources.
2. Front-load the product innovation process to explore thoroughly alternative solutions while there is maximum design space.
   Toyota has developed processes through disciplines like *Kentou* (study phase) and *Mizen Boushi* (proactive problem prevention) to significantly reduce the chaos of the FEI.
3. Create a level product innovation process flow.
   Many innovation development processes are common across products/solutions. Toyota's TPS system uses adaptive tools with repetitive processes to not only eliminate waste (muda) but also eliminate "unevenness" (mura) and "overburden" (muri).
4. Utilize rigorous standardization to reduce variation and create flexibility and predictable outcomes.
   Innovation management or product development is always working to balance creativity with reproducibility. Toyota accomplishes this by creating high-level system flexibility balanced with standardized lower-level tasks. They employ three levels of standardizations:
   i. ***Design standardization:*** a common architecture comprised of reusable modules and shared components.

ii. ***Process standardization:*** a templated product and building system based on standard manufacturing processes.

iii. ***Engineering skill set standardization:*** designed to deliver flexibility in program planning and staffing.

People Sub-System

5. Appoint a chief engineer to integrate development from start to finish.

Also called a program manager, most companies assign project management responsibilities to this role. Toyota differs in that they have a functional, skilled chief engineer who is solely responsible for managing the entire development process from start to finish. They are accountable and responsible for things like:

- Determining which engineering or manufacturing solution the effort will use in the event of an impasse on the team.

6. Organize to balance functional expertise and cross-functional integration.

Toyota achieves through the chief engineer role along with modular engineering team and the use of the "obeya" (big room) to enhance cross-functional engagement and focus.

7. Develop towering competence in all engineers.

Engineering by definition is a deeply technical profession. Many organizations attempt to train engineers to understand business concepts and, in the process, potentially dilute their depth of technical ability. Toyota encourages engineers to deepen and not broaden their skills and aligning their activities through a chief engineer or program manager.

8. Build in learning and continuous improvement.

Company's leaders emphasize the importance of organizational learning and continuous improvement.

Tools & Technology Sub-System

9. Build a culture to support excellence and relentless improvement.
   - Culture in general is big part of Toyota's product strategy. Beyond learning and continuous improvement, their entire commitment to quality and excellence is actively passed down from leaders from one generation to the next and celebrated company-wide regularly.
10. Adapt technologies to fit the organization's people and processes.
    - Using technology to solve a problem without first understanding the scope of the problem leads to even poorer performance.
11. Align the organization through simple visual communication.
    - While culture and standardization are key elements of Toyota's success, without clear and concise ways to communicate, it is difficult to keep an organization aligned. Toyota relies on *Hoshin Kanri* (Policy Deployment) to ensure that the company's strategic goals continue to progress at all levels. **The 7 Steps of Hoshin Kanri Planning**
      1. Establish the Vision and Assess the Current State. . . .
      2. Develop Breakthrough Objectives. . . .
      3. Define Annual Objectives. . . .
      4. Cascade Goals Throughout the Organization. . . .
      5. Execute Annual Objectives. . . .
      6. Monthly Reviews. . . .
      7. Annual Review.
12. Use powerful tools for standardization and organizational learning.
    - Finally, without tools to manage learnings and standards across programs and time it is difficult to sustain innovation. This type of continuous improvement is one of the core tenets of *Kaizen* (good change or improvement), Toyota, through an evolutionary process, has developed powerful tools to capture, distill and communicate learnings and standards from program to program at all levels and through time.

The benefits and limitations of lean product innovation are shown in Table 4.8.

TABLE 4.8 The benefits and limitations of lean product innovation

Benefits	Limitations
• Focus of process is on transformation of information, not on heavy-handed governance. • An even-driven approach simplifies collaboration and enables design optimization. • Emphasis on proactively managing risks to schedule cost, performance, and quality. • Can be scaled to any size project. • Simple, often visual tools are used to capture learning, track progress, set priorities, and solve problems.	• Requires dedicated and experienced workers, both to suggest system improvements and to positively respond to system changes. • Requires a change in organizational structure and culture. A unified and committed project culture is necessary with an appropriate and supporting organizational structure. • Requires strong supplier management. The focus of Lean product innovation or "just in time" delivery requires good communication and coordination of suppliers. • Requires a willingness and ability across the organization to accept changes in project goals and direction.

4.3.11 Lean Startup

Lean startup is an approach to building new businesses based on the belief that entrepreneurs must investigate, experiment, test, and iterate as they develop products. The concept of Lean startup originated in the early 2000s and evolved into a methodology around 2010. It was developed by Silicon Valley entrepreneurs Blank (2010) and Ries (2011).

Proponents of the methodology say Lean startup principles ensure that entrepreneurs develop products that customers actually want, rather than attempting to build businesses based on untested ideas. Proponents also describe this mentality as "fail fast, fail cheap" because the Lean startup process is designed to limit the time and money invested in product ideas when entrepreneurs test and prove their potential value.

The Lean startup methodology calls for entrepreneurs to start their business ventures by searching for a business model and then testing their ideas. Feedback from potential customers is then used to adjust their ideas as they move forward. The Lean startup methodology also advocates for entrepreneurs to continually engage in this activity loop – exploring and developing hypotheses that they then test among customers to elicit feedback, something known as validated learning. Entrepreneurs use that customer feedback to re-engineer their products.

There are six critical elements for the proper execution of the Lean startup methodology:

1. **Build-measure-learn:** As described by Ries (2011), and shown in Figure 4.23, "the lean startup process can be summarized in three words: Build, Measure, Learn. Simply put, you need to build a prototype to measure what works and learn what your customers want. Repeat This process as many times and as fast as you can. And hopefully from this you will build a sustainable business."
2. **The Business Model Canvas:** The Business Model Canvas (BMC), first developed by Osterwalder et al. (2010), is a simple yet effective visual strategy tool that organizations, big and small, use for business model innovation. The BMC is an important tool used by Lean startup businesses with its emphasis on entrepreneur-focused business planning.

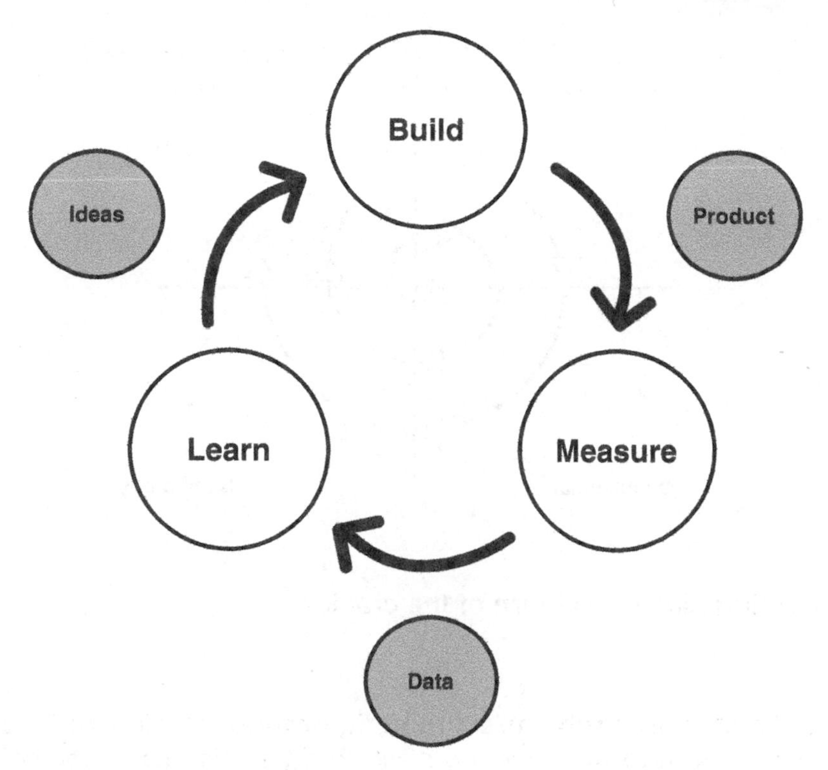

FIGURE 4.23 The build-measure-learn cycle.
Source: Adapted from Ries (2011).

The importance of the organization's business model in the context of strategy and innovation is very key. Business models, if incorrectly defined and/or do not support the innovation strategy and management, the technology strategy, and the product strategy, will not achieve the goal of value creation – which is ultimately what it's all about for profit-making firms. The growth of the Internet, the ease of accessibility, and globalization, among others, have created the impetus for firms to incorporate innovation into every facet where value can be captured and translated into profit. Refer to chapter 2.4.4, for more detail on the Business Model Canvas.

3. **Learning plans: One turn of the crank:** As Ries (2011), the author of *The Lean Startup,* wrote, "Validated learning is the process of demonstrating empirically that a team has discovered valuable truths about a startup's present and future business prospects." A learning plan describes how key hypotheses are to be tested. Completing a learning plan is "one turn of the crank," shown in Figure 4.24. The plan is categorized by these four quadrants:

   - Market – patient, consumer.
   - Organization – staffing, budgeting, structure.
   - Commercial – investment, profitability.
   - Technology – technical, innovations/platforms.

   All quadrants are not required for each learning plan. Only cover those areas that are critical to project success or failure. Some attributes of a good learning plan are:

   - Good plans fail fast/learn quickly; probability should generally rise as phases progress.
   - Are there ways to bring proof forward? Can you learn about low-probability events earlier?
   - Design periodic proof or pivot points on problematic issues.

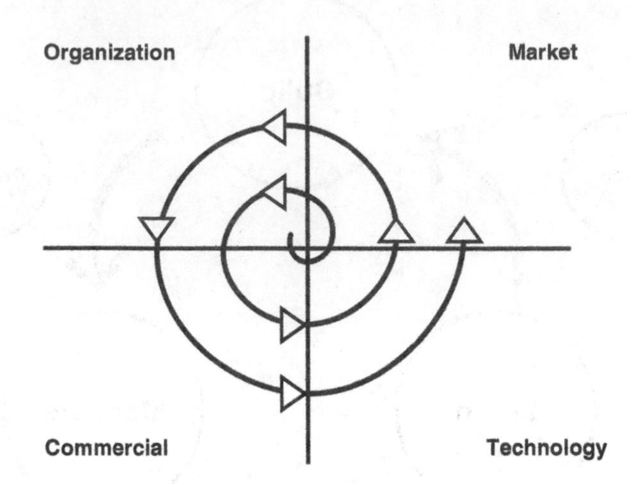

FIGURE 4.24 The learning plan – "one turn of the crank."
Source: Ries (2011).

4. **Three stages of a startup: Problem/solution fit, product/market/fit & scale.**
   Stage 1: Problem/solution fit. *Is there a clear pain (or gain) worth solving for a customer?* It is fundamental that there be a real "pain or gain" for the target customer. Too many early-stage ideas fail because the project team or leadership are enamored with a new technology or the elegance of the solution.
   Stage 2: Product/market fit. *Does the product meet the market need?* Not only is this stage critical to validating the product features and how attractive it will be in the market, but it is also important to decide on the business model used in the go-to-market.
   Stage 3: Scale. *Accelerating growth to achieve economies of scale.* Are the right production, sales, service, and support structures in place? Is the business model becoming profitable?
5. **Minimum viable product:** A minimum viable product (MVP) has just enough features to satisfy early customers and provide feedback for future product development. "The lesson of the MVP is that any additional work beyond what was required to start learning is waste, no matter how important it might have seemed at the time" (Ries, 2011).

   Gathering insights from an MVP is less expensive and faster than developing a product with more features, which increases costs and risk if the product fails – for example, due to incorrect assumptions. The term was coined and defined by Frank Robinson in 2001 and then popularized by Steve Blank (Blank & Dorf, 2012; Blank 2016).
6. **Pivoting.** Entrepreneurs have found that the extreme uncertainty of a new product or service usually requires many course corrections, or "pivots," to find a successful formula. Eric Ries suggests 10 forms of pivoting:
   - **Zoom-in pivot.** In this case, what previously was considered a single feature in a product becomes the whole product. This highlights the value of both focus and an MVP, delivered quickly and efficiently.
   - **Zoom-out pivot.** In the reverse situation, sometimes a single feature is insufficient to support a customer set. In this pivot, what was considered the whole product becomes a single feature of a much larger product.
   - **Customer segment pivot.** Your product may attract real customers, but not the ones in the original vision. In other words, it solves a real problem but needs to be positioned for a more appreciative segment.

- **Customer need pivot.** Early customer feedback indicates that the problem solved is not very important or money isn't available to buy. This requires repositioning, or a completely new product, to find a problem worth solving.
- **Platform pivot.** This refers to a change from an application to a platform or vice versa. Many founders envision their solution as a platform for future products but don't have a single killer application just yet.
- **Business architecture pivot.** Many years ago, Geoffrey Moore observed two major business architectures: high margin/low volume or low margin/high volume. You can't do both at the same time.
- **Value capture pivot.** Changes in how a startup captures value can have far-reaching consequences for business, product, and marketing strategies. The "free" model doesn't capture much value.
- **Engine of growth pivot.** Most startups these days use one of three primary growth engines: the viral, sticky, and paid growth models. Picking the right model can dramatically affect the speed and profitability of growth.
- **Channel pivot.** The mechanism by which a company delivers its product to customers is called the sales or distribution channel. Channel pivots usually require unique pricing, feature, and competitive positioning adjustments.
- **Technology pivot.** Using a completely different technology to solve the problem. This is most relevant if the new technology can provide superior pricing and/or performance to improve.

A case study of the application of Lean startup is presented in Appendix B, Case 2.

4.3.12 Jobs To Be Done (JTBD)

Some would say JTBD is an innovation process or framework, yet others would classify it as a demand-generation tool. This is in large part because of two separate but overlapping origins of JTBD theory. One originates with Christensen and Raynor (2003), who viewed the value of JTBD as understanding why customers make buying decisions. This is the demand generation perspective. The other originates with Ulwick (2016), who views the reason people buy products as to get a job done, and you study that job, and make it the unit of analysis. This approach uses a more rigorous process that leads to clearly defining the product customers want to get a job done. In both approaches, JTBD is widely practiced and therefore is included as a valuable innovation practice.

As an example of demand generation, Efosa Ojomo (2021) wrote a blog post entitled, *The Power and Simplicity of Jobs to Be Done Theory*. In this post, he shared an anecdote that illustrates that while most all frameworks or processes focus on the functional aspects of product innovation, you can use JTBD to also focus on the social and emotional aspects.

Developed by the late Professor Clayton Christensen and his longtime collaborator Bob Moesta, the Theory of Jobs to Be Done (or Jobs Theory) is a useful framework that helps innovators better understand consumer behavior. It highlights the fact that people don't simply buy products or services; they *hire* them to make progress in specific circumstances (what we call their Job to Be Done, or for short, "job").

Jobs Theory goes beyond superficial categories to expose the functional, social, and emotional dimensions that explain why consumers make their choices. The functional dimension is simple: the product or service must *function* in a way that helps the consumer overcome a struggle. The job's social dimension is about how others view the consumer using the product or service. And the emotional component of the job is about how the product or service makes the consumer feel. To perfectly nail the job, innovators should focus on understanding the job's functional, social, and emotional dimensions.

For example, several years ago, Godrej, an Indian conglomerate, developed chotuKool, a new refrigeration product to bring cooling to the hundreds of millions of Indians who lived without refrigeration. After extensive research, Godrej learned that most people it would target with this new product didn't have access to electricity, couldn't afford to purchase food in bulk, and couldn't afford most refrigerators on the market. chotuKool was designed specifically for this market.

chotuKool is a small thermoelectric refrigerator powered by a rechargeable battery, a laptop-like power supply. The product weighs around 16 pounds, has a capacity of 35 liters, and can maintain a temperature of 10 degrees Celsius (50 degrees Fahrenheit). It consumes only half the electrical power a conventional refrigerator uses and consists of only 20 components and no moving parts (depending on the size, conventional refrigerators could contain more than 200 components). Finally, at roughly $60, chotuKool costs a fraction of the least expensive refrigerator on the market. From all indications, this product should have been a slam dunk. But socially and emotionally, chotuKool struggled to nail the job. Before long, chotuKool was seen as the refrigerator you purchased if you couldn't afford conventional refrigerators. In fact, upon further research, Godrej learned that many people purchased conventional electricity-based refrigerators and left them unplugged due to erratic electricity supply. For them, the fact that they owned a refrigerator, even though it wasn't often plugged in, was a status symbol. It made them feel good and elevated them socially. Though it did not solve the functional component of the job, it nailed the social and emotional ones.

Ultimately Godrej learned from its rocky beginning and went on to develop a refrigerator that satisfied each component of its customer's Job to Be Done. Other innovators should similarly take note: by understanding the functional, social, and emotional components of their customers' jobs and nailing each, they'll increase their odds of success tremendously.

So how does JTBD actually work? Zbignev Grecis (2015) posted the image depicted in Figure 4.25 and the associated descriptions.

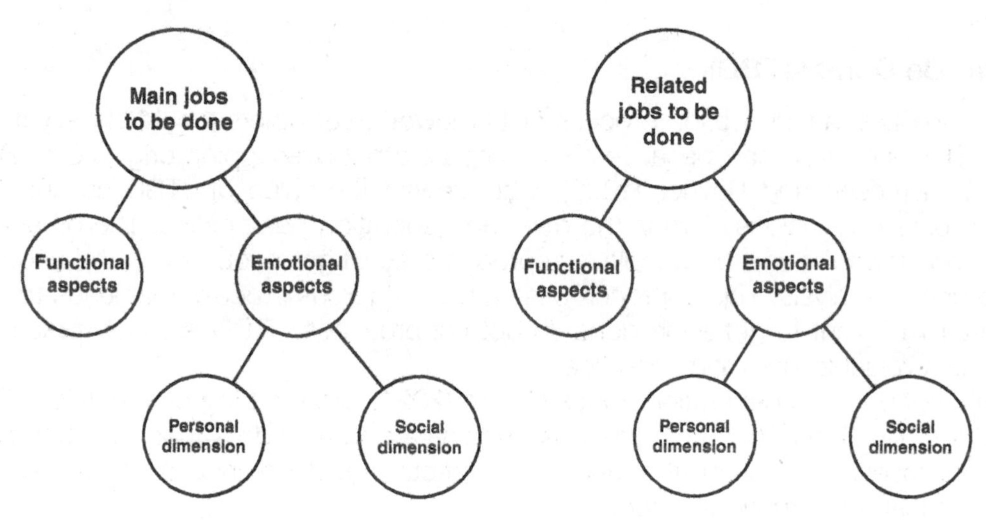

FIGURE 4.25 The two types of jobs

There are two different types of jobs:

1. **Main jobs to be done** describe the task customers want to achieve.
2. **Related jobs to be done,** which customers want to accomplish in conjunction with the main jobs to be done.

Within these two types of jobs, there are:

- **Functional job aspects:** the practical and objective customer requirements.
- **Emotional job aspects:** the subjective customer requirements related to feelings and perception. These can be further broken down into ***personal dimensions,*** how the ***customer feels*** about the solution, and ***social dimensions,*** how the ***customer believes they are perceived*** by others while using the solution.

To help illustrate JTBD further, let us work with the following example, modification of an off-road vehicle, shown in Figure 4.26.

FIGURE 4.26 Modification of an off-road vehicle

One JTBD is to modify your off-road vehicle, in this case, a Jeep, to perform better while off-road but maintain on-road drivability. We'll narrow this very broad JTBD further by focusing on upgrading the tires.

Most stock base-model Jeeps come with P245/70R17 tires, which translates to approximately a 31″ by 10″ tire on a 17″ wheel for those more familiar with standard units. Tire manufacturers who focus on tire performance only often do not capture the anticipated market share because they focus heavily, if not solely, on the performance of their tires – how durable they are and how well they perform in the target environment (paved road, dirt, mud, rock, etc.).

Size, Durability, and Performance are the functional aspects of a tire. If you spend most of your time using your Jeep in muddy or wet environments, your functional tire class would be M/T or Mud Terrain. In this example, we want to maintain drivability on the road, so we would probably focus more on the MTX or Mud/All-Terrain class of tire. But what about the emotional aspects of tire selection?

In a recent poll of nearly 2300 Jeep enthusiasts in the United States, 68% of respondents indicated that Durability and Performance were their primary decision factors, followed by 21% who cited Price. Only 8% indicated that they were most influenced by how "cool" the tires would look on their Jeep. These data, collected through a survey, would seem to indicate that durability and performance are by far the market force to be addressed. However, does that represent customer decision-making accurately?

The *cool factor* is often more important for some Jeep enthusiasts because they focus heavily on both how the tire appearance makes them *feel* about their Jeep and how others *perceive* their Jeep. How would you determine this from the data shared previously? This factor comes to light with a quick view of actual purchase data. An off-road tire store in the same area as the surveyed Jeep enthusiasts was presented with the survey data and responded, "It always comes down to looks." Their sales history proved out that more than 80% of the time, when presented with either two options where one is measurably more expensive or one is measurably lower in quality/performance, the high-cost or lower-performing tire is selected if it rates higher in appeal.

4.4 COMPARING PRODUCT INNOVATION PROCESS MODELS

An increase in the understanding of what underpins successful product innovation has led organizations to evolve and adapt process methods to meet the needs of specific organizational contexts and product categories. Researchers and practitioners have recognized that the "one-size-fits-all" approach to product innovation processes does not work. In the preceding sections, we have overviewed a number of the key models that have been developed and applied over recent years. Each model has specific benefits and limitations. Although an application of one of these models in its purest form may be appropriate in some situations, in most cases, a blend of elements from a number of models is most appropriate. The following is a comparison of selected processes.

4.4.1 Agile vs. Lean

The difference between Agile and Lean is simple to understand, but most people feel they are somehow equivalent. They are not.

Lean is designed to reduce waste and improve operational efficiency, especially related to repetitive tasks, as often seen in manufacturing. The value of the Lean approach to product innovation is its focus on a core set of principles or guidelines underpinning the process of developing new products. It is not a defined process focused on the activities and tasks required to develop a new product successfully. Section 4.3.10 summarizes the 12 guidelines for Lean product innovation, as initially applied by Toyota.

Agile, specifically in the form of Scrum, is designed to execute tasks over a short time frame, with frequent customer involvement, and to be able to make changes quickly. The structure, process, and roles are all well defined as they are applied to the development of a product or product component. In a nutshell, Agile is a time-focused, iterative philosophy to build a product step by step (incrementally), delivering it in smaller pieces. One of its main benefits is adapting and changing at any step (depending on feedback, market conditions, corporate obstacles, etc.) and supplying only relevant products to the market.

As can be noted, they have nothing to do with each other per se – you do not need to innovate new products to be Lean, and you don't have to be operationally efficient to be Agile. Further, they are not exclusive of each other. They can be used together in the same organization.

4.4.2 Agile vs. Stage-Gate®

Stage-Gate® is not a project management or micro-planning model. Rather, it is a comprehensive and holistic idea-to-launch system and a macro-planning process. It is cross-functional (i.e., it involves technical product developers, as well as marketing, sales, and operations). It places significant emphasis on "gates" that form the basis of an investment decision model, asking key questions: (1) is the organization doing the right project, and (2) is the project being done right?

By contrast, Agile (Scrum) was originally designed to rapidly develop working software. In practice, the development stage consists of a number of sprints, where each sprint or iteration produces a working product (executable code or software that works) that can be demonstrated to stakeholders (i.e., customers). An iteration may not add enough functionality to warrant a market release, but the goal is to have a demonstratable product at each iteration's end. Multiple iterations are usually required to release a product. A sprint typically takes two to four weeks.

Robert Cooper has provided a good explanation of the features of both Stage-Gate® and Agile (refer to Table 4.9). He refers to the commonly held belief that Stage-Gate® is appropriate for "hardware" products while Agile is appropriate for software, concluding that the two methods are mutually exclusive. Cooper asserts, "Agile and Stage-Gate® are not substitutes for each other. Rather, Agile is a useful micro-planning tool or project management tool that can be used within the Stage-Gate® process to accelerate certain stages – probably stages 3 and 4" (Cooper, 2015).

TABLE 4.9 Stage-Gate vs. Agile

Characteristic	Stage-Gate	Agile
Type of model	Macro-planning	Macro-planning, project management
Scope	Idea to launch, end to end	Development and testing stages only
Organizational breadth	Cross-functional – technical, marketing, manufacturing	Technical (software code writers, engineers, IT staff)
End points	A launched new product in the market-place	Developed and tested software product
Decision model	Investment model: go/kill model involves senior management governance	Largely tactical: actions needed for the next sprint

4.4.3 Integrated Product Development (IPD) vs. Other Process Models

IPD, as the name implies, provides a framework designed to emphasize the integration of functions, roles, and activities involved in product innovation. It has been defined as "a philosophy that systematically employs an integrated team effort from multifunctional disciplines to [effectively and efficiently] develop . . .new products that satisfy customer needs" (Kahn et al., 2013). One of the features that has evolved into the IPD model is "learning and continuous improvement."

Logically, the macro-planning features and decision-making basis of the Stage-Gate® model; the micro-planning and flexibility of the Agile model; the focus on reduction in wasted time and effort associated with Lean; and the overall integration of product innovation as a learning company are potentially complementary and not mutually exclusive. Merging elements of each model into a true fit-for-purpose product innovation process model, with an overall focus on learning and continuous improvement, is a sign of advanced product innovation practice.

4.4.4 The Key Question: Which Process to Use?

The answer is: ***there is no clear-cut answer***.

- Most of the processes discussed in this chapter have common elements:
- They seek to add structure to product innovation projects.
- They emphasize the value of the cross-functional team.
- They focus on the importance of the voice of the customer.

In one sentence, each process has its specific point of emphasis:

- Stage-Gate® strongly emphasizes a structured and disciplined approach.
- Agile promotes flexibility and speed of delivery.
- Lean focuses on the importance of removing waste (resources, time).
- Integrated product development focuses on the integration of functions, role, and activities.
- Design thinking is based on human-centered problem-solving.
- Jobs to be done focuses on underlying customer needs.

So, which process should you use? While there is no clear-cut answer, here are some general tips:

1) Consider your organization's culture, resources, and skill sets. Organizations with a rigid structure and lack of flexibility will probably find difficulty in applying an agile approach and may be better suited to the discipline of Stage-Gate®.
2) High risk and new to the company products. These may have the potential for high returns, but balanced against high risk, are more suited to a Stage-Gate® process. This is particularly the case in the pharmaceutical industry. As discussed earlier, the number of gates can be adjusted according to the level of risk and need for speed to market.
3) Iterative development and modifications. Where iterative development based on previous products is required, Agile is often the preferred process. This is particularly the case in the IT and software industry.
4) A range of products in your portfolio. Where the product innovation portfolio contains a range of products from low to high risk, new to the company to iterative development, current technology to requirement for new technologies, etc., selection of the most appropriate process for each type of product is vitally important. DO NOT force fit all projects into the same process.

So, is there a RIGHT or BEST process?

The answer is clearly NO.

The best organizations will choose the process that suits their specific needs. They modify the process as required and will use a combination of processes. Why not a Stage-Gate® foundation with an Agile approach within certain stages, underpinned with Jobs to be Done for identification of customer needs, Design Thinking to provide human-centered problem solving, and underpinned with Lean principles?

And, above all, a LEARNING ORGANIZATION that is focused on continuous improvement.

4.5 PRODUCT INNOVATION PROCESS CONTROL AND MANAGEMENT

Each product innovation process discussed in this chapter has significant strengths in basic structure and underlying principles. However, organizations can implement them poorly, with a frequently seen problem of the process taking precedence over outcomes – the process takes on a life of its own. For example, delivery teams will often produce required artifacts, such as requirements documents or architecture documents, solely to pass through the process gate.

A well-defined control and management (governance) responsibility vested in a senior manager, or a management team is an excellent vehicle for ensuring the overall effectiveness of the product innovation process. That is, the right amount of process, efficiently and effectively focused on delivering the right outcomes. The *product* is the focus of a product innovation process, not the process itself.

4.5.1 What is Product Innovation Governance

The Project Management Institute (PMI) defines governance as applied to program management as (emphasis and applicability to product innovation in bold):

> "[that which] covers the **systems and methods** by which a program and its strategy are **defined, authorized, monitored,** and **supported** by its sponsoring organization. Program Governance refers to the practices and processes conducted by a sponsoring organization to **ensure** that its programs are **managed effectively and consistently** (to the extent feasible). Program Governance is achieved through the actions of a **review** and **decision-making** body that is charged with endorsing or approving recommendations made regarding a program under its authority." (PMI Standard for Program Management 3rd Edition, 2013:51)

From a product innovation governance perspective, the following questions need to be adequately answered:

- Is the product innovation process tailored to the specific needs of the organization and its products or services? Is the process well communicated, understood, and accepted across the whole organization?
- Have measurable targets for new product or service outcomes been determined and agreed upon? Are these communicated and commonly understood by all people involved in the product innovation process?
- Are there specific metrics designed for each stage in the product innovation process (for example, actual spend vs. budget, milestones on time, overall development cycle time)? Are these used as a basis for learning and continuous improvement?
- Is the appropriate balance of management authority and individual responsibility defined and in place?
- Do the decision protocols and processes enable effective and timely decision-making? Do any significant and unnecessary "roadblocks" exist that add to frustration and delays?
- Is the product innovation process reviewed regularly, based on inputs from a cross-section of the organization and from the output and process metrics?
- Are specific processes and instances in place to resolve potential disagreements among team members, especially when cross-functional teams participate in the same projects?

4.5.2 The Role of the Board of Directors and Senior Management

McNaughton (2023) discusses the importance of supportive and robust governance in fostering innovation. He emphasizes the importance of the Board and senior executives in "setting expectations and policies, organizing the firm, delegating decision-making and hiring the right people to support innovation. In Table 4.10, McNaughton provides a summary of how the Board and senior executives influence innovation.

TABLE 4.10 How do the board and CEO influence innovation

The Board	The CEO and other C-suite executives
• Receives instruction from owners about investment objectives. • Sets parameters for risk tolerance and monitors risk. • Participates in strategy setting including the portfolio of innovation. • With management chooses how innovation will be organized. • Monitors and audits performance. • Oversees compliance. • Appoints CEO and monitors performance. • Delegates decision rights. • Nominates board members. • Sets up sub-committees or advisory boards (e.g., on innovation, or digital transformation).	• Sets a vision for innovation. • Defines how the company will capture value from innovation. • Chooses how innovation will be organized. • Chooses innovation processes/methods. • Defines roles and assigns responsibility. • Fosters an innovation culture. • Champions and models innovative behavior. • Sets policies and incentives and establishes management practices that support innovation. • Allocates resources and sets budgets. • Identifies and overcomes obstacles and resistance to innovation. • Defines how to measure innovation, sets targets, and monitors performance.

Source: McNaughton, 2023 / John Wiley & Sons.

4.6 IN SUMMARY

- This chapter has described a range of new product models. Each has its strengths and is generally more applicable at certain stages of the full product innovation process. The Stage-Gate® provides an idea-to-launch process, while most other models focus on specific stages throughout this overall process.
- Using a well-developed product innovation process(es), the best companies have significantly higher product success rates on average (76%) compared to the rest of the companies' success rate (51%).
- Product innovation is a risk vs. reward process. The application of disciplined processes and practices is designed to reduce the level of uncertainty and improve the likelihood of product success.
- Costs associated with new product innovation generally increase dramatically throughout the product innovation process – especially during final design, prototyping, and scale-up to commercialization. It is vitally important to put significant effort into the early stages of product innovation (FEI, front end of innovation) to ensure that an informed decision is made about the project's likelihood of success before costs increase.
- Stage-Gate®, IPD, waterfall, Agile, Lean, concurrent engineering, systems engineering, jobs to be done, and design thinking all have merits in application to specific organization and product situations. It is important to fully understand the principles of each model and apply the most appropriate model to the specific organizational context. This often requires a blend of two or more models.
- All process models have the following principles in common:
  - A focus on strategic alignment.
  - Knowledge-based decision-making to reduce the risk of product failure.
  - An emphasis on stakeholder input to design decisions.
  - The use of cross-functional teams.
  - A structured framework, understood and applied across the organization.
- Organizations that are truly successful in product innovation understand the fundamental principles of new product success. They learn from others and are constantly striving for continuous improvement.
- Fundamental to the successful implementation of product innovation is clarity of definition around the intent of the innovation strategy and of specific product innovation projects. This clarity of definition is provided in what is called the product innovation charter (PIC).
- A business sustainability strategy and principles need to be built into every product innovation project. These factors relate to reducing waste, recycling where/when possible, and reusing materials. Furthermore, it includes operating within the widely accepted triple constraint of people, planet, and profit.

4.A PROCESS PRACTICE QUESTIONS

1. Mary works as a product development engineer at an automotive parts supplier. Her boss has been complaining about the long lead time in getting new products to market. He believes that the iterative changes, inherent in most of the new products, could be carried out much more efficiently. What type of product development process would you suggest that Mary recommend to her boss to reduce time to market?
   A. Waterfall
   B. Integrated
   C. Agile
   D. Stage-Gate
2. In the early stages of the new products process there is a lot of uncertainty. These stages are often referred to as
   A. Concept generation.
   B. Front end of innovation.
   C. Business case development.
   D. Concept evaluation.
3. The sign of a mature product development process within an organization is the ability of the organization to:
   A. Include and integrate stakeholders and senior management throughout the development process.
   B. Use iterative and risk-limiting steps to facilitate effective and efficient new product development.
   C. Continuously fill the pipeline with new products.
   D. Develop its own best practices from the various models and experiences.
4. The first step in the classic waterfall process is requirements. The last step is known as:
   A. Development.
   B. Maintenance.
   C. Verification.
   D. Implementation.
5. Key elements of Agile new product process include the scrum, the scrum master, the scrum team, the sprint, the product backlog, and:
   A. Stakeholders.
   B. Project manager.
   C. Product owner.
   D. Product champion.
6. A company is developing a product that is basically a relatively simple line extension of existing products. It knows the market well and the risks associated with product failure are low. What type of Stage-Gate process would be most likely appropriate for this development?
   A. A 5-stage process with heavy emphasis on initial business analysis.
   B. A 5-stage process with significant market research built into the decision-making.
   C. A relatively short 3-stage process focusing on speed to market.
   D. A 3-stage process with a strong emphasis on pre-launch test marketing.
7. In the Jobs-to-be-done (JTBD) framework there are two types of jobs that are defined: Main Jobs and Related Jobs. Both types of jobs have multiple aspects. Which of the following is **not** an aspect of either a Main or Related job-type:
   A. Emotional.
   B. Financial.
   C. Functional.
   D. Developmental.

8. Gates are defined as decision points based on deliverables, criteria, and outputs. Outputs include:
   A. Financial statements.
   B. Yes or no decisions.
   C. The highs and lows of impact and probability.
   D. Go, kill, hold, recycle decisions.
9. Lean processes are focused primarily on:
   A. Putting more discipline into processes.
   B. Making processes more Agile.
   C. Removing waste.
   D. Encouraging cross-functional integration.
10. The framework, functions, and processes that guide activities in project, program, and portfolio management and provide guidance, decision-making, and oversight is called. . .
   A. Governance.
   B. Project management.
   C. Team leadership.
   D. General management.

Answers to practice questions: Product innovation process

1. C
2. B
3. D
4. B
5. C
6. C
7. B
8. D
9. C
10. A

REFERENCES

Agile Manifesto. (2001). *Agile manifesto for software development*. https://agilemanifesto.org/

Blank, S. (2010). *Not all those who wander are lost*. Cafepress.

Blank, S. (2016). *SyncDev methodology*. Sync.Dev. http://www.syncdev.com/minimum-viable-product/.

Blank, S. and Dorf, B. (2012). *The startup owner's manual: The step-by-step guide for building a great company*. K&S Ranch.

Booz, Allen, and Hamilton (1982). *New products management for the 1980s*. Indiana University.

Brown, T. (2008). *Change by design: How design thinking transforms organizations and inspires innovation*. Harper Collins.

Cagan, M. (2017). *Inspired*, 2e. Silicon Valley Product Group.

Christensen, C.M. and Raynor, M.E. (2003). *The innovator's solution*. Harvard Business School Press.

Cooper, R.G. and Kleinschmidt, E. (1986). An investigation into the new product process: Steps, deficiencies, and impact. *Journal of Product Innovation Management 3* (2): 71–85. https://doi.org/10.1016/0737-6782(86)90030-5.

Cooper, R.G. (2001). *Winning at new products*. 3e. Basic Books.

Cooper, R.G. (2014). What's next after stage-Gate®. *Research-Technology Management 57* (1): 20–31. https://doi.org/10.5437/08956308X5606963.

Cooper, R. G. (2015). *Stage-Gate® and agile development: Debunking the* myths. http://www.stage-gate.com/resources_stage-gate_agile.php

Cooper, R.G. and Sommer, A.F. (2016). Agile-stage-gate: New idea-to-launch method for manufactured new products is faster, more responsive. *Industrial Marketing Management 59*: 167–180. https://doi.org/10.1016/j.indmarman.2016.10.006.

Cooper, R.G. (2017). *Winning at new products*, 5e. Basic Books.

Cooper, R.G. and Sommer, A.F. (2018). Agile-stage-gate for manufacturers. *Research-Technology Management 61* (2): 17–26. https://doi.org/10.1080/08956308.2018.1421380.

Cross, N. (1982). Designerly ways of knowing. *Design Studies 3*: 221–227. https://doi.org/10.1016/0142-694X(82)90040-0.

Espinoza-Orias, N., Cooper, K., and Lariani, S. (2018). Integrated product innovation at Nestlé. In: *Designing sustainable technologies, products and policies* (ed. E. Benetto, K. Gericke, and M. Guiton), 89–103. Springer https://doi.org/10.1007/978-3-319-66981-6_50.

Fricke, E., & Schulz, A. P. (2005). Design for changeability (DfC): Principles to enable changes in systems throughout their entire life cycle. *Systems Engineering, 8(9)*. https://doi.org/10.1002/sys.20039

Grecis, Z. (2015). *8 things to use in "jobs-to-be-done" framework for product development*. UX Collective https://uxdesign.cc/8-things-to-use-in-jobs-to-be-done-framework-for-product-development-4ae7c6f3c30b.

Gurtner, S., Spanjol, J., and Griffin, A. (2018). *Leveraging constraints for innovation*. Wiley https://doi.org/10.1002/9781119390299.

Knapp, J., Zeratsky, J., and Kowitz, B. (2016). *Sprint: How to solve big problems and test new ideas in just five days*. Bantam Press (GB).

Knudsen, M.P., von Zedtwitz, M., Griffin, A., and Barczak, G. (2023). Best practices in new product development and innovation: Results from PDMA's 2021 global survey. *Journal of Product Innovation Management 40* (3): 257–275. https://doi.org/10.1111/jpim.12663.

Haberfellner, R., De Weck, O., Fricke, E., & Vössner, S. (1986). *Systems engineering: Fundamentals and applications*. Birkhäuser.

Kahn, K.B., Evans Kay, S., and Slotegraaf, R.J. (2013). & Uban. In: *The PDMA handbook of new product development*, 3e (ed. S.). John Wiley & Sons https://doi.org/10.1002/9781118466421.

Liedtka, J. (2020). Putting technology in its place: Design thinking's social technology at work. *California Management Review 62* (2): 53–83. https://doi.org/10.1177/0008125619897391.

Liedtka, J. and Ogilvie, T. (2011). *Designing for growth: A design thinking tool kit for managers*. Columbia Business School Publishing.

Mascitelli, R. (2011). *Mastering lean product innovation: A practical, event-driven process for maximizing speed, profits and quality*. Technology Perspectives.

McKim, R.H. (1972). *Experiences in visual thinking*. Brooks/Cole Publishing Co.

McNaughton. (2023). Innovation governance. In: *The PDMA handbook of innovation and new product development*, 4e (ed. L. Bsteiler and C.H. Noble), 121–136. John Wiley and Sons.

Morgan, J.M. and Liker, J.K. (2006). *The Toyota product innovation system, integrating people, process and technology*. Productivity Press.

Naveh, E. (2004). The effect of integrated product innovation on efficiency and innovation. *International Journal of Production Research* 43 (13): 2789–2808. https://doi.org/10.1080/00207540500031873.

New York Times. (1985, December 2). Demise of the Sgt. York: Model turns to dud. https://www.nytimes.com/1985/12/02/us/demise-of-the-sgt-york-model-turns-to-dud.html

Ojomo, E. (2021). *The power and simplicity of jobs to be done theory*. https://www.christenseninstitute.org/blog/the-power-and-simplicity-of-job-to-be-done-theory/

Osterwalder, A., Pigneur, Y., and Smith, A. (2010). *Business model generation: A handbook for visionaries, game changers, and challengers*. Wiley.

Pahl, G., Beitz, W., Feldhusen, J., and Grote, K.H. (2007). *Engineering design: A systematic approach*, 3e. Springer https://doi.org/10.1007/978-1-84628-319-2.

Phillips, R., Neailey, K., and Broughton, T. (1999). A comparative study of six Stage-Gate approaches to product innovation. *Integrated Manufacturing Systems 10* (5): 289–297. https://doi.org/10.1108/09576069910371106.

Pichler, R. (2013). *Agile product innovation with scrum: Creating products that customers love*. Addison-Wesley Signature Series (Cohn).

Ries, E. (2011). *The lean startup: How today's entrepreneurs use continuous innovation to create radically successful businesses*. Crown Business.

Rittel, H. and Webber, M. (1973). *Dilemmas in general theory of planning*. *Policy Sciences* 4 (2): 155–169. http://www.jstor.org/stable/4531523.

Royce, W.W. (1970). Managing the development of large software systems. In: *Proceedings of IEEE WESCON*, vol. 26, 328–388. IEEE.

Sutherland, J. (2014). *Scrum: The art of doing twice the work in half the time*. Currency.

Ulwick, A.W. (2016). *Jobs to be done: Theory to practice*. Ideation Press.

Winner, R. I., Pennell, J. P., Bertrand, H. E., & Slusarczuk, M. M. G. (1991). *The role of concurrent engineering in weapons system acquisition*. Institute for Defense snalyses Report R-338.

5

Product Innovation

Design & Development

Required to facilitate efficient and effective evolution of a product from initial idea through to a developed, manufacture and "market-ready" form.

Product Development and Management Body of Knowledge: A Guidebook for Product Innovation Training and Certification, Third Edition.
Allan Anderson, Chad McAllister, and Ernie Harris.

WHAT YOU WILL LEARN IN THIS CHAPTER

Product innovation makes use of a wide range of methodologies, concepts, and techniques, which we often simply refer to as *tools*. Tools are used as products move through the different phases of the product innovation process, evolving an idea into a product concept that then moves through deeper design and development, resulting in a product to launch to markets. Some tools are specific to certain phases of the product innovation process, while others are applied in many areas.

In this chapter, we focus on tools that are more specifically applied to the design and development phases of the product innovation process. Each of these phases benefits significantly from applying tools to reduce uncertainty and ambiguity and to assure conformance with the design's customer requirements and technical specifications. The most commonly used tools are described, their advantages, and where they can be applied at the various phases of product design and development.

PDMA's 2021 Best Practices Study shows that the "best companies" are more likely to use design and development tools than the "rest" of the companies (Knudsen et al., 2023).

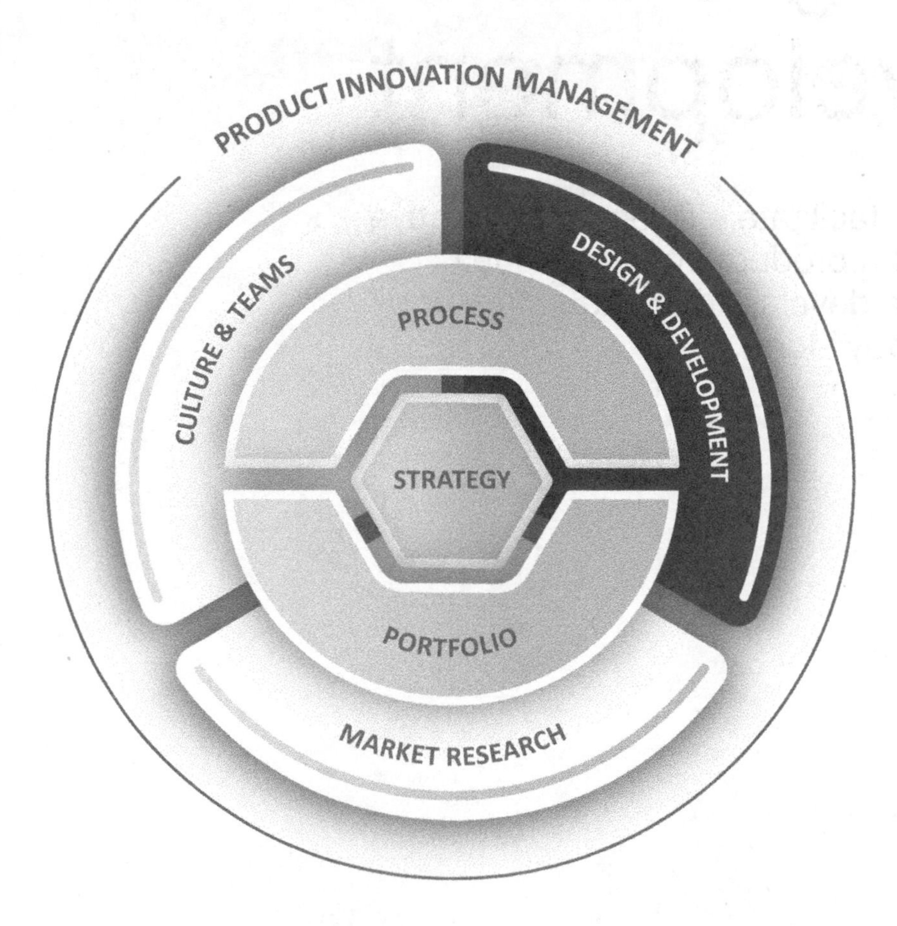

THE CHAPTER ROADMAP

This chapter is organized into two main parts: (1) Design and (2) Development. The first part describes design tools for constructing and refining a product concept. These tools are useful for moving from an initial idea or innovation insight to a clarified and validated product concept. The design tools are organized into primary categories: Ideation, Concept Design, and Embodiment. The customer's needs are increasingly understood during the progression from Ideation to Embodiment, and the product elements that create value for the customer are identified. This results in a detailed description of a product concept and early prototypes.

The second part describes development tools, organized by initial design specifications, detailed design specifications, and fabrication and assembly. Moving from the initial design specifications to fabrication and assembly adds specificity to the detailed product concept, enabling the product to be developed, implemented, or manufactured. Also, related tools are addressed for Usability testing, Performance testing, Quality Assurance, Manufacturing, and Sustainability.

The level of detail needed at any stage of the design and development process depends on several factors, including the industry, product form (e.g., software, physical manufacturing, and services), product complexity, newness to company, etc. For your product work, depending on such factors, you will find some of the tools applicable and others that are not. Further, organizations will vary on what activities they consider design versus the activities they consider part of development. The line between design and development is not absolute. The categories discussed in this chapter provide a means of organizing the tools.

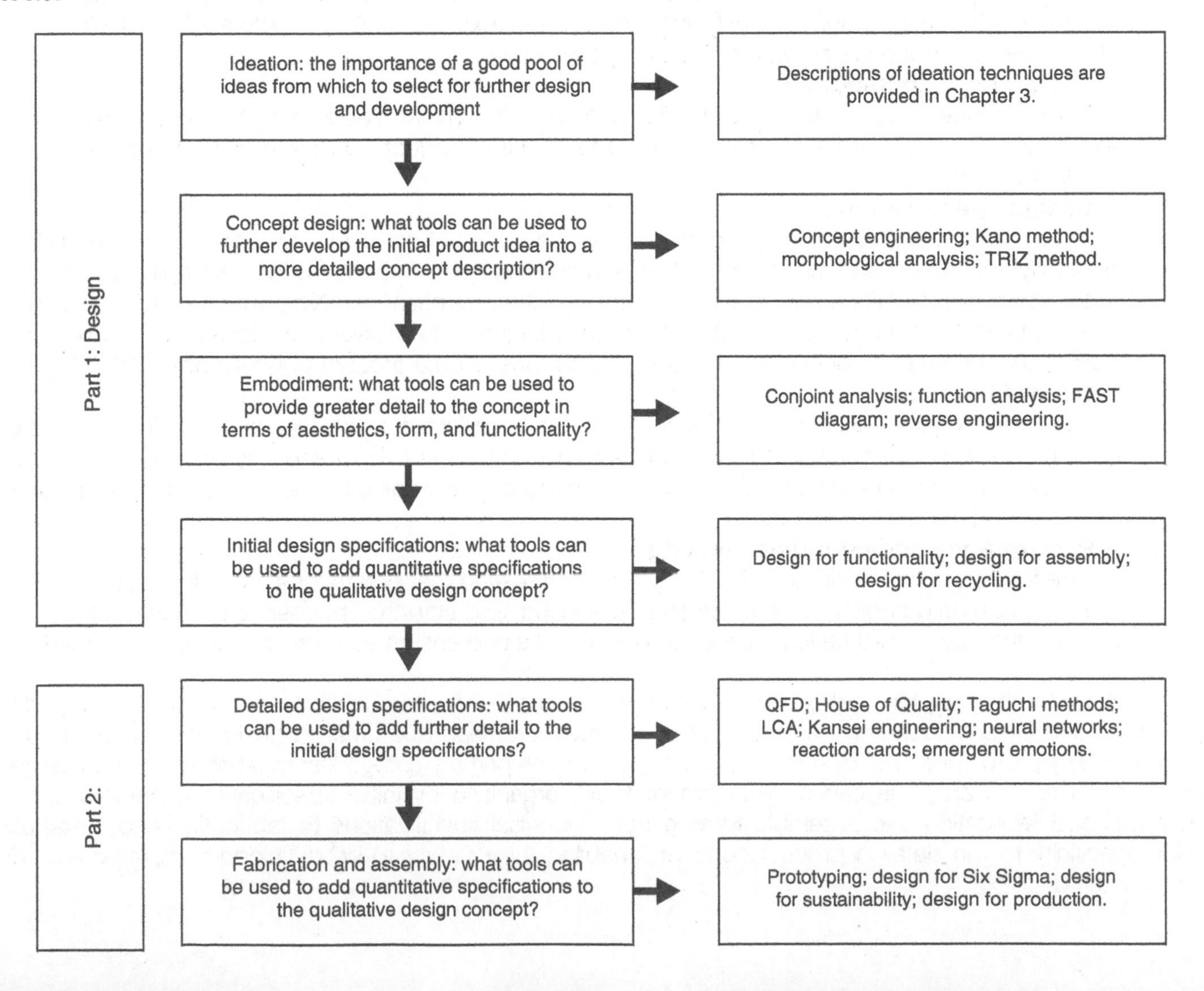

5.1 DESIGN AND DEVELOPMENT PROCESS STEPS

Following is a list of activities included in most product design and development processes:

i. **Idea generation or Ideation**

As discussed in Chapter 3, idea generation forms the basis for new products to be included in the new products portfolio. Idea selection is based on a range of well-defined criteria, including strategic fit, competition, resource availability, sustainability, and return on investment. These new product ideas may be totally new to the organization or modifications to existing products.

ii. **Basic concept description**

To create a product concept description, ideas are expanded on with greater detail related to a defined target market and its specific requirements for the key design components:

- Aesthetics
- Functionality
- Quality

iii. **Embodiment**

The basic product concept is further refined, and detail is added using Embodiment tools. This involves the preparation of visual and functional representations of the concept in the form of sketches, wireframes, models, and prototypes.

iv. **Quality assurance**

Quality Assurance is a planned and systematic set of actions necessary to provide adequate confidence that the product optimally fulfils customers' expectations. A quality assurance plan should be developed in parallel with design specifications.

v. **Design for manufacturing**

For manufactured products, design for manufacturing involves optimizing the product's design for its manufacturing and assembly process, merging the product's design requirements with its production method.

vi. **Design specifications**

The previous activities move from idea to detailed product concept. The product concept is generally descriptive and mostly qualitative. It is necessary but insufficient for developing the product. Design specifications add specificity using quantitative measures. Note that the term "design specifications" is not used in all organizations or industries. The PDMA BoK uses this term for the activity that enables product development to occur based on a product concept description.

vii. **Usability testing**

On-going user input is an essential element of the product design process, providing feedback and suggestions for improvement in design features. The use of visual and physical representations of the product concept can add significantly to the quality and value of user feedback. User testing is conducted to gain this feedback.

viii. **Performance and endurance testing**

Performance and endurance testing is a type of testing that determines a product's performance, such as durability, speed, scalability, handling, and stability. The testing process includes manual and automated tests designed to reveal faults and ensure your product is fit for the market.

As shown in the chapter roadmap, we have organized the design and development tools into two parts: (1) Design and (2) Development. Part 1 describes the tools that lead to a detailed design of a product concept and early prototypes. The design tools are organized into primary categories: Ideation, Concept Design, and Embodiment. Part 2 describes development tools, organized by initial specifications, detailed specifications, and fabrication and assembly. Moving from the initial specifications to fabrication and assembly adds specificity to the detailed product concept, enabling the product to be developed, implemented, or manufactured.

PART 1: DESIGN

5.2 INTRODUCTION TO DESIGN

For the purposes of this book, we are defining the design process as the earlier activities of product innovation that focus on the evolution of a product from initial insight or idea to a well-defined and understood product concept. Consequently, the output of design is a product concept with sufficient detail for a development team to begin their work (addressed below in Part 2: Development).

This evolutionary process requires consideration of a range of factors, including:

- Consumer needs and wants.
- Aesthetics.
- Functionality.
- Quality.
- Material and component availability.
- Cost vs. what price the consumer is willing or able to pay.
- Capital cost and return on investment.
- Competition.
- Manufacturability.
- Environmental impact.

Information on these factors enables the design process to progress through the application of appropriate design tools. Complementing these design tools are ideation tools described in Section 3.2.1. Also, Chapter 5, Marketing Research, describes consumer and market research tools that contribute to product design and development.

5.2.1 Ideation

A successful product starts with a good idea. The greater the pool of quality ideas, the greater the likelihood of developing successful products. Although these new ideas can happen by accident, a structured process using sound ideation tools will likely lead to a sustained source of good ideas.

A selection of ideation tools was discussed in Section 3.2.1 as the foundation of the new product portfolio. Several contexts for new product idea generation exists. They include improvement to an existing product, a line extension, a new to-the-organization product, or even a new-to-the-world product. Although most ideation tools can be applied to each of these contexts, some are better suited to specific areas of focus. A few examples are presented below.

Product improvements and line extensions would benefit from:

- SCAMPER, where participants focus on specific action verbs to help them think differently about a product and avoid making assumptions limiting their past thinking – Substitute, Put to another use, Eliminate, etc.
- Mind-mapping, where participants graphically outline ideas for product improvement from a defined starting point of product aesthetics, functionality, and quality properties, and encourage mapping ideas from this starting point.
- Story boarding asks participants to develop a story about the customer's use of a product to better understand areas where the product could be improved.
- Crowdsourcing utilizes the power of social media to engage with users and potential users to submit ideas for product improvements and line extensions.

New to the organization product would benefit from:

- Brainstorming, where a large pool of ideas is generated and subsequently evaluated against portfolio-specific screening criteria, as discussed in Chapter 3.
- Brainwriting, where participants write down their ideas and share these with others in a group for further discussion and evolution of these ideas.
- Crowd sourcing is also useful here to generate new product ideas that an organization has not yet pursued.

New to-the-world products would benefit from ideation approaches that seek to gain a better understanding of trends and unarticulated customer needs. These include:

- Scenario building involves the development of future-focused scenarios and then asking what new product would create customer value in these scenarios.
- Delphi utilizes expert input to predict future trends for markets, technologies, or products.
- Ethnography seeks to better understand users or potential users in their normal environment, leading to the identification of needs through direct observation that users are either unwilling or unable to articulate. These needs then lay the foundation for the new product idea.
- Big data is becoming more widely used to understand current product usage and potential future trends. As discussed in Section 3.2.1, the challenge is "mining the nuggets" of truly valuable information from the mass of available data.

The ideation stage usually results in a relatively brief description of the product idea with very little detail on its form, function, benefits, features, etc.

"Anyone can dream up great ideas, but an idea is nothing until it is realized, be it as a website, a physical product, an app, or a user interface" – Jens Martin Skibsted, Danish designer.

5.2.2 Concept Design

Ideation often provides a very simple description for a new product opportunity. To enable greater focus and clarity of direction for those involved in the further development of the product, more detail, substantiated by customer input, is required.

The next stage in the product design process involves the development of a more detailed concept description. This provides greater detail about the new product idea enabling:

- Clarity and alignment for all members of the product team and those associated with the project.
- A vital means of explaining the proposed product to potential customers and other stakeholders, both to seek their views on product benefits and features, and their advice on possible changes or improvements before incurring additional development expenses.
- A more detailed basis for feasibility analysis.

In Section 2.7.1, we described the three levels of a product: (1) core, (2) tangible, and (3) augmented, refer to Figure 5.1.

1. ***Core product*:** The benefits the target market will derive from the product.
2. ***Tangible product*:** The physical and aesthetic design features that give the product its appearance and functionality.
3. ***Augmented product*:** These are benefits that may be provided as an addition to the product, either free of charge or for a higher price.

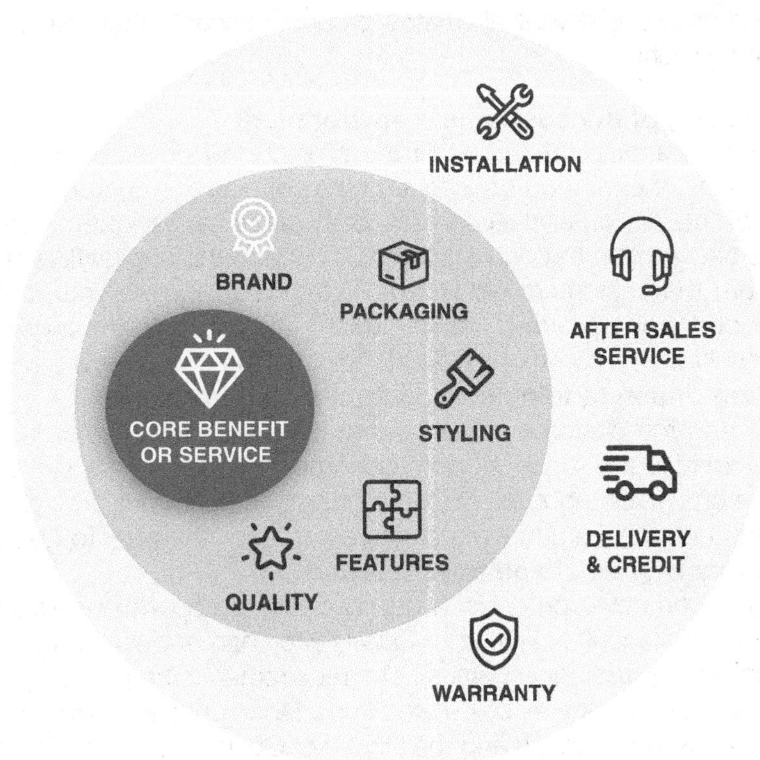

FIGURE 5.1 The three levels of a product

Concept Example

IDEO, the global design company, is recognized for its human-centered product design. A classic example of a concept description, IDEO created is for a new shopping cart, considering factors such as maneuverability, shopping behavior, child safety, maintenance cost, and replacement cost.

"The nestable steel frame lacks sides and a bottom to deter theft, and holds removable plastic baskets to increase shopper flexibility, help protect goods, and provide a method to promote brand awareness. A dual child seat uses a swing-up tray for a play surface, and a hole provides a secure spot for a cup of coffee or a bunch of carnations. . . . One of the unique—and potentially patentable—features of the cart is the design of its steerable back wheels. Normally fixed straight for stability and familiarity, an easy sideways effort allows the wheels to turn left or right. Pushing the cart forward puts the wheels straight again."

The three levels of the product provide a useful framework for evolving the concept description. A wide range of tools is available for this. A number of these tools are described in Chapter 6, Market Research. These include focus groups, ethnography, customer site visits, social media, and a range of multivariate techniques. Following is a selection of other commonly used tools applied to creating and refining a product concept.

Concept Engineering

The Concept Engineering (CE) method, developed in collaboration between the Center for Quality Management (CQM), CQM member companies, and MIT, provides a stepwise, customer-centered process for developing product concepts, Burchill and Walden (1994). The method determines the customer's key requirements to be included in the design, and proposes several alternative product concepts that satisfy these requirements. The method involves five stages of work.

Stage I: Understanding of the customer's environment

This involves understanding the scope of the project and developing a roadmap that will guide the exploration activities, the method of collecting the voice of the customer, and the development of a common image of the customer environment and use of the product. It requires being immersed into the customer's use context through customer visits and the application of contextual inquiry. The voices collected from the customers will be translated into requirements that identify the expected features to be included in the product design. This stage requires a triangulation of perspectives in the design team, resulting in a common understanding of the customer's environment.

Stage II: Converting understanding into customer requirements

This stage converts the customers' voices into requirement statements. The method of *priority marking* is used to identify the most critical requirements. Generally, a set of 24 customer requirements is collected, from which eight are selected based on customer priority for further development in the next stage. Also, all requirements are clustered around "themes" to guide the creative inquiry.

Stage III: Operationalizing what you have learned

This stage involves the development of a quality chart and operational definitions for the customer requirements selected in stage II. A House of Quality-like chart shows the relationship matrix between customer and technical requirements (metrics). The requirements are prioritized using any of the following methods: The Kano questionnaire, self-stated importance questionnaire, or the critical requirements questionnaire. Both the Kano methods and the House of Quality are discussed later in this chapter.

In addition, this stage involves the creation of general metrics for the requirements. An attempt to minimize the number of metrics per requirement is usually made with two metrics per requirement. The reason is that if a metric correlates with many requirements (e.g., four), it suggests that it is too abstract and, thus, difficult to interpret. Validity and easiness of use of the metric are the two main criteria to evaluate.

Stage IV: Generating concepts

This stage transitions from requirements to the *solution space*. First, it is desirable to generate as many diverse concepts as possible. For this, decomposing the design problem into sub-problems is helpful, as this facilitates considering solutions to specific components of the product or system. There are several methods for decomposition – such as functional analysis, Fast diagrams, process flows, and metrics tree diagrams. This stage ends with clear concept descriptions for the concepts carried to the next stage.

Stage V: Selecting the final concept

This stage selects a final concept using numerical analysis and scoring of the most viable concepts already envisioned. Concepts are systematically reviewed, using a scoring process to screen the concepts. They are evaluated against customer requirements and organizational and business constraints. This leads to selecting the best concept that creates the most value for the customer and the organization.

Similar to a Retrospective from an Agile project, the final step of Concept Engineering is to increase learning by reflecting on the overall process and identifying opportunities for improvements.

User Experience (UX) and User Interface (UI) Design

> "No product is an island. A product is more than the product. It is a cohesive, integrated set of experiences. Think through all of the stages of a product or service – from initial intentions through final reflections, from first usage to help, service, and maintenance. Make them all work together seamlessly."
>
> —*Don Norman, inventor of the term "User Experience"*

User experience (UX) refers to any interaction a user has with a product or service. UX design considers each and every element that shapes this experience, how it makes the user feel, and how easy it is for the user to accomplish their desired tasks. A UX designer's primary goal is for each user to interact positively with a product or service. Whether the interaction solves a problem, provides entertainment, or helps the user find critical information, the experience should satisfy the user.

User Interface (UI), on the other hand, refers to the interfaces users engage with. The UI design process may include buttons or widgets, text, images, sliders, and other interactive elements. UI designers ensure that every visual element, transition, and animation included within a product or service sets the stage for a cohesive, positive experience. Refer to Table 5.1 for a comparison of UX and UI design applied to software and electronic products.

TABLE 5.1 UX vs. UI design for software and electronic products

UX design	UI design
Feel The feel of the overall experience within the product.	**Look** How the product's interfaces look and function.
Prototyping Creates wireframes and testable prototypes that form the basis of a website or service's user experience.	**Design** Finalizes the products and designs for actual user engagement.
High-level Takes a high-level view of the product, ensuring the collective user flow is fully realized and consistent.	**Details** Works on individual pages, buttons, and interactions, making sure they are well-done and functional.

Consideration of both UX and UI, and the tools used to generate their understanding, are essential during all stages of product innovation, starting with the detailed concept description. Detailed design features should evolve as more knowledge of the desired UX and UI is gained.

Following is a selection of commonly used tools that are used to assist the development of the product concept and, in turn, the overall UX design of the product.

Kano Method

Widely used by experienced product managers and innovators, the Kano method has proved useful for identifying customer needs and latent demands, determining functional requirements, developing concepts as candidates for further product definition, and analyzing competitive products or services within a product category.

The Kano method classifies the product requirements into three broad categories: Basic requirements, performance requirements, and excitement requirements.

- **Basic requirements** fulfill basic functions in each product in the category; these are essential – without them, there is user dissatisfaction.
- **Performance requirements** provide real benefits and enhance the utility of the product. These reflect functional performance attributes and are needed for differentiation.
- **Excitement requirements** (generally referred to as the WoW! effect) are generally unexpected. They delight the customer and provide superior satisfaction. Their presence increases the overall experience and they do not provide dissatisfaction if not present.

The Kano method is operationalized by identifying the set of user requirements, known as customer requirements (CRs), that users look for in products or services. In addition to the above categories, Kano describes four types of elements of a product:

1. ***Attractive quality elements:*** Customers are satisfied when present but do not feel dissatisfied when these requirements are absent.

2. ***One-dimension quality elements*:** These elements show a proportional relationship between functionality and satisfaction. More functionality implies higher levels of satisfaction.
3. ***Must-be quality elements*:** These are the elements that the user expects at a minimum. They do not lead to satisfaction when fulfilled. However, they cause dissatisfaction when not fulfilled.
4. ***Indifferent quality elements*:** These elements provide neither satisfaction nor dissatisfaction regardless if they are fulfilled.

Succinctly, the Kano method is used to narrow the set of requirements that will make the product competitive in the market within a specific product category. Applying it involves a customer survey and analyzing the results to classify customer requirements as attractive, one-dimension, must-be, or indifferent. Further analysis can prioritize the requirements to include in the release of a product, accounting for development time and cost, support of product vision, technical feasibility, etc. Further details are available at Kano method websites, such as www.kanosurveys.com.

Concept Scenarios

This technique generates scenarios as a way to learn how product concepts would work in real-life situations. Product design teams collect sketches, illustrations, photos, and descriptions of activities to help develop a customer journey. Scenarios detail the actual actor, describe the context and the actor's goal, and describe how the actor achieves the goal. The technique aims to explore how the product concept can meet customers' expectations.

Every concept scenario implies the following steps:

1. Identify possible concepts generally developed with other methods, such as the Kano or morphological analysis.
2. Create and identify a scenario around each possible concept: actor, context, and process to achieve customers' goals.
3. Evaluate and reformulate the product concept as you go through the scenario.
4. Illustrate the scenarios by reflecting on all activities, users, and other participants involved in real-life situations where the concept is being used.
5. Discuss the added value of the concept within the imagined situations and user context.

This approach streamlines product design by reducing the time to identify critical customer requirements, describe the product functionality within the user experience in mind, communicates new product features, and provides guidance for concept testing.

Courtyard by Marriott

An example of the application of concept scenarios is the design for the Courtyard by Marriott. This mid-priced hotel was designed for business travelers. The concept scenario included specific working desk spaces, couches, and small gathering places for business interviews, business centers, restful beds, and green views from the bedroom windows to the exterior. This concept was tested by assessing the perception of the target markets, recommendations were provided to enhance design (e.g., addition of closets), and later a conjoint analysis technique was applied to evaluate the relative weight that each hotel component had on potential customer satisfaction.

Morphological Analysis

This technique is useful when the designer attempts to build possible concepts with the information gathered through previous exploration. The approach generates a system-level solution that meets the needs and expectations of potential users. It aims to identify possible *elements* common to several possible *solutions* known as *design parameters*.

A typical morphological analysis has the following steps:

1. Identify user-centered categories or dimensions critical to the product design.
2. Use these categories to organize possible concept ideas. These ideas are branched from each category.
3. Create a morphological chart with categories/dimensions on the horizontal, and below each category, a series of concept ideas (see Figure 5.2).
4. Combine possible ideas from each category into potential solutions to enter the product design stage.
5. Identify specific criteria to allow for a comparison of the adequacy of the solutions found.
6. Discuss the possible solutions among the product innovation team for final assessment.

Figure 5.2 presents an example showing the application of a morphological analysis in the creation of concepts for a hotel using five critical design dimensions of hotel design: Adaptability (ability to adapt to changing conditions), Community (providing a sense of sharing), Technology (current, available technology), Engagement (understanding and meeting customer needs), and Well-being (comfortable, happy, and safe environment). Each bulb represents a possible idea within the design dimensions (adaptability, community, etc.). The links between bulbs represent possible conceptual connections leading to four possible solutions or combinations of ideas that satisfy all critical design dimensions.

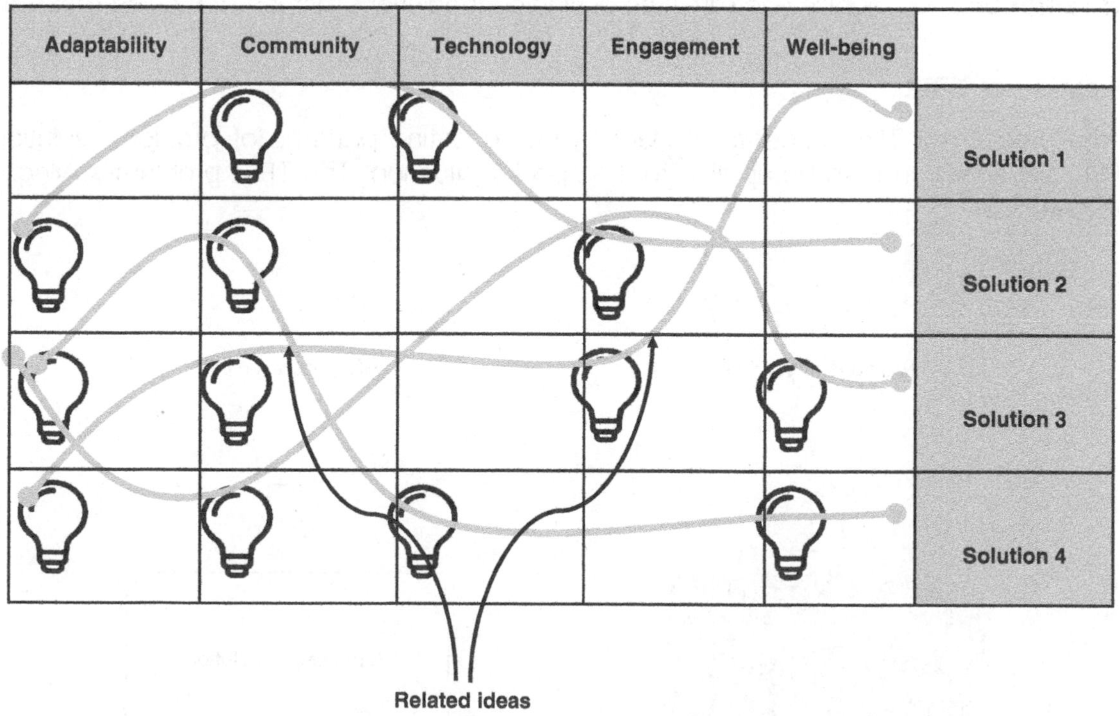

FIGURE 5.2 Example of morphological analysis

TRIZ Method

TRIZ is a problem-solving method based on logic and data, not intuition, which accelerates the project team's ability to solve problems creatively. TRIZ also provides repeatability, predictability, and reliability due to its structure, algorithmic approach, and database of solutions to known problems. TRIZ is the Russian acronym for the "Theory of Inventive Problem Solving" (TIPS). G.S. Altshuller and his colleagues in the former U.S.S.R. developed the method between 1946 and 1985 by examining patents. Their work concluded that:

- in excess of 95 percent of all patents used only seven inventive tools,
- less than 5 percent come from breakthroughs in science and brand-new ideas,
- many of the exceptional patents improved performance by resolving contradictory requirements,
- if patents were categorized by what they did functionally rather than by industry, the same problem was solved many times with a small number of techniques.

This primary research lays the foundation for the fundamental principles of TRIZ:

- Problems and solutions are repeated across industries and sciences. By classifying the "contradictions" in each problem, you can predict good creative solutions to a problem.
- Patterns of technical evolution tend to be repeated across industries and sciences.
- Creative innovations often use scientific effects outside the field where they were developed.

TRIZ contains four paradigm shifts, which can be used with creative right-brained thinking and analytical left-brained thinking to create a new system. These are:

- ***Functionality*:** The system possesses a Main Useful Function (MUF), and any components that don't contribute toward the MUF are ultimately harmful.
- ***Resources*:** Identify anything in the system not being used for maximum potential.
- ***Ideality*:** Invision the final ideal solution where all functions are achieved without causing harm or problems.
- ***Contradictions*:** Recognize and eliminate design contradictions instead of making trade-offs.

The application of TRIZ

Much of the practice of TRIZ consists of learning the repeating patterns of problems-solutions. These general TRIZ patterns can then be applied to the specific situation. The TRIZ problem-solving method is shown in Figure 5.3.

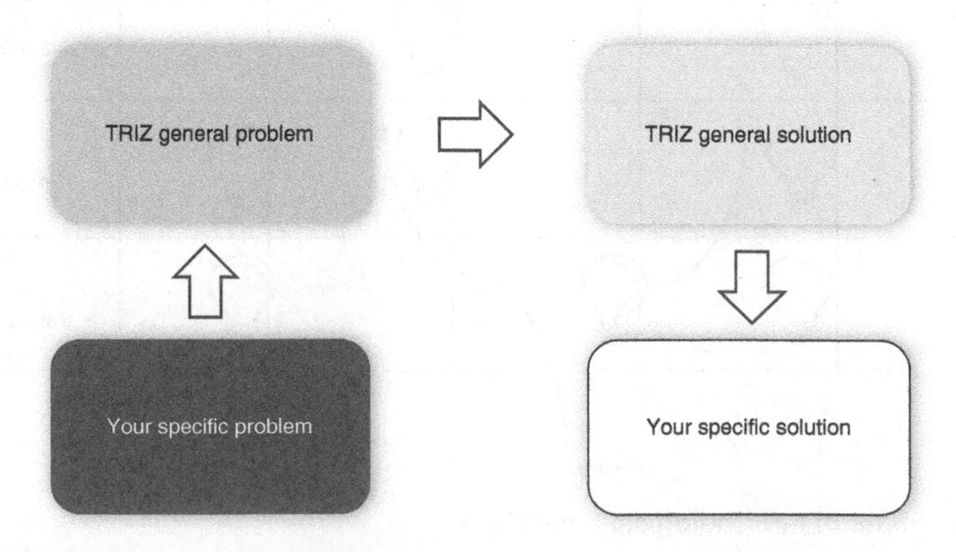

FIGURE 5.3 TRIZ problem-solving method

Tools used in TRIZ

Altshuller (1984) defined inventive problems as those which contain conflicting requirements, which he called contradictions. Further, he found that the same fundamental solutions were used repeatedly, often separated by many years. He reasoned that their task would have been simpler if later inventors knew of earlier solutions. He, therefore, set about extracting, compiling, and organizing such knowledge. He defined 39 basic properties (refer to Table 5.2) and 40 principles (refer to Table 5.3) for solving problems containing contradictions in any two-of-39 properties. This he gave in the form of a contradiction table of size 39 × 39 with each cell giving up to four principles (and examples from patent database) that may be used to eliminate the contradiction.

TABLE 5.2 TRIZ 39 engineering parameters

1. Weight of moving object	15. Duration of action by a moving object	27. Reliability
2. Weight of stationary object	16. Duration of action by a stationary object	28. Measurement accuracy
3. Length of moving object	17. Temperature	29. Manufacturing precision
4. Length of stationary object	18. Illumination intensity	30. External harm affects the object
5. Area of moving object	19. Use of energy by moving object	31. Object-generated harmful factors
6. Area of stationary object	20. Use of energy by stationary object	32. Ease of manufacture
7. Volume of moving object	21. Power	33. Ease of operation
8. Volume of stationary object	22. Loss of Energy	34. Ease of repair
9. Speed	23. Loss of substance	35. Adaptability or versatility
10. Force	24. Loss of Information	36. Device complexity
11. Stress or pressure	25. Loss of Time	37. Difficulty of detecting and measuring
12. Shape	26. Quantity of substance/the matter	38. Extent of automation
13. Stability of the object's composition		39. Productivity
14. Strength		

TABLE 5.3 Altshuller's 40 principles

1. Segmentation	11. Beforehand cushioning	21. Skipping	31. Porous materials
2. Taking out	12. Equipotentiality	22. 'Blessing in disguise' or 'Turn Lemons into Lemonade'	32. Color changes
3. Local quality	13. 'The other way round'	23. Feedback	33. Homogeneity
4. Asymmetry	14. Spheroidality – Curvature	24. 'Intermediary'	34. Discarding and recovering
5. Merging	15. Dynamics	25. Self-service	35. Parameter changes
6. Universality	16. Partial or excessive actions	26. Copying	36. Phase transitions
7. "Nested doll"	17. Another dimension	27. Cheap short-living objects	37. Thermal expansion
8. Anti-weight	18. Mechanical vibration	28. Mechanics substitution	38. Strong oxidants
9. Preliminary anti-action	19. Periodic action	29. Pneumatics and hydraulics	39. Inert atmosphere
10. Preliminary action	20. Continuity of useful action	30. Flexible shells and thin films	40. Composite materials

TRIZ has now evolved into an international science that is widely applied across a range of disciplines and industries. The complexities of TRIZ preclude a detailed explanation of its application in this publication.

An excellent overview of TRIZ theory, together with examples of application, is provided by Ladewig (2007), along with publications from Haines-Gadd (2016) and Gadd (2011).

TRIZ applied to product innovation

Although many people may find TRIZ too complex in its finest detail, a rudimentary application using Altshuller's 40 principles can be extremely useful in the idea generation stage and early concept development, both for new products and for product improvements. See examples in Table 5.4.

TABLE 5.4 Examples of application of Altshuller's 40 principles to product creation, sourced from The Triz Journal

Principle	Solution
Segmentation (divide and object into independent parts)	Individually wrapped cheese slices
Local quality (provide different packaging for different uses)	"Adult" editions of Harry Potter books
Universality (make an object perform multiple functions)	Chocolate spread sold in glasses (with a lid) that can be used for drinking afterward
Nested doll (nest one object inside another)	Store within a store (coffee shops in a bookstore)
Another dimension (tilt or re-orient object)	Squeezable ketchup bottles that sit on their lids

5.2.3 Embodiment Design

The embodiment design stage builds on the basic concept description based on additional user input, manufacturing criteria, and economic constraints. It extends the design, often in the form of one or more types of prototypes, providing additional meaning to stakeholders. This can lead to more meaningful feedback for further design enhancements. A few examples are below.

A sketch is a freehand drawing. It is very cheap and fast and often done during brainstorming sessions. Its main goal is to illustrate the product or aspects of the product.
A mock-up is a model of what your final product will look like. Product mock-ups are frequently used to present a final product in a real-life context.
A wireframe is a detailed black-and-white layout of the product. They provide specifics about how the product is laid out or constructed. Wireframes are commonly used for software products and websites. You plan the layout of elements (images, buttons, and text) at this stage. A simple example of is presented in Figure 5.4.

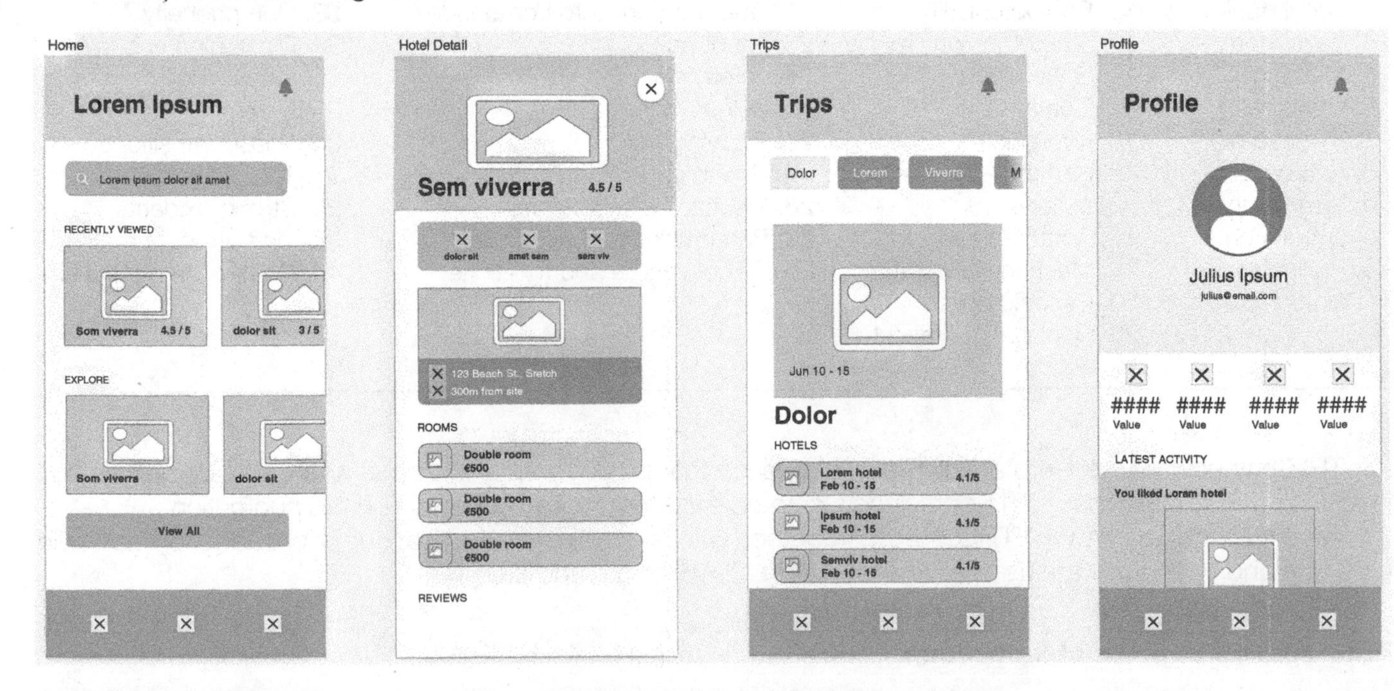

FIGURE 5.4. Simple example of a wireframe for a travel app.
Image source: https://www.sketch.com/blog/wireframe-examples/. Used with permission.

A Prototype

Prototyping is a design method that builds a representation of a product concept. Its main purpose is to assess how well the product features satisfy user expectations. As such, prototyping is an example of a problem-solving approach to design. A prototype approximates the form of a design idea that allows the product development team to verify the product operations and functions, its components, layout, functionality, appearance, and behavior. Prototypes range from concept sketches to fully functional artifacts.

Prototyping fidelity is the degree to which the design allows for a reasonable understanding of the product in use. If the design team is in the initial stages of product conceptualization, low-fidelity prototypes are suitable. In the latest stages, high-fidelity prototypes are needed. In addition, medium fidelity prototypes allow for evaluating the interactive aspect of consumer experience, exposing possible flaws before finalizing the product release.

There are several types of prototypes based on the level of comprehensiveness or learning needed, as well as the stage of the product innovation process (see Figure 5.5).

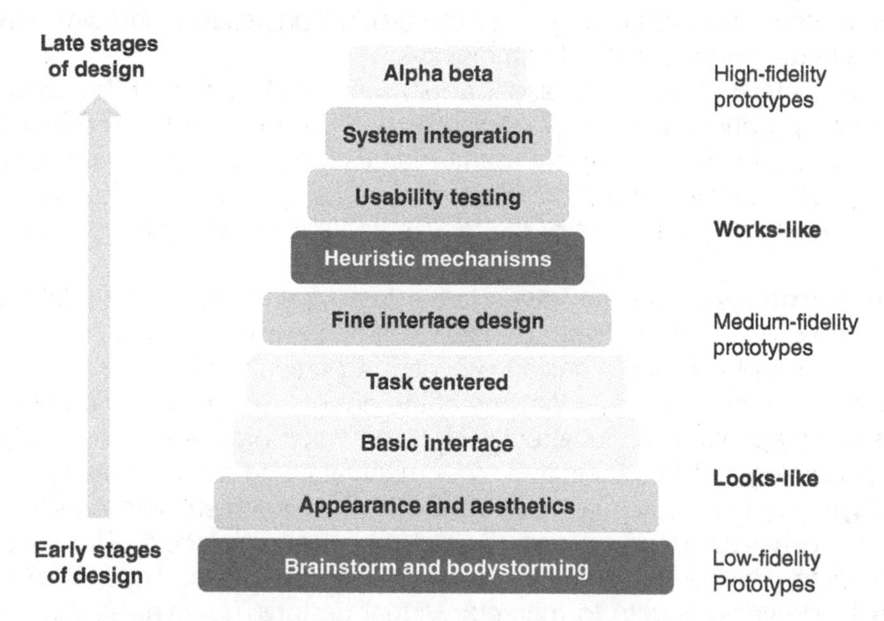

FIGURE 5.5 Testing objectives and types of prototypes according to the design process.
Source: Rodriguez (2017) / Createspace Independent Pub.

Prototypes can be classified based on two dimensions: physical-analytical and focused-comprehensive. A clay mock-up is a physical prototype, while a simulation model is analytical in essence. Focused prototypes allow for the evaluation of few attributes (e.g., paper prototypes), while comprehensive ones allow for greater scrutiny and analysis of several variables simultaneously (e.g., automobile prototype).

Prototypes are valuable throughout the design process, as the product evolves to its final form. Specific purposes and applications of prototypes are described below.

***Paper prototyping*:** This is the most common basic form to represent concepts and ideas. This prototype does not represent the technical characteristics of the design. Mainly, it is used to explore idea generation and brainstorm possible solutions using sketching to depict the ideas. Storyboards are one form of a paper prototype that can represent the customer interface through a series of storyboard frames.

***Functional prototyping*:** The purpose of these prototypes is to test how the product works and delivers the expected functions. Usability testing allows the identification of possible design errors at early stages when all functions have not yet been finalized. This method favors identifying specific tasks for the user, completion of the task by the user, identifying areas for improvement, and modifying the prototype as needed. Overall, experimentation allows for evaluating user-centered design requirements, including ergonomic and cognitive considerations and the user's experience.

***Experience prototyping*:** This type of prototyping applies to the design of product or service experiences. The method requires a description in a structural manner of the different stages of product/service delivery and the critical interactions between the prototype and the user. In the service industry, service maps or blueprints are used to identify the "critical moments" or moments of truth that may create an imbalance or misspecification of the value delivery. In the case of products, this method studies the dynamic relationship between the user and the usage setting that generates functional and emotional concerns. Experience prototyping allows users, designers, and others involved in the innovation process to be active participants within the "experiential setting" and social context as they discover and generate rich interpretations.

***Alpha prototype*:** This prototype may not be entirely functional and is used for testing purposes. Typically, the organization uses these prototypes internally in a controlled environment to support them. During testing, features, functions, and subsystems are scrutinized, and the product's performance is assessed. The design is close to the production version, and simplified production processes are used (e.g., machine instead of molding).

***Beta prototype*:** Beta prototypes represent a fully functional version of the product and are evaluated before the preproduction stage. This prototype is used to test the product in its actual usage environment with customers, assemblers, parts manufacturers, and component suppliers. The main purpose is to assess reliability, and products are made using actual production manufacturing facilities. This prototype allows an analysis of the requirements and adjustments needed before the product is finalized.

***Pre-production prototype*:** This prototype is the final type of physical model that specifies all the elements of the design in addition to specifications of part production and components before entering the manufacturing phase of the design process. A particular interest is in confirming the necessary technical requirements and specifications of the design. Design for assembly and manufacturing is validated at this stage with careful attention to production processes, assembly times, parts integration and outsourcing, balance lines, and other manufacturing adjustments.

***Virtual prototyping*:** This method combines the virtual environment with engineering design, allowing the designer to assess design sensitivity and optimization process. Before the final product is built, virtual objects provide the ability to test "what-if" scenarios. Thus, it reduces development time and time to deliver products to markets. Virtual prototyping involves the creation of 3D CAD (computer-assisted design) models. These models allow for the analysis of the form and shape, how parts and components fit, rendering, and assembly. In addition, this method requires input devices to sense user interaction and motion, output devices to replace users' sensory input with computer-generated input, and software to handle real-time processing, rendering capabilities, and simulation tests. Virtual prototyping is mainly used by industrial designers as they aim to visualize their concepts.

***Rapid prototyping*:** For physical products, rapid prototyping (RP) has become critical in reducing product development cycles and time to market without compromising quality. Rapid prototyping is the physical modeling of a design using machine technology in an additive and/or subtractive manufacturing process. 3D printing is an example of an additive process where the prototype is built layer by layer. A wide range of materials (such as ABS filament, liquid plastics, resins, and metals) can be used in additive manufacturing. Milling is an example of subtractive manufacturing, which removes material using a high-speed cutting head. Computer Numerical Control (CNC) is commonly used to control subtractive manufacturing equipment. Both additive and subtractive manufacturing processes for rapid prototyping rely on CAD software to construct a virtual model of the object before it is physically formed.

Figure 5.6 shows the application of a number of prototypes throughout the development of a model car for 8- to 10-year-old children. Each technique evolves greater detail into the final design.

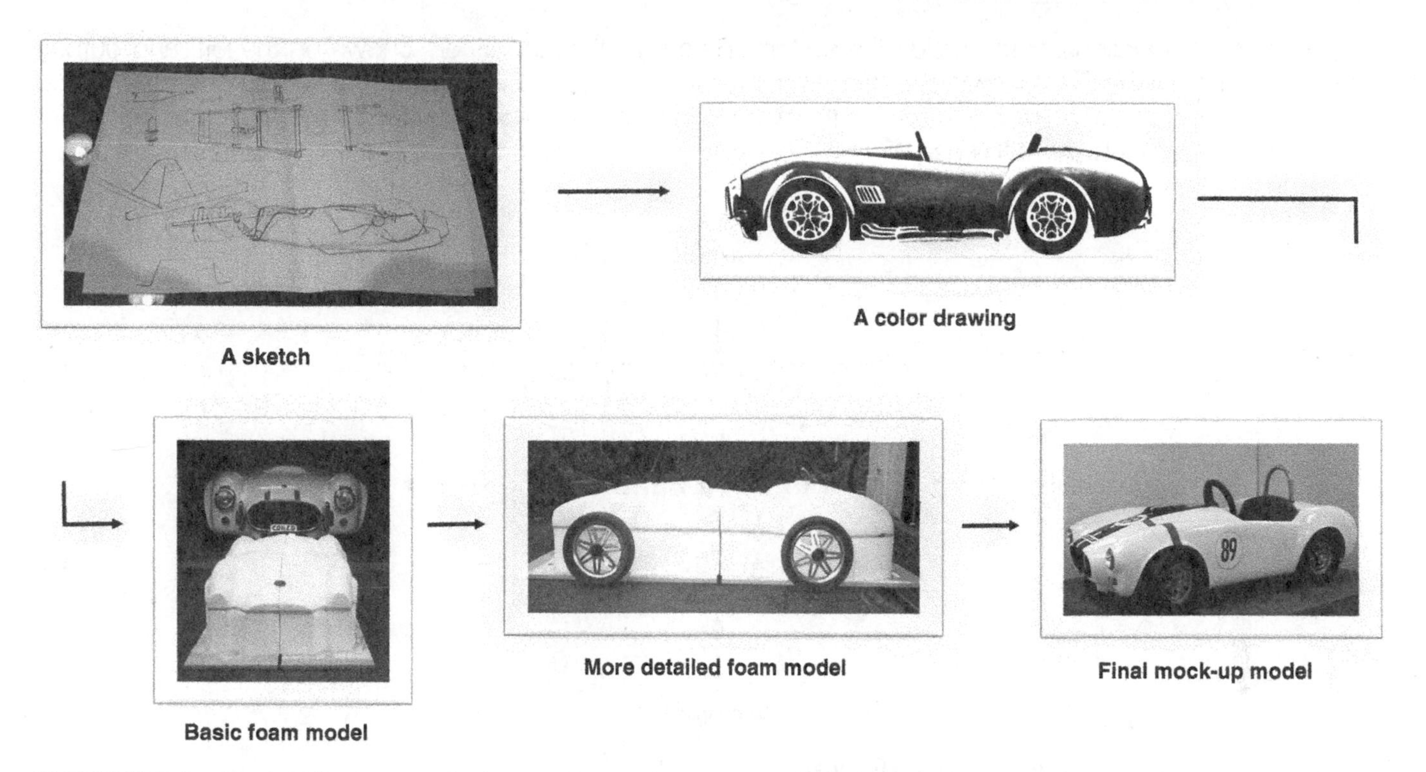

FIGURE 5.6 Embodiment techniques throughout a design process.
Source: Used with permission Elizabeth Irwin, Massey University.

5.2.4 Tools to Assist Embodiment Design

The embodiment design stage is both iterative and evolutionary. A wide range of tools can be used to add greater detail to the product concept and its design. Following are some tools useful for embodiment design.

Consumer Research Tools

Consumer research tools that can be applied to product concept development and refinement are summarized in Chapter 6, Market Research. They provide insights for creating, refining, and evolving an initial product concept. These include:

- Focus groups
- Consumer panels
- Sensory panels
- Surveys
- Multivariate techniques – multidimensional scaling, factor analysis, conjoint analysis.

Functional Analysis

Functional analysis is the foundational method when applying value analysis methodology. The central objective of value analysis is to provide performance at the lowest cost. Functional analysis considers that a product is represented by a system composed of several functions interrelated under an operational logic. Functional analysis divides a system into smaller parts called functional elements. It is used to create and examine the functions of a product concept. There are different ways to record and display the functions which make up a system design. A Functional Flow Block Diagram is a popular graphical method. It uses a rectangular box to represent each function (see Figure 5.7). Arrows represent flows or states of any type to and from the function. The flows connect to other functions or to outside the system. By convention, inputs are shown on the left, and outputs are shown on the right. The function box itself transforms the inputs to the outputs. Mechanisms are the entities which perform the function but which are not themselves transformed. They are normally shown by arrows at the bottom and typically represent the use of a device. Controls are inputs that command, limit, or direct the operation of the function and are normally shown at the top.

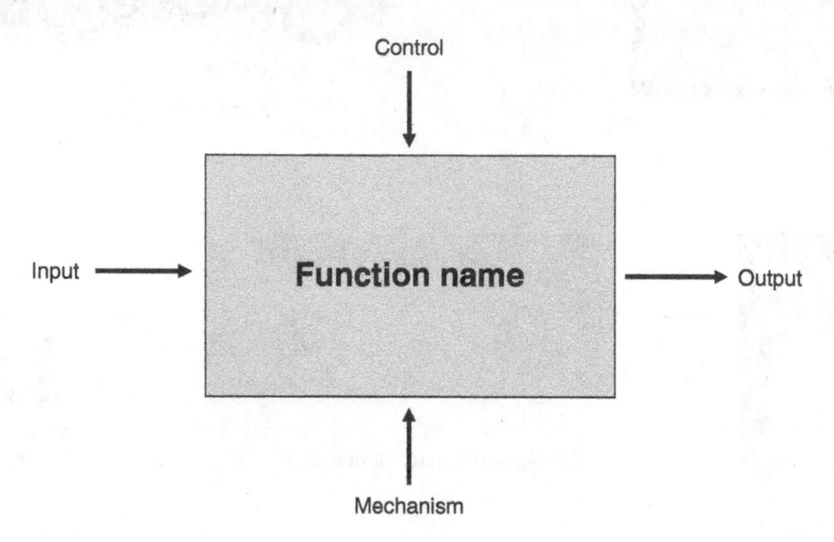

FIGURE 5.7 Functional flow block diagram

A function is defined as a relation verb–noun, i.e., stack plates. The active verb "stack" has an "action" over the noun "plates." The assumption behind functional analysis is that the customer purchases a product because of the functions it performs.

According to functional analysis, functions can be classified as basic and secondary, intrinsic or extrinsic, and use or aesthetic, described in relation to a television:

A basic function answers "no" to the following question: If I didn't have to perform this function, would I still have to perform any of the other functions listed? Once the basic function has been identified, all other functions become secondary. Being a secondary function does not imply being irrelevant. On the contrary, secondary functions support the basic function so the product can achieve its purpose.

Intrinsic functions are those that happened inside the boundaries of the product, while **extrinsic functions** refer to those that occur while the user interacts with the product, for example:

- User pushes channel change button: extrinsic function.
- As a result, several operations occur within the television set: intrinsic functions.

Use functions refer to something that needs to be accomplished for the product to perform: user presses start button on TV remote.

Aesthetic functions are accomplished through any of the five senses: sight, smell, taste, touch, and sound. For example:

- Surround sound.
- Sharp color contrast.

Fast (Function Analysis System Technique)

FAST is a technique that builds on the results of a functional analysis. It aids in thinking about a problem objectively and in identifying the scope of the project or product by showing the logical relationships between functions. Organizing the functions into a function-logic, FAST diagram enables participants to identify all the required functions. The FAST diagram can be used to verify if, and illustrate how, a proposed solution achieves the project or product's needs and identify unnecessary, duplicated, or missing functions. Developing a FAST diagram is a creative thought process supporting communication between team members.

The development of a FAST diagram helps teams to:

- Develop a shared understanding of the project.
- Identify missing functions.
- Define, simplify, and clarify the problem.
- Organize and understand the relationships between functions.
- Identify the basic function of the project, process, or product.
- Improve communication and consensus.
- Stimulate creativity.

The diagram is built following How-Why logic, as shown in Figure 5.8. This logic requires functions to be ordered from left to right. From left to right, we answer the question: How is this function carried out? The response is by several functions. When the analysis is done from right to left, we answer the question: Why do we need this function? The approach from both ends – left to right (How) and right to left (Why) – allows for a robust diagram and configuration of functions, enhancing posterior analysis and opportunities for redesign and product improvement.

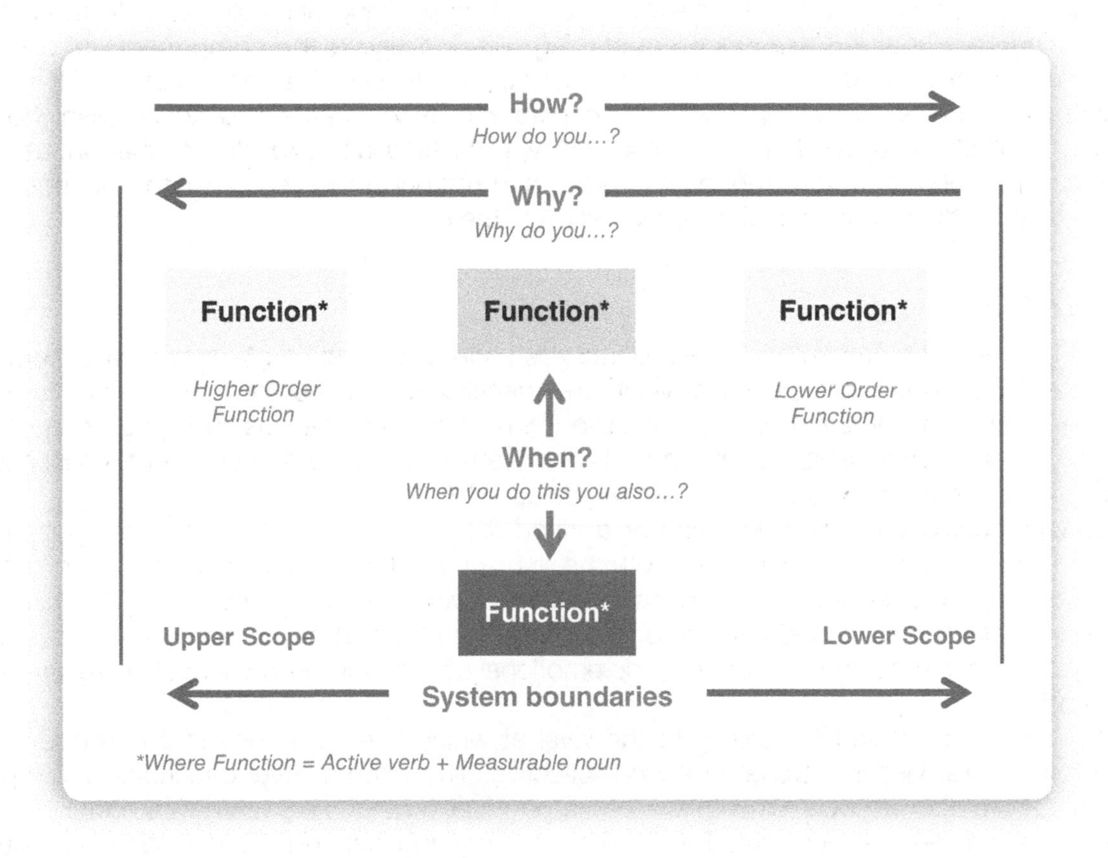

FIGURE 5.8 How-Why logic of a FAST diagram

Consider a mouse trap as an example, shown in Figure 5.9. Functions are ordered from left to right, starting on the left side. The highest order function, which is the main job the customer wants the product for, starts on the left. A mouse trap's function is to "eliminate mice." Then answer the question: How is this function carried out? The response is by several functions that work together. When the analysis is done from right to left, we answer the question: Why do we need this function? The approach from both ends – left to right (How) and right to left (Why) – allows for a robust examination of product functions, creating a thorough analysis that enhances opportunities for product improvement.

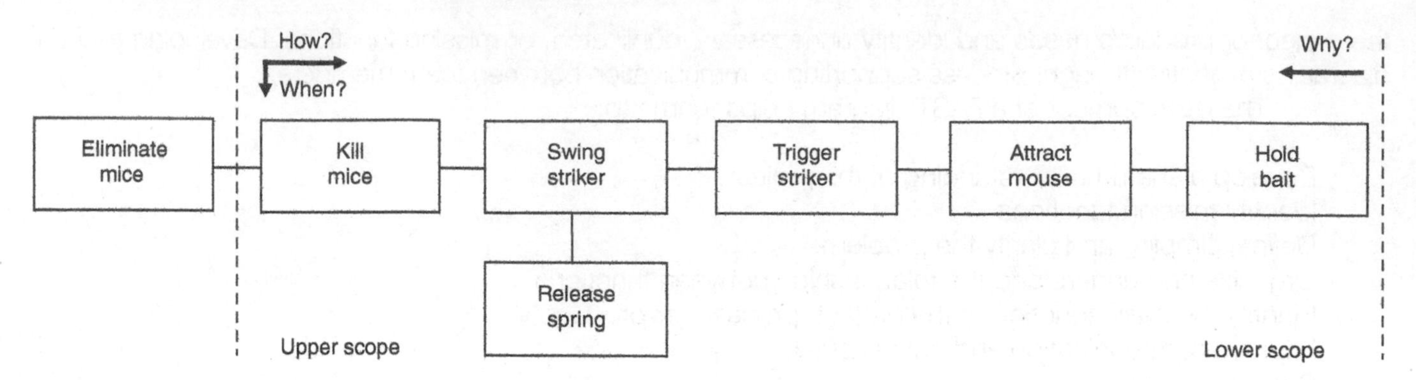

FIGURE 5.9 Mouse trap example of FAST

Reverse Engineering

Reverse Engineering (RE) is the implementation of value analysis (VA) tear-down processes to formulate ideas for product improvement. This method is based on dissembling products, systems, components, and data to identify the functions embedded that allow for comparison of competitors' products and production processes.

In the context of product concept refinement, reverse engineering aims to stimulate the creative process by focusing on identifying the functions, starting with identifying the components of a system and their understanding at a deep level. New conceptualizations are possible, approaching RE as imitation-type thinking where parts and components are copied with few variations to accomplish the expected performance. Another RE approach is the research-type thinking used to identify the language through "signals" (creativity clues) that the embodiment emits and regroup these signals into new abstractions or components, resulting in a more valuable configuration of the product.

Emotional Design

Consumers build mental representations of products as soon as they start interacting with them. The sensory information that consumers perceive allows the identification of the product's functions which contribute to the creation of meaning and an affective connection with the design in the consumer's mind. Designers rely on emotional design to elicit moods and feelings that create positive emotional associations and a feeling of trust in the product, thus improving its usability.

For example, consider the first time you or a friend received an iPhone. Apple recognized that your emotional connection with the iPhone starts with the unboxing of the product. The designers worked hard to create packaging and an unboxing experience like receiving a much-anticipated gift. This included a precise amount of friction designed into the box so that when the top of the box is removed, you have a pleasing sensory experience. In this example, the emotional connection begins even before interacting with the actual product.

Emotions can be classified according to the level at which users process them. The levels of information processing are: visceral, behavioral, and reflective. The visceral level describes a low processing and basic motor reaction associated with the user's physical senses, i.e., aesthetics, color, or smell. The behavioral level is a midlevel processing that involves memory and learning associated with product usage, i.e., functionality and usability. The reflective level entails feeling, emotions, and cognition that determine understanding, interpretation, and reasoning. At this level, the product connects with the sense of self and identity.

Several methods have been developed to help designers assess the impact of emotional design on the user's preference and intention to purchase. A brief description of the more commonly encountered methods follows:

Kansei engineering: This method identifies the relevant design elements (color, size, and shape) embedded in a product as determined by user preferences. It aims at the development or improvement of products and services by translating the customer's psychological feelings and needs into the domain of product design. The method requires the identification of Kansei words, generally adjectives to present users' emotions and feelings. Developing product concepts for a "sporty" car, a "flashy" smartwatch, and an "exotic" travel experience are examples where Kansei engineering applies.

***Sentiment analysis*:** This technique is used to classify and understand people's opinions in product review blogs or social networks. It allows for the identification of opinion polarities (positive, neutral, or negative) expressed on product features through automated processing. In the past, the method made use of Naive Bayes probabilistic classifier, but today, it is more common to see Artificial Intelligence approaches used to infer sentiment.

***Neural networks*:** This approach creates non-linear models to examine the complex relationship between input variables (product features) and output variables (user perceptions). They are constructed using a deep learning algorithm and trained using known preference combinations. For example, if a company has previous research showing user preference for different sizes of bottles as well as user preference for a different color of bottles, along with other factors, a neural network can be constructed to reveal the best combination of factors with the highest preference.

***Microsoft reaction card*:** This method assesses the emotional response and desirability (visual appeal) of a design or product. Participants describe a design based on 118 product-reaction words by selecting cards that are relevant to the product or design. The method concludes with an explanation of why the card is attached to the design. The method uses cluster and frequency analysis and word cloud processing.

***Emergent emotions*:** This method considers that emotions are dynamic, ongoing, and part of a recursive process. The user's pattern of response to a design is driven by the appraisal results. These include emotional responses and the desirability of product features resulting from emotional impact and differentiation. The method uses neural network and other non-linear dynamic modeling within the context of artificial intelligence to explain consumers' emotional processes, as illustrated in the Geneva Expert System on Emotions (GENESE).

PART 2: DEVELOPMENT

5.3 INTRODUCTION TO DEVELOPMENT

The division between where design stops and development begins is arbitrary and varies by organization. The structure of this chapter was chosen to emphasize that in the more successful organizations, a well-understood and sufficiently accurate product concept is designed before more expensive development activities begin. Some organizations view the distinction between design and development as moving from the customer space to the engineering/developer space.

- ***Customer***: The product concept is created and articulated based on an understanding of the customer's problem or desired objective. The design of the product concept is largely expressed in terms that the customer understands and appreciates. It reflects the customer's language about their needs and describes the solution to their problem or how what they want will be accomplished.
- ***Engineering/developer***: For the tangible product or intangible service to become a reality, specifications and requirements must be created, components designed and made, software written, assembly conducted, etc. The engineers and developers need to know what must be made for the product concept to become *real* and later made available to customers.

The tools and concepts presented below in Part 2: Development section are those that engineers and developers can use to transform the previously designed product concept into the specifications, drawings, requirements, etc., they need to develop the product through to commercialization.

The distinction between designing the product concept and designing and developing the product is specific to each organization, and you may see different ways to categorize the tools and the process itself. Further, many of the tools summarized below as Development concepts traditionally have *design* in their name, such as *Design for functionality*. We have retained the traditional name instead of changing it; so we don't use *Develop for functionality*.

We want you to remember that a product concept should first be designed before additional, and more expensive engineering and developer resources are engaged to develop the product.

5.3.1 Initial Specifications

Where the concept description provides a description (qualitative presentation) of the product concept's benefits and features, the product design specifications provide the quantitative requirements. So, for example, where the concept description may describe the size of a new product as being able to fit into a coat pocket, the product design specifications would specify the actual physical dimensions required.

Product design specifications are intended to clarify the product design and to provide quantification and objectivity. They enable the communication of the product design requirements to other design team members and to progress the product's development from design through to manufacture. Following are specific areas of focus for design specifications.

An example of the development of a product concept and design specifications is presented in Appendix B, Case 3.

Design for Functionality (DFF)

Functionality determines the final performance of a product. Does the product do what it was intended to do, or more importantly, what the user wants it to do? Functional design is the process of responding to the needs or desires of the people who will be using the product in a way that allows those needs and desires to be met.

Design for Manufacturing and Assembly (DFMA)

DFMA focuses on reducing time to market and production costs by prioritizing both the ease of manufacturing the product and the simplified assembly of its parts into the final product. It is important that this is carried out early in the design and development process, and involves a cross-functional team with requisite knowledge of manufacturing and assembly. Further discussed in Section 5.3.7.

Design for Maintenance (DFM)

Maintenance of a product implies monitoring its actual condition, maintaining it, and allowing for recovery of the system as changes occur (such as wear and tear, corrosion, useful life deterioration, and other changes over its operating lifetime). Inadequate maintenance affects the functionality of the product, the economy, and the safety of the entire system. Overall, DFM focuses on safety, ergonomics, and assembly of the product.

Products should be designed to be maintained, repaired, and/or replaced with parts or components. Decisions regarding the selection of materials, assemblies, parts, devices, and components determine the maintainability or the capacity of the system to be inspected, restored, and serviced when components fail as they reach their operational life expectancy.

Design for Recycling (DFR)

The notion of DFR is an aspect of design for sustainability (addressed later in Section 5.3.8). DFR implies the use of materials that allow for reusing or reprocessing production waste, products, and parts composing products. The methods for DFR center on reusing and reprocessing products. Several guidelines are followed during DFR, such as ease of disassembly, material compatibility, material separation, and opportunities to reprocess parts and components. Consequently, synergies should be sought when applying DFR, DFM, and DFMA.

Design for Usability (DFU)

Usability is a measure of how well a specific user in a specific context can use a product to achieve a defined goal effectively, efficiently, and satisfactorily. A design's usability depends on how well its features accommodate users' needs and contexts.

Design for usability encompasses a product's functionality, serviceability, maintainability, ease of operation, reliability, safety, aesthetics, operating context and environment, and customizability.

Design for Serviceability (DFS)

Design for serviceability focuses on the ability to diagnose, remove, or replace any part, component, assembly, or subassembly of a product while performing service repairs and troubleshooting. Serviceability facilitates manipulation and operation while bringing the product to its original working specifications, putting it in operation quickly, and avoiding downtime.

Nowadays, many products are designed with instrumentation that alerts when service is required in advance of possible system failures. Intelligent capabilities alert users of possible failures and even have self-corrective mechanisms to maintain operation as the user waits for the unit to be serviced.

5.3.2 Quality Function Deployment (QFD)

Quality Function Deployment (QFD) is a technique for deriving product specifications from customer needs. It tends to be used in complicated manufacturing environments, such as the design of an automobile. However, regardless of your product environment and if you use QFD formally or not, you will find it useful to understand the principles of QFD. All product projects must somehow bridge or translate customers' needs and wants to the information designers/engineers/developers need to create the product. While this is an obvious aspect of creating a product, it is also where priorities are easily confused, and misunderstandings occur. QFD contains the principles to make the bridging from customer to development explicit. This leads to clarity and improved communications among the project team.

QFD is a structured method employing matrix analysis for linking "what the market requires" to "how it will be accomplished in the development effort" (Hauser & Clausing, 1988). This method is most valuable during the stages of development where a cross-functional team is used to focus on how customer needs translate into product specifications and features that will address these needs. The most commonly used example of QFD is the House of Quality, shown in Figure 5.10.

The basic steps in building the House of Quality:

1. Identify customer needs and wants and rate these needs and wants on a defined scale.
2. Identify product design attributes and define the direction for improvement in the design attributes.
3. Relate the customer attributes to the design attributes.
4. Determine the importance of each product design attribute.
5. Analyze the interrelationships between the product design attributes.
6. Conduct an evaluation of competing products.
7. Identify customer attributes for further improvement and target values for design attributes.

Let's use the example of the design of a mobile phone to demonstrate the application of the House of Quality. This example, shown in Figure 5.10, is purely hypothetical and is not based on any real product example. It is intended to demonstrate the basic use of the House of Quality. Greater levels of complexity can be added, depending on the specific focus of its application.

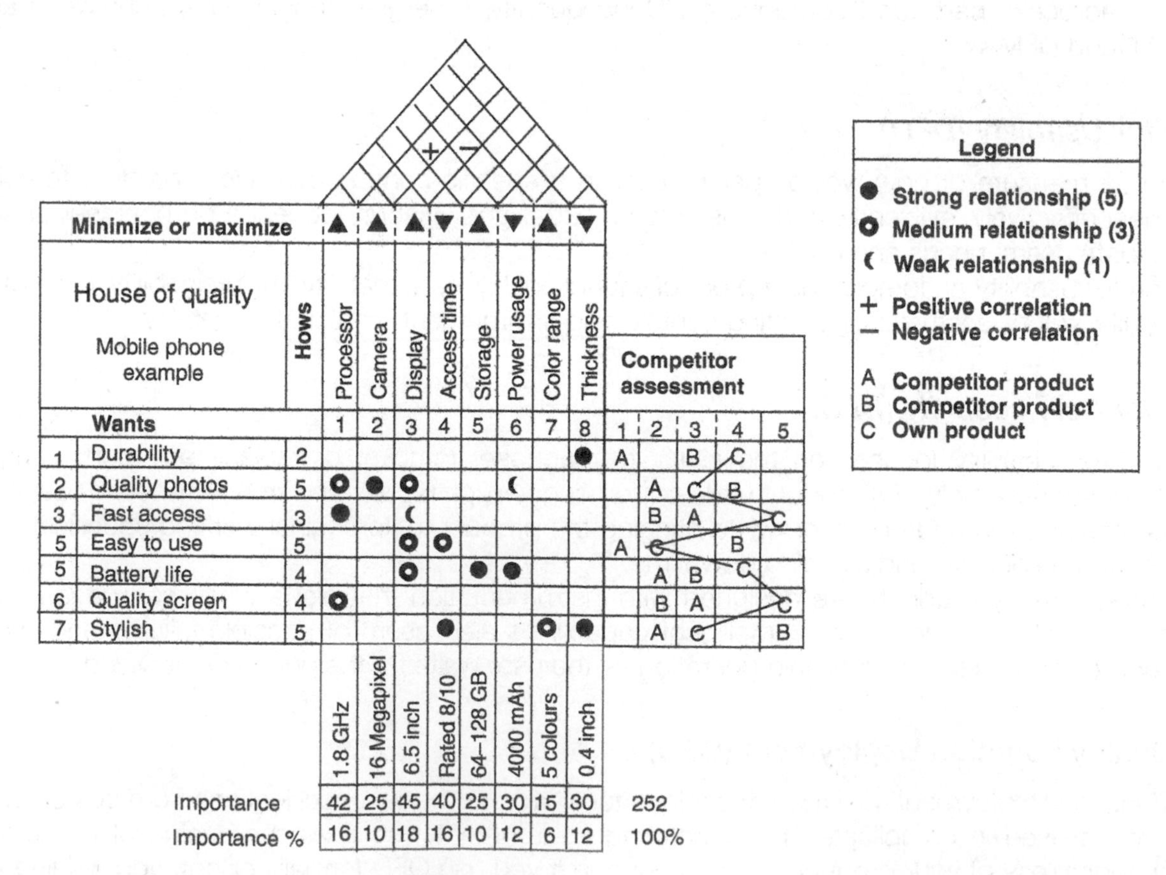

FIGURE 5.10 House of quality example

Step 1. Identify customer needs and wants

- Known as the *whats*, these are product requirements in the customer's terms, using market research, such as interviews, surveys, or focus groups. They provide answers to the questions:
- "What does the customer expect from the product?"
- "Why does the customer buy the product?"
- Salespeople and technicians are also important sources of information – both in terms of these two questions and in terms of product failure and repair.
- Often these are expanded into secondary and tertiary needs/requirements.
- After identifying the needs, they are rated according to priority/importance to the customer.

For our mobile phone example, the following key customer needs and wants have been identified:

- ***Durability***: How resistant to damage is the phone in various situations?
- ***High-quality photos***: Can high-quality photos be produced in a range of conditions?
- ***Fast access***: How quickly can various functions be accessed?
- ***Easy to use***: How easy is it to navigate the various functions?
- ***Long battery life***: How long does the battery last under normal operating conditions?
- ***High-quality screen***: How clearly are images defined under various conditions?
- ***Stylish design***: Is the design "on-trend" with its look, feel, and functionality?

The customers' wants, with their associated importance rating, are included on the left side of the House of Quality in Figure 5.10. Durability is rated 2, quality photos 5, and so on, where a higher rating demonstrates greater desirability.

Step 2. Identify design attributes/requirements and the direction for improvement

Known as the *hows*, design attributes are expressed in the language of the designer/engineer and represent the technical characteristics (attributes) that must be deployed throughout the design, manufacturing, and service processes. These must be measurable because the output will be controlled and compared to objective targets.

In our mobile phone example, the design attributes, or "hows," are included in the central part of the House of Quality: Processor, Camera, Display, Access time, Storage, Power usage, Color range, and Thickness. The direction for improvement in the design attributes is shown by solid arrows above each column. So, for example, the upward pointing arrow for the "processor" indicates a desire for an increase in this attribute, while a downward arrow for "thickness" indicates a desire to reduce thickness.

The roof of the House of Quality shows, symbolically, the interrelationships between design attributes. It can show areas where there are conflicting targets of attributes, requiring compromise. For example, in Figure 5.10, there is a positive relationship between the camera and storage, indicating that as the quality of the camera increases, the storage requirement will also increase. There is a negative relationship between display and power usage.

Step 3. Relate the customer attributes to the design attributes

- Symbolically we determine whether there is no relationship, a weak one, a moderate one, or a strong relationship between each customer attribute and each design attribute.
- The purpose is to determine whether the final design attributes adequately cover customer attributes.
- The lack of a strong relationship between a customer attribute and any design attribute shows that the customer attribute is not adequately addressed or that the final product will have difficulty in meeting the expressed customer need.
- Similarly, if a design attribute does not affect any customer need, then it may be redundant, unnecessary, or the designers may have missed some important customer need.
- Consequently, the relationship analysis can prevent the product from being overbuilt (more than customers want) or underbuilt (less than customers want).

For example, in Figure 5.10, we see that for quality of photos there is a medium relationship with processor, a strong relationship with camera, a medium relationship with display, and a weak relationship with battery capacity.

Step 4. Determine the importance of each product design attribute.
The importance of each design attribute (the "hows") is calculated by adding the product of design rating and the rating for each of the "wants." In Figure 5.10, the "Processor" importance is $(5\times3)+(3\times5)+(4\times3) = 42$. The relative importance of each design attribute is calculated as a percentage of the sum of the importance scores. For "Processor" this is 42/252 or 16%.

Step 5. Analyze the interrelationships between the product design attributes
As shown in Figure 5.10, there is a strong relationship between the "processor" and "fast access" and a moderate relationship between "display" and "easy to use." There is no relationship between "access time" and "durability."

Step 6. Conduct an evaluation of competing products

- This step includes identifying importance ratings for each customer attribute and evaluating existing products/services for each of the attributes.
- Customer importance ratings represent the areas of greatest interest and highest expectations as expressed by the customer.
- Competitive evaluation helps to highlight the absolute strengths and weaknesses of competing products.

The right side of Figure 5.10 shows a comparison of two competitor products against the "own product" based on the customer "wants." Although the "own product" is relatively strong on "screen quality," "battery life," "fast access," and "durability," it is relatively weak on "ease of use" and "quality of photos." This points to a focus on the camera and displays design attributes for further improvement.

Step 7. Identify design attributes for improvement and develop targets

- This is usually accomplished through in-house testing and then translated into measurable terms.
- The evaluations are compared with the competitive evaluation of customer attributes to determine the inconsistency between customer evaluations and technical evaluations.
- For example, if a competing product is found to better satisfy a customer attribute, but the evaluation of the related design attribute indicates otherwise, then either the measures used are faulty, or the product has an image difference that is affecting customer perceptions.
- On the basis of customer importance ratings and comparison with competitor products, targets and directions for each design attribute are set.

So, for our mobile phone example in Figure 5.10, we have already noted that the current design is weak in its ease of use. Referring to House of Quality, we see that Display and Access Time are key design attributes influencing ease of use. Both of these design attributes have relatively high levels of importance across all customer attributes (18% and 16%, respectively). Based on technical design input, specific targets are set with a specific focus on key design attributes (6.5 inches for the display and a panel rating of at least 8/10 for access time).

Benefits and Limitations of QFD

Benefits include:

- It involves a team approach, bringing designers and engineers/developers together, leading to consensus, and promoting cross-functional discussion.
- As organizations grow, it is common for product designers/engineers/developers to lose customer focus. QFD focuses the product innovation team on customer requirements.

- It provides a structured basis for defining product design specifications and engineering design requirements from customer requirements.
- As development unfolds and changes in the design are suggested, given appropriate feedback from market testing, QFD prevents losing sight of the original product concept, customer requirements, and priorities.

Limitations include:

- It can be cumbersome (large numbers of whats and hows create large tables) and slow to go through the complete QFD process.
- It can feel tedious, and people may lose sight of what they are seeking to achieve in the product design.
- As consumer needs evolve swiftly, effort must be focused on identification of the correct design specifications. This is critical when new technologies appear to support the product category where the product is competing.

5.3.3 Design of Experiments (DOE)

Design of experiments (DOE) is used to find cause-and-effect relationships. It is a systematic method to determine the relationship between factors affecting a process and the output of that process. DOE can be applied to both a product and its manufacturing process. It provides the basis for fully understanding the product or process.

- Which variables have the greatest effect?
- How can I expect the product or process to change if one of the input variables is altered?
- Are there any interactions among the variables?
- What are the optimal variable conditions?

Experimental design can be an extremely complex subject to fully understand, requiring a sound knowledge of statistics. Figure 5.11 shows the major levels of experimental design, related to the popular Taguchi method. The Taguchi method is used to design efficient and reliable products at the level of quality customers expect. The advantage of the Taguchi method is that it requires relatively basic statistics to apply experimental design compared to other approaches.

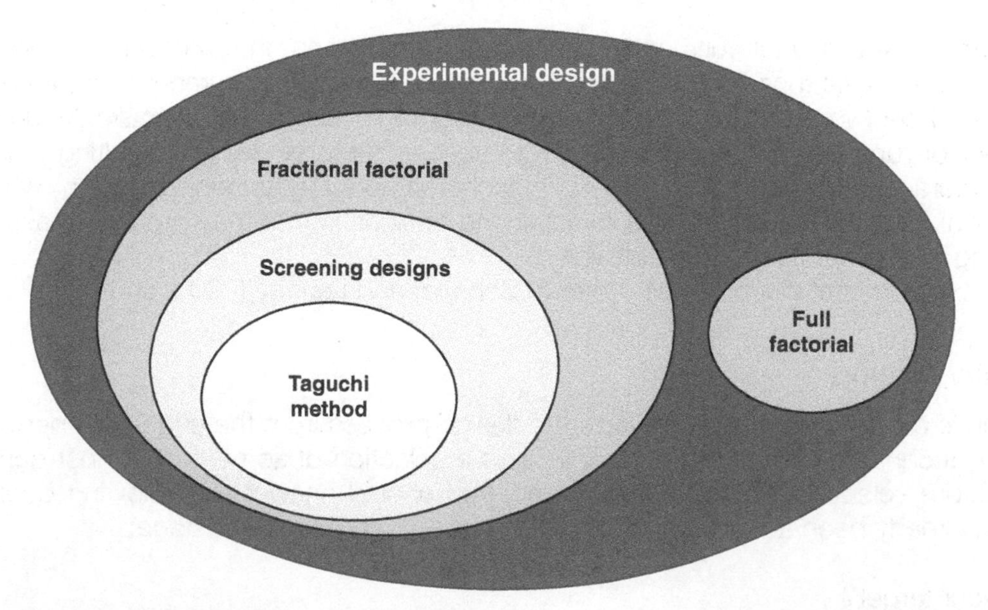

FIGURE 5.11 Experimental design

Let's examine the simple process of making bread to understand the basics of experimental design. In bread making, the proofing process is the final rise of the dough before baking. Three variables, time, temperature, and yeast weight, affect the volume of the loaf of bread during the proofing process.

A traditional and straightforward way of determining the effects of each variable would be to change the variable levels, one at a time, and observe the impact on loaf volume. This will certainly provide some useful information but misses one valuable consideration – the interaction among the variables. For example, both temperature and time have an impact on yeast activity; they are interdependent. The application of experimental design addresses this issue.

Table 5.5 shows how a full factorial design might be applied to the bread proofing example. In setting up this design, the following steps would be taken:

1. The input variables were specified – time, temperature, and yeast weight.
2. High and low levels were defined for each input variable.
3. A number of "experimental runs" were defined (all combinations of the three variables at the high and low levels). This requires eight runs.
4. All eight experimental runs are carried out. That is, eight different loaves of bread. The volume of each loaf is measured, and the results are analyzed.

TABLE 5.5 Factorial design to study the effect of proofing time and proofing temperature on specific volume of bread

Run number	Proof time (min)	Proof temp (°C)	Yeast weight (g)
1	60	30	1
2	120	30	1
3	60	50	1
4	120	50	1
5	60	30	2
6	120	30	2
7	60	50	2
8	120	50	2

Although eight experimental runs may not be a problem, as the number of variables increases, the number of experimental runs for a full factorial may be prohibitive. The number of experimental runs = 2^n, where n is the number of variables. So, 4 variables would require 16 runs, 5 would require 32 runs, and so on.

The number of runs can be reduced by using fractional factorial design, resulting in less information regarding the interaction among variables. The Taguchi method is a screening design, which is a type of fractional factorial. The advantage of Taguchi screening designs is that they provide a taste of the subject without requiring a university degree in statistics.

For further experimental design details, refer to Schmidt and Laundy (2005) and Roy (2001).

5.3.4 Usability Testing

Usability testing is critically important to inform the design process from the end user's perspective. Several usability testing techniques are available. Following is a selection of some of the most commonly applied techniques. Before selecting the technique to use, you should answer the following questions, most of which will have already been addressed in the earlier concept development stage:

- Who is your target?
- What specific information do you need to aid in the further design process?

- How critical is it to get the best information? Is "close enough" good enough?
- How much money can you afford to spend?
- How much time can you afford to get the results?

Guerrilla Testing

Guerrilla testing is the simplest form of usability testing. Basically, guerrilla testing means going into a public place such as a supermarket or a park to ask people about a product prototype. They are asked to perform a quick usability test. It is a low-cost and relatively simple testing that enables real user feedback.

When to use: Guerrilla testing works best in the early stages of the product development process when you have a tangible design (wireframes or basic prototypes) and want to know if you are heading in the right direction with the design. Guerrilla testing is also good for collecting personal opinions and emotional impressions about ideas and concepts. But Guerrilla testing, by definition, might not represent your product's target audience. Consequently, it may not be right to test niche products.

Laboratory (lab) Usability Testing

As the name suggests, lab usability testing is testing run in special environments (laboratories) and supervised by a moderator. A moderator is a professional who is looking to obtain feedback from live users. During a moderated test, moderators facilitate test participants through tasks, answering their questions, and replying to their feedback in real-time.

When to use: Lab usability testing works best when you need to have in-depth information on how real users interact with your product and what issues they face. It will help you investigate the reasoning behind user behavior. The fact that this testing is moderated versus unmoderated enables you to collect more qualitative information. At the same time, lab testing can be expensive to organize and run because you need to secure an environment, attract test participants, and hire a trained moderator who is skilled in facilitating (not leading) responses and in observing, interpreting, and reporting on participant reactions.

Unmoderated Remote Usability Testing

Unmoderated remote usability testing occurs remotely without a moderator. It offers quick, robust, and inexpensive user testing results to guide product design. Test participants are asked to complete tasks in their own environment using their own devices and without a moderator present, which leads to the product being used naturally. The cost of unmoderated testing is lower; however, this type of testing offers less detailed testing results.

When to use: Unmoderated remote usability testing works best when you need to obtain a large sample to prove critical findings from your initial moderated research. In other words, you have a particular hypothesis that you want to validate on a large segment of your users. Unmoderated remote usability testing will help you test a particular question or observe user behavior patterns.

Contextual Inquiry

Contextual inquiry is an interview and observation method for general customer research that provides insights into their wants and needs as well as usability insights. Test participants (real users) are first asked a set of questions about their experience with a product and then observed and questioned while they work in their own environments. This provides the design team with information about the user experience from real users.

When to use: Contextual inquiry is useful for getting rich information about users in their workspace, personal preferences, and habits. It is particularly useful at the beginning of the design process, during the development and refining of the concept description. It helps the product team design a well-tailored experience or improve the experience for existing products.

Online Testing

Online concept testing is the process of evaluating product concepts and providing feedback on design features. It can include both quantitative and qualitative research via surveys and online communities. The main quantitative technique is an online concept survey. You can reach thousands of targeted and engaged respondents through online surveys, providing rich and robust data for analysis. In the survey, the concept test would involve consumers reviewing a text description or a visual representation of the concept. Then, the audience would be prompted to answer questions or to discuss their impressions of your idea. Qualitative techniques can either use an online community – a carefully curated space to engage respondents in a group setting, as well as one-to-one tasks – or run a series of in-depth interviews over video chat, in which you have the ability to share concepts on the screen.

When to use: Online testing is a way to collect feedback from test participants scattered around different parts of the globe at a lower cost compared to other usability testing. It allows you to reach several thousand users to obtain general feedback on ideas or concepts or to engage with specific communities of users to obtain rich information from targeted users. These communities can continue to provide input as the product concept evolves in detail and future versions are launched.

Alpha, Beta, and Gamma Testing

Alpha testing is an internal check done by the in-house development or QA team. It is commonly used for software products to discover software bugs before external users test the software in beta testing. Software behavior is assessed and verified in controlled conditions, often using automated and manual tests.

Beta testing has also been called pre-release testing or field testing. It is conducted by a limited number of users external to the organization, called *Beta-Testers*, before the official product delivery. The main purpose of beta testing is to verify software compatibility with actual user environments (different software and hardware configurations, types of network connection, etc.) and to get the users' feedback on software usability and functionality. This is testing in an uncontrolled environment to understand how the product responds under real-world conditions.

Gamma testing is the final stage of the testing process conducted before software release. It makes sure that the product is ready for market release according to all the specified requirements. Gamma testing focuses on software security and functionality. During gamma testing, the software does not undergo any modifications unless the detected bug is of high priority and severity. Only a limited number of external users perform gamma testing. The checking includes the verification of certain specifications, not the whole product. Feedback received after gamma testing is considered as updates for upcoming software versions. In place of gamma testing, some organizations perform alpha testing again after a beta test and then launch the product.

5.3.5 Performance and Endurance Testing

Performance and endurance testing is a type of testing that determines a product's performance, such as durability, speed, scalability, handling, and stability. The testing process includes manual and mechanical endurance tests designed to reveal faults and ensure your product is fit for market. Ongoing application of performance and endurance testing throughout the design process, with cross-functional team involvement, is critical to minimizing the costs of product failure when it reaches the market.

Failure Mode and Effects Analysis (FMEA)

Failure modes and effects analysis (FMEA) is a step-by-step approach for identifying possible failures in a design, a manufacturing or assembly process, or a product. *Failure modes* means the ways, or modes, in which something might fail. Failures are any errors or defects, especially ones that affect the customer and can be potential or actual. *Effects analysis* refers to studying the consequences of those failures.

FMEA can be applied to a range of products and services, including automobiles, pharmaceuticals, healthcare, banking, and many others:

- When an existing process or product is being applied in a new way
- Before developing control plans for a new or modified process
- When improvement goals are planned for an existing process or product
- When analyzing failures of an existing process or product

Basic steps to complete an FMEA:

1. Compile a list of failures that could happen in each step of the process.
2. Measure the impact of every failure by asking: What happens when this failure occurs?
3. Determine how severe the failure is by rating it on a ten-point scale. A 1 is a defect the customer doesn't even notice; a 10 is a catastrophe.
4. Rate the likelihood of failure using a ten-point scale. 1 indicates no chance; 10 is guaranteed failure.
5. Establish the probability of discovering the error before the customer does, using a 10-point scale. If a failure is easy to discover, it is rated a 1. Undetectable failures are assigned a 10.

Sometimes FMEA is extended to FMECA (failure mode, effects, and criticality analysis) although today, criticality is included in most definitions of standard FMEA.

Endurance Testing

Performance and endurance testing is a type of testing that determines a product's performance, such as durability, speed, scalability, handling, and stability. The testing process includes manual and mechanical endurance tests designed to reveal faults and ensure your product is fit for market.

Mechanical endurance testing: Electronics, automotive parts, and other heavy-use items endure extreme stress levels daily through repeated and prolonged use. When a product is subjected to this repeated stress, it can be prone to failure even if it never reached its static stress limitations. As an example, the number of times a car door can be opened and closed before failing is determined with mechanical endurance testing. Although calculating static mechanical stress limits is useful, it fails to evaluate a product's performance over its expected lifetime, as it can't account for damage stemming from prolonged use in different environmental conditions. Mechanical endurance testing is thus needed to measure this damage over a product's lifespan.

Many types of mechanical endurance testing can be performed to emulate the entire life cycle of a product, depending on industry standards and unique product needs. Some of the most common types of testing include fatigue, push-pull-press, and open-close testing:

- ***Fatigue testing***: Experts assert that fatigue causes between 50 and 90 percent of all mechanical failures. While sources vary on the exact percentage, fatigue analysis of engineering material is vital to ensuring a finished product's longevity. In fatigue testing, engineers determine localized and progressive structural damage caused when a material experiences cyclic loading. Tests measure the product's resistance to cyclic stress over several stress levels.
- ***Push-pull-press testing***: Also known as tensile, compression, and impact testing, push-pull-press tests calculate a material's resistance to heavy impact from tension or compression. Tests measure

the force required to pull, push or press an object to its breaking point, which determines the maximum amount of tensile stress a material can undergo before failure. These tests are particularly useful in industries such as mechanical and structural engineering.
- ***Open-close testing***: Hinges on doors, windows, flip phones, garage door openers, and many other products need to be tested extensively to ensure durability over many years. Open-close automation testing machinery quickly simulates years' worth of use.

Software endurance testing: Sometimes referred to as long duration or SOAK testing, determines whether a system can sustain a continuous expected load. Endurance tests can last anywhere between a few hours or a few weeks, depending on typical application usage patterns. The following are some examples of software endurance testing:

- ***Test memory leakage***: Checks are done to verify if there is any memory leakage in the application, which can cause the crashing of the software or operating system over time.
- ***Test connection closure between the layer of the system***: If the connection between the layers of the system is not closed successfully, it may stall some or all modules of the system.
- ***Test database connection close successfully***: If the database connection is not closed successfully, it may result in a system crash.
- ***Test response time***: The system is tested for the response time of the system as the application becomes less efficient as a result of the prolonged use of the system.

5.3.6 Quality Assurance

Quality assurance is a program for systematically monitoring and evaluating the various aspects of a project, service, or facility to ensure that standards of quality are being met. Quality assurance creates clear expectations for everyone. A quality assurance process sets clear expectations and standards from the start of the project. There's less room for error or misunderstandings when people know what the expectations are early in the project.

Six Sigma

Six Sigma aims to reduce variations in a business and manufacturing process via dedicated improvements in the various processes. This requires a sustained commitment from all members of the team. Applications of Six Sigma that focus on the design or redesign of products and their enabling processes to meet customer needs and expectations are known as Design for Six Sigma (DFSS). The phases or steps of DFSS are not universally recognized or defined – companies and Six Sigma training organizations have defined DFSS differently. Many times a company will implement DFSS to suit their business, industry, and culture; other times, they will implement the version of DFSS used by the consulting company assisting in deploying Six Sigma. Because of this, DFSS is more of an approach than a defined methodology.

Design methodologies for six sigma

Several methodologies exist for six sigma design applications. Following is a brief description of four methodologies, which are also summarized in Table 5.6. The differences between these methodologies are difficult to appreciate fully. For further information, the American Association of Quality has helpful resources (https://asq.org).

DMAIC

DMAIC is a data-driven quality strategy used to improve processes. It is an integral part of a Six Sigma initiative, but in general, it can be implemented as a standalone quality improvement procedure or as part of other process improvement initiatives such as Lean. The DMAIC methodology should be used when a product or process is in existence at your company but is not meeting customer requirements or is not performing adequately. DMAIC is an acronym for the five phases that make up the process:

- **Define** the problem, improvement activity, opportunity for improvement, project goals, and customer (internal and external) requirements.
- **Measure** process performance.
- **Analyze** the process to determine the root causes of variation and poor performance (defects).
- **Improve** process performance by addressing and eliminating the root causes.
- **Control** the improved process and future process performance.

DMADV

DAMDV is focused on the process of designing a new product, service, or process.

The five phases of DMADV are:

- **Define** the project goals and customer (internal and external) requirements.
- **Measure** and determine customer needs and specifications; benchmark competitors and industry.
- **Analyze** the process options to meet customer needs.
- **Design** the detailed process to meet customer needs.
- **Verify** the design performance and ability to meet customer needs.

IDOV

IDOV is a specific methodology for designing new products and services to meet Six Sigma standards. Its main difference from DMADV is a focus on meeting the financial goals of an organization. IDOV is a four-phase process:

- **Identify** customer needs and strategic intent.
- **Design** the detailed design by evaluating various design alternatives.
- **Optimize** the design from a productivity (business requirements) and quality point of view (customer requirements) and realize it.
- **Validate** by piloting the design, updating as needed, and preparing to launch the new design.

DMEDI

DMEDI is a creative approach to designing new robust processes, products, and services. It is focused on obtaining significant competitive advantages, or quantum leaps over current environments. DMEDI projects typically require more time and resources to complete. Its five-stage process is:

- ***Define*:** The problem, improvement activity, opportunity for improvement, project goals, and customer (internal and external) requirements are defined and understood.
- ***Measure*:** The measure phase in DMEDI requires more examination than in DMAIC because there is little, if any, existing process definition. To meet the need for clearly defined customer requirements, a quality function deployment (QFD) or house of quality is used in multiple iterations and phases to properly define a product or service that the customer truly desires.
- ***Explore*:** The Explore phase is focused on delivering a conceptual design for a new process. By contrast, the Analyze phase of the DMAIC process looks to break down data for an existing process to identify defect root causes. The DMEDI phase is conceptual; the DMAIC phase is tangible.
- ***Develop*:** DMEDI develops an optimal design based largely on meeting customer desires. In contrast, DMAIC delivers a rational future-state design based on statistical or mathematical proof.
- ***Implement*:** This phase validates the capability of the proposed process to meet or exceed project objectives and identify problems. This is done through a permanent, full-scale pilot deployment rather than the temporary pilot, small-scale of the DMAIC methodology.

TABLE 5.6 Design for Six Sigma methodologies

DFSS Methodologies			
DMAIC	**DMADV**	**DMEDI**	**IDOV**
Define	Define	Define	Identify
Measure	Measure	Measure	Design
Analyze	Analyze	Explore	Optimize
Improve	Design	Develop	Validate
Control	Verify	Implement	

5.3.7 Design for Manufacturing and Assembly (DfMA)

Following is an overview of the key elements of the Design for Manufacturing and Assembly process. Readers are referred to Ulrich and Eppinger (2016) for more extensive coverage.

Design for Manufacturing and Assembly (DfMA) is an amalgam of two processes, *design for manufacturing* (DfM) and *design for assembly* (DfA).

Design for Manufacturing (DfM) is a design method to reduce the complexity of manufacturing operations and the overall cost of production, including the cost of raw materials.

Design for Assembly (DfA) is a design method to facilitate or reduce the assembly operations of parts or components of a product.

Designing for Manufacturing and Assembly (DfMA) is a critical part of the product development cycle. It involves optimizing the design of a product's manufacturing and assembly process and merging the design requirements of the product with its production method. Employing DfMA reduces the cost and difficulty of producing a product while maintaining its quality.

Overview of the DfMA Process

Ulrich and Eppinger (2016) suggest a five-step process:

1. Estimate the manufacturing costs
2. Reduce the costs of components
3. Reduce the cost of assembly
4. Reduce the costs of supporting production
5. Consider the impact of DFM decisions on other factors.

Basics of Effective DfMA

Standardization: Standardization cuts costs by reducing inventory and scale-up impacts. Here are some ways to think about part standardization:

- Design parts that can be reused within a product or shared between product lines.
- Standardize hardware within a product and across products to reduce inventory needs.
- Make design modular to simplify product changes or redesigns.
- Use standard components instead of custom-made ones, where possible.
- Consider the future availability of parts over the expected life of the product in the market.

Design simplicity: Simplifying the design cuts down on time, inventory, and cost to manufacture the product:

- Minimize assembly steps and inventory by making multifunctional parts.
- Use designed-in alignment or quick securing methods like snap fits. Fastening techniques like bolting or gluing take longer to secure and require more inventory.
- Test the design with working prototypes, for example, using 3D printed prototypes.

Alignment and compliance: Errors in alignment can damage parts or equipment, reducing yield or even causing line shutdown.

- Resolve assembly issues by analyzing how tolerances will impact assembly.
- Design integrated part features to help alignment during assembly.
- Add tapers or chamfers to guide components during assembly insertion steps.

Set-up time reduction: Reduce set-up time by reducing the number of operations required per part or assembly.

- Reduce the number of setups or rotations required per part or assembly,
- Assess where the manufacturing line could be upgraded with improved tools or workstations.

DfMA Performed Throughout the Product Innovation Process

DfMA should form an integral part of the product innovation process. The importance of cross-functional involvement in product innovation has been emphasized throughout this book. Too often, manufacturing input is left until too late in the product innovation process. Effective DfMA requires manufacturing input from an early stage and throughout the process:

- In concept development to ensure manufacturing capability and capacity is available.
- In financial feasibility to contribute to manufacturing cost estimates.
- In detailed design specifications to provide advice on design detail that will ensure optimal product quality and minimal costs in manufacturing.
- In scale-up to manufacture through intimate knowledge of the product's design and commitment to its success.

5.3.8 Design for Sustainability

The concept of design for sustainability applies a combination of environmental, social, and economic priorities and considerations during design and development, and over the life of a product. Product conceptualization and creation, the first step in the design process, has a major role in defining how sustainability is incorporated in contrast to the final stages of the product development process. Design for sustainability is design with the intention of achieving sustainable outputs of the product or service and requires a holistic view and systems perspective. The three pillars of sustainability – known as the triple bottom line (TBL) – stand for: people (social equity), profit (economy), and planet (environment) (see Section 2.9). Sustainability occurs when organizations are committed to designing products based on the notion that a profitable organization depends on a healthy society and both are sustained in a healthy environment.

The Principles of Sustainable Product Design

The principles of sustainable product design are:

- Use non-toxic, sustainably produced, or recycled materials with low environmental impact.
- Use manufacturing processes to produce products that are energy efficient.
- Build longer-lasting and better-functioning products.
- Design products for reuse, easy disassembly, and recycling.
- Use life cycle analysis tools to design more sustainable products.
- Shift the consumption mode from personal ownership of products to the provision of services that provide similar functions. Some examples are Interface Carpets (carpet tiles do not require much trimming during installation and create less waste), Xerox (copier leasing rather than purchase), and Zipcar (car sharing).
- Materials should come from sustainably managed renewable sources that can be composted when their usefulness is exhausted.

Sustainability Guidelines

Sustainability guidelines are specific listings of specific issues to take into account during the development process. They suggest aspects to take into account and are considered a checklist that ensures awareness of sustainability issues in the development process. Some of these guidelines are:

SPSD framework

SPSD stands for sustainable product and service development. The approach implements sustainable product and service development throughout a product's life. It focuses on transforming the offering of products into services to reduce manufacturing.

ARPI framework

ARPI stands for analysis, report, prioritize, and improve. The approach ensures the implementation of eco designs and includes an assessment of the environment, reporting on these, ranking the considerations, and suggesting actions for improvement. An example of ARPI application in the electrical and electronics industry is provided by Simon et al. (2000).

MDE framework

This framework is a guide to material, design, and ecology (MDE). It emphasizes the selection of materials and their impact on product methods, function design, market requirements, price, and environmental impact.

Product sustainability index (ProdSI)

The ProdSI represents the level of sustainability built into a product using the TBL. Each cluster includes several factors assessed through specific metrics. A possible list of clusters and factors is shown in Table 5.7. Clusters are given a weight (from 0 to 100), and factors within each cluster are evaluated on a scale of 1–10, where numerals closer to 10 represent a higher level of compliance with the criteria.

TABLE 5.7 Clusters and factors to calculate Product Sustainability Index (PSI)

Clusters	Factors	
Product's Environmental Impact	Life cycle (useful life span) Environmental effect (toxicity, emissions) Residues	Ecological balance and efficiency Regional and global impact (CO_2, ozone depletion)
Product's Social Impact	• Operational safety • Health and wellness effects • Ethical responsibility • Social well-being (quality of life, peace of mind, life satisfaction)	• Employee safety and health • Education
Product's Functionality	• Service life - durability • Modularity • Ease of use • Maintainability/serviceability • Upgradability	• Ergonomics • Reliability • Safety • Functional effectiveness
Resource Utilization	• Energy efficiency and power consumption • Use of renewable source of energy • Material utilization and efficiency (content and hazardous material)	• Water use and efficiency • Installation and training costs • Operational cost (labor cost, energy, capital, etc.)
Product's Manufacturability	• Manufacturing methods • Assembly • Packaging	• Transportation • Storage
Product's Recyclability/ Manufacturability	• Disassembly • Recyclability and recovery	• Disposability • Remanufacturing/reusability

Source: Jawahir et al. (2015) / John Wiley & Sons.

Analytical Tools for Sustainability

Life cycle assessment (LCA)

This method has been used for eco-design for more than 30 years. LCA provides quantitative data on the product's environmental impact from cradle (extraction of materials) to grave (end of life). The method has four phases:

1. Determination of goals and scope of the LCA,
2. Inventory of energy and material inputs across all stages of the product's life,
3. Evaluation of the environmental impact associated with inputs and outputs during the life cycle, and
4. Interpretation of results to reach corrective decisions.

Additional details of the LCA process are shown in Figure 5.12.

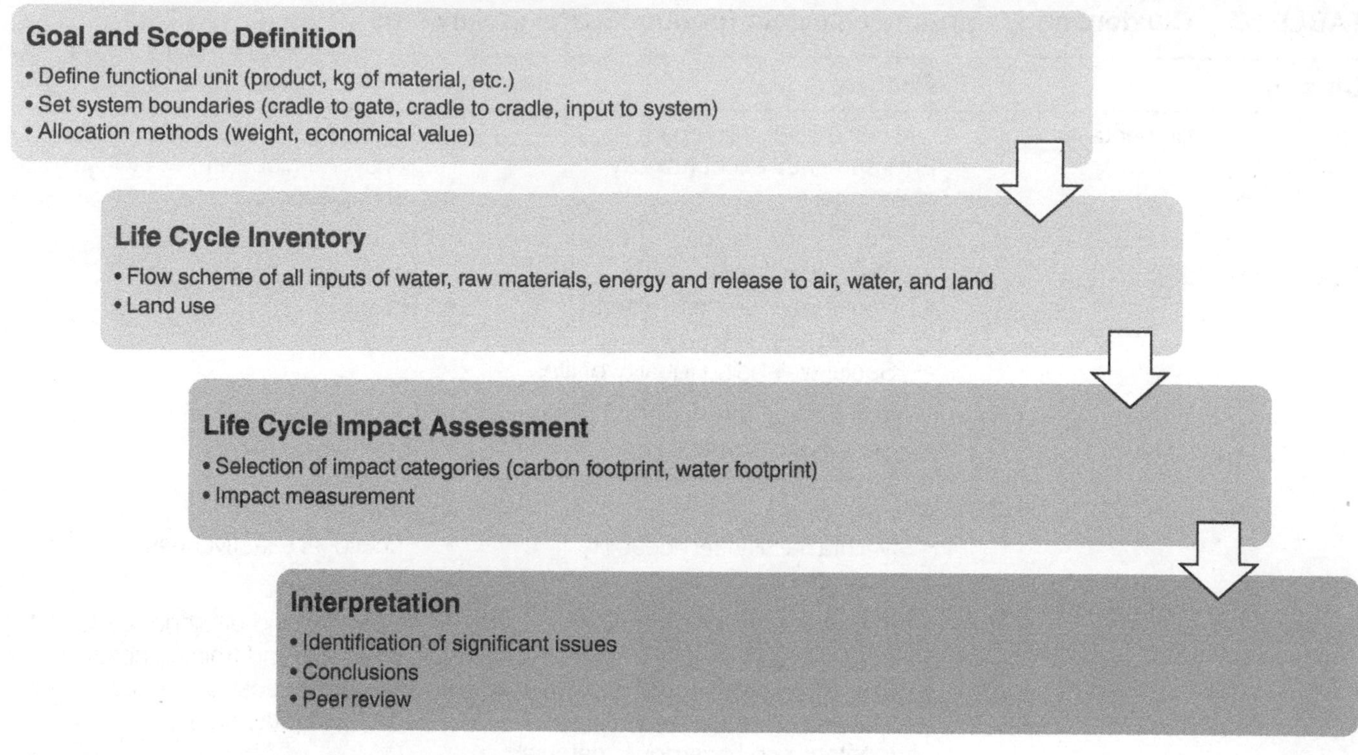

FIGURE 5.12 Outline of the process for life cycle assessment (LCA)

Simplified life cycle assessment (SLCA)

This method is a simplification of the LCA. Specific information and parameters during the evaluation of the environmental impact phase (i.e., inventory data) are not examined in order to reduce the amount of data gathered.

Life cycle costing (LCC)

This method involves the analysis of all the costs associated to produce a product, including the activities created by all the actors throughout the product's life (supplier, manufacturer, and consumer). External costs and end-of-cycle costs are neglected in the conventional LCC method. As such, the method is a cradle-to-grave cost analysis. All costs are used to estimate the net present value (NPV) through a discounted cash flow. Furthermore, the annual cost of owning, operating, and maintaining assets over their entire life known as Equivalent Annual Cost (EAC) is calculated.

Quality function deployment for environment (QFDE)

This method integrates QFD, benchmarking, and LCA to conduct environmental impact studies of products and their components. The QFDE follows the logic of axiomatic design, which considers that every design should follow four domains in sequence:

1. Customer domain,
2. Functional domain,
3. Physical domain, and
4. Process domain.

The QFDE method starts by connecting the typical customer requirements of the QFD with sustainable requirements to identify functional requirements. Later, these functional requirements assist in identifying key design parameters. Typically, the method has four phases:

1. Association of customer requirements (CRs) to engineering metrics (EM), known as technical or functional requirements,
2. Relationship of EMs to the product components,
3. Estimation of design changes in components or subsystems after the benchmarking process, and
4. Evaluation of the impact of changes in EMs on environmental quality requirements.

The framework for sustainable design using QFDE is shown in Figure 5.13.

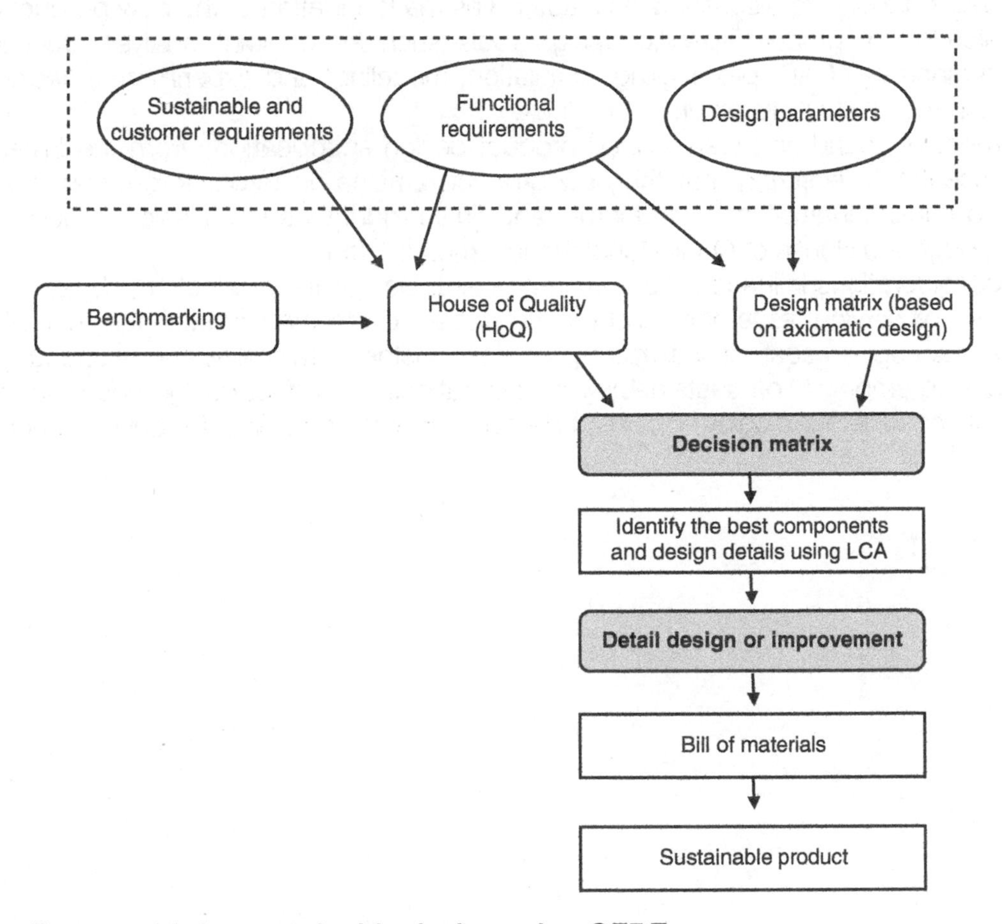

FIGURE 5.13 Framework for sustainable design using QFDE.
Source: Hosseinpour et al. (2015) / Emerald Publishing Limited.

5.4 IN SUMMARY

- Using appropriate design and development methodologies and tools is key to product innovation success. This chapter focused on methodologies applied during the design and development phases of product innovation. The design and development phases evolve the product from initial idea through to final design specification, ready for implementation or manufacture and launch. Market research tools, which also contribute to constructing a product concept, are discussed in Chapter 6.
- Idea generation (ideation) lays the platform for successful new products. Several techniques were discussed – from mind-mapping, storyboarding, personas, and ethnographic methods, to a day in the life of the customer. All these techniques allow for deep-dives into the consumption experience. Further ideation tools are discussed in Chapter 3 as the foundation of the new product portfolio.
- The application of product-specific design tools such as conjoint analysis, functional analysis, reverse engineering, CAD, prototyping, simulation, modeling, and experimental design are essential resources in the product development toolbox.
- Development of sound and well-defined product design specifications from the initial product concept is essential to ensuring that the new product remains on target throughout the development process to implementation or manufacture and final product launch. Design specifications are presented through the House of Quality and Taguchi methodology.
- Several considerations influence the design process. Design for production, design for functionality, and design for service, assembly, and maintenance provide the product designer with critical perspectives that support cost minimization, product robustness, reliability, and reaching quality targets.
- The increasing emphasis on sustainability considerations in product design necessitates the application of tools and methodologies, including life cycle assessment, QFD for environment, and life cycle costing.

5.A DESIGN & DEVELOPMENT PRACTICE QUESTIONS

1. Product design specifications are primarily intended to:
   A. Translate qualitative design features into quantitative parameters.
   B. Identify the core benefits of a product.
   C. Identify customer needs.
   D. List the tangible features of a product.
2. The House of Quality relates the voice of the customer to the voice of:
   A. Marketing.
   B. Advertising.
   C. Engineering.
   D. Manufacturing.
3. Which ideation tool used in new product development generates a graphical output linking various pieces of information, ideas, and thoughts?
   A. Mind-mapping.
   B. Brainstorming.
   C. Brainwriting.
   D. Delphi technique.
4. Which diagrams are constructed using How-Why logic?
   A. FAST.
   B. Functional Analysis.
   C. Conjoint.
   D. Bubble Charts.
5. Target product design specifications must use a set of metrics to ensure they are met during the design work. Suppose you are designing a new automobile. Which of the following is an acceptable product design metric?
   A. The seats are comfortable.
   B. The car accelerates from 0 to 60 miles per hour in less than 6 seconds.
   C. The towing capacity is adequate for most people to use a boat trailer.
   D. There are a wide range of colors and fabrics available for the interior.
6. DFF, DFP DFA, DFM, DFU are approaches to determine. . .
   A. Concept definition.
   B. Product design specifications.
   C. Manufacturability.
   D. Embodiment design.
7. The product concept description should include which three elements?
   A. Customer needs, environmental factors, customer use reports.
   B. Qualitative description, quantitative parameters, and technical deliverables.
   C. Core benefits, tangible attributes, and augmented features.
   D. Competitive benchmarks, concept description, qualitative measures
8. What is the pattern in TRIZ problem-solving matrix?
   A. Identify the specific problem first, then the general problem, and identify a general solution before a specific solution.
   B. Identify the specific problem first, test a solution with a targeted customer group, generate more prototypes, select specific solution.
   C. First use a general problem with general solution, then select the specific problem to solve.
   D. Identify a cross-functional team that can test various product solutions with customers and select the highest rated solution.

9. Jack is a consultant to a manufacturing company. He has been asked to evaluate the components and parts of the product as well as the whole design (system view) and the method of assembly used (manual, automatic, fixed automation, and robotic assembly) to reduce the number of parts used. What specific design tool should Jack use?
   A. Design for manufacturing.
   B. Design for assembly.
   C. Design for functionality.
   D. Design for life.
10. What specific design tool uses the following methods: Think-aloud laddering, Quantification Theory I (QTI), PLS analysis, and Genetic and Fuzzy logic?
    A. Taguchi.
    B. Six Sigma.
    C. Kansei.
    D. TRIZ.

Answers to practice questions: Product design and development

1. A
2. C
3. A
4. A
5. B
6. B
7. C
8. A
9. B
10. C

REFERENCES

Altshuller, G. (1984). *Creativity as an exact science*. Gordon & Breach. https://doi.org/10.1201/9781466593442

Burchill, G. and Walden, D. (1994). Mutual learning: Industry/academia collaboration for improved product development. *Center for Quality Management Journal 3* (2): 23–39.

Gadd, K. (2011). *TRIZ for engineers: Enabling inventive problem solving*. John Wiley and Sons, Ltd. https://doi.org/10.1002/9780470684320.

Haines-Gadd, L. (2016). *TRIZ for dummies*. Amazon.

Hauser, J.R. and Clausing, D. (1988). The house of quality. *Harvard Business Review 66* (3): 63–73.

Hosseinpour, A., Peng, Q., and Gu, P. (2015). A benchmark-based method for sustainable product design. *Benchmarking: An International Journal 22* (4): 643–664. https://doi.org/10.1108/BIJ-09-2014-0092.

Jawahir, I.S., Wanigarathne, P.C., and Wang, X. (2015). Product design and manufacturing processes for sustainability. In: *Mechanical engineers' handbook* (ed. M. Kutz), 414–443. John Wiley & Sons https://doi.org/10.1002/9781118985960.meh207.

Knudsen, M.P., von Zedtwitz, M., Griffin, A., and Barczak, G. (2023). Best practices in new product development and innovation: Results from PDMA's 2021 global survey. *Journal of Product Innovation Management 40* (3): 257–275. https://doi.org/10.1111/jpim.12663.

Ladewig, G.R. (2007). TRIZ: The theory of inventive problem solving. In: *The PDMA toolbook 3 for new product innovation* (ed. A. Griffin and S. Somermeyer), 3–40. John Wiley and Sons https://doi.org/10.1002/9780470209943.ch1.

Rodriguez, C.M. (2017). *Product design and innovation: Analytics for decision making*. Createspace Publishers.

Roy, R.K. (2001). *Design of experiments using the Taguchi approach: 16 steps to product and process improvement*. John Wiley & Sons Inc.

Schmidt, S.R. and Launsby, R.G. (2005). *Understanding industrial designed experiments*, 4e. Air Academy Press.

Simon, M., Pool, S., Sweatman, A. et al. (2000). Environmental priorities in strategic product development. *Business Strategy and the Environment 9* (6): 367–377. https://doi.org/10.1002/1099-0836(200011/12)9:6<367::AID-BSE262>3.0.CO;2-D.

Ulrich, K.T. and Eppinger, S.D. (2016). *Product design and development*, 6e. McGraw-Hill Education.

6

Product Innovation

Market Research

Required to provide market-related information and data to underpin decision-making in all aspects of strategy development, portfolio management, the product innovation process, and life cycle management

Product Development and Management Body of Knowledge: A Guidebook for Product Innovation Training and Certification, Third Edition.
Allan Anderson, Chad McAllister, and Ernie Harris.

WHAT YOU WILL LEARN IN THIS CHAPTER

Market research is critical to informed decision-making at all levels of product innovation management. This chapter presents a range of market research methods that can assist decision-making at various stages of product innovation. The strengths and weaknesses of each technique are discussed, together with how each contributes to specific innovation management decisions.

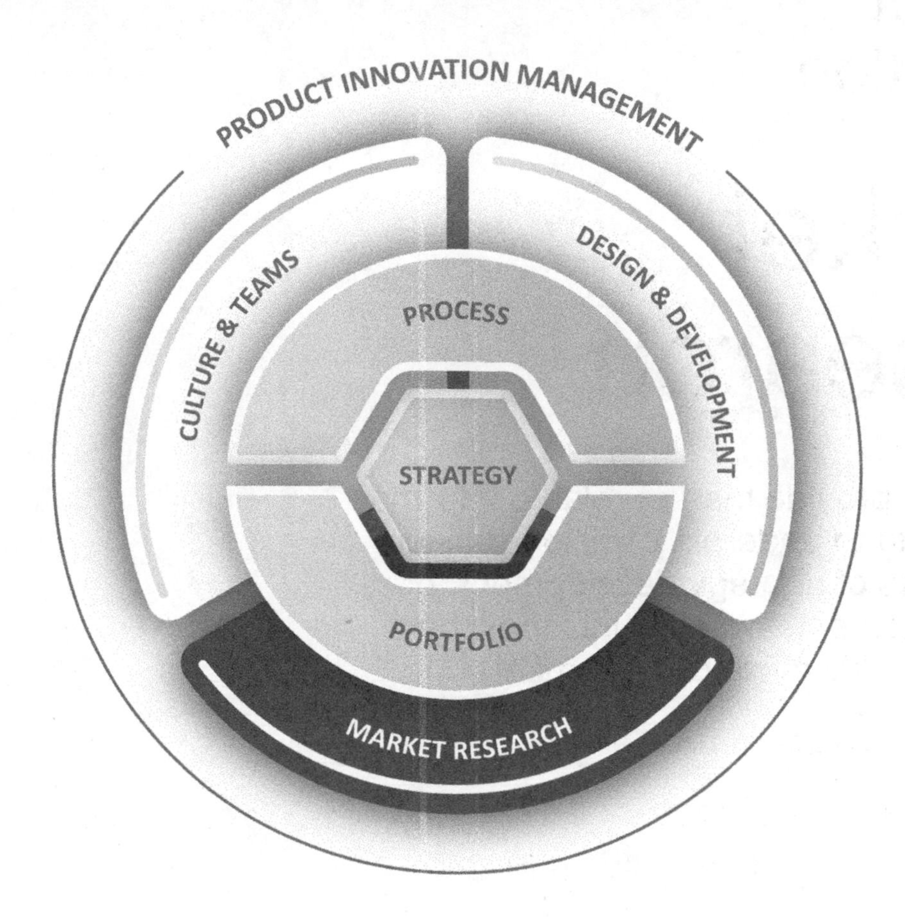

THE CHAPTER ROADMAP

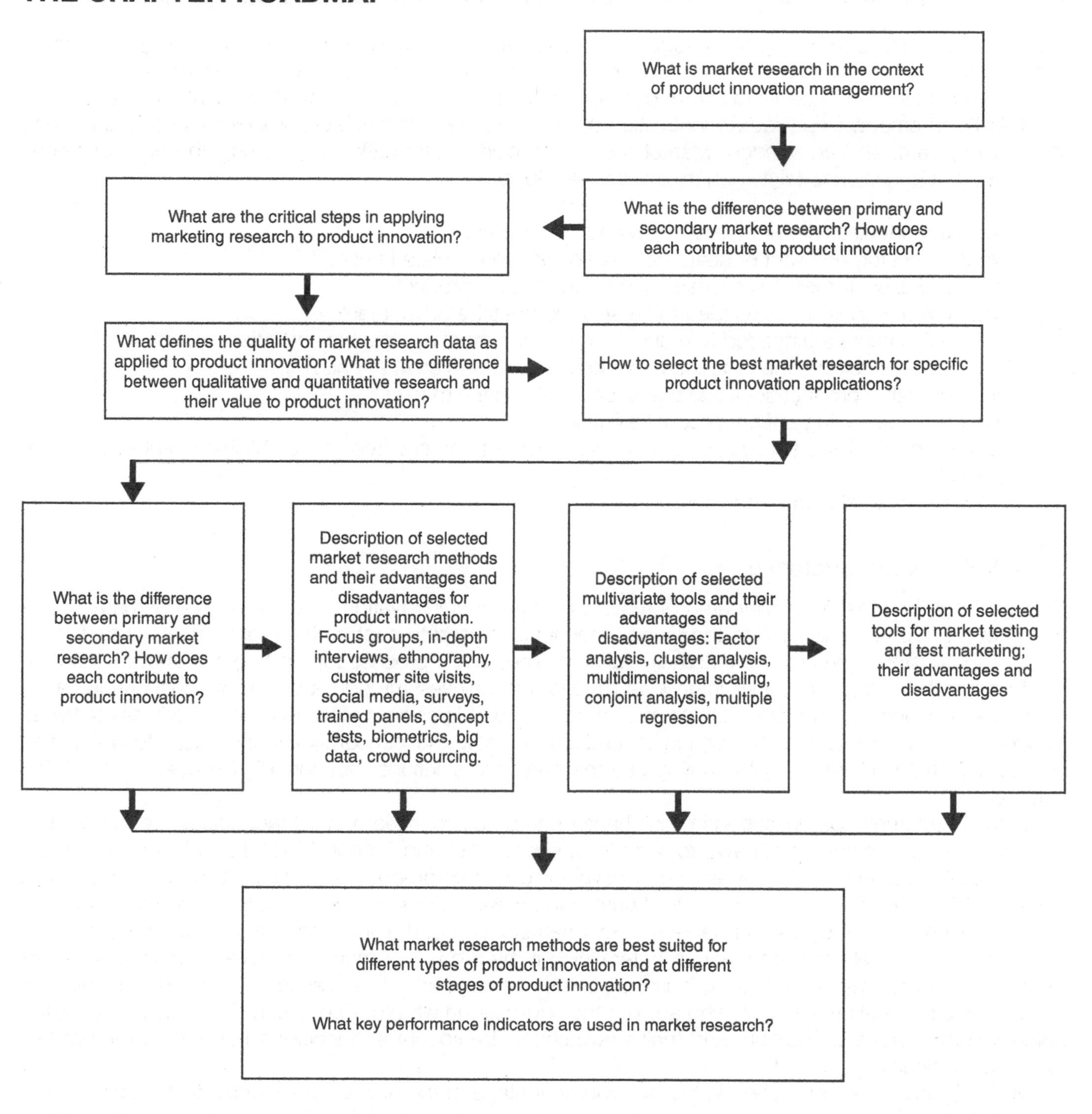

6.1 HOW MARKET RESEARCH CONTRIBUTES TO PRODUCT INNOVATION

Understanding and addressing the needs of stakeholders, typically with a focus on customer needs, is essential to developing successful new products and improving existing products. A wide range of market research techniques is available to inform decision-making throughout the product innovation process.

This book emphasizes product innovation as a risk vs. reward decision process founded on sound information, data, and knowledge. Market research provides information for making the right decisions and, in turn, reduces uncertainty and minimizes risk. Specifically:

- What opportunities are out there – now and in the future?
- What do customers want or need – articulated vs. unarticulated needs?
- What drives customers to purchase and re-purchase a product?
- What value proposition is needed to create a competitive advantage?
- What refinements are needed to make the product a more acceptable and desirable solution?
- Will the customer buy the product, how often, where, and at what price?
- Are there other product solutions that already exist in the market?
- What is the advantage of our product solution?
- Can the product solution be protected and advantage be secured by establishing intellectual property rights?
- Is the product solution sustainable?

6.1.1 Voice of the Customer

One of, if not the most important area of research for product innovation, is voice of the customer (VoC), a market research term applied to the range of techniques used to capture the requirements and/or feedback of the customer (internal or external). VoC is defined as "a process for eliciting needs from consumers that uses structured in-depth interviews to lead interviewees through a series of situations in which they have experienced and found solutions to the set of problems being investigated. Needs are obtained through indirect questioning by coming to understand how the consumers found ways to meet their needs, and, more important, why they chose the particular solutions they found" (Belliveau et al., 2002, Chapter 11).

Many of the methods described in this chapter can be categorized as VoC techniques. For a review of VoC techniques as a source of new product ideas, see Cooper and Dreher (2015). In addition to researching the final customer or consumer, when developing a new product or service, it is also essential to consider other key stakeholders in the final purchase decision. For example, most food products have traditionally been sold through supermarkets. Supermarkets are not the final consumers of the products, but they play an extremely important role in determining if the product is made available to consumers. Their views are as important as the final consumers'. In the case of medical devices for personal use, it has generally been a medical professional who advises the end-user on what device to use. They are critical stakeholders in the purchase decision, and their views should be sought and included throughout the product innovation process.

VoC research is not only used in the development stage of the new product innovation process but is often used to develop customer or market segmentation schemes using demographic, psychographic, behavioral, and lifestyle data to create customer personas for launch and marketing communications. Further, VoC research may be revisited, and additional insights sought or confirmation of previous insights at any stage of product development.

6.1.2 Six Critical Steps of Market Research

VoC and market research, in general, is a process. Naresh (2009) provides a list of six steps that underpin most market research processes, see Figure 6.1.

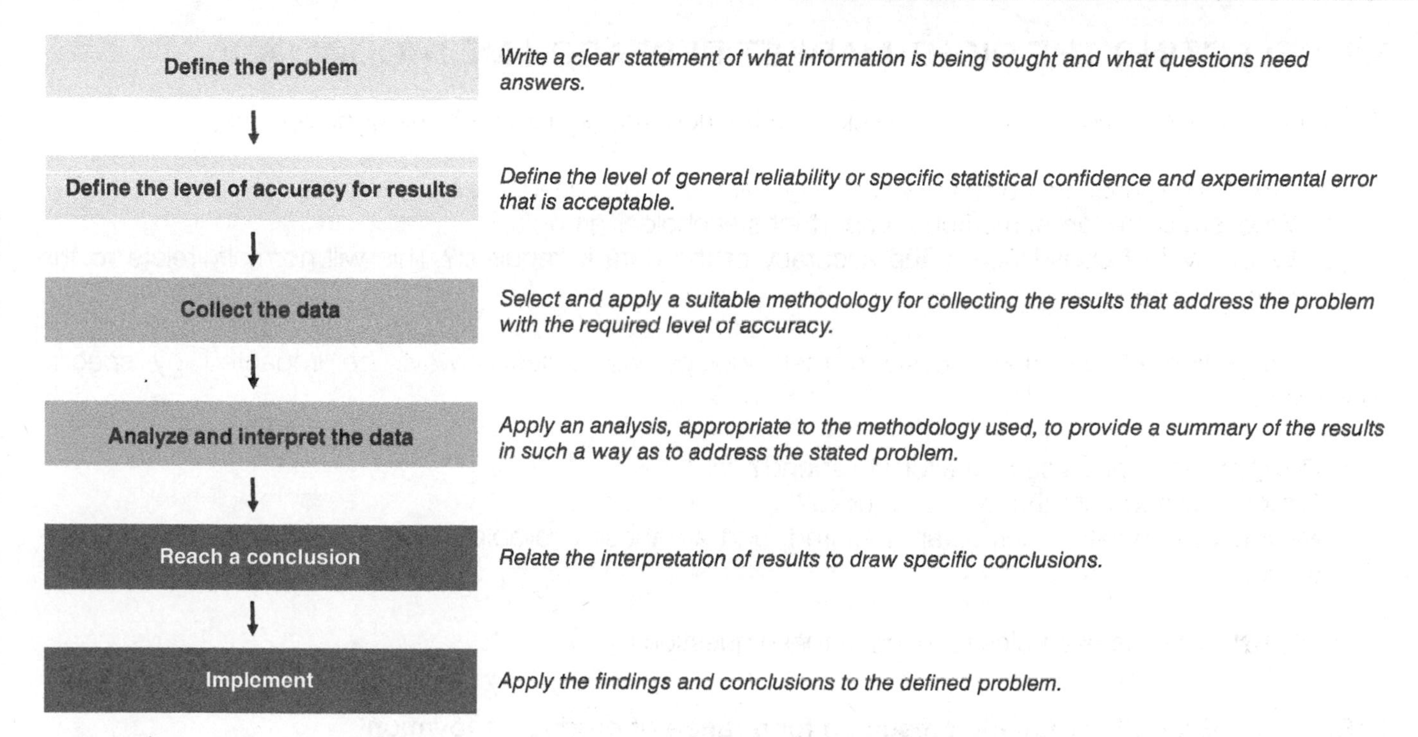

Define the problem
Write a clear statement of what information is being sought and what questions need answers.
Define the level of accuracy for results
Define the level of general reliability or specific statistical confidence and experimental error that is acceptable.
Collect the data
Select and apply a suitable methodology for collecting the results that address the problem with the required level of accuracy.
Analyze and interpret the data
Apply an analysis, appropriate to the methodology used, to provide a summary of the results in such a way as to address the stated problem.
Reach a conclusion
Relate the interpretation of results to draw specific conclusions.
Implement
Apply the findings and conclusions to the defined problem.

FIGURE 6.1 Six steps that underpin market research processes

6.2 SELECTING THE RIGHT MARKET RESEARCH METHODOLOGY

The three critical questions to answer in selecting the right market research methodology are:

1. What information is required, or what specific questions need to be answered?
2. Who is/are the target market or important stakeholder group(s)?
3. What level of confidence in the accuracy of the data is required? This will normally relate to the risks involved.

The selection of a market research methodology will almost always be impacted by specific constraints:

- ***Budget:*** how much can we afford to spend?
- ***Time:*** how soon are the results required?
- ***Resources:*** what resources are required, and are these available within the budget and time constraints?

Table 6.1 provides examples for each of these questions.

TABLE 6.1 Examples of market research for a range of product innovation

What information is required?	Who is the target market?	What level of accuracy?
What is the best design for a new gymnasium website?	Customers and potential customers of the gymnasium.	Risks are low. Feedback with reasonable accuracy is sufficient.
What is the potential market for a new yogurt flavor to be included in an existing, widely marketed brand?	Household shoppers and yogurt consumers	Although risks are relatively low due to current production and market knowledge, significant launch and inventory cost are involved. A high level of accuracy is required.
Does a newly designed can opener meet the needs for people with arthritis?	People who are afflicted with hand arthritis or similar issues that limit hand strength and flexibility.	A new product to the company, there are significant risks due to lack of specific product and market experience. A high level of accuracy is required.
What improvements can be made to a new robotic pool cleaner?	Swimming pool owners both home and commercial.	Low level of risk at the early stage of development due to existing product experience. Low to moderate accuracy required.
Is there a market for a radically new 3D interface for home televisions?	Television users. Most likely those users who are technically competent and inclined to try new products.	Low risk at this early stage of concept development and testing. But risk will almost certainly increase through development. Low accuracy is required initially leading to high accuracy as investment increases.
Does a prototype for a new medical device to reduce sleep apnea meet market needs?	Sleep apnea patients, medical practitioners.	High risk due to the need for government regulatory approval. High level of accuracy from all key stakeholders.

A high-level process leading to the selection of an appropriate market research methodology is shown in Figure 6.2.

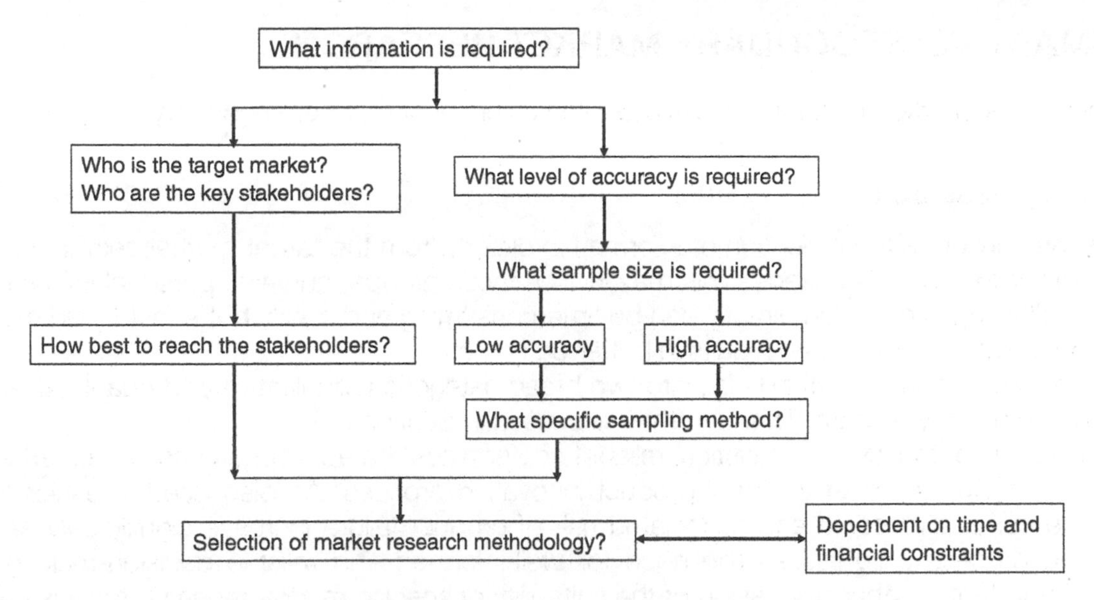

FIGURE 6.2 High-level process for market research methodology selection

The remainder of this chapter describes specific market research methods and their appropriateness at different stages of the product innovation process in terms of:

- The information they can provide.
- How suitable they are for specific stakeholder groups.
- The level of accuracy and statistical reliability.
- Time required to use the method.
- The resources and overall cost involved.

6.3 PRIMARY VS. SECONDARY MARKET RESEARCH

At the highest level, market research can be classified as either primary or secondary.

6.3.1 Primary Research

Primary research involves the collection of information directly from the target market segment and specifically for an organization's own needs. This may involve focus groups, surveys, personal interviews, observation, etc. Collecting information directly can be time-consuming and costly but should yield more specific and useful data, which in turn often results in better outcomes.

Primary market research methods fall into two broad categories, qualitative and quantitative, based on the nature of the underlying data. This is further discussed in Section 6.4.1.

To a large degree, the level of statistical reliability determines the appropriateness of a market research technique at specific stages of the new product innovation process. As discussed in earlier sections of this book, the cumulative costs and associated risk of product failure increase significantly as the product innovation process progresses. The need for well-informed and reliable decision-making becomes increasingly important. Further discussion of the suitability of specific market research tools is presented in Section 6.10 and summarized in Table 6.18.

6.3.2 Secondary Research

Secondary research relies on the collection of information from studies previously performed and published by other individuals, groups, or agencies. This method is significantly less costly and can be performed more rapidly, provided the research sources are readily available. Finding appropriate research data is often the most time-consuming part of secondary research.

Secondary market research involves searching for sources of existing data that were previously collected by someone else and can include:

- Government statistics.
- Syndicated data.
- Industry or trade publications.
- Trade fairs and conferences.
- Newspapers and magazines.
- Organization annual reports.
- Research publications.
- Scientific journals.
- Scholarly articles.
- Patent and trademark (intellectual property) databases.
- Open-source databases, wikis.
- Internet websites, white papers, blogs, and discussion forums.

A comparison of primary and secondary market research is presented in Table 6.2.

TABLE 6.2 Comparison of primary and secondary market research

Comparison	Secondary market research	Primary market research
Cost	Usually inexpensive. Often free but sometimes requires payment for specific information.	Can range low to high cost depending on the scope and method used.
Focus and quality of information	Mainly generic. But still offers a general overview and trending as a basis for more targeted primary research.	Allows for specific focus on the information required and the accuracy of that information.
Currency of information	Often dated if drawn from public sources. More up-to-date information may be sourced at a cost from specific providers.	As current as the completion of the primary research and its analysis.
Timeframe	Relatively fast, especially in the age of online databases targeted at specific topics.	Varies according to the scope and quality of information required and the method employed.
Resources required	Any person with some level of online search training and familiarity with the information domain.	Requires in-house or outsourced expertise, especially where more complex methods are employed.
Risk minimization	In general, not sufficiently reliable as a basis for high-risk decision-making.	Dependent on the market research method employed to provide a level of confidence in the information for decision-making, particularly during the later stages of the product innovation process.

6.4 THE QUALITY OF MARKET RESEARCH DATA

Market research can be categorized according to the type of data that is gathered, either quantitative or qualitative, and the extent to which the data are representative of the target market.

6.4.1 Qualitative vs. Quantitative

Qualitative and quantitative market research are defined as follows:

> ***Qualitative Market Research:*** Research conducted with a small number of respondents, either in groups or individually, to gain an impression of their beliefs, motivations, perceptions, and opinions. Frequently used to gather initial consumer needs and obtain initial reactions to ideas and concepts. Results are not representative of the market in general or projectable to that market.
> ***Quantitative Market Research***: Consumer research, often surveys, is conducted with a large enough sample of consumers to produce statistically reliable results that can be used to project outcomes to the general consumer population. Used to determine importance levels of different customer needs, performance ratings of and satisfaction with current products, probability of trial, repurchase rate, and product preferences. These techniques are used to reduce the uncertainty associated with many other aspects of product innovation.

Ottum (2004) provides a summary of the differences between quantitative and qualitative market research, as shown in Table 6.3.

TABLE 6.3 Comparison of qualitative and quantitative market research

Qualitative market research	Quantitative market research
Words and images	Numbers
"Soft" data	"Hard" data
More exploratory	More confirmatory
Great for understanding unmet needs	Great for optimizing the new product's appeal
Analyzed by looking for themes and deeper meaning	Analyzed using statistics

6.4.2 Qualitative and Quantitative at Different Stages of the Product Innovation Process

Both qualitative and quantitative methods can add value to decision-making throughout the product innovation process. But it is always important to remember that quantitative methods, particularly those using statistical sampling, are better where high-risk decisions need to be made.

Qualitative techniques remain extremely valuable in:

- Providing context and direction at the start of the process where costs and risks are low – qualitative research can provide insights about customers' problems that quantitative research cannot,
- Providing context or hypotheses for quantitative studies,
- Following quantitative methods to help give context to results, quantitative techniques should be applied, where possible, as project costs and risks increase.

6.4.3 Target Market Representation

When the market research data are to be used to draw statistically reliable conclusions representing an entire population or target market, appropriate sample selection must be employed. Sampling methods are mainly divided into two categories: probability sampling and non-probability sampling. In the first case, each member has a fixed, known opportunity to belong to the sample, whereas in the second case, there is no specific probability of an individual being a part of the sample. A comparison of the two sampling methods is shown in Table 6.4.

TABLE 6.4 Comparison of probability and non-probability sampling

Comparison	Probability sampling	Non-probability sampling
Definition	A sampling technique in which the subjects of the population get an equal opportunity to be selected in a representative sample.	A method of sampling where each member of the population does not have an equal opportunity to be selected in the sample.
Also known as	Random sampling	Non-random sampling
Basis of selection	Strictly random	Arbitrary, convenience, or purposeful
Results	Unbiased	Biased
Method	Objective	Subjective
Inferences	Statistical	Analytical
Hypothesis	Tested	Generated

6.4.4 Sampling Methods

Random Sampling: The simplest method of sampling for quantitative market research is random sampling. A random sample is defined as a subset of a statistical population in which each member of the subset has an equal probability of being chosen. A simple random sample is meant to be an unbiased representation of a group. However, while simple, it is often not practical to use.

Advantages of random sampling include representation of the target population and elimination of sampling bias.

Disadvantages include extreme difficulty to achieve in practice, and the cost and time involved.

In order to overcome some of the disadvantages of pure random sampling, a number of other sampling methods can be applied to maintain a high level of precision but at significantly reduced time and cost. These include systemic, stratified, and cluster sampling.

Systemic Sampling: Samples are chosen at regular intervals depending on the size of the population and how large a sample is needed. Be sure there isn't a pattern in the original population that creates a bias in the sample selection.

Stratified Sampling: The population is divided into strata according to some variables that are thought to be related to the variables of interest. A sample is taken from each stratum. This is intended to reduce sampling error because if the strata really are related to the variables of interest, then each stratum is more homogeneous, and it has less variation in the target variables.

Cluster Sampling: The population is divided into clusters, and a sample of clusters is taken. This tends to increase sampling error because clusters tend to be similar; if they were identical, there would be no point in taking more than one observation within the cluster because they would all be identical. The loss of precision is related to the variability within the clusters, which is only known after the sample is taken. In single-stage cluster sampling, the whole cluster is the sample. In multistage cluster sampling, random sampling is done within the cluster in one or more stages.

6.4.5 Sample Size and the Statistical Basis of Probability Sampling

Statistical formulas, tables, or easy-to-use online calculators are used to determine a sample size sufficient to establish the statistical reliability of research results.

The following information is needed for sample size calculation:

- ***Margin of error (confidence interval):*** What range around the calculated survey result can be tolerated, e.g., survey result ±5%.
- ***Confidence level:*** What is the confidence that the actual results fall within the confidence interval required?
- ***Variance:*** What variance is expected in the results? This is often estimated from population statistics or past studies.

General sample size relationships:

- The lower the required confidence interval (margin of error), the larger the sample size required.
- The higher the required level of confidence, the larger the sample size required.
- The higher the variance in the population with regard to what is being surveyed, the larger the sample size required.
- Selecting a sample size that is too small will lead to lower accuracy and confidence in the results.

6.5 MARKET RESEARCH METHODS

6.5.1 Focus Groups

A focus group is a qualitative market research technique where 8–12 market participants are gathered in a physical or virtual room for a discussion under the leadership of a trained moderator. The discussion focuses on a consumer problem, product, or potential solution to a problem.

Typical characteristics of a focus group:

- Screening questions are used to select 8–12 participants.
- It is often conducted in a specialized facility with a trained moderator.
- Specialized facilities are set up with a table and chairs, a one-way mirror, and the capacity to audio and video record the group.
- Observers can view proceedings from an adjacent room through a one-way mirror.

Considerations in applying focus groups:

- Avoid conducting a single focus group. Three or more is best; though not statistically reliable, multiple groups do provide greater confidence in the findings.
- Ensure the moderator has the necessary background and training to handle the topic.
- First, use proper planning and preparation – avoid bringing a random group of people together for an unstructured chat.

Screen out focus group veterans – people who regularly participate in focus groups.

- Avoid conducting a single focus group. Three or more is best; though not statistically reliable, multiple groups do provide greater confidence in the findings.
- Ensure the moderator has the necessary background and training to handle the topic.
- First, use proper planning and preparation – avoid bringing a random group of people together for an unstructured chat.
- Screen out focus group veterans – people who regularly participate in focus groups.
- A focus group is not a quantitative technique – statistical conclusions cannot be made.

The strengths and weaknesses of focus groups are summarized in Table 6.5.

TABLE 6.5 Strengths and weaknesses of focus groups

Focus group strengths	Focus group weakness
• Interaction among participants provokes discussion and can provide fresh insights and in-depth understanding. • Comments come directly from representatives of the market — unfiltered through questionnaires or analysis. • Questions can be changed quickly in response to participants' comments. • Behavior of participants can be observed, especially in usage studies.	• Group dynamics can suppress some individuals' contributions or lead to dominance by others. • Comments by participants can be open to interpretation. • Conclusions do not apply directly to people outside of focus group — findings may not represent the larger market segment. • Quality of results is heavily influenced by the skill of the moderator. • Conducting groups is limited to participant availability at a set time and place.

The value of focus groups in product innovation decision-making is that when run well, they can provide great insights throughout the new product innovation process:

- What opportunities are out there, and what are customers' ideas on gaps in the market?
- What do customers want or need? Focus groups are not as good at identifying needs that the customer has yet to recognize or cannot articulate. Results are likely to be within the current product or market paradigm.
- What value proposition should be built into the new product? Focus groups are used to better understand the features that drive value and should be built into the product.
- What refinements should be made to make the product more acceptable? Focus groups are used on an ongoing basis to provide direction for product refinement. Lead user groups or customer panels, comprising customers who have an active interest in the product category and who lead the market in purchasing, are particularly useful (Thomke & Von Hippel, 2002).

6.5.2 In-depth Interviews

This qualitative research method involves conducting longer intensive interviews probing and exploring a specific topic, one on one, with individual participants. The research gathers detailed insights, perspectives, attitudes, thoughts, behaviors, and viewpoints on a problem, idea, program, situation, etc.

The strengths and weaknesses of in-depth interviews are summarized in Table 6.6.

TABLE 6.6 Strengths and weaknesses of in-depth interviews

In-depth interview strengths	In-depth interview weaknesses
• Provides more detail, context, language, emotions, and relationships for each area explored. • Designed to be more open-ended, provide emergent information and insight, reveal hidden relationships, connections, complexity, and help explain quantitative findings. • Can be a more relaxed, personal, private, safe, and individual-based conversation that provides richer, deeper insight. • Valuable prior to quantitative studies to provide initial framework for the quantitative research purpose and questions. It also provides richer, authentic language for use in quantitative studies. • Valuable when investigating or exploring the reasons behind quantitative research results.	• Time-consuming and costly. • Subject to participant availability at a set time and place. • Must be conducted by a professional trained in interviewing techniques. • Can be biased and subject to interpretation. • Findings may not represent the larger market segment. • Typically conducted with a smaller number of participants than other types of research due to increased time, effort, and expense.

The value of in-depth interviews in product innovation is that they can provide great insights throughout the new product innovation process:

- What opportunities are out there, and what are customers' ideas on gaps in the market? In-depth interviews generally provide a more relaxed environment that facilitates probing questions to dig deeper into respondents' ideas.
- What do customers want or need, especially unarticulated needs? The more relaxed environment and opportunity to develop a closer relationship with respondents can provide a better understanding of unarticulated needs.
- What is the value proposition to be built into the new product? An in-depth interview provides the opportunity to build on new ideas and further develop the value proposition.
- What refinements should be made to make the product more acceptable? In-depth interviews can be used, either with the same respondents or different respondents to provide direction for on-going product refinement.

- They work especially well in business-to-business (B2B) research where individuals with specific knowledge or experts are needed, or in business-to-consumer (B2C) when the scenario is sensitive or private. If appropriate to the study design, conducting phone or video conference interviews can offer flexibility and lower costs.

6.5.3 Ethnography

Ethnography is defined as a descriptive, qualitative market research methodology for studying the customer in relation to their environment. Ethnographic market research (EMR) refers to a type of research that helps organizations understand the customer in terms of cultural trends, lifestyle factors, attitudes, and how social context influences product selection and usage. Researchers spend time in the field observing customers and their environment to acquire a deep understanding of their lifestyles and cultures as a basis for better understanding their needs and problems.

In contrast to focus groups, ethnography utilizes a variety of techniques and forums to present a complete picture of consumers and how products and services fit into their daily lives. Further reference and explanation of the application of EMR to product innovation are provided by Miller et al. (2004) and Katz (2010).

Ethnographic market research can be applied on-site or in-home. On-site EMR takes place wherever the consumer is utilizing the product or service; for example, in a restaurant, store, office, or even the car. Conducting place-based research allows the researcher to interview and observe as the behavior is carried out and provides an opportunity for follow-up questions as needed. Immersion tours of markets can be conducted on location to research entire markets vs. individual consumers. This is especially helpful in geographic expansion.

In-home EMR is similar to on-site and is conducted in the participant's home environment. The research can include one or multiple family members and can last for several hours or longer. The researcher is immersed in the home environment and observes, asks questions, and listens to obtain insight into consumer trends, reactions, and problems. Consumers go about solving those product or service-based issues. In-home sessions provide insight into customers' unmet needs, how to improve products, what new items are needed, and how changing needs affect usage.

The strengths and weaknesses of ethnography are summarized in Table 6.7.

TABLE 6.7 Strengths and weaknesses of ethnography

Ethnography strengths	Ethnography weaknesses
• Provides a unique opportunity to identify attributes, situations, and states that are truly valued by customers and reveal unidentified dissatisfiers or hidden problems. • Can be a very useful means of identifying unarticulated needs that participants may be unwilling or unable to identify during a traditional interview, focus group, or survey, especially if the new product is unfamiliar, the situation private, or sensitive. • Can facilitate the discovery of unarticulated needs can form the basis for radically new products.	• It can take a long time to carry out, especially if the researcher is seeking a broad representation of the target market. • More resources are often needed compared to other methods. • It relies on the researcher's interpretation of what is observed. • It has no basis in statistical reliability.

The value of ethnographic market research to product innovation decision-making is derived from its usefulness in understanding customer needs and product usage.

- What opportunities are out there, particularly new product opportunities? Ethnographic research seeks to get the researchers to walk in the customers' shoes — to really understand the issues and problems that get to needs that new products can satisfy. It is a high-empathy method.
- What do customers want or need? Ethnography is an excellent technique for getting to know the customer and getting to the heart of their needs, which they may be unable or unwilling to articulate.

- What refinements should be made to make the product more acceptable? Ethnographic research can be used to test a product prototype, observe customer reaction to it, and watch the application of the prototype. For example, it can examine if the product is used in the intended way and to discover what problems the customer has in using the product. Ethnography can also be used with products already launched in the market for new and improved versions or line extensions.

6.5.4 Customer Site Visits

Customer site visits are defined as "a qualitative market research technique for uncovering customer needs. The method involves going to a customer's work site, watching as a person performs functions associated with the customer needs a firm wants to solve, and then debriefing that person about what they did, why they did those things, the problems encountered as they were trying to perform the function, and what worked well" (Kahn et al., 2004, Chapters 15–16).

Similar to in-depth interviews, customer site visits are one of the market research methods used most commonly by business-to-business (B2B) organizations. It involves one or more associates from a vendor directly interviewing, observing, and interacting with one or more customers (or potential customers) of that vendor at the site of product use.

The strengths and weaknesses of customer site visits are summarized in Table 6.8.

TABLE 6.8 Strengths and weaknesses of customer site visits

Customer site visit strengths	Customer site visit weaknesses
• Face-to-face communication on customer premises can provide a large amount of contextual information. • Site visits often provide the opportunity to see the product in use, observe first-hand its strengths and weaknesses and discuss directly with the user what improvements might be made or desired. • Visits are useful for technical staff involved in product innovation to better understand the customer's needs and how these can best be translated into product design specifications. • Visits by product innovation team members promote better communication and understanding of requirements leading to greater team cohesion. • Conducting site visits also often help build stronger customer relationships.	• Site visits can be expensive and time-consuming, especially if travel is involved. • Where the organization is supplying a number of customers, comments and suggestions from one or two of these may not represent the views of the majority. Sample size is important to get a more reliable view. • The quality of the information gathered is highly dependent on asking the right questions of the right people. So, for example, an interview with a marketing representative of the customer organization may provide limited, or unreliable, technical information. • Sending an unqualified representative from the organization with limited product knowledge and interview skills can have a detrimental impact on customer relationships and lead to unreliable information.

Considerations in Applying Customer Site Visits

- Be sure to inform sales representatives of the visit and enlist their support.
- Ensure the customer representatives who host the visit are decision-makers or influencers with the requisite knowledge to provide the information sought.
- Build in a tour beyond the meeting room to see first-hand where and how the product will be used, if possible.
- Ask customers to identify problems. If they suggest solutions, listen, acknowledge, and accept their input.
- Take samples, visual aids, and anything that can enable a clearer understanding.

The value of customer site visits to product innovation decision-making is the ability to gather in-depth market and technical information:

- What opportunities are out there? Particularly with B2B companies, where there may be levels of buyers, users, and influencers in the customer organization who can provide insights into new technologies or competitors' products.
- What do customers want or need? Site visits often focus on current needs, problems, and required improvements. These are specific issues that the customer is best placed to articulate. Visits can also reveal unidentified needs that the customer is either unable or unwilling to articulate and requires lines of questioning and observation to identify the customer's needs beyond the current product(s).
- What refinements should be made to make the product? The outcome is often to make the product more acceptable, especially where detailed concepts or prototypes form the basis for the refinement discussion.

6.5.5 Social Media

Social media has matured and has opened up new ways of interacting with and sourcing information from the marketplace. Media such as Facebook, Twitter, YouTube, Pinterest, Instagram, WhatsApp, LinkedIn, Snapchat, Reddit, blogs, and various discussion forums all offer a medium for engaging with and listening to customers. In 2022, Facebook alone reported nearly 3 billion users worldwide. Social media marketing usage rate for marketing purposes among US companies has risen from 86 in 2013 to 92% in 2022 (Statista, 2022). In addition to these more global social media platforms, a number of other country or region-specific platforms exist. For example, Vkontakte in Russia, Viadeo in France, Line in Japan, Kaokao Talk in South Korea and in China, with a range of "home-grown" platforms and networks like WeChat and Weibo. When using social media for marketing communication and research, it is vitally important to select the most appropriate platform for the target market.

Social media is regularly used for co-creation in the ideation and design phases and for driving awareness and promotion in the commercialization and launch phases of product innovation. Social listening is a type of VoC market research that monitors digital conversations and analyzes what customers are saying about specific topics. A well-reported example of how social media listening supports the new product innovation process is how, after analyzing social media conversations, McDonald's began offering its all-day breakfast menu in key markets.

Clorox brand Brita reports using social listening to learn that millennials regularly complained about a roommate or partner who drank the last of the Brita water and did not refill the pitcher. They launched a new product, Brita Stream, targeted at millennials. It filters as it pours, eliminating waiting for the water to filter when the user finds it empty.

TABLE 6.9 Strengths and weaknesses of social media

Social media strengths	Social media weaknesses
• It can provide direct and immediate contact with current and potential customers. • If selected carefully, specific social media allows targeting to a very narrow audience. • It provides the opportunity to engage with a loyal following of supporters or lead users, either as a basis for ongoing idea generation or for input during the product design process (refer to crowdsourcing later in this chapter).	• It can be subject to a high degree of bias based on the composition of people using a specific social medium and those who actively engage in this medium. • It is difficult to maintain control and achieve a clear focus on the problem or question. • Although it may be possible to receive a large response to a survey through social media, true statistical confidence still may not be assigned to the results as the sample may not accurately reflect the population. • Social media reciprocity (the practice of automatically liking and sharing within a network of friends and followers) and customer turbulence, creating rapid changes in attention such as the viral effect, may bias or cloud results

In addition to social listening, companies worldwide apply integrated social media programs to their marketing and product management. Nike, for example, uses Twitter, Facebook, Instagram, and Pinterest along with its own platforms. Nike created Nike+ to be a virtual training club. As of 2022, its primary platform was the Nike SNKRS App, which is positioned as "Your Ultimate Sneaker Source."

Strength and weaknesses of social media are presented in Table 6.9.

The value of social media in product innovation decision-making is its facility to reach large numbers of targeted customers quickly and easily. This makes it an excellent market research tool for product innovation:

- What opportunities are out there? Social media provides an excellent source of information on what people are doing and thinking, both individuals and organizations, in real time. This can lead to new opportunities directly or by inference.
- What do customers want or need? Social provides the product developer with a broad canvas of information related to the general market – and in some cases, even to specific market segments. Note that social media is largely unstructured and needs caution in interpretation.
- What refinements should be made to make the product more acceptable? Specific and targeted social media groups form an excellent basis for online lead users to provide ongoing testing and input to a product's design and use.

There are multiple opportunities for innovation using social media analytics. Social media can serve as a source of consumer insights, complementing traditional market research methods. The speed with which social media data are generated and analyzed offers opportunities to be proactive in leveraging the data (Moe & Schweidel, 2017).

Examples of Social Media Applied to Product Innovation

Non-profit Compassion International seeks donors to fulfill its mission of caring for children in poverty. It used social media metrics to identify social causes receiving the most attention and charitable donations. At the top of the list was sustainable water projects. Compassion went on to co-develop a water filter for villages that only had access to dirty water supplies. This resulted in new donors joining Compassion's mission.

OBO, a New Zealand-based manufacturer of special field hockey equipment, has used a suite of social media tools to inform its design process for new and improved products. It focuses on user groups; both lead users at the high-performing international level and general users at the club level. More detail on OBO's use of social media is presented in Appendix B, case 4.

6.5.6 Surveys

Surveys can be broadly defined as the polling of customers or stakeholders to identify their level of satisfaction with an existing product or to discover needs for new products or services. Surveys are a basic building block and foundational to market research in product innovation. Surveys are ubiquitous. They can be, and very often are, used in combination with other market research techniques.

A survey consists of:

- A fixed set of questions asked of a sample of respondents.
- A sample that is large enough for the study intent and selected to consist of certain characteristics.

Is a survey a qualitative or quantitative research approach?
This depends on both the sample selection and the questionnaire design.

Sample size selection
Probability sampling, as described in Section 6.4.3, provides the basis for statistical analysis providing a level of confidence in the results. Non-probability sampling generally provides for a cheaper and more rapid survey, albeit without the benefits of statistical confidence.

Questionnaire design

Questions used in surveys are many and varied. There are a number of ways of classifying questions. For example:

- ***Open-ended questions:*** These cannot be answered with a simple yes or no answer or a selection from a multiple-choice answer list. The respondent is asked to elaborate on their response.
- ***Closed-ended questions:*** Provide a number of optional answers to a question, a yes or no, or a rating, for example, "like a lot to dislike a lot."
- ***Nominal questions:*** Answers are required in defined categories. For example, religious affiliation, sex, city where you live, etc.
- ***Likert scale questions:*** Respondents are presented with a statement and asked to indicate their agreement with the statement on a defined scale, usually 5, 7, or 9 points. For example:
  - Very satisfied
  - Moderately satisfied
  - Neither satisfied nor dissatisfied
  - Moderately dissatisfied
  - Very dissatisfied
- ***Ranking scale (or ordinal) questions:*** Respondents rank items or choose from an ordered set. For example, respondents could be shown five product benefits and asked to rank them 1 to 5 in order of importance, with 1 being the most important benefit.

It is important to note that the question style significantly impacts its suitability for statistical analysis. The critical consideration is the nature of the scale used for the responses, mainly whether the intervals between the individual points on the scale can be considered equidistant and whether the data is normally distributed. If the answer is "yes" to these two criteria, then parametric statistical analysis can be applied. This simply means that basic descriptive statistics, such as the sample's mean, mode, and standard deviation, can be extended to the larger population. If the answer to the two criteria is "no," then the parametric analysis is unsuitable, and non-parametric statistical analyses should be applied, for example, rank order correlation or chi-square. The important point is to be aware of the pitfalls of using inappropriate data for statistical analysis of survey data. If in doubt, consult a qualified statistician.

TABLE 6.10 Strengths and weaknesses of surveys

Survey strengths	Survey weaknesses
• Relatively easy to administer, depending on scope. • Cost-effective, but cost depends on survey mode. • Can be administered remotely via online, mobile devices, mail, email, kiosk, or telephone. • Conducted remotely can reduce or prevent geographical dependence • Capable of collecting data from a large number of respondents. • Numerous questions can be asked about a subject, giving extensive flexibility in data analysis. • With survey software, advanced statistical techniques can be utilized to analyze survey data to determine validity, reliability, and statistical significance, including the ability to analyze multiple variables.	• Respondents may need to be incentivized to participate in a survey. • Respondents may not feel encouraged to provide accurate, honest answers. • Respondents may not feel comfortable providing answers that present themselves in an unfavorable manner. • Respondents may not be fully aware of their reasons for any given answer because of lack of memory on the subject, or even boredom. • Surveys with closed-ended questions may have a lower validity rate than other question types. • Data errors due to question non-responses may exist. The respondents who choose to respond to a survey question may be different from those who chose not to respond, thus creating bias. • Survey question answer options could lead to unclear data because certain answer options may be interpreted differently by respondents. For example, the answer option "somewhat agree" may represent different things to different participants and have its own meaning to each individual respondent.

Surveys are a valuable market research tool for product innovation decision-making. They are frequently used as the data collection tool for other more specific research approaches, such as Conjoint or Kano analysis. Surveys can be used to provide insights, directly or indirectly, into any of the following:

- What do customers want or need?
- What refinements should be made to make the product more acceptable?
- What drives customers to purchase and re-purchase a product?
- What value proposition should be built into our new product?
- Will the customer buy the product: how often, where, and at what price?

The strengths and weaknesses of surveys are presented in Table 6.10.

6.5.7 Consumer Panels

Consumer panels are defined as groups of consumers in specific sectors or segments, typically recruited by professional research companies and agencies. The panel participants are used as respondents to answer specific research questions relating to product testing, taste testing, or other sensory testing; concept testing; concept sorting, and many other areas. Often, they are a specialist panel that may take part in numerous projects in a category. Consumer panels are particularly useful for short, quick surveys, where the emphasis is on a sample of those with specialized knowledge or selection criteria rather than a representative sample of the general population. Consumer panels fall into two broad categories:

- **Untrained panels** representing the attitudes, beliefs, perceptions, and behaviors of the target market. Although not necessarily statistically representative of the target market, these panels can at least provide valuable input to evaluating and designing desired features and functionality of a new product.
- **Trained panels** comprise individuals who are trained to assess, generally in some quantitative way, the specific attributes of a product. The application of trained panels to sensory testing is further discussed in the following section.

Strengths and weaknesses of consumer panels are summarized in Table 6.11.

TABLE 6.11 Strengths and weaknesses of consumer panels

Consumer panel strengths	Consumer panel weaknesses
• Untrained panels provide valuable insights into consumer preferences and suggestions for product improvement. But should not necessarily be considered representative of the entire market. • Trained panels are invaluable in some industries, such as food and cosmetics, where instrumental measurements are either unavailable or unable to provide the required information. • Consumer panels can be quickly formed, especially with the assistance of a research firm that specializes in panels.	• Untrained panels, although useful as indicators of consumer preference and attitudes, may not be statistically representative of the target market. • The use of trained panels to provide preference data should be avoided. Their training ceases to make them representative of the target market.

6.5.8 Sensory Testing

Sensory testing is widely used in consumer products. It can be used throughout the product innovation process to explore concepts in the early phases through to testing prototypes or validating product performance prior to launch.

Sensory testing methods are divided into two broad classes: affective and analytical methods. Affective methods use consumer panels, described in the previous section.

The most common analytical methods of sensory testing are discrimination (or difference) and descriptive methods. Discrimination tests can be used to determine if products are different or if one product has more of a selected characteristic than another. Descriptive methods are used to provide more comprehensive profiles of a product by asking panelists to identify the different characteristics within the product and quantify characteristics. Trained panelists must be used for descriptive methods. Trained panelists are recommended for both discrimination and descriptive methods.

Some of the more commonly used analytical tests include:

- ***Triangle:*** The **triangle test** uses three samples to determine if an overall difference exists between two products. The three samples include two that are identical and one that is different. The panelist is asked to identify the odd sample.
- ***Duo trio:*** In a **Duo trio test,** panelists are presented with a reference sample and two other samples. Panelists are asked to identify the sample that is the same as the reference.
- ***Paired comparison:*** In a **paired comparison test**, two different samples are presented, and one asks which of the two samples has most of the sensory property of interest, e.g., which of the two products has the sweetest taste.

The most commonly used descriptive method is Quantitative Descriptive Analysis (QDA®). The QDA® methodology requires trained panelists to assess the product against defined sensory descriptors on line scales. Statistical analyses can be performed to determine product differences. Results can also be presented graphically in the form of a "spider web," as shown in the comparison of two wines in Figure 6.3.

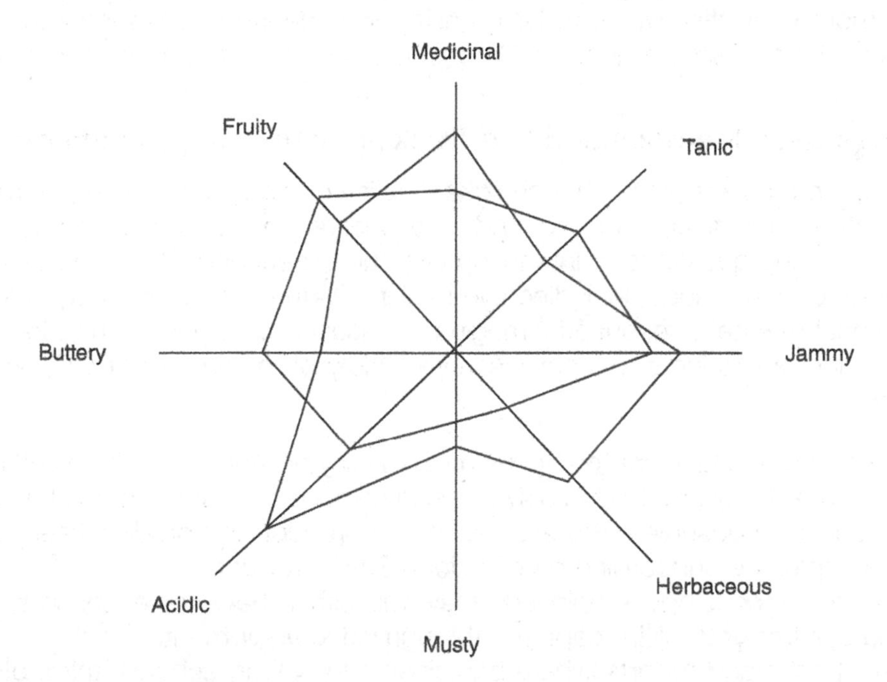

FIGURE 6.3 QDA® spider web of two wines

6.5.9 Trained Panels

As discussed in sensory methodologies above, trained panels are often required to provide quantitative and statistically reliable data. Trained panels are especially used in formulated products, such as food and cosmetics, where instrumental measurement of sensory characteristics is sometimes difficult or unavailable. In such cases, trained panels can be invaluable for objectively evaluating changes to product characteristics (for example, taste or texture) resulting from formulation or process changes. Panel training is not an insignificant exercise, involving considerable time and expertise in trainers. Training increases the sensory acuity of panelists and provides them with a rudimentary knowledge of procedures used in sensory evaluation. Training also develops the ability of panel members to detect, recognize and describe sensory stimuli. Basically, the training process involves:

- Selection of panel members based on their ability to recognize the sensory property or properties under investigation.
- Training the selected panelists in the use of appropriate descriptors consistent across the entire panel.
- Training the selected panel members to quantify the level of the property using reference samples at defined levels.

The American Society for Testing and Material (1981) provides greater detail on the guidelines for selecting and training sensory panel members.

6.5.10 Concept Tests and Concept Sorts

Concept tests and concept sorts are quantitative research surveys that evaluate customer acceptance of a new product or service idea (referred to as a concept). These tests are used in the early phases of the new product innovation process prior to the development phase. Concept sorts help identify and rank which product concepts are strongest when a variety of options for development exist. Technology has made concept testing easy for participants and more convenient. Participants can complete concept test surveys online, with flexible scheduling, anywhere around the world. Testing and sorting concepts prior to development is more cost-effective at reducing risk and expense prior to incurring increasingly larger development costs as the process progresses. Proof of concept can be the result of concept testing.

6.5.11 Biometrics-based Methods and Eye Tracking, Virtual and Augmented Reality

Biometrics is an emerging market research technique. It relies on specialized tools, biometric technologies, and applications to study biological, cognitive, and emotional responses toward various products and services without actually asking questions or interfering with the experience. Biometric data collection can be obtained using a number of methods, including eye tracking, virtual reality, and augmented reality. Specific tools such as functional magnetic resonance imaging, measuring changes to the blood flow in the brain, are used to test a range of products and services in a variety of media formats. Basically, the biometrics process is as follows:

- A sample is recruited and brought to a focus group facility or market research setting.
- Each participant may be asked to don galvanic skin response sensors on one hand.
- They are then asked to observe a representation of a product, the product itself, or an aspect of the product, for example, the onboarding process for a SaaS product.
- Depending on the technology employed, they will either have face sensors attached for their responses and eye tracking will be captured through video equipment.
- The equipment tracks and reports where participants look first, second, third, etc., and provides a visual scan overlaid on an image of the product being tested. It is used to answer questions on consumers' reactions to various stimuli, online products, and services, websites, apps, images of products, packaging, and messaging. It is widely used in software, retail product packaging, marketing, and advertising.

Eye tracking and biometrics are typically used with high-fidelity prototypes or a live website, focusing on a user's attention and engagement with a particular set of design features. The resulting data adds another layer of understanding to the broader user experience with the product.

Examples of companies using Biometrics market research include Expedia, in learning what excites potential customers as they plan a holiday online. Also, Walmart filed a patent application in 2018 for a system and method for a biometric feedback shopping cart handle, collecting biometric feedback from sensors on shopping cart handles for monitoring consumer biometrics like pulse, temperature, and stress, and comparing against a baseline.

Virtual reality (VR) testing is a growing segment of the market research field and is conducted using specialized equipment, including a headset and/or gloves with tracking sensors that create three-dimensional (3D) simulations and enable participants to interact in a realistic environment. VR is used in conjunction with tools already used in market research, such as eye tracking, consumer behavior analysis, and market environment simulations. VR allows companies to perform product use testing without developing actual prototypes, using VR simulations instead. As a result, financial risks can be minimized. It also enables researchers to observe and examine consumer behavior without costly trials in the marketplace.

Augmented reality (AR) is similar to VR. Whereas VR replaces the participant's real world with an entirely separate reality, AR overlays elements of a new reality into the participant's present environment. AR is sometimes referred to as mixed reality and has the added convenience that it can be viewed and interacted with on device screens such as computers, displays, tablets, phones, and watches.

VR and AR platforms are already currently used by companies like Coca-Cola, Heineken, and Nike in their social channels and networks and in the broadcasting of large commercial sporting and entertainment events, promoting immersive experiences. These are the same platforms that allow researchers and developers to gain in-depth insight by offering an immersive experience with a high level of interaction, key elements for market research.

6.5.12 Big Data

Big data is described as "a collection of large and complex data from different instruments at all stages of the process which go from acquisition, storage, and sharing, to analysis and visualization" (Pisano et al., 2015). "The main driver of the interest in big data is the potential usefulness of it for informing marketing decisions, including those across consumer areas such as problem recognition, consumer search, purchase behavior, consumption, post-purchase evaluation, and post-purchase engagement." (Hofacker et al., 2016).

Big data and the big data analytics industry have matured significantly. The act of gathering and storing large amounts of information for analysis has been widely used, growing as storage technology improved. The big data concept gained momentum in the early 2000s when industry analyst Doug Laney articulated the now-mainstream definition of big data as the three Vs (Laney, 2001):

1. ***Volume:*** Organizations collect data from a variety of sources, including business transactions, social media, and information from sensor or machine-to-machine data. In the past, storing it would have been a problem – but new technologies (including cloud-based services) have eased the burden.
2. ***Velocity:*** Data stream in at unprecedented speeds and must be dealt with in a timely manner. RFID tags, sensors, and smart metering are driving the need to deal with torrents of data in near-real time.
3. ***Variety:*** Data come in all types of formats, from structured, numeric data in traditional databases to unstructured text documents, email, video, audio, stock ticker data, and financial transactions.

Zikopoulos et al. (2015) provide a useful analogy with gold mining to explain the meaning and potential of big data:

> "In the old days miners could readily spot nuggets or veins (high value per byte data) because they were visible to the naked eye. But there was more gold out there that wasn't readily visible. Trying to find it would have required the mobilization of millions of people. Today miners work differently. Gold mining leverages new generation equipment that can process millions of tons of dirt (low value per byte data) to find 'nearly-invisible to the naked eye' strands of gold. And with modern equipment those small strands can be extracted from the mass of dirt and processed into gold

bars (high value data). Today there is a mass of data, residing in different forms in a range of different locations. The challenge for the future is to locate this data and process it into a form that is useful for a specific purpose."

The strengths and weaknesses of Big Data are presented in Table 6.12.

TABLE 6.12 Strengths and weaknesses of Big Data

Big data strengths	Big data weaknesses
• Big data is cost-effective as many organizations already have big data analytics in place. • It improves indirect customer involvement. • Big data enables managers to capture customers' explicit and implicit knowledge. • It can be applied to any phase of product innovation. • Data capture is in real time and can be generated much faster than traditional studies. • It can refocus attention away from internal product-focused innovation and turns attention to innovation around the customer experience. • It is *real* customer behavior, versus what consumers *say* they will do in an interview or survey.	• Role of big data in the product innovation process is directly affected by how data-driven the organization is overall. • To organize and manage big data successfully, organizations need to have established innovation ecosystems and build data alliances with stakeholders, including partners, suppliers, and other entities with common interests. • Data scientists need to be available to be part of product projects. • Stimulating and mining customer interaction via public software platforms may generate data that can also be mined by competition.

6.5.13 Crowdsourcing

Crowdsourcing is a means of collecting information, goods, services, ideas, funding, or other input into a specific task or project from a large and relatively open group of people, either paid or unpaid, most commonly via technology platforms, social media channels, or the Internet. Many companies and organizations also use their own websites as a means of crowdsourcing new product ideas and improvements to existing products.

LEGO has a dedicated website and online community for fans and customers to contribute their own product ideas. LEGO uses the platform to co-develop with customers. The platform has rapidly grown and has more than 1.5 million users at the time of this writing. The platform is named LEGO Ideas. It is designed to motivate and incentivize participation by allowing users to submit LEGO build ideas, vote for their favorite ideas, state how much they would pay for it, and explain why they like it so much. If more than 10,000 people support the idea, then it goes to the official Lego review board, which decides whether or not to put it into production. If the idea is commercialized, the creator even receives name recognition on the product package and a 1 percent royalty on worldwide sales. LEGO Ideas can be found at ideas .lego.com.

Anheuser-Busch, in 2012, ran a crowdsourcing project to create a new beer. It varied slightly from the typical consumer-led crowdsourcing projects as the initial recipes were created during a competition involving the brewmasters at Budweiser's 12 breweries. However, more than 25,000 consumers were involved in the subsequent taste tests to decide the winning brew, so the wisdom of the crowd was involved during the development process. The winning recipe, Black Crown, went on sale in 2013.

The strengths and weaknesses of crowdsourcing are presented in Table 6.13.

TABLE 6.13 Strengths and weaknesses of Crowdsourcing

Crowdsourcing strengths	Crowdsourcing weaknesses
• Crowdsourcing brings together communities around a common project or cause. • It is an efficient way of solving time-intensive problems. • It creates deeper engagement by communities, who resonate and build loyalty to the product or solution. • A wider view can often lead a different perspective. • The numbers involved can come close to full market representation, providing some level of statistical confidence in the results. (But care should be exercised ~~in~~ to prevent over-reliance on crowdsourced data. It is important to use other inputs in addition to crowdsourcing to validate and successfully incorporate the crowd-sourcing input into the new product process.)	• Results can be easily skewed based on the crowd being sourced. • Lack of confidentiality or ownership on an idea. Potential for plagiarism. (However, some organizations use NDAs to protect against this.) • It has the risk to miss the best ideas, talent, or direction and fall short of the goal or purpose. • Reciprocal voting can occur and that may distort value of research results.

6.6 MULTIVARIATE RESEARCH METHODS

Multivariate research, testing, and analysis are used in product innovation and management to explore the relationship that exists between the numerous variables that impact new product success.

In this type of research, a dependent variable is examined in association with one or more predictor or independent variables. Multivariate methods provide a more accurate view of the interactions, correlations, and trade-offs that play out when looking at the possible ranges of product and market attributes and characteristics. It can point out which variables, attributes, and characteristics are highly correlated and can detect potential problems and risks to the product value proposition when decisions around choices need to be made.

A wide range of multivariate techniques can be applied to product innovation. Most of these require an understanding of statistics and the support of an expert. While they involve more complex design and analysis, they provide the potential to gain deeper and more valuable customer insights.

Several multivariate techniques used in product innovation management are briefly outlined below.

6.6.1 Principal Components and Factor Analysis

Deriving Information from Data

Data is an indispensable variable in our life. The train timetable, the text message you received today, and the supermarket item's price are all data. These data do not have any meaning in themselves without a context attached to them. This context is what transforms data into information. In other words, information is structured data, having logical reasoning that makes it coherent and can drive decisions.

In today's world, where we make decisions based on this structured data, there is a potential danger of having too much information and even the risk of losing some important information. The phenomenon of having too much information is often referred to as the *curse of dimensionality*. Two techniques that address this issue, through what is referred to as *dimensionality reduction*, are *Principal Component Analysis (PCA)* and *Factor Analysis*. These techniques help us overcome the curse of dimensionality while minimizing the loss of information.

The main focus of principal components and factor analysis is to summarize the information in a large number of variables into a smaller number of factors.

PCA identifies the smallest number of factors or components necessary to explain as much (or all) of the variance, in a dataset, as possible. It reduces the unnecessary features present in the data by creating or deriving new dimensions (also called components). These components are a linear combination of the original variables. This way, PCA converts a larger number of correlated variables (i.e., breaks down the data) into a smaller set of uncorrelated variables.

Factor analysis is also used to decrease the larger number of attributes into a smaller set of factors. When analyzing data with many predictors, some of the features may have a common theme amongst themselves. The features that have similar meaning underneath could be influencing the target. A factor is, therefore, a common or underlying element with which several other variables are correlated. Also, these latent variables are not directly observable and hence are not measurable by themselves with a single variable.

There are two types of factor analysis, exploratory and confirmatory:

Exploratory factor analysis identifies the underlying factors in a set of variables where the researcher does not have a hypothesis.

Confirmatory factor analysis is used to confirm the effects and correlation among a set of variables, to test a pre-determined hypothesis relating to the relationship among the variables.

A Simple Example of Factor Analysis

In surveying patient experience at a doctor's surgery, three responses, "spends time with me," "answers my questions," and "looked me in the eye," were shown by factor analysis to be highly related and could be summarized by the factor or key theme, **"cares about me."** This theme can then be used in the promotion of the surgery as it captures the key elements that patients value.

The Difference between Principal Components Analysis and Factor Analysis

Despite the obvious similarities between the two techniques, factor analysis differs from principal components analysis in its approach to factor extraction. Factor analysis analyzes the covariance among variables to produce either correlated or uncorrelated factors, and principal components analysis analyzes the variance of the variables to produce only uncorrelated components. In principal components analysis, extracted components are linear combinations of the raw data, and in factor analysis, extracted factors are an explanation of the raw data and the relationships within them.

6.6.2 Multidimensional Scaling

Multidimensional scaling (MDS) provides a means of visualizing, or mapping, the level of similarity of individual cases of a dataset. It is particularly useful in visually representing products that are perceived as similar by consumers. The distribution of products in a multidimensional space can provide an indication as to the product dimensions considered important by consumers. It also provides an indication as to where gaps exist in the current product offering.

The basic MDS process is as follows:

1. Select the objects to be researched. An example is products within a specific product category, with a view to developing and launching a new product in this category.
2. Make a list of all possible paired combinations of the existing products. Techniques are available, such as factor analysis, to reduce the number of pairs if the list becomes too long.
3. Ask a sample of the target market to rate the similarity or substitutability of products in each pair. A sample size of 30–50 is common.
4. Analyze the paired comparison scores using MDS software to generate a visual map representing the relationships among the products.

The dimensions of this visual map represent the bundles of key attributes that are important to consumers in their decision on similarity or substitutability. Usually, the number of dimensions is kept to two or three for simplicity of communication. See Figure 6.4.

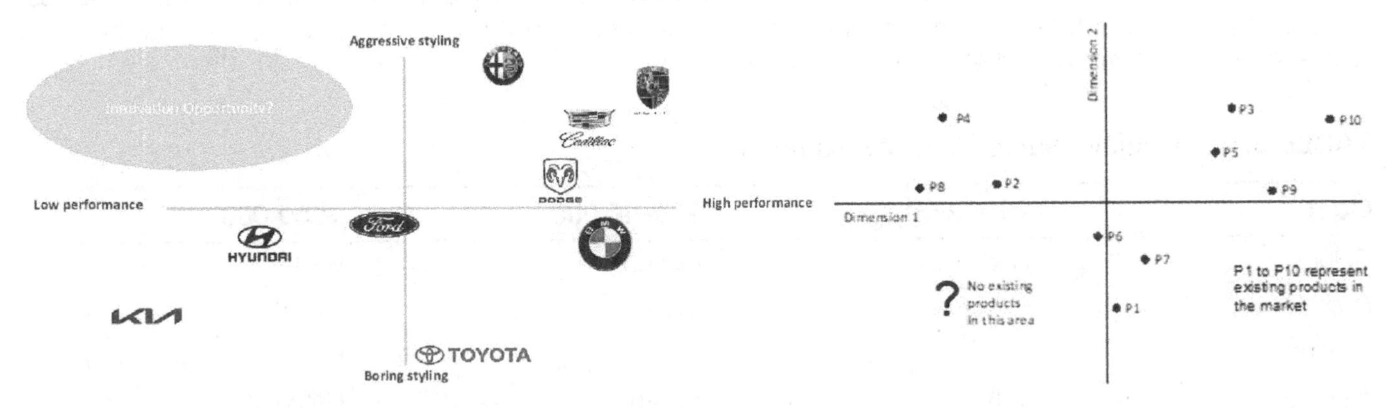

FIGURE 6.4 Multidimensional scaling example

6.6.3 Conjoint Analysis

Conjoint analysis is a multivariate statistical technique used in product innovation to determine how people value different attributes (features, functions, benefits) that make up an individual product or service.

A conjoint survey captures the relationship between different attributes of a product and how preferable they are both comparatively and when combined.

One of the primary purposes for using conjoint analysis is to gain strategic insights and making better business decisions relating to product pricing, product feature development, branding, package design, and marketing messaging validation.

Conjoint surveys don't let respondents say that everything is important. It systematically varies product attribute levels to create competing, realistic product profiles, and then records what people choose. The objective of conjoint analysis is to determine what combination of a limited number of attributes is most influential on respondent choice or decision-making.

The basic conjoint analysis process is as follows:

1. Potential attributes of the new product are defined with different levels or features of these attributes (e.g., battery life of 1 day, 5 days, 10 days).
2. The different levels or ranges of attributes are set randomly into combinations.
3. A sample of consumers (usually 30–150) is asked to rate the different combinations of attributes.
4. Results are analyzed to identify the most influential attributes in consumer decision-making.

Conjoint Analysis Example

Table 6.14 presents a simple example based on the evaluation of the "right" combination of vehicle attributes to assist in new vehicle design. When searching for the "right" vehicle, the consumer typically considers a multitude of variables: type, year, mileage, price, color, accessories packages, fuel type, fuel consumption, etc. Given a large number of variables and potential options for each variable, the consumer faces a complicated decision. The perfect choice, based on the ideal variable and options, is almost always impossible. Trade-offs have to be made. Conjoint analysis doesn't let respondents say that everything is important. It systematically varies product attribute levels to create competing, realistic product profiles and then records what people choose.

Even in this simple vehicle example with three options for six different attributes, the complete number of possible combinations is very large. Statistical design processes are used to reduce this number of combinations to make the process more manageable for respondents. Conjoint surveys present samples of possible combinations, called choice sets, for respondents to rate, as shown in Table 6.14. Using survey results, it is possible to calculate a numerical value that measures how much each attribute and level influenced the respondent's choices. Each of these values is called a preference or utility score.

Preference scores can be used to develop simulation models which help to identify the specific features and pricing that balance value to the customer with cost to the organization and forecast potential demand in a competitive market situation.

TABLE 6.14 Conjoint survey: simple example

Cost	<$30,000	$30–40,000	>$60,000
Color	Red	Black	White
Fuel	Petrol	Diesel	Electric
Seating	2	4	6
Type	SUV	Sedan	Hatch
Engine	2l	3l	4l
Accessories	Full	Moderate	Minimal

6.6.4 Multiple Regression Analysis

Multiple regression analysis is often used in product innovation to analyze survey-based data. It provides detailed insight that can be applied to new products or improve products or services when there are factors, key drivers, and product attributes that can impact the product's perceived value proposition. It is used more often than simple linear regression in product innovation due to the number of factors that can impact new product success. It can be used to identify which variables have an impact on the topic of interest and is used to predict the value of a variable based on the known value of two or more other variables (predictors).

In the analysis, the results are often plotted on an axis that displays the relationship between the datasets. Multiple regression analysis is widely used to predict, optimize, support, or validate decisions, avoid risk or prevent mistakes, and provide new insight into unarticulated relationships in the product variables being studied.

6.6.5 Multivariate Analysis Strengths and Weaknesses

The strengths and weaknesses of multivariate techniques are summarized in Table 6.15.

TABLE 6.15 Strengths and weakness of multivariate techniques

Multivariate technique strengths	Multivariate technique weaknesses
• Provide insights into the interrelationships among variables which does not exist with univariate techniques. • Techniques including multidimensional scaling and factor analysis can be powerful tools to uncover unarticulated needs and identify gaps in current product offerings.	• Complex both in terms of design and analysis requiring a high level of statistical understanding. • Require specific programs, which required specialized expertise, for analysis. • For the data to be meaningful, large sample sizes are required.

6.7 PRODUCT USE TESTING

Product use testing measures the performance of a product under specific conditions replicating actual consumption, operation, manipulation, or handling. It is also referred to as field testing or user testing. Product use testing focuses on the product meeting consumer needs and requirements prior to launching the product in the market.

Product use testing should not be confused with test marketing or market testing which are discussed in Section 6.8.

6.7.1 In-home Use Test vs. Central Location Tests

In-home use test (IHUT) is where a product is utilized or consumed and tested in the participant's home setting. An IHUT is especially useful when the product is used frequently, two or more times per week, or if the product is a food, beverage, or beauty product and can be consumed during different use occasions. It is also useful if the usage is a sensitive or private situation. The product is shipped to participants, who log their usage experiences and responses into online software. Real-life product use in the home or other intended use environment is important to the reliability of research results.

Central location test (CLT) is where testing is carried out in a specially designed testing location. Individual participants or a group of participants meet in a central location where the product is tested under simulated use conditions.

A summary of how to choose between IHUT and CLT is presented in Table 6.16.

TABLE 6.16 Choosing between IHUT and CLT

Factor	IHUT vs. CLT
Time available	CLT delivers quicker results.
Market representation	IHUT provides for deeper insights into realistic product usage.
Natural setting	IHUT has a significant advantage over CLT in that it is carried out in a real-life setting.
Broader audience	Where you are looking for responses from a broader audience, for example a food product that would be consumed by a whole family, then IHUT is preferred over CLT.
Attracting participants to a central location	Some target consumers may find it difficult to get to a central location, for example the elderly, people caring for small children, low income etc. IHUT would be preferred.
Control over product usage	Where product use and testing requires significant instruction and supervision, CLT is preferred.
Confidentiality	CLT allows for greater confidentiality, possibly through use of no-disclosure agreements.

6.7.2 Alpha, Beta, and Gamma Testing

Alpha, beta, and gamma testing is a form of market research primarily used in the software or technology industry to test a new product during development and just prior to launch. This form of market research may not be strictly quantitative, in the sense of providing a specific level of statistical confidence, due to the lack of statistically based sample selection. It does, however, provide the level of detailed feedback that only comes from the use of the product in its final, or near-to-final, form and functionality.

Alpha Testing

Alpha testing is initial usability testing, which is normally done by in-house testers, such as a quality control group. On rare occasions, alpha testing is done by the client or an outsider. Once the alpha testing version is released, it's then called the Alpha release.

Beta Testing

Beta testing is done by end users before product delivery. The users give feedback or report defects. Change requests are made and fixed prior to full product launch. The version released after beta testing is called Beta release. Beta testing can be considered pre-release testing. Beta test versions of software are now distributed to a wide audience partly to give the product a real-world test and partly to provide a preview of the next release. The main objective behind beta testing is to get feedback from different customer groups and check the product's compatibility with different kinds of networks and hardware.

Gamma Testing

Gamma testing (or a gamma check) is performed when the application is in its final state and ready for release to customers. The gamma test is the last opportunity to identify critical errors in the product. Security, performance, and usability are generally the focus of gamma testing. No feature development or enhancement of the software is planned or undertaken as a result of this test, and only tightly scoped bug fixes are written at this stage. Gamma testing is less common now, as greater pressure has been imposed on speed to market and automated testing tools have improved.

6.8 MARKET TESTING AND TEST MARKETING

6.8.1 The Difference Between Market Testing and Test Marketing

These terms can be confusing. Both tests are designed to provide customer feedback from specific target markets at later stages of product development. The difference is that ***market testing*** focuses on an existing product in a new market or market segment, while ***test marketing*** focuses on a new product in a market.

Market Testing

This testing is carried out to determine the market potential for an existing product in a specific area where the product did not exist earlier. This includes testing new market segments, considering target users, geographies, demographics, or any new market attribute that was not a target in the initial product launch. The test is done to ascertain whether the product will be successful or not in the new market and to analyze its sales potential.

Test Marketing

Test marketing is carried out by companies who are planning to launch a new product. The test is carried out on a small scale to determine the response of their customers to the new product, frequently collecting data on all four aspects of the marketing mix (product, price, promotion, place). This test is conducted before the product is launched. It provides valuable customer feedback on areas for product improvement as well as the overall sales potential.

The strengths and weaknesses of market testing and test marketing are shown in Table 6.17.

TABLE 6.17 Strengths and weaknesses of market testing and test marketing

Strengths of test marketing and market testing	Weaknesses of test marketing and market testing
• Provides information that significantly increases the probability of making the correct decision. • Significantly reduces the probability of wasted capital and other expenditures on an unsuccessful market launch or expansion. • All elements of the distribution and marketing plan can be tested and validated. • Data can be used to improve sales forecasting for full launch.	• Can be time-consuming and expensive. • May lead to delays in full expansion launches. • Could provide competitors with an early insight into potential market plans and allows them more time to launch a competitive response.

6.8.2 Specific Methods of Test Marketing and Market Testing

Simulated Test

Customers are selected and surveyed on their brand familiarity and preferences in a specific product category. These customers are exposed to promotional material and other marketing mix variables related to the product in a staged or artificial marketplace. Researchers expose customers to advertising and other marketing mix variables to gauge their purchase intent.

Controlled Test

A panel of stores is selected to stock the new product under real market conditions. Shelf position and number of product facings are controlled, and sales are measured at checkout. A sample of customers is later interviewed or surveyed to give feedback on the product.

Sales Wave Research

Used to test consumer reaction to new products prior to full-scale commercialization; new products are placed in consumer homes to determine the reaction to them, and the rate at which the products are re-purchased is tracked.

6.9 MARKET RESEARCH AND VOICE OF THE CUSTOMER: INTEGRAL TO PRODUCT INNOVATION

Cooper (2023) claims that "in general, marketing activities are the most poorly executed activities of the entire new product process, rated far below corresponding technical (engineering, design, R&D) activities. Moreover, relatively few resources are spent on marketing actions (except for the launch), accounting for less than 20 percent of total projects. A market focus is relevant throughout the entire new product project," as shown in Figure 6.5.

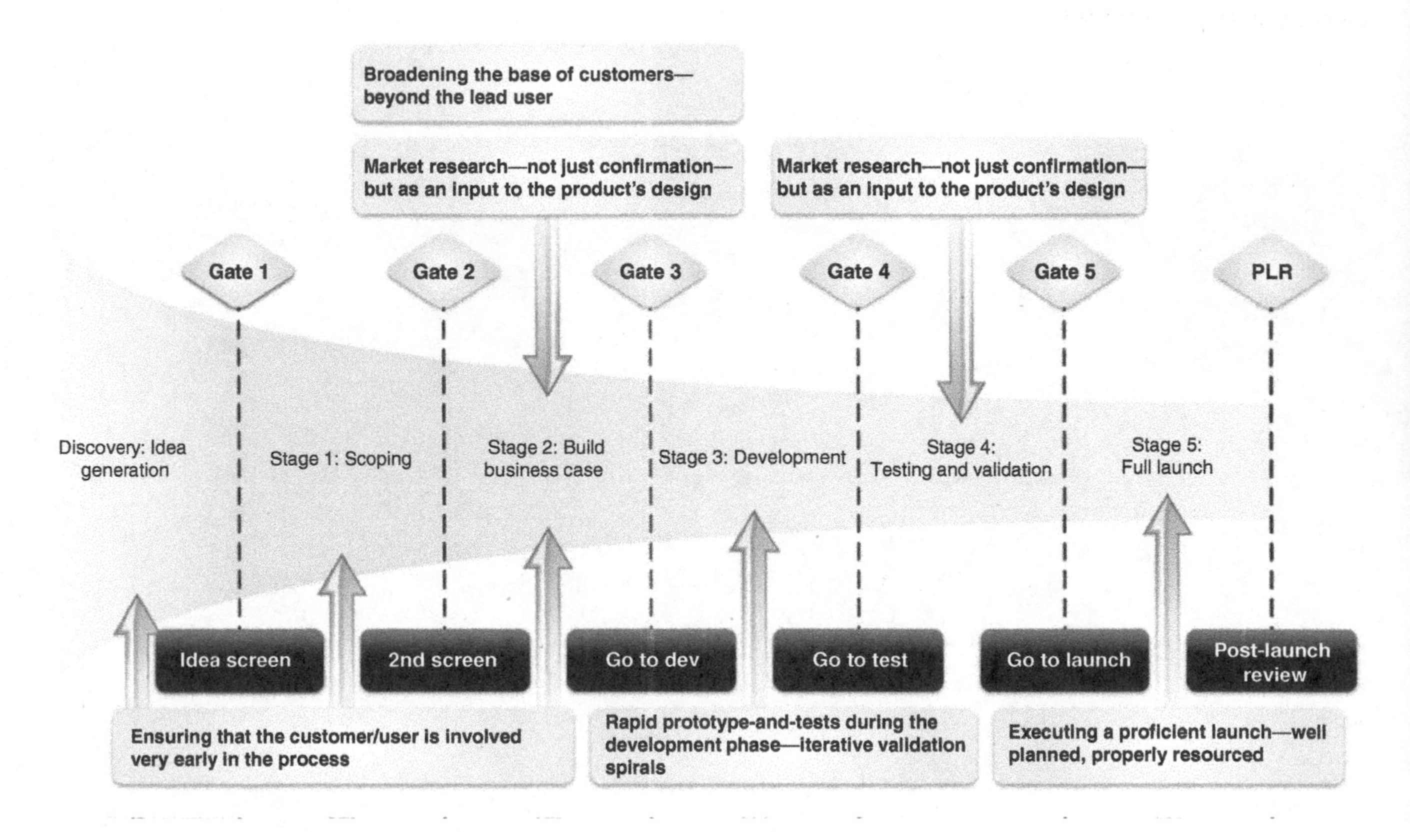

FIGURE 6.5 A strong customer focus includes key actions from beginning to end in the innovation process. From *What separates the winners from the losers and what drives success.*
Source: Cooper, (2023) / Stage-Gate International.

Cooper (2023) goes further to point to a high degree of customer participation at the prototype development stage (feedback, testing, opinions, and comments), is much more likely to result in market success. He claims that "smart project teams build in a series of deliberate iterative steps whereby successive versions of the product are shown to the customer to seek feedback and verification, as shown as the curved arrows in Figure 6.6. Not only do iterations reduce market uncertainties, they also can be used to reduce technical uncertainties by seeking technical solutions in an experimental, iterative fashion."

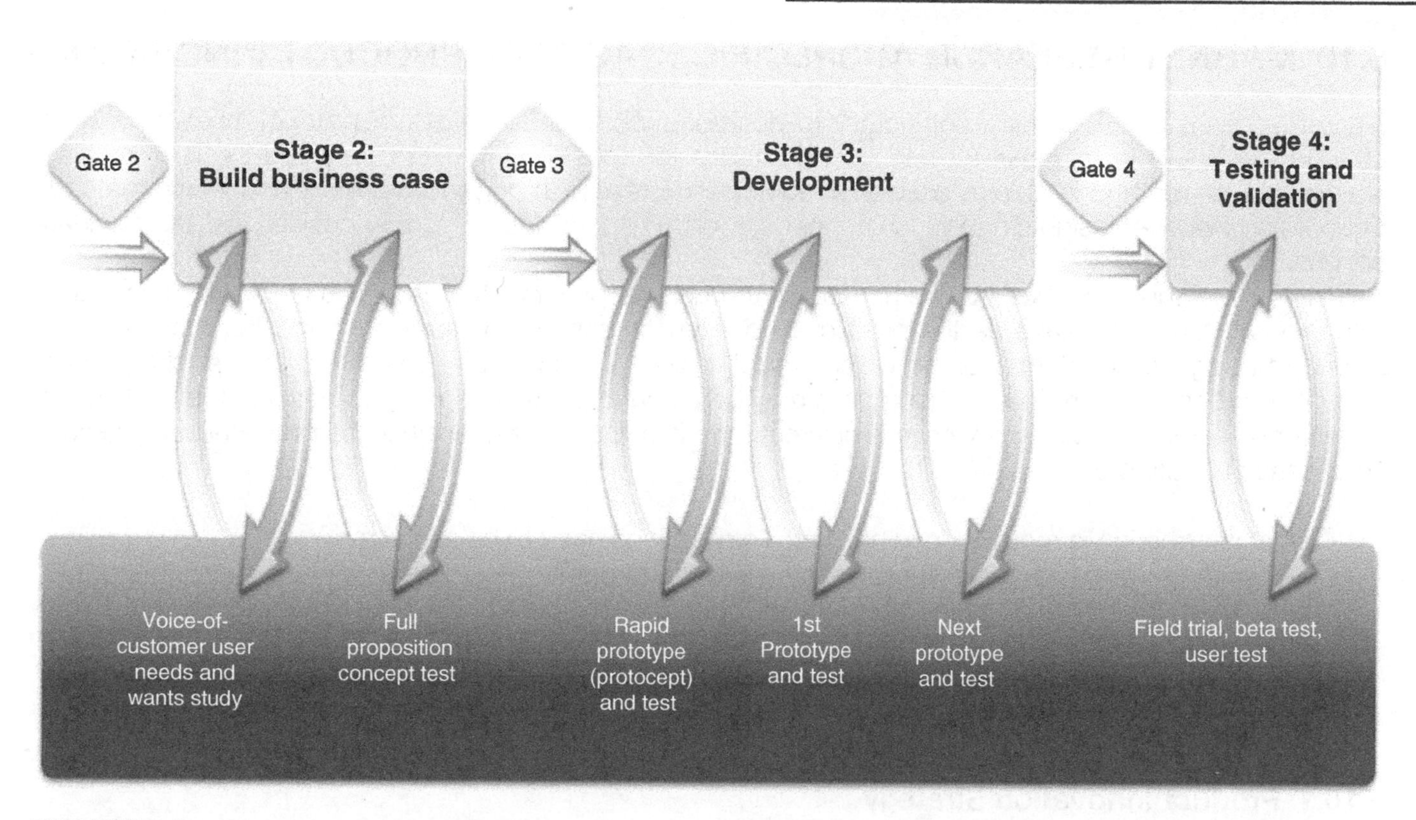

FIGURE 6.6 Iterative or spiral development – a series of "build-test-feedback-revise" iterations with customers – gets the product right. From *What separates the winners from the losers and what drives success.*

Source: Cooper, (2023) / Stage-Gate International.

6.10 MARKET RESEARCH AT SPECIFIC STAGES OF PRODUCT INNOVATION

Market research is an essential contributor to decision-making at all stages of product innovation. Each method discussed in this chapter has its advantages for specific stages and required decision-making. No one tool or method is able to answer all questions. It is also important to understand which methods explore and discover insights and which methods confirm and validate or narrow possibilities and eliminate options.

Throughout this book, we have emphasized the general principle that the risks and costs associated with product failure increase as the product innovation process progresses. This implies that the quality and reliability of the information on which critical decisions are made needs to increase as the product innovation process progresses. Although this may be a sound general principle to apply in the selection of a market research method, every case is different, and it is important to answer the three critical questions addressed in Section 6.2:

- What information is required, or what specific questions need to be answered?
- Who is the target market or important stakeholder group(s)?
- What level of confidence in the accuracy of the data is required? This will normally relate to the risks involved.

The following is a guide to some of the methods best suited to each phase of product innovation.

6.10.1 Product Innovation Strategy

This phase generally requires forward-looking research. What are the trends? Where are the new potential markets? Which markets are likely to grow and which to decline? What customers do we already have relationships with? What customer problems do we understand well?

The use of a range of methods is important in strategy development. Comparing and cross-referencing various sources lays the foundation for sound decision-making during this phase. Secondary research sources can provide a good source of information, particularly where relevant trend analysis is provided. SWOT, PESTLE, and Delphi methods, as described in Chapter 2 Strategy, provide the most appropriate methods for targeted primary research.

6.10.2 Identifying Opportunities

This phase is often called the front end of innovation or the discovery phase. Opportunities may come in the form of totally new products, modifications, or improvements to existing products, or line extensions to existing product lines or platforms. The primary focus is identifying these opportunities and early-stage evaluation of their potential. Various forms of primary and secondary qualitative market research methods are well suited.

Secondary methods can be good sources of new ideas, including Internet searches, trade fairs, trade journals, patents, suppliers, social media channels, and government and trade statistics.

Primary qualitative methods are particularly useful for identifying opportunities, especially insights competitors have missed. Methods include:

- A range of ideation techniques, crowdsourcing, and big data analysis.
- Focus groups involving customers or other stakeholders, distribution channel members, sales force representatives, and organization staff.
- Social media sites, both general social media sites, specific discussion forums, and blogs.
- Customer site visits, particularly for B2B product innovation.
- Ethnography for identifying unarticulated needs that can lead to product ideas.
- Multivariate analyses identify gaps in the current market offerings or the underlying attributes valued by customers and which can be formed into the new product concept.

6.10.3 Evaluating Ideas and Early-stage Business Analysis

The selection of opportunities for inclusion in the product portfolio is discussed in Chapter 3, Portfolio Management. The focus during this phase is the collection and analysis of key market information required relating to market size and sales potential, competitor and competitor products, target market characteristics, and the price the customer is likely to pay. This information provides the basis for early-stage financial analysis and for ongoing refinement of the financial analysis.

Secondary research can provide some general information on aspects of the target market, sales potential, and competitive landscape.

Primary research methods include:

- Interviews
- Focus groups
- Surveys
- Concept sorts

6.10.4 Concept Development and Testing

During this stage, a more detailed concept description is sought, leading to product design specifications (refer to Chapter 5.3). Customer and other stakeholder input are required to identify the key benefits, attributes, and functionality desired in the product.

Primary options for research methods include focus groups, lead-user groups, online discussion forums, customer site visits, surveys, concept testing and concept sorts, sensory testing, recipe formation and testing, and specific multivariate techniques such as conjoint analysis.

Secondary research into competitive solutions includes various online sources and intellectual property databases for existing patents and trademarks. More detailed and robust secondary research on market potential and the business case is important to justify the increased cost of development in the next stage.

6.10.5 Prototype Development and Product Use Testing

During this stage, the concept takes physical form with possible functionality. Project costs start to rise significantly, and it is increasingly important to make the right decisions regarding product benefits, form, and function. The availability of samples or prototypes that display form and possibly function provides a far better basis for seeking customer feedback than the concept description (words and images) used to describe the product in the previous stage. Demonstration of the prototype during research will certainly add significantly to customer understanding of the product and, in turn, result in more valuable and reliable feedback.

At this stage, the accuracy and reliability of information become more and more important. Consideration should be given to primary quantitative research, including surveys, and alpha testing, with qualitative tools (interviews, focus groups, and customer site visits) to help explain the quantitative results. Virtual reality, 3D prototype models or visuals, images, and video are useful as stimuli for research when building iterative physical prototypes is cost-prohibitive.

6.10.6 Pre-launch Product and Test Marketing

At this stage, the product is developed and manufactured to its final commercial form and functionality. Next comes the decision on whether to proceed with full product launch. Although obtaining the most reliable information to avoid product failure is a critical consideration, this often has to be balanced against the need for speed to market and the cost of testing. If speed to market is an absolute priority, or the risk of product failure is relatively low, then a beta test or full launch may be appropriate. Where there is significant potential for damage to a brand or financial loss, then a test market may be justified.

6.10.7 Post-launch Product Testing and Market Research

Once the product is launched, and throughout its life cycle, market research often plays an important role in determining its performance and success. Market research provides information for decision-making around line extensions, market expansion, and understanding competitive response. Some examples include the use of customer panels or lead user groups that provide answers to specific time-sensitive questions or longitudinal studies over a product's lifetime, competitive benchmarking and tear-down analysis, B2B customer satisfaction surveys, etc. Many companies purchase syndicated industry data, scan-track data, or participate in industry studies to learn more about their customers, markets, and market share over time.

A summary of appropriate market research methods in each phase of the product innovation process is provided in Table 6.18. This is intended to provide a guide to the most appropriate methods based on the information required and the risks involved in decision-making in each phase. The final selection of market research methodology will depend on a range of factors:

- Is the product new to the organization, a line extension, or an improvement?
- How familiar is the organization with the product and market? What level of prior knowledge resides in the organization?
- How much new capital expenditure will be required on plant and equipment?
- What level of new human resource commitment is required?
- What existing brand value could be impacted by the success or failure of the new product?
- What is the likelihood and timeframe for competition?

So, for example, a company planning to enter the electric vehicle market may have limited relevant prior knowledge currently within the company. Significant financial commitment will be required for R&D, and specialized human resources in the early phases of product innovation. Large capital expenditures will be required for prototyping and for final manufacture. There will almost certainly be significant risks related to brand equity. Also, costs associated with launch, market support, and after-sales service will be high, relative to the current product range. This all adds up to significant risks throughout the product innovation process, demanding high-quality and reliable market information. This points to the need for quantitative research methods that provide statistically reliable information from the specific target market.

On the other hand, a food manufacturing company specializing in fermented milk products is planning to develop a line extension to its existing yogurt range. The company has significant knowledge of, and experience in, the yogurt market. Very little new capital equipment expenditure is likely to be required. Product launch and after-sales maintenance of the market can be handled with existing resources. And there is unlikely to be a negative reaction to the extent that it would negatively impact the brand. All pointing to the need for qualitative market research, providing input to the new product's sensory profile and functionality. Although statistically reliable market data may be helpful, the costs of this type of research are probably not justified, given the low risks involved.

TABLE 6.18 Market research at different phases of product innovation

Product innovation phase	Information required	Level of risk	Appropriate research methods
Strategy development	Future trends in products and markets.	Relatively high. Strategic decisions can lead to significant investments in R&D, equipment, and human resources.	Although largely qualitative, it important to cross-compare information from a range of sources leading to objective analysis. Methods include: • Secondary research, Surveys, SWOT, Delphi, PESTLE
Opportunity identification	Ideas for new products or product improvement. Information on a target market – who is the customer, demographics, psychographics? What are their needs and desired benefits?	Relatively low. Initial project costs are generally low and no commitment to capital at this stage. Proceeding to the next stage of concept evaluation should not be costly with limited commitment to resource and project continuation.	Mainly qualitative: • Secondary research, Social media, Focus groups, Customer site visits, Lead user groups, Ethnography, Multivariate tools
Opportunity evaluation	Who is the target market and how large is the market? What is the economic potential? How well does the product fit within the new product portfolio?	Relatively low, but at this stage you are heading down the road to project commitment where costs and project risks can rise rapidly.	Qualitative with some attempt to quantify: • Secondary research, Focus groups, Social media: lead users, Customer site visits, Surveys
Concept development	What features, functionality, and aesthetics does the customer want or need in the product? This information provides the basis for concept development and product design, leading to product design specifications.	Low to medium. But a commitment at this stage leads to much higher financial commitment in more expensive design and prototyping.	Mainly qualitative: • Focus groups, Social Media, Lead users, Online discussion forums, Customer site visits
Prototyping and testing	Is the product design heading in the right direction? Now that the customer can start to see the tangible product, is it meeting their expectations – their needs and wants? What changes are required? Customer feedback at this stage can provide more confidence in the financial analysis and the business case for commercialization.	Very much dependent on the project. Risk may be regarded as high where a commitment at this stage is leading to potential capital investment and high costs of commercialization. Further business analysis may be carried out to confirm earlier evaluations.	Mainly qualitative. But with a focus on quantitative, particularly where risks associated with commercialization are high: • Secondary research, Focus groups, Customer site visits, Surveys, Conjoint analysis

(Continued)

TABLE 6.18 (Continued)

Product innovation phase	Information required	Level of risk	Appropriate research methods
Pre-launch testing	Information is required from the target market on product acceptance (possibly relative to competitors), sales potential, pricing. All required to firm up on the business case for commercialization.	Dependent on the project. High risk, where significant expenditure will be required for commercialization. The trade-off between uncertainty leading to product failure and the pressure for speed to market must be considered. Low risk where commercialization costs are low and product/market knowledge is high.	Qualitative: • Beta testing, Market testing, Test marketing, Probability sampling could be applied to Test Marketing to increase the statistical reliability of the data. Thus, providing greater confidence in the business case analysis.
Post-launch and life cycle management	Information determining product success and measuring distribution, sales, market share, demand, competitive response, customer user satisfaction are required. Data on potential for line extensions, improvement, or enhancements, market or distribution expansion may be required, and research methods are similar to the opportunity identification step.	High. Commitment of ongoing human and operational resources are required to launch and maintain products in market. Firms risk long-term financial sustainability by ignoring changing consumer needs, demand and the competitive landscape.	Quantitative: • Primary sales, distribution data, Syndication data Qualitative: • Longitudinal and satisfaction surveys, Secondary research, Industry data, Economic and consumer trends, Competitive analysis, Social media analysis

6.11 MARKET RESEARCH METRICS AND KEY PERFORMANCE INDICATORS

While primary research is designed to help answer unique and specific questions that guide product innovation and management decisions, there are also many common metrics used in market research for product innovation, development, and product life cycle management that measure, monitor, analyze, and present results in a comparable and efficient way to track product success. These common measures are ubiquitous across industries and categories and provide a core set of transferable measurements across businesses.

The following are common product innovation and management research metrics with a brief and basic description of what is measured:

- ***Acquisition effort:*** The extent to which your product or service is accessible to your customer.
- ***Awareness:*** Product and brand: the extent to which customers are familiar with your product or brand.
- ***Brand development index:*** Sales of your brand compared with its average performance in all markets.
- ***Brand image:*** How customers think and feel about your product and brand.
- ***Convenience:*** The extent to which your product or service makes your customer's life easier, saves time or effort, etc.
- ***Customer attitudes:*** The extent to which your customers have a favorable or unfavorable attitude toward your product or service.
- ***Distribution:*** The extent to which your product is available in the market.
- ***Ease of use:*** The extent to which your product or service is simple to operate, consume, engage, or interact with.
- ***Engagement:*** The extent to which your customers interact in a relationship with your company, brand, product, or service.
- ***Installed base:*** The number of units that are actually in use over a particular time frame.
- ***Market penetration:*** The percentage of your target market that you have reached at least once in a specific period of time.
- ***Market share:*** The percentage of the total market held by your company, brand, product, etc.
- ***Market size:*** A measure of the total market potential (in terms of sales, profit, number of potential buyers, units sold, volume, etc.) for a company, product, or service.
- ***Net promoter score:*** The likelihood someone would recommend your product or service to a friend.
- ***Percent of all commodity volume (%ACV):*** The percentage of the total annual sales volume aggregated from all retailers where your product can be sold.
- ***Pride (to own, to serve):*** The extent to which your product or service contributes to a positive sense of self for your customer.
- ***Satisfaction:*** The extent to which your product or service meets the needs of your customer; how happy customers are with a company's products, services, and capabilities.
- ***Usage and purchase intent:*** The extent to which someone says they will use or purchase your product or service.
- ***Willingness to pay:*** The highest price at which a customer says they will definitely buy your product or service.

As a specific example of the application of market research metrics, Table 6.19 presents four of the most important SaaS (Software as a Service) sector metrics (Bernazzani, 2022).

TABLE 6.19 Four of the most important market metrics for the SaaS sector

Metric	Description
Customer churn rate	Measures how much business you've lost within a certain time period. It is one of the most important metrics in tracking the day-to-day vitality of your business, and can help you better understand customer retention across specific date or time periods.
Revenue churn	Reveals how much revenue was lost in a given period. For subscription-based companies. Choose a specific period of time – one year, one month, all time, etc. – and be consistent in entering the metrics from that time for accuracy.
Customer lifetime value (CLV)	The average amount of money that your customers pay during their engagement with your company. Find your customer lifetime rate by dividing the number 1 by your customer churn rate. For example, if your monthly churn rate is 1%, your customer lifetime rate would be 100 (1/0.01 = 100).
Customer acquisition cost (CAC)	Shows exactly how much it costs to acquire new customers and how much value they bring to your business. To calculate CAC, divide your total sales and marketing spend (including personnel) by the total number of new customers you add during a given time.

6.12 IN SUMMARY

- Market research includes a range of techniques that are essential to product innovation to enable informed decision-making throughout the new product innovation process and management of the product life cycle.
- The two basic categories of market research are primary and secondary. Primary research is original research conducted by an organization specifically to meet its objectives. Secondary research relies on the collection of information from studies previously performed and published by other individuals, groups, or agencies.
- Market research techniques can be further classified as either qualitative or quantitative. In general, quantitative techniques provide a greater level of statistical confidence in the research. Qualitative techniques can provide context and deeper insights into customers' problems and unmet needs.
- In selecting which market research technique to use, it is important to recognize the costs and risks associated with a specific decision. Qualitative techniques, although valuable throughout product innovation, are particularly helpful in the early stages of idea generation and concept development. In the later stages of product innovation, costs and risks will generally increase, thus requiring greater confidence in the market research and justifying the additional costs of quantitative research.
- Ongoing application of market research throughout the new product innovation process significantly increases the chances of product success. Specific techniques are more appropriate at different stages of the process. No single technique is perfect, and the use of a range of techniques and cross-referencing the results is the best approach.
- Crowd sourcing and Big Data has gained prominence as a key source of market research for product innovation. They provide large volumes of data that can often compensate for the lack of statistical reliability, although care should be taken in over-reliance on their statistical accuracy.

A. MARKET RESEARCH PRACTICE QUESTIONS

1. A product manager reviewed technical publications, electronic databases, and websites prior to designing a research project that directly contacts prospective customers. This review of already-published materials is an example of:
   A. Market testing.
   B. Voice of the customer.
   C. Portfolio management.
   D. Secondary market research.
2. During new product development, companies often use to measure end user satisfaction with a product and to determine whether the company can deliver the total quality product as promised.
   A. Lead user research.
   B. Product use testing.
   C. Secondary research.
   D. Quality function deployment.
3. You are responsible for validating the value proposition of adding a number of new features to your current product line. A way to explore the association between adding one or more new features and the perceived value to the user would be to:
   A. Color code the feature sets so that they are easily identified by the user.
   B. Add fake sets of features to see if respondents identify the actual product features.
   C. Run a volumetric study of the most profitable product prototype to forecast sales.
   D. Conduct a conjoint study varying feature sets and price points to optimize the new feature set and product price.
4. Factor analysis, multidimensional scaling, conjoint analysis, and multiple regression analysis are examples of:
   A. Qualitative research techniques.
   B. Multivariate research techniques.
   C. Voice of customer techniques.
   D. Ethnographic research techniques.
5. Qualitative market research techniques should be considered when?
   A. Context and direction are needed.
   B. Deep customer insights are sought.
   C. Customers' unmet needs have yet to be discovered.
   D. All of the above.
6. A software company is developing an update to a specific app for online shopping. The company has extensive experience with its target market through previous product launches. Even if there are minor faults with the app, these can be quickly remedied after launch without significant negative customer response. Speed to market is critical. The company is seeking customer reaction to the improved features and functionality of the new app. What market research approach would you recommend?
   A. Focus groups.
   B. In-house alpha testing followed by beta testing.
   C. Test marketing.
   D. Factor analysis.
7. Research that involves collecting information specifically tailored to your needs (such as focus groups or surveys) is known as____.
   A. Primary research.
   B. Secondary research.

C. Qualitative research.
D. Quantitative research.

8. A disadvantage of focus groups in marketing research is ____.
   A. Their high cost to implement focus groups.
   B. The long process of recruiting participants, conducting the focus groups, and providing analysis.
   C. The complexities of executing focus groups.
   D. They are not good at identifying needs customers have yet to recognize.
9. A toy manufacturing company is developing a "powered vehicle" for 10–12-year-old children. The company has extensive experience in the toy market, but mainly with under 5-year-olds. The main risks associated with the project relate to a lack of previous engagement with the 10–12-year group along with meeting their specific requirements. Getting the product right is more important than speed to market. What market research technique would you recommend?
   A. Consumer panels selected from the target age group.
   B. Alpha testing.
   C. Lead users.
   D. Crowdsourcing.
10. Social media is particularly good for reaching ____.
   A. Potential new customers.
   B. Lead users.
   C. Customers in foreign markets.
   D. Those needing customer service.

Answers to practice questions

1. D
2. B
3. D
4. B
5. D
6. B
7. A
8. D
9. A
10. B

REFERENCES

American Society for Testing and Material (1981). *Guidelines for the selection and training of sensory panel members (ASTM-STP 758)*. ASTM.

Belliveau, P., Griffin, A., and Somermeyer, S. (ed.) (2002). *The PDMA toolbook 1 for new product development*, *Vol. 2*. John Wiley & Sons.

Bernazzani, S. (2022). *15 metrics every SAAS company should care about*. HubSpot Blog https://blog.hubspot.com/service/saas-metrics.

Cooper, R.G. (2023). What separates the winners from the losers and what drives success. In: *The PDMA handbook of innovation and new product development*, 4e (ed. K.B. Kahn), 3–44. John Wiley and Sons.

Cooper, R.G. and Dreher, A. (2015). *Voice of the customer methods: What is the best source of new product ideas?* Stage-Gate International https://www.stage-gate.com/new-product-development-process/voice-of-customer-methods-what-is-the-best-source-of-new-product-ideas/.

Hofacker, C., Malthouse, E.C., and Sultan, F. (2016). Big data and consumer behavior: Imminent opportunities. *Journal of Consumer Marketing 33* (2): 89–97. https://doi.org/10.1108/JCM-04-2015-1399.

Kahn, K.B., Castellion, G., and Griffin, A. (ed.) (2004). *The PDMA handbook of new product development*, 2e. John Wiley and Sons.

Katz, G. (2010). *Rethinking the product development funnel*. Massey University. http://stream.massey.ac.nz/mod/url/view.php?id=1423336.

Laney, D. (2001). *3D data management: controlling data, velocity, volume and variety*. Gartner.

Miller, C., Perry, B., and Woodland, C.L. (2004). Ethnographic market research. In: *The PDMA toolbook 2 for new product innovation* (ed. P. Belliveau, A. Griffin, and S. Somermeyer), 297–330. John Wiley and Sons.

Moe, W. and Schweidel, D. (2017). Opportunities for innovation in social media analytics. *Journal of Innovation Management 34* (5): 697–702. https://doi.org/10.1111/jpim.12405.

Naresh, K.M. (2009). *Marketing research: An applied orientation*. Prentice Hall.

Ottum, B. (2004). Quantitative market research. In: *The PDMA handbook of innovation and new product development*, 2e (ed. K.B. Kahn, G. Castellion, and A. Griffin), 279–301. John Wiley and Sons.

Pisano, G., Verganti, R., and Deschamps, E. (2015). A study on the impacts of business model design and innovation on firm performance. *Entrepreneurship Research Journal 5* (3): 181–199. https://doi.org/10.1352/2326-6988-3.3.199.

Statista. (2022). *Social media marketing usage rate in the United States from 2013 to 2022*. https://www.statista.com/statistics/203513/usage-trands-of-social-media-platforms-in-marketing/

Thomke, S. and Von Hippel, E. (2002). Customers as innovators: A new way to create value. *Harvard Business Review 80* (4): 74–81.

Zikopoulos, P., Eaton, C., DeRoos, D. et al. (2015). *Big data beyond the hype: A guide to conversations for today's data center*. Paul McGraw-Hill.

7

Product Innovation

Culture and Teams

Essential to forming and maintaining an innovative environment that enables, encourages, and rewards product innovation processes and practices.

Product Development and Management Body of Knowledge: A Guidebook for Product Innovation Training and Certification, Third Edition.
Allan Anderson, Chad McAllister, and Ernie Harris.

WHAT YOU WILL LEARN IN THIS CHAPTER

While strategies, processes, and tools are important for product innovation success, ultimately, it is people that really matter – the culture, teams, and leaders. In this chapter, we discuss the importance of developing the right culture of innovation; the roles and responsibilities of management; what is required to achieve a high-performing team, team development, and team leadership; and what team structures are appropriate in specific situations.

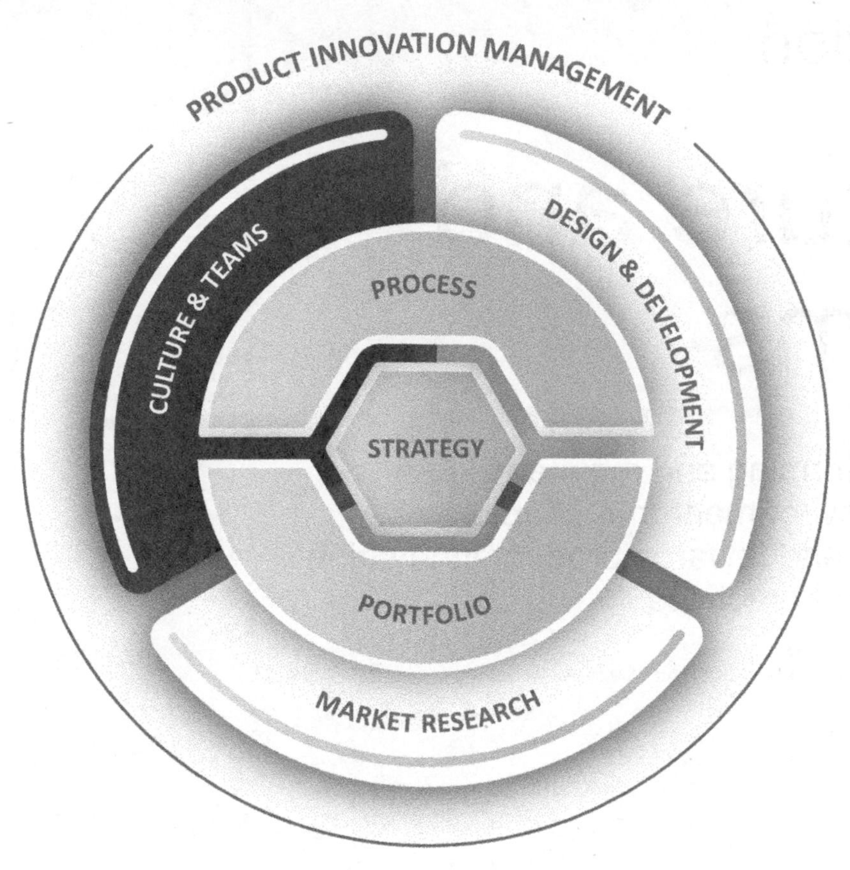

CHAPTER ROADMAP

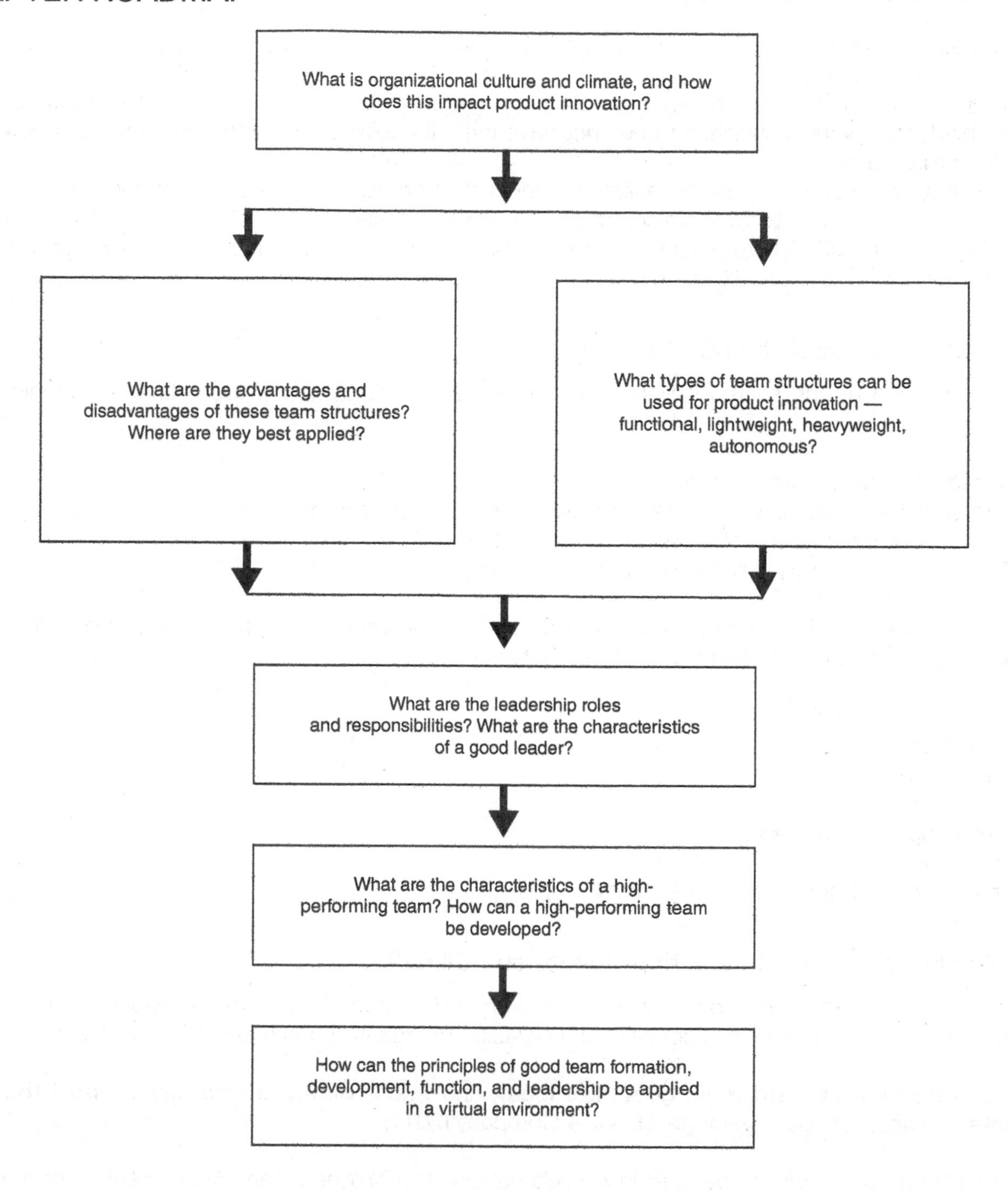

7.1 PRODUCT INNOVATION CULTURE AND CLIMATE

"Culture eats strategy for breakfast." This quote was attributed to the management guru Peter Drucker in 2006 and made famous by Mark Fields, who later became chief executive of Ford Motor Company.

Put simply, this means that no matter how strong your strategic plan is, its implementation will be seriously hindered if your organization does not have the right culture supported and shared by everyone across the organization.

While strategy and process are essential to product development and product management, this alone will not lead to the organization's sustained success. People are ultimately what make an organization successful. The culture and climate ultimately provides the framework within which the strategy and process can be positively and successfully implemented.

7.1.1 What are Organization Culture and Climate?

Culture is defined as the shared beliefs, core values, assumptions, and expectations of people in the organization. Culture:

- Reflects the organization's values.
- Is observable in customs, rites, ceremonies, stories, heroes, and patterns.
- Infers how work is accomplished.
- Expresses survivability: what we do naturally to survive.

Climate is defined as the set of properties of the work environment, perceived directly or indirectly by employees, that has a major effect on employee behavior:

- Quality of leadership
- Communication
- Responsibility
- Trust
- Fair recognition and rewards
- Opportunity
- Employee involvement and participation

7.1.2 The Importance of Innovation Culture and Climate

The importance of an innovation culture is strongly advocated across most leading organizations. Following are quotes from senior executives of a selection of organizations that lead in product innovation:

> "Organizations whose **strategic goals are clear**, and whose **cultures strongly support those goals**, possess a huge advantage." Booz & Company (2011)

> "Innovation is so deeply embedded in Apple's culture. The boldness, ambition, belief there aren't limits, a desire to make the very best products in the world. It's the strongest ever. It's in the DNA of the organization." Tim Cook, CEO of Apple.

> "I encourage our employees to go down blind alleys and experiment. We've tried to create tools to reduce the cost of doing experiments so that we can do more of them. If you can increase the number of experiments you try from a hundred to a thousand, you dramatically increase the number of innovations you produce." Jeff Bezos, Amazon Founder.

"What does it take to create this kind of innovative impact? It starts with great hiring. . .ensuring a 'culture fit' is critical. . .sharing Airbnb's core values, and comfortable with risk and enjoy adventure. . .environmental elements that support the innovative culture include:

- Collaborative workspaces, fun spaces encouraging people working across disciplines.
- Think globally, act locally.
- A flat organization where risk-taking is encouraged. Ask for forgiveness rather than permission." Katie Dill, Airbnb.

"I came to see, in my time at IBM, that culture isn't just one aspect of the game—it is the game. In the end, an organization is nothing more than the collective capacity of its people to create value. Vision, strategy, marketing, financial management—any management system, in fact—can set you on the right path and carry you for a while. But no enterprise—whether in business, government, education, health care, or any area of human endeavor—will succeed over the long haul if those elements aren't part of its DNA." Lou Gerstner, former CEO of IBM.

Although the Toyota production system (TPS) is largely credited for the organization's success, Toyota views employees not just as pairs of hands but as knowledge workers who accumulate *chie* – the wisdom of experience – on the organization's front lines. Toyota, therefore, invests heavily in people and organizational capabilities, and it garners ideas from everyone and everywhere.

Venture capitalist Ben Horowitz wrote in, "What you do is who you are," that a great culture alone does not get you a great company (Horowitz, 2019). The point he's making is that if the market doesn't want what you're selling, culture isn't going to get you to success. That said, culture is to business as nutrition is to an athlete. If you're a talented athlete, you may be successful, even with poor nutrition. However, if you also have a solid nutritional regimen, you are much more likely to be a world-class athlete.

Going beyond Horowitz's nutrition analogy, a strong cultural identity breeds strong intellectual and emotional alignment. When people are aligned in mission, vision, and values, much more is possible. Organizations with award-winning cultures demonstrate this alignment, such as Google, Adobe, HubSpot, Zoom, and many others (Peart, 2020).

The critical importance of culture to product innovation success is also strongly supported by the results of the PDMA's CPAS survey carried out in 2012 (Markham & Lee, 2013). Although it is universally accepted that culture plays a very important role in product innovation success, there is no "magic bullet" for creating a culture of innovation. The "right" culture is very much personal to the organization. What works well in one organization may be completely wrong for another.

One factor contributing to this is employee engagement. Gallup has researched aspects of workplace engagement for decades (Harter, 2022). They define employee engagement as the involvement and enthusiasm of employees in their work and workplace. Employee engagement in the US has varied from 28 to 36% since 2001, meaning that only about a third of the workforce, on average, feels engaged at work. Engaged employees are needed to support an innovation culture.

As an example, a study of 150 American companies demonstrated widespread gaps between employee views of their work environment and engaging, creative, innovation team climates (Denning, 2015).

- Only 5% of survey respondents felt highly motivated to innovate,
- More than 75% reported their ideas were poorly reviewed and analyzed,
- One in seven (16%) did not believe intellectual property was viewed as a critical business function, and
- Nearly half (49%) felt they would not receive any recognition or benefit for developing successful ideas.

There are several factors and initiatives that feature in most successful innovation cultures:

- Organizational values with a clear focus on product innovation.
- A business strategy that clearly emphasizes the role and importance of product innovation.
- A clear sense of goals and direction within the business strategy is communicated, understood, and shared across the organization.
- Trying and failing is better than not trying at all. "Ask for forgiveness, not permission."
- Individual performance objectives are clearly related to overall organization innovation objectives, and good performance is recognized in appropriate ways.
- Fit with the innovation culture is a key recruitment criterion.
- Effective communication should be encouraged internally and externally.
- Constructive conflict is encouraged. A vigorous debate is better than passive agreement.
- As far as possible, make work enjoyable and rewarding – not just financially, but personally.
- Work is engaging, and leaders encourage both professional and personal growth.
- Constructive conflict is encouraged to support idea generation and problem-solving.

7.2 MANAGEMENT RESPONSIBILITIES

In this section, the roles of management are discussed with respect to several areas crucial to successful new product innovation. These include:

- Strategy,
- Product innovation processes,
- Organization and teams, and
- The product itself.

As you explore culture and team organizational structures across companies, you will find that roles, responsibilities, and titles vary greatly. For example, a program manager, portfolio manager, and senior product manager may all perform the same duties in different companies. The structure and titles of an innovation team are often heavily influenced by the pedigree of the leadership team itself. If the team is heavily sales-based, titles will tend to be more market-facing such as Director of Innovation. If the leaders are more operational in their background, titles such as Program Manager or Portfolio Manager will seem more tactical.

7.2.1 Product Innovation Strategy Roles

As discussed in Chapter 2, corporate strategy is built on the foundation of mission, vision, and values. Further, all other strategies, including innovation strategy, are derived from corporate strategy, as shown in Figure 7.1. Corporate strategy is determined by a corporate executive team. A typical executive team includes senior representatives from key corporate functions, such as finance, marketing, sales, manufacturing or development, and technology. A company's board of directors may be directly or indirectly associated with the executive team, with the CEO (chief executive officer) linking the two groups.

Titles and specific roles and responsibilities will again vary significantly from organization to organization based on the size, industry, geography, and more.

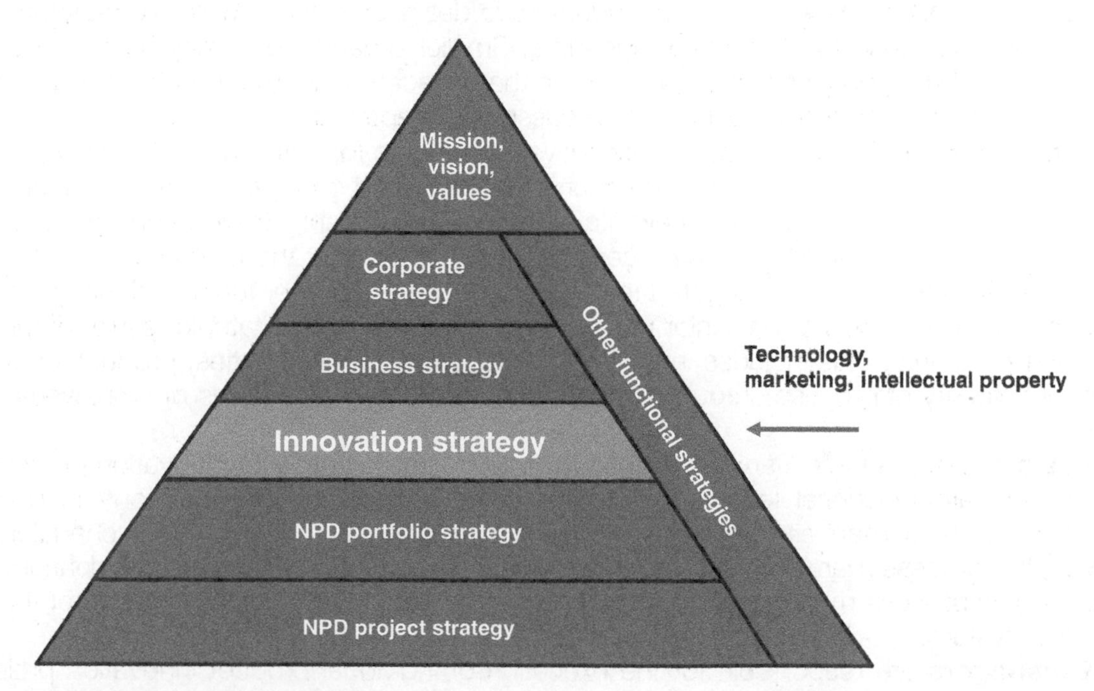

FIGURE 7.1 Hierarchy of strategies

Business Strategies are typically defined by senior leadership team members, or in the case of companies with multiple business units, leaders of those respective units. Some typical titles involved in the creation of a Business Strategy include but are not limited to:

- President
- Executive Vice President, Senior Vice President, or Vice President
- Director

Innovation Strategies expand upon the questions answered in the Business Strategy and are often developed by a cross-functional team. As with all strategies, alignment is key, and it is typically the senior leadership team that ensures that alignment. The objective is for innovation to be integrated consistently across organizational objectives.

Functional Strategies, including development/engineering, marketing, sales, manufacturing, procurement, and finance, are developed by the functional leader and a senior team from the specific functional area. Normally, goals are implemented by these functional leaders to support the implementation of higher-level strategies and to achieve tactical objectives.

Product Strategies will normally be developed by the product team led by a senior product person, such as the VP of Product, who coordinates with the senior leadership team to ensure alignment. Whereas the overall business and innovation strategies may change very little on an annual basis, product strategies reflect changing trends and market opportunities. The product strategy includes life cycle management and product or brand management of individual products and platforms. These concepts are discussed in Chapter 1, Product Innovation Management.

7.2.2 Product Innovation Process Roles

Organizations that successfully launch repeatable innovations typically include several specific management roles. In some cases, these innovation leaders have designated titles, while in many organizations, these roles are filled by business or functional leaders. Smaller organizations may have one person fulfill many roles. Titles and designations vary depending on the project management methodology (e.g., Stage-Gate®, waterfall, Agile, or hybrid approaches, as discussed in Chapter 4).

First, the **process champion** is a senior executive responsible for establishing the new product innovation process. A process champion works to ensure the quality and consistency of the implementation of product innovation processes. Training of new staff members and talent development to support innovation also fall within the responsibility of the process champion. Note that the facilitation of product innovation processes and training may be delegated to the process owner or other functional managers.

The **process owner** is typically a senior manager responsible for the strategic results of the organization's innovation programs. This includes innovation strategy alignment activities, product innovation process throughput, quality of process outputs, and active participation by all levels of employees across the organization.

A **process manager** is necessary to successfully implement any product innovation process. Usually, process managers are functional leaders responsible for ensuring that product portfolio decisions are implemented in an orderly manner. In addition to ensuring the adequacy of approved schedules, budgets, and resources, the process manager often facilitates innovation training, brainstorming, ideation, and post-launch reviews. The process manager will gather and analyze data to support the metrics of the organization's innovation system.

Project managers are responsible for the execution of individual product innovation projects. They follow the accepted methodology for project management and ensure that project milestones are delivered on time and on budget. Depending on the product innovation scale and scope of work, project managers may serve only in a supervisory role or work on the project's technical or marketing aspects.

7.3 PRODUCT INNOVATION TEAM STRUCTURES

7.3.1 Team Membership

Individual team members are responsible for implementing the work of an innovation project. Team members should be trained in both product innovation process implementation and demonstrate functional knowledge and expertise to accomplish the project's work. ***The most successful innovation companies use cross-functional teams throughout the product innovation life cycle***.

Team Diversity

Research into the impact of diversity on financial performance dates back several decades or more. Harvard Business Review reported in the article, *"How and Where Diversity Drives Financial Performance,"* that companies which were above average in total diversity (measured as the average of six dimensions: migration, industry, career path, gender, education, and age) were 19% points higher in innovation revenue and 9% points higher in EBIT (Lorenzo and Reeves, 2018).

Abbie Griffin has made many contributions to the innovation discipline and researched serial innovation teams in organizations. Some of her research is shared in the book she co-authored titled, "Serial Innovators: How Individuals Create and Deliver Breakthrough Innovations in Mature Firms" (Griffin et al., 2012). The authors identified eight types of thinkers who are beneficial to innovation teams: linear, spatial, connection, holistic, critical, concrete, divergent, and convergent. Consequently, diversity in thinking styles also improves innovation performance.

While more diverse teams tend to deliver better financial results, diversity alone is not sufficient. Diversity without sound organizational and managerial processes is not enough. Rather it is the addition of diversity to sound leadership practices that appears to deliver an added benefit.

Multi-disciplinary and Cross-functional Teams

High-performing, multi-disciplinary teams improve the quality of the product and decrease project development times. Cross-functional communication brings issues to light sooner and facilitates collaboration. Hand-offs between functional teams at milestones reduce knowledge transfer and should be minimized. Successful innovations are delivered by product innovation teams that combine R&D, engineering, operations, and marketing skills throughout the development project. An ideal cross-functional team shares the following group characteristics:

- Includes all necessary functional representatives,
- Ensures team member assignments are continuous from project initiation to market launch,
- Provides appropriate communication tools,
- Establishes clear project and team objectives with expected performance outcomes, and
- Indicates functional, project, and career alignments.

7.3.2 Project Team Structures

Typical project team structures used in innovation range from a model with deep, functional expertise to autonomous teams tasked with designing and developing new-to-the-world products. Wheelwright and Clark (1992) first identified four common product innovation team models:

- Functional team,
- Lightweight team,
- Heavyweight team, and
- Autonomous team.

Each team model is described below.

Functional Teams

Functional teams are typically built from an organizational hierarchy. As indicated in Figure 7.2, product innovation team members are drawn from individual functions, such as engineering, manufacturing, and marketing. Each team member is responsible for a portion of the product innovation project that corresponds to their functional expertise. One or more functional managers very loosely coordinate team members' work, and the work often involves hand-offs among the functions.

A functional team can be used successfully in product innovation under the following situations:

- In developing fundamental research for deployment across a broad set of product lines, and in which deep functional expertise and knowledge are critical to development,
- In entrepreneurial and small businesses, where there are few resources and few projects worked simultaneously, and,
- In organizations that seek incremental product improvements as part of a defender strategy so that functional competency outweighs the need for multi-disciplinary activities.

Team member performance and tasks are generally maintained by the functional manager. Functional teams are also used for very low-risk product improvements, especially in slow-moving industries.

Organizations emphasizing production efficiency over innovation are more likely to use functional teams.

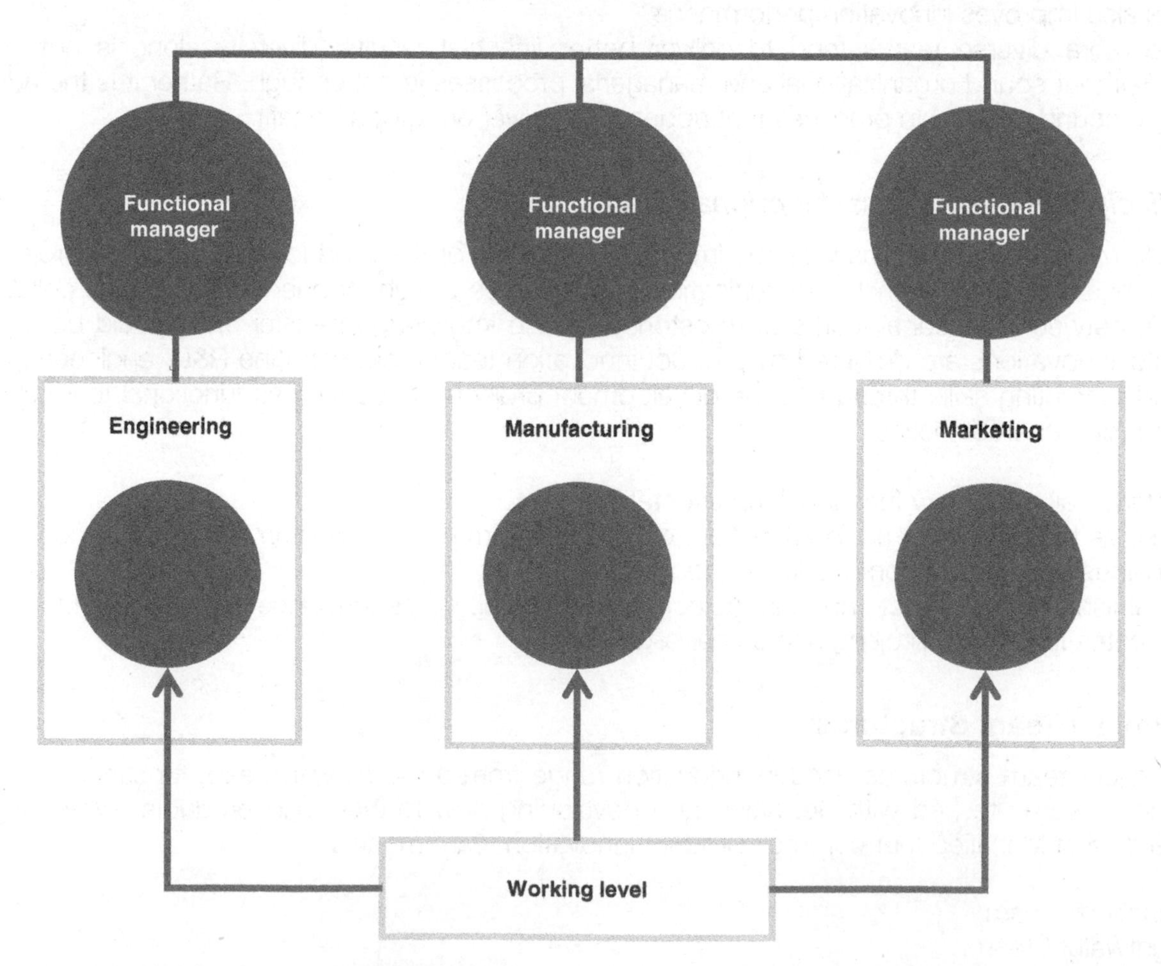

FIGURE 7.2 Functional team structure

Lightweight Teams

As illustrated in Figure 7.3, and as compared to a functional team, a lightweight team adds a degree of coordination for the new product innovation project. Functional liaisons are identified, and a project manager is named. The project manager may also be a significant independent contributor to the project work in a lightweight team. Team members continue to report to their own functional manager with the added responsibility of the innovation project. The project manager does not have authority over the team members and serves primarily as a communication liaison for project information.

Lightweight teams are often used for minor product improvements in which the development work requires coordination among functions. Functional team members have deep expertise regarding the product and features but may not be able to participate in the project throughout the full design and development life cycle.

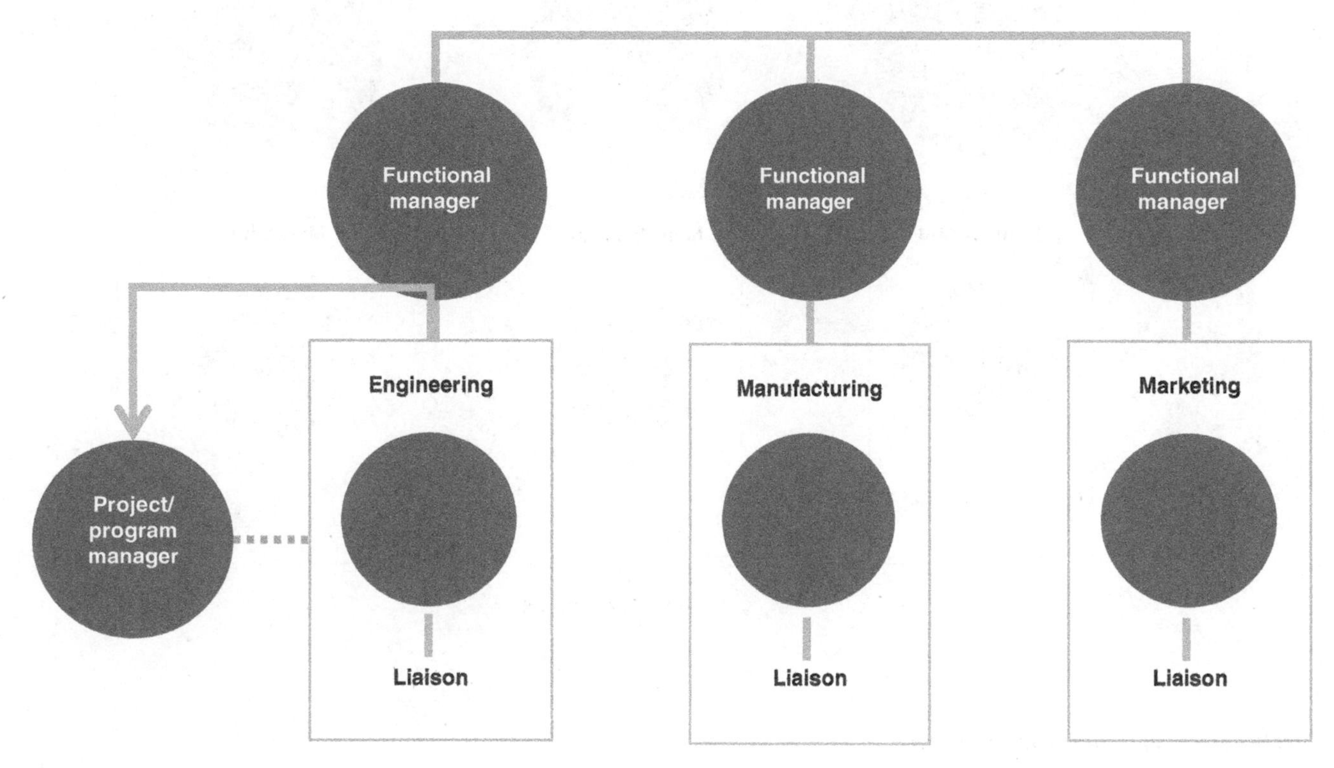

FIGURE 7.3 Lightweight team

Heavyweight Teams

In contrast to the lightweight team, a heavyweight team has a greater focus on the project than on functional alliances. A key driver for project work comes from the market as the product innovation team converts a concept into a commercially viable product or service. As shown in Figure 7.4, a cross-functional core team is assembled with members from every discipline necessary to accomplish the product innovation work. The project manager/leader, in contrast to lightweight teams, does have authority while also providing coordination and communication, both internally and externally. The project manager tends to be a more senior leader in the organization.

Because the role of the project manager is clearly delineated and the innovation work is more complex, the project manager formally directs the work of individual team members. In the heavyweight team model, the project manager has responsibility over the work of staff members, but the functional managers retain career management and ultimate authority for their performance.

While communication, coordination, and collaboration are highly focused on the product innovation project in a heavyweight team, this organizational structure is not right for every project. The heavyweight team structure should be used when the technical or marketing development is complex, involving new applications, customers, and markets. Teams can be large, with each core team member supervising another sub-team within their function. Heavyweight teams are more resource intensive than functional workgroups and lightweight teams, requiring skilled leadership that can focus and energize team members across a broad spectrum of functions and disciplines. In many cases, the heavyweight team members are dispersed geographically, and the team leader will deploy additional tools and techniques to manage the virtual team (see Section 7.6).

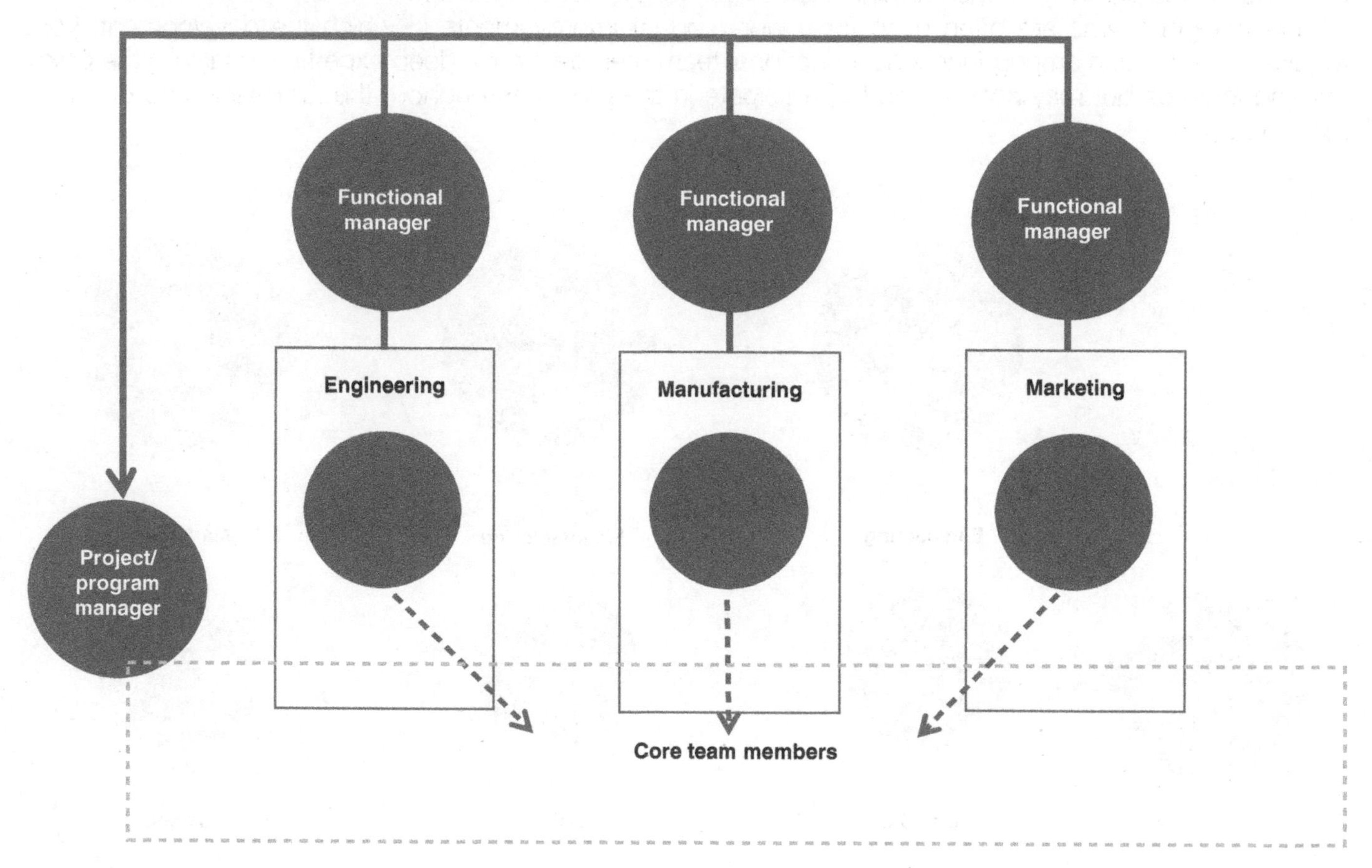

FIGURE 7.4 Heavyweight team

Autonomous Teams

The word autonomous means *independent and self-governing*. Thus, autonomous teams are used in new product innovation for major, long-term ventures. Such teams are sometimes called *tiger teams* or known as *skunk works projects*, which was coined at Lockheed Aircraft Corp. They aim to model the structure of an entrepreneurial startup firm within corporate boundaries. Clayton Christensen recommended the autonomous team structure for radical, disruptive innovation (Christensen & Raynor, 2003).

Autonomous teams, as shown in Figure 7.5, are led by a senior executive as the project manager and remove team members from their home function to form the stand-alone project venture. The project leader has full and complete authority and responsibility for the team and the success of the new product innovation effort. Often the venture team is housed in a separate location, away from the organization's headquarters or operational facilities, to yield higher independence and autonomy for the development team.

A key advantage of autonomous teams is the laser-like focus on the purpose and mission of the project. Often these teams work on new-to-the-world products with disruptive technologies entering (or creating) brand-new markets. This type of work is typically energizing to team members, and they often remain assigned to the maintenance and growth of the new product line over its subsequent life cycle, including the development of next-generation products and services.

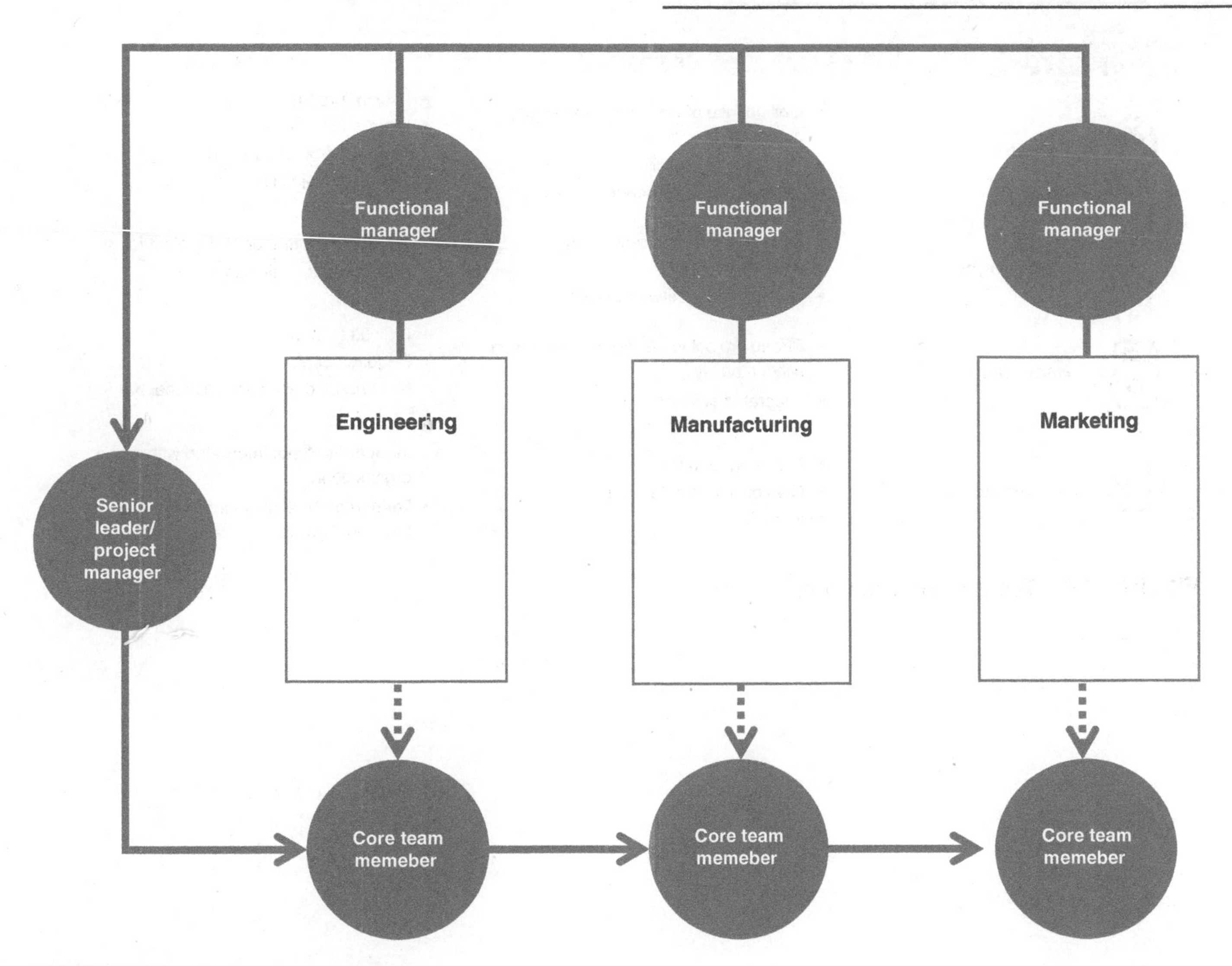

FIGURE 7.5 Autonomous team

Team Structure Comparison

Each organizational structure for new product innovation projects has strengths and weaknesses (see Figure 7.6). The correct team structure should be chosen for the needs of the innovation project. However, this is less common than it should be, as many organizations use the same team structure for a wide variety of project types. Instead, senior leaders should work with the portfolio management team, the product innovation process owner, and the product innovation process manager to identify the appropriate team structure for each innovation project.

In general, projects requiring depth of knowledge and expertise with little customer interaction are better suited to execution by a functional work team or a lightweight team. As uncertainty and the complexity of technology and business development increase, team structures such as the heavyweight and autonomous teams become more valuable.

Strengths of one team structure can be weaknesses of others. For example, having a deep knowledge of a single product technology is a strength for the work of a functional team, but a heavyweight or autonomous team needs to have open, customer-focused perspectives to yield broadly applied and creative solutions. Career congruency is well-established for lightweight teams but may be less certain for individuals working on a complex product innovation project under a new venture.

Type of development team	Strengths	Weaknesses
Functional	• Optimal use of resources, expertise, depth, scale economies • Control accountability • Career path congruence	• Lack of breadth • Rigid and bureaucratic • Task not project oriented • Slow and disjointed • Turf expertise driven
Lightweight	• Improved communications and coordination • Less idle time between tasks	• Weak project leader and project focus • Frustrating to individuals
Heavyweight	• Strong project focus, commitment, and accountability • Integrated solution	• Difficult to staff • Requires depth • Must break down into functional barriers
Autonomous	• Focus on results • Own business objectives • Innovative	• Independent, not integrated with rest of organization • Team members may lose connection to functional group

FIGURE 7.6 Team structure comparison

7.4 TEAM DEVELOPMENT

7.4.1 Defining a High-performing Team?

Katzenbach and Smith (1993) defined a team as "a small number of people with complementary skills who are committed to a common purpose, set of performance goals, and approach for which they hold themselves mutually accountable." High-performing teams typically grow and develop through several well-known stages and processes. Team leaders work with the product innovation process champion and product innovation process owner to ensure that innovation team members have the right skills and that the team structure is appropriate for the scale and scope of the project work. Innovation teams are successful when the team climate assures strategic alignment, engagement, and empowerment as key arenas.

> ***Strategic alignment:*** Team members need to understand how the project is connected to, and driven by, business objectives. The common purpose of the project is aligned with and contributes to the overall goals of the organization.
>
> ***Engagement:*** Motivated team members feel a sense of pride in their work and camaraderie with their fellow teammates. Rewards and recognition for both individual and team contributions drive improved performance.
>
> ***Empowerment:*** Empowered team members are more creative and make decisions that lead to better product designs. Open dialogue is encouraged, and team members' views are considered in decision-making for innovation projects. The team itself is empowered to do the work in the best way possible.

As illustrated in Figure 7.7, team success factors of strategic alignment, engagement, and empowerment are supported by additional elements. For example, strong leadership is required for an innovative team climate, as are interaction and involvement in which all team members have an equal voice in team decisions.

Self-esteem, one of the values in Maslow's hierarchy of needs, is reinforced by management and leadership for team members and is shared through rewards and recognition. Open communication is extremely important for innovation teams so that the team members can speak freely to share ideas and concepts.

In addition to empowerment at the team level (shown in the central core of Figure 7.7), individual **empowerment** means that each person is treated equally and that the team climate is mutually supportive of each team member, promoting psychological safety. *Effective processes* are also important for successful innovation teams. These processes include standards for knowledge sharing and meetings, as described in Section 7.6 on virtual teams. *Trust and diversity* are both key characteristics to building an engaged and empowered team because creativity will flourish when team members share trusting relationships and when different experiences contribute to unique solutions. Productive *conflict management* (see Section 7.4.4) is a common characteristic among high-performing innovation teams since some conflict can drive creativity, with task-based conflict resulting in better outcomes, but unhealthy conflict can hinder the accomplishments of a team.

Finally, as indicated with strategic alignment, a successful innovation team requires a project goal that is clear and succinct. Common goals unite team members and are especially important when working with team members in functional groups or who are dispersed as in a virtual team.

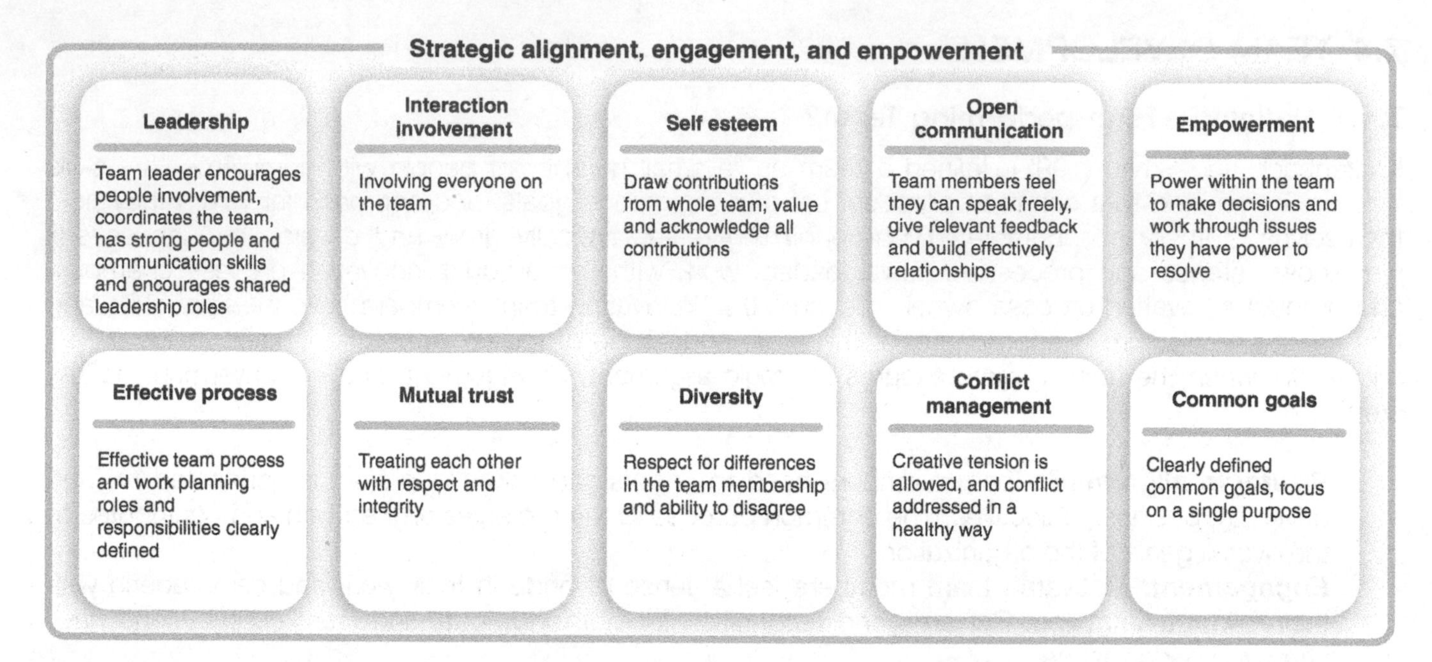

FIGURE 7.7 Framework for a high-performing team

7.4.2 Team Formation

Psychologist Bruce Tuckman (2001) developed a model of the growth stages of high-performing teams in the 1960s. His research indicates that teams naturally advance sequentially through a series of activities and emotions to reach their highest collaborative working potential. A team that is disrupted by changes in membership, for example, will backtrack to the initial stage and again advance sequentially through the team formation phases. These stages are *forming, storming, norming, performing, and adjourning* (refer to Figure 7.8). Note that if the team is ongoing, as in manufacturing or marketing, the stage of *reforming* is substituted for *adjourning*. Because product innovation teams are typically project-oriented, the adjourning stage is most common. However, as the model suggests, allowing teams to remain intact has significant performance advantages over teams that are rebuilt for each project.

Forming: In this stage, most team members are positive and courteous to one another. Some are anxious, as they may not fully understand what the team will do. Others are excited about the task ahead. The leader plays a dominant role at this stage because team member roles and responsibilities are generally not defined. Typically, this stage is not lengthy because as people start to work together, they try to get to know their new colleagues. In many instances, team members are appointed to innovation teams based on skills or product expertise and may assume a set of expectations for their role on the team, which can, in part, lead to storming.

Storming: The storming stage is characterized by conflict, and it is the primary responsibility of the team leader to manage these disruptions in a healthy way. People assigned to the team may be frustrated with the pace of development on the team or confused by the apparent duplication of roles. Often the problem statement is not clearly defined, and the storming stage is used to clarify the team's purpose and goals. Many teams, especially with weak or inexperienced leadership, can get stuck in the storming stage.

Norming: Gradually, the team moves into the norming stage. This is when people start to resolve their differences, appreciate colleagues' strengths, and respect the leader's authority. During the norming stage, team members establish their own ways of working together and agree upon standard practices. One way in which teams can reach an agreement is to have a clear vision for the project, project goals, and team processes documented in a team charter.

Performing: The team reaches the performing stage when hard work leads, without interpersonal friction, to the achievement of the team's goals. The team structures and processes established by the team are working well. The leader delegates more work to individual team members and concentrates on developing individual team members and skills for the overall group. Team members feel comfortable with each other and enjoy being part of the team. Project work is completed at a rapid pace, and learning is at a high level.

Adjourning: As indicated, in all temporary teams, the stage of adjourning arrives as the team's work on the project is complete. In product innovation projects, the product is launched and turned over to standard business operations. Team members are free to be assigned to other projects or are returned to their home organizations. In some cases, team members are reformed into new business divisions for ongoing support of the new product.

Forming
Appropriate team members join
May be a chaotic state as people find their roles
Team is positive and courteous to one another; some are anxious, and others are excited
Leader plays a dominant role

Storming
Conflict, frustrated with pace of development, or confusion with roles
Leader manages these disruptions in a healthy way

Performing
Team structures and processes are established and working well, team feels comfortable with each other and work rapidly
Leader delegates work to individuals and develops team members and skills

FIGURE 7.8 Stages of team formation

7.4.3 Work Styles

As a team is formed and develops, personal and professional work style preferences can impact the team's effectiveness. Work styles are especially important for team members to understand and respect in cross-functional, multi-disciplinary teams (Gallup, 2023). Individual bias can hinder trust and increase conflict if team members fail to recognize and value diversity within the team. For example, a classic point of conflict arises in innovation teams between engineering and marketing representatives. Engineers and marketers typically use their own jargon and are assumed to approach problems from very different perspectives. Engineers may assume marketers make only qualitative decisions, while marketers may assume the engineers are too slow and plodding in their analysis. Other business discipline representatives face similar stereotyping based on personality assumptions.

Using a work style assessment during the forming stage of a project team can help team members to overcome bias and assumptions. Work style assessments, like DiSC®, provide a tool and common vocabulary for team members to improve communication and collaboration. In particular, DiSC® creates a team profile index that allows team members to discuss problems, motivators, and stressors while guiding enhanced dialogue for creativity (Scullard & Baum, 2015).

The DiSC® work assessment tool describes everyone's preferred working style. Unlike other assessment tools (such as the Myers-Briggs or Big Five Personality Traits), DiSC® focuses on a team member in a professional setting. The four primary categories of work styles, refer to Figure 7.9, include the following:

D-Dominance: Individuals with a "D" work style tend to prefer a fast pace of work. They are quick to make decisions and may be perceived as overly demanding. These team members are action-oriented.

I-Influence: Team members with an "I" work style are highly energetic and talkative. They will build social relationships easily and seek new people to engage with. Some team members may view those with the "I" style as "all talk, no action"; however, these team members add enthusiasm to the work.

S-Steady: Innovation team members with the "S" work style are considered even-tempered and calm, valuable traits when the new product innovation work is chaotic or unstructured. These team members are more accommodating than many other people and will easily show empathy for others. While other team members may view their work pace as moderate, these individuals help to stabilize uncertain project activities.

C-Conscientious: The final category of DiSC® work styles are analytical and reserved. These team members need to have a complete dataset to evaluate before making a rational decision. They are often perceived as unemotional but pride themselves on accurate, detailed work. Others may view their attention to detail as a hindrance to project work.

Regardless of the tool used, it is important to understand the skillsets present on your team and how those skills will be most likely employed through personality and behavioral lenses. In most cases, product innovation project work is best accomplished with a balance of team member work styles. Both a diversity of product experience and a diversity of work styles can lead to increased creativity that yields more novel product solutions.

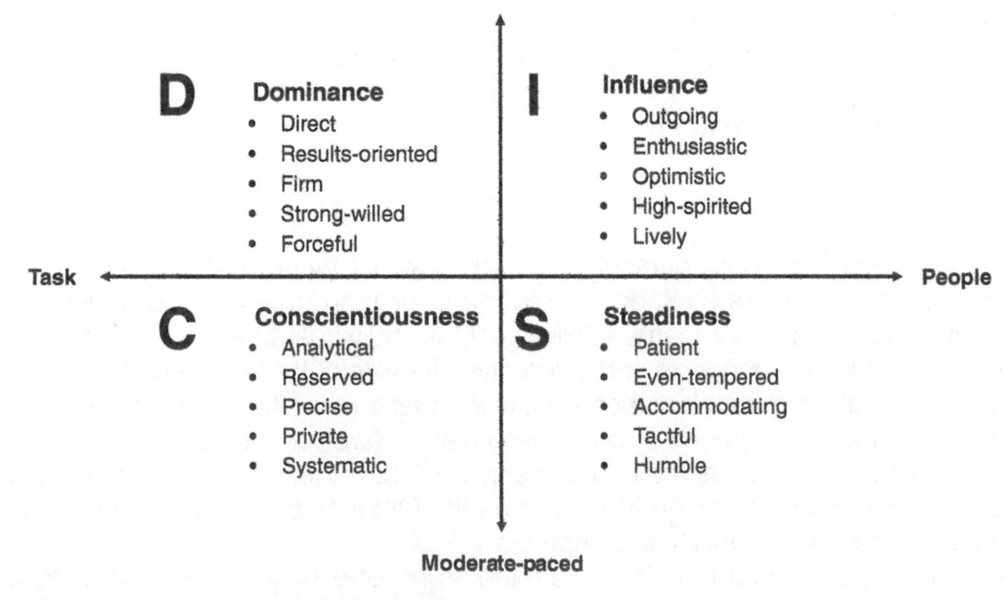

FIGURE 7.9 DiSC® work style assessments

7.4.4 Conflict Management

While dialogue and problem-solving are usually considered *healthy conflict*, disruptions to teamwork and the ability of group members to collaborate can hinder effective and productive project accomplishments. Conflict that is not addressed directly or dismissed as unimportant can fester, arising later to disrupt relationships and the team's productivity. Potential causes of conflict on an innovation team include (Kerzner, 2013):

- Resources, both people and financial,
- Equipment and facilities,
- Capital investment,
- Budget and costs.
- Technical opinions and trade-offs,
- Functional priorities,
- Organizational procedures and policies,
- Regulatory limitations,
- Scheduling,
- Project and functional responsibilities,
- Project constraints and boundaries, and
- Work style differences.

It is important to recognize that conflicts are a characteristic of high-performing teams. In these teams, the conflicts are associated with tasks and disagreements about the best way to accomplish a task or objective. Such disagreements, in teams that are also characterized as having high mutual accountability and a focus on delivering the expected results, lead to better decisions and outcomes.

A traditional approach to conflict management is known as the Thomas-Kilmann model (Kilmann & Thomas, 1978). The two dimensions, shown in Figure 7.10, are cooperativeness and assertiveness. Cooperation is viewed as a *concern for others* and is characterized by behaviors to encourage acceptance of views from other team members. Assertiveness is also known as *concern for self* and focuses on getting one's own views accepted by others. By combining various levels (low, medium, high) of cooperativeness and assertiveness, several conflict management styles are evident.

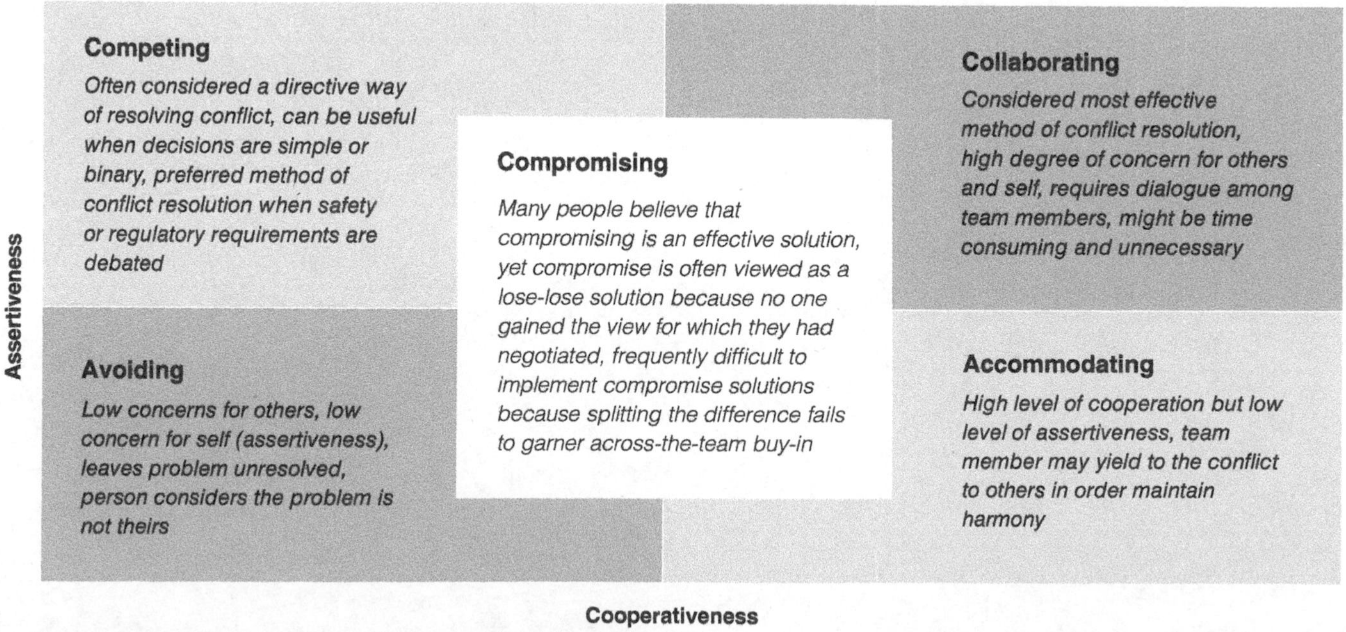

Competing
Often considered a directive way of resolving conflict, can be useful when decisions are simple or binary, preferred method of conflict resolution when safety or regulatory requirements are debated

Collaborating
Considered most effective method of conflict resolution, high degree of concern for others and self, requires dialogue among team members, might be time consuming and unnecessary

Compromising
Many people believe that compromising is an effective solution, yet compromise is often viewed as a lose-lose solution because no one gained the view for which they had negotiated, frequently difficult to implement compromise solutions because splitting the difference fails to garner across-the-team buy-in

Avoiding
Low concerns for others, low concern for self (assertiveness), leaves problem unresolved, person considers the problem is not theirs

Accommodating
High level of cooperation but low level of assertiveness, team member may yield to the conflict to others in order maintain harmony

Assertiveness

Cooperativeness

FIGURE 7.10 Thomas–Kilmann conflict model

Avoiding: Avoiding is characterized by low concern for others (cooperativeness) and a low concern for self (assertiveness). In many cases, the avoiding style leaves the problem unresolved. In other situations, a person may assume they are not part of the problem and leave the situation for others (who are closer to the problem) to engage in generating a solution.

Accommodating: In conflict resolution, the accommodating style is characterized by a high level of cooperation but a low level of assertiveness. In this situation, a team member may yield the conflict to others to maintain harmony in the team.

Compromising: At an intermediate level of cooperativeness and assertiveness is compromising. In typical interactions, many people believe that compromising is an effective solution, yet compromise is often viewed as a lose-lose solution because no one gained the view for which they had negotiated. It is frequently difficult to implement compromise solutions because splitting the difference fails to garner across-the-team buy-in.

Competing: With a high level of assertiveness and a low level of cooperativeness, competing is often considered a directive way of resolving conflict. Using the competing style to resolve team conflicts can be useful when decisions are simple or binary. It is also the preferred method of conflict resolution when safety or regulatory requirements are debated.

Collaborating: Many people believe that collaborating is the most effective method of conflict resolution because it involves high degrees of concern for others and concern for self. Collaboration requires dialogue among all team members so that each can express their own view of the situation. A drawback of collaboration for conflict resolution is that it is time-consuming and may not be necessary for every discussion item.

Managing conflict in project teams requires effective communication and negotiation. The team charter documents the process for escalation of conflicts, especially regarding decisions of resourcing and funding.

7.5 TEAM LEADERSHIP

As previously discussed, executives in senior management provide and set the direction for innovation through strategy and portfolio decisions. The product innovation process champion, process owner, and process manager also provide direct leadership for innovation teams. Innovation team leaders manage both the human resources of a project and the logistical goals of the project, such as scope, schedule, and budget. Effective leaders are emotionally intelligent and support the team as *servant leaders* rather than *task dictators*.

7.5.1 Roles and Responsibilities

The team leader provides direction, guidance, and support to a group of individuals working toward a specific goal. Effective leaders know the team members' strengths, weaknesses, and motivations. Team leader roles include:

- Providing purpose (what the team should achieve),
- Building a star team, not a team of stars,
- Establishing shared ownership for the results,
- Developing team members to their fullest potential,
- Making work interesting and engaging,
- Motivating and inspiring team members,
- Leading and facilitating constructive communication,
- Monitoring progress without micromanaging.

7.5.2 Communication

Team selection, its development, and ongoing operation are significantly impacted by a range of factors, both internal and external. Effective communication plays a significant role in the performance of the team and the interactions with the team leader.

Team performance is influenced by clear communication of the following:

1. Culture and environment of the organization; it's the values and behaviors that are encouraged for high performance.
2. Structure of the organization, including the roles and expected relationships among various functions.
3. Processes in place to promote and enhance team performance, such as the utilization of a team charter document to outline team member expectations.
4. The skills and capabilities represented in the team and how these can work together to achieve desired project outcomes.
5. Appropriate rewards with recognition that drive both individual and team motivation.
6. Leadership at all levels, including senior management, and how these are engaged and provide direction and support.
7. Collaboration and cooperation within the core team and how this extends, where necessary, to other functions and sub-teams, as indicated in Figure 7.11.

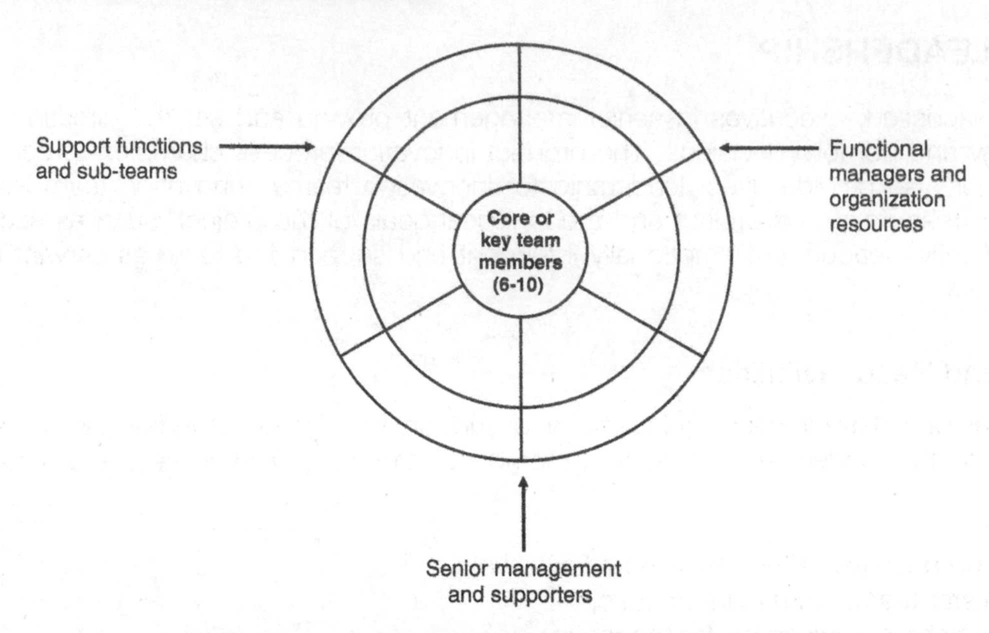

FIGURE 7.11 Team communication network structure

7.5.3 Emotional intelligence

Leaders high in emotional intelligence (EI) are twice as effective as leaders low in emotional intelligence. As Goleman (1998) has noted, technical competency and IQ are necessary but insufficient qualities for strong leadership performance. Emotional intelligence is comprised of self-management components and elements directed toward managing relationships (see Figure 7.12).

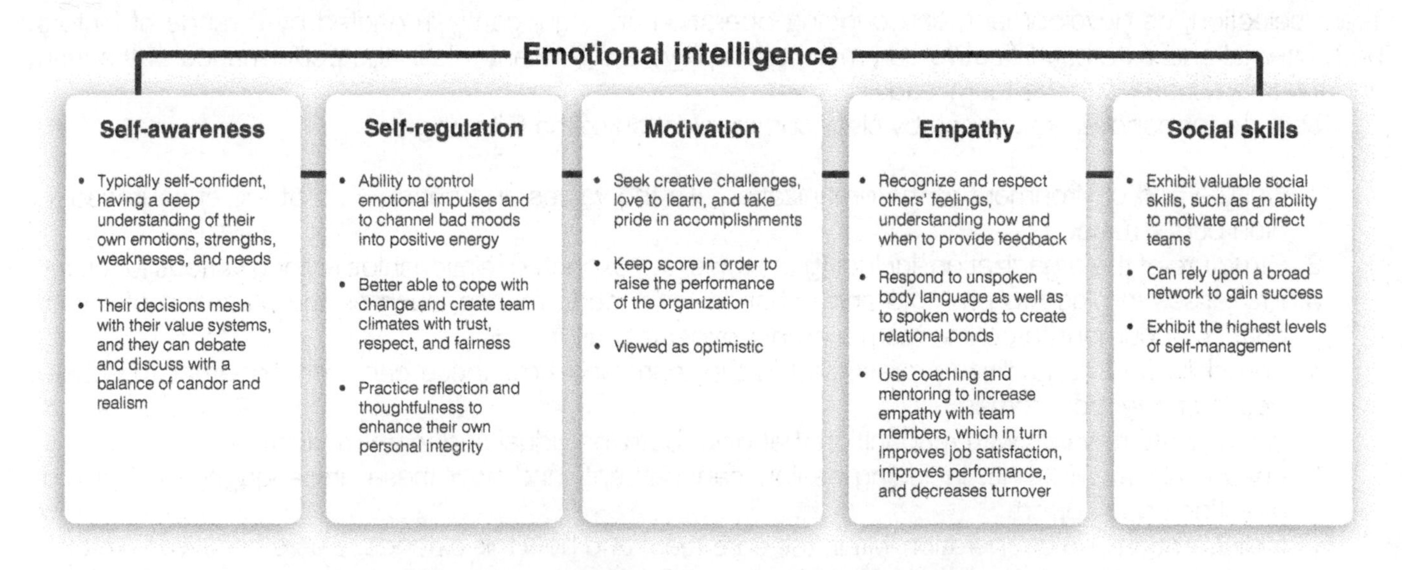

FIGURE 7.12 Elements of emotional intelligence for innovation leaders

Self-awareness is the first self-management component in EI. Leaders with a high degree of self-awareness are typically self-confident, having a deep understanding of their own emotions, strengths, weaknesses, and needs. Their decisions mesh with their value systems, and they can debate and discuss with a balance of candor and realism.

Self-regulation is another trait of a leader with high EI. Self-regulation is the ability to control emotional impulses and to channel bad moods into positive energy. Leaders with self-regulation are better able to cope with change and create team climates with trust, respect, and fairness. Such leaders practice reflection and thoughtfulness to enhance their own personal integrity.

Motivation is a characteristic of self-management leading to high EI. Motivated leaders seek creative challenges, love to learn, and take pride in their accomplishments. Leaders with high self-motivation also keep score to raise the performance of the organization and are typically viewed as optimistic.

Empathy and social skills are the two elements of EI that are supported by the traits of self-management yet are reflected in effective relationships. Leaders with empathy recognize and respect others' feelings, understanding how and when to provide feedback. An empathetic leader will respond to unspoken body language as well as to spoken words to create relational bonds. Innovation leaders use coaching and mentoring to increase empathy with team members, which in turn improves job satisfaction and performance while decreasing turnover.

Social skills may not traditionally be valued as leadership competencies. However, leaders with high EI exhibit valuable social skills, such as an ability to motivate and direct teams. In building rapport with various functions, team members, suppliers, distributors, and customers, leaders with high social skills can rely upon a broad network to gain success. Individuals who exhibit the highest levels of self-management are usually also adept at social skills.

7.6 VIRTUAL TEAMS

So far we have covered culture, leadership, and team structures but with an underlying assumption of geographic proximity. In other words, traditional team theory implies co-location. In this section, we will explore how Virtual Teams (sometimes referred to as remote or distributed teams) impact the concepts discussed thus far.

Remote work in the United States has been growing steadily for many years now. Figure 7.13 shows that even prior to the global COVID pandemic which began in 2020, remote work has been growing rapidly (Strap, 2020).

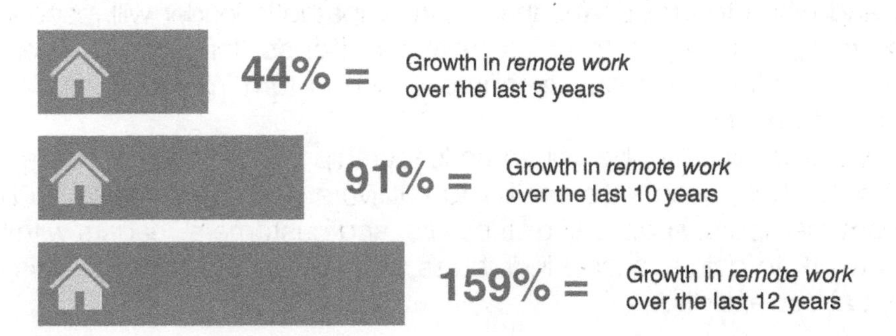
44% =
Growth in *remote work* over the last 5 years
91% =
Growth in *remote work* over the last 10 years
159% =
Growth in *remote work* over the last 12 years

FIGURE 7.13 Pre-COVID trends through 2017

What's even more interesting is the rapid growth in remote work, with a 417% increase projected by 2025 compared to prepandemic levels (Ryan, 2023).

While it is beneficial to co-locate product innovation teams, it is often impractical and/or infeasible to house a team in one central location. As we've already seen, remote work is on the rise, and the development of effective online collaboration tools has been accelerated by the COVID pandemic to the point where even product innovation jobs, once thought to be an in-person role exclusively, are now increasingly available as remote or virtual positions.

Much has been written on the topic of virtual vs. co-located teams. It is important to focus on the aspects of team leadership that are most important relative to the context of the team. Simply stated, all teams have the same challenges, but some are more prevalent in the virtual context. Specifically, communication and team cohesion need more attention in virtual teams.

One of the most cited downsides to virtual teams involves cultural, ethnic, or language barriers related to communication (Hardenbrook & Jurgens-Kowal, 2018). To help mitigate these challenges, the Virtual Team Model (VTM) provides five elements and 16 operational practices to bridge communication gaps in a virtual team (see Figure 7.14).

7.6.1 Initiation and Structure

Virtual teams will go through the same development stages as discussed in Section 7.4.2. Some of the requirements that differ for virtual teams revolve around team cohesiveness and commitment to a shared vision. Most leaders today have experience in, and are skilled at, leading and motivating a team which they have physical access to. In-person teams need to be functionally capable, committed to success, and aligned on vision. The most common practice is to have a team-building exercise where the team is required to interact physically, meet each other, building shared experiences – this is most often not an option for a virtual team. Therefore, some of the initiation and structural components of team building differ slightly. Some of the more important virtual team model initiation and structure components include:

- ***Hiring for purpose:*** For a virtual team, it is important to select team members who share the mission and vision of the project statement. Because these individuals are often working independently,

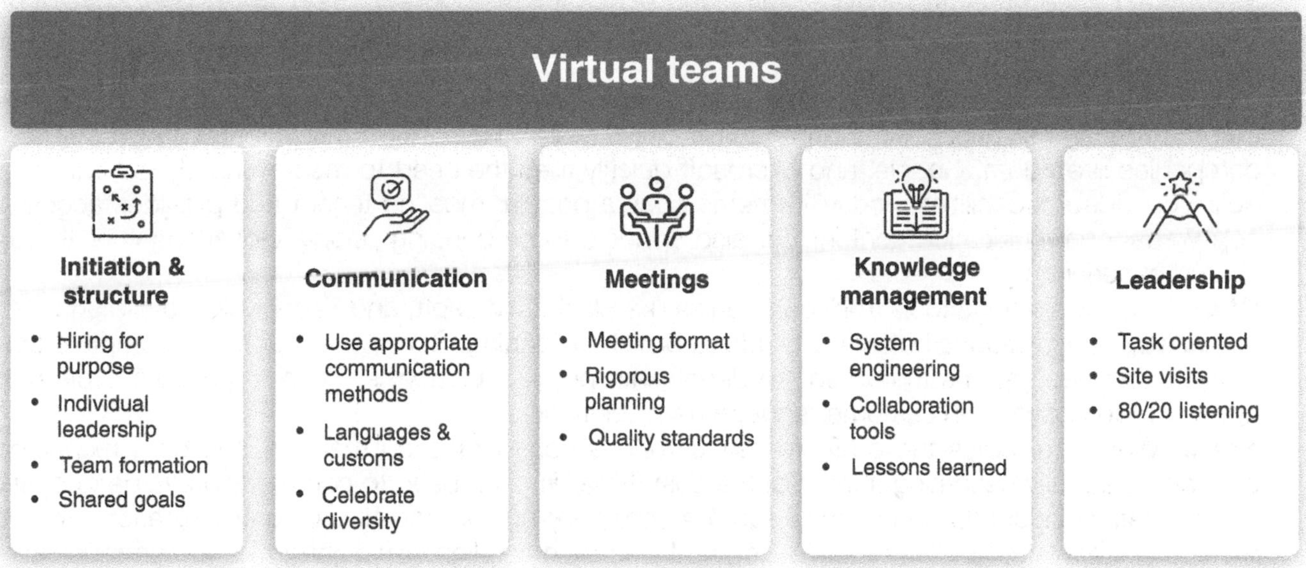

Virtual teams

Initiation & structure
- Hiring for purpose
- Individual leadership
- Team formation
- Shared goals

Communication
- Use appropriate communication methods
- Languages & customs
- Celebrate diversity

Meetings
- Meeting format
- Rigorous planning
- Quality standards

Knowledge management
- System engineering
- Collaboration tools
- Lessons learned

Leadership
- Task oriented
- Site visits
- 80/20 listening

FIGURE 7.14 Virtual Team Model

hiring for purpose assures their alignment with strategic objectives so that decisions are naturally coordinated with the project and organizational goals.

- ***Individual leadership:*** Dispersed team members may not interact with their leader daily. They need self-motivation to schedule and complete project tasks. Virtual team members represent the project to their local management in local markets and thus require both generalist and specialist knowledge. Such team members can work on in-depth problems within their field of expertise but also need broad education and experience to represent the project, the product, and the brand locally.
- ***Team formation:*** As indicated in Section 7.4.2, teams usually advance through stages of forming, storming, norming, performing, and adjourning. Phases such as storming and norming can be more difficult without face-to-face dialogue. For larger projects or complex innovation work, a face-to-face kickoff meeting can facilitate the necessary conversations to build trust among team members. If a face-to-face kickoff meeting is impossible, frequent video conferencing accompanied by social team-building activities serves as a secondary substitute. Building trust is an important outcome of the team formation phases.
- ***Training in online tools:*** As the use of online tools for virtual meetings increased, so has the capability of these tools, including many interactive features. Training in the use of these tools can significantly enhance the value of virtual meetings.
- ***Shared goals:*** Team members working from a common office or lab space receive frequent feedback on their progress. However, dispersed team members often work alone in isolated spaces. Rather than relying on natural conversations and interactions of a co-located team, virtual teams are successful by focusing on the common goal. The *hallway conversations* can be replaced with video and chat collaboration tools like Slack. Individuals accustomed to virtual work often perform better in an asynchronous communications modal. Team leaders should reiterate the shared purpose at every opportunity, regardless of whether they're leading in-person or virtual teams.

7.6.2 Communication Practices

Usually, face-to-face communication is the richest form of discussion because body language, facial expressions, and gestures make up as much as 80% of communication. Even with the high-quality video conferencing available today, many components of face-to-face communication are lost among dispersed team members. Five operational practices within the Virtual Team Model can improve the outcomes of dispersed project work teams:

- ***Video:*** Prior to the pandemic of 2020, distributed teams relied heavily on voice and asynchronous communications solutions. As the pandemic drove more than 50% of the global workforce remote, companies like Zoom, Google, and Microsoft quickly filled the need to meet remotely but with high-definition video capabilities. Today, *cameras on* is a popular meeting tenant and provides access to non-verbal communication content but also appears to be bringing people closer together through shared experience.
- ***Chat:*** Communications tools from companies like Slack, Microsoft, and Facebook provide an analog for *hallway conversations* but with the added benefit of logging. Some, like Slack, are complete communications ecosystems that integrate deeply into process platforms, development life cycle management tools, and even customer engagement solutions.
- ***Email:*** While email was the de-facto asynchronous communication method, it is both technologically and culturally becoming a thing of the past. Email is often used to communicate large quantities of information, nuanced concepts, or as a record-keeping tool for official communications. It is by no means obsolete, but it is rapidly being replaced by other, lighter-weight forms of communication, such as WhatsApp and WeChat.
- ***Language and customs:*** While English is often chosen as the primary language for a dispersed team, individual skills, comfort, and competency may vary. Team members who share a different primary language may feel uncomfortable speaking during team meetings. The team leader sets the tone for acceptance and patience if team members use awkward phrasing or poor grammar. Communicating via chat often relieves social and emotional pressure brought on by hesitation to "find the right word" or accents.
- ***Celebrate diversity:*** Teams need internal trust to succeed. Intellectual trust is sufficient for co-located teams demonstrating their educational background and experience (Rosenfeld et al., 2011). However, when the innovation challenge involves more risk, emotional trust surpasses cognitive capability as a measure of success. Dispersed teams establish emotional trust by building shared goals and celebrating their diversity. Virtual teams can inspire deeper respect and trust for one another by telling stories of their local cultures, such as special holidays or sporting events. Personal relationships lead to increased trust and team cohesiveness. Here too, technology is meeting the demand. Google's calendaring tool, for example, provides pre-defined events that can be added to individual or team calendars with the click of a mouse for cultural, religious, sports, and more.

7.6.3 Virtual Meetings

In many projects, meetings are used for one-way communication to inform the project leader of status. Meetings in a virtual team also serve as a primary exchange for ideas, concepts, and planning future work. Four operational practices associated with meetings yield significant performance improvements for globally dispersed teams:

- ***Medium format:*** To keep team members interested and engaged in meetings, the format of the meeting varies. For teams spread across many time zones, consider rotating the meeting time. It should remain predictable in its rotation but can also accommodate a variety of start times. For example, for weekly meetings, perhaps the first Monday of each month is at 3pm GMT −5, whereas the second Monday of each month is at 3pm GMT −8.
- ***Video:*** Most participants of virtual meetings have wondered at some point if other participants were paying attention in the meeting and fully engaged in the discussion and decision-making. Conferencing can be used to ensure team members are engaged and able to communicate effectively. Asking team members to turn on and leave their webcams on during virtual meetings results in several benefits. Participants tend to give more attention to meetings when webcams are used. It is also easier

to gauge it team members are engaged in the meeting or distracted. A well-timed question can help bring distracted members back to the topic of the meeting. Also, research shows that when webcams are required to be used in newly forming virtual teams, trust, rapport, and cohesion develop faster (Olson et al., 2014).

- ***Purposeful planning:*** Traditional face-to-face meetings for co-located teams can be spur-of-the-moment sessions. However, preplanning is a best practice for all meetings, regardless of geography. Virtual teams require preplanning. For periodic meetings, the agenda is distributed in advance. The project shared goals are discussed at each meeting, reinforcing the team structure. Minutes of the meeting, verifying follow-up actions and responsibilities, are distributed within a couple of days of the meeting. If the video conference is recorded, a link to the meeting is provided with the meeting minutes. Most collaboration tools available today provide pre-defined containers to facilitate and store meeting artifacts, including the agenda, minutes, and other items relevant to the project meeting.
- ***Quality standards:*** Because virtual teams most often work asynchronously, individual work product quality is even more important. In-person teams can often compensate for defects with ad-hoc meetings or tightly coordinated work efforts. These are also available to the virtual team but are sometimes harder to facilitate, so increasing the emphasis on initial quality, describing, and discussing what a job or deliverable looks like is the key to sustained success in a virtual team environment.

If not organized well and run efficiently, virtual meetings can be extremely boring and unrewarding. Some simple tips to make these meetings more productive include:

- Go into the meeting with a plan and key objectives to be addressed.
- Keep meetings short.
- Strike a balance between talking and listening.
- Add participatory elements.
- Use visual aids.
- Make time to socialize.
- Use virtual icebreakers for new team members.

7.6.4 Knowledge Management

An important element of innovation work is to transfer lessons learned to both operational and next-generation R&D teams. Knowledge management is the element within the Virtual Team Model framework that describes what and how to document innovation learnings from dispersed teams. Effective use of knowledge management speeds next-generation product innovation and creates a data repository to drive product innovation process in portfolio analysis. Three operational practices make up the knowledge management element of the Virtual Team Model:

- ***Systems engineering:*** Systems engineering rigorously documents project work to minimize errors in hand-offs and to ensure customer requirements are fully incorporated in the design. Systems engineering is especially useful for virtual teams because version control is maintained across project work that might be completed anywhere in the world.
- ***Collaboration tools:*** Innovation work is most successful when the team members collaborate with each other, with customers, and with suppliers and vendors. Crowdsourcing tools are an example of collaboration software used with customers. Cloud communication and project management tools aid the virtual team in knowledge transfer and project progress updates, including project scope, schedule, and budget. Software and collaboration tools are continually being updated, with some updates being compulsory. As with all tools and processes, change management is required to ensure consistent use and efficient execution.
- ***Lessons learned:*** Agile project teams conduct a retrospective at the conclusion of each sprint to improve their work together before moving to the next sprint. This practice has expanded beyond

traditional agile methodologies and is now considered a best practice for all frameworks. A simple framework for a lessons-learned discussion/retrospective is to ask what the team should continue doing, what should be stopped, and what should be added.

7.6.5 Leadership of Virtual Teams

While leadership skills are universal, some special abilities are necessary to manage a virtual team. Individual leadership and motivation are necessary skills for team members, a part of initiation and structure in the Virtual Team Model. Additional operational practices within the leadership element include focus on the project work, visiting team member locations, peer visits, and deep listening.

- ***Task-oriented:*** Leadership in a face-to-face team involves motivation and monitoring. Management in a co-located team environment is often indirect and personal relationships are easily established. Virtual team members are normally more self-motivated, and they are driven by the challenge of the work itself. They are used to working autonomously, so the leader's role in a virtual team involves more coordination of work, but it does not remove the need for motivation and monitoring. Performance measurements focus on task completion, timing and integration of piece work, and overall project goal alignment. Recall that a large part of the leader's responsibility on a virtual team is to ensure and reiterate sharing of the project goals and vision.
- ***Site visits:*** In some scenarios, site visits are possible, and in others, they are not economically feasible. Take, for example, an organization that simply works from home, but most of the employees are within driving distance of one another. This practice is increasingly common in the United States. Conversely, there are many startups that have only one employee in each country, which may make the prospect of site visits less likely. Regardless, anytime leaders and team members can meet in person, personal relationship values increase, and this is reflected in the efficiency and effectiveness of team efforts.
- ***80/20 listening:*** Finally, leadership is most effective when leaders spend more time listening rather than talking. Especially for virtual teams, where team members are technically skilled, self-motivated, and work autonomously, a project leader serves the team best with deep listening. Pareto's law of getting 80% of the benefit by addressing the highest impact 20% of issues is at the heart of 80/20 listening. Team leaders need to clearly understand the challenges faced by remote team members and assist as quickly as possible. Many one-way, directive communications from the leader to the team can be handled by email or chat tools, thus preserving meeting time for active listening and problem-solving.

7.7 INCENTIVIZING PRODUCT INNOVATION

7.7.1 Individual or Team Performance Metrics

Chapter 1.10 focused on organizational metrics for product innovation and their application to continuous improvement. While organizational metrics can provide the foundation for individuals and teams involved in product innovation, they need to be tailored to the behaviors that lead to successful product innovation and continuous improvement. Metrics for individual and innovation team performance should not be focused on outcomes alone. Punishment for lack of success can inhibit risk-taking. That is not to say that ongoing failure should be tolerated. Understanding the reasons for failure and taking remedial action is far more valuable to the individual, the team, and the organization. ***Failure*** is an expectation when developing new products and services. Markets and technologies change, especially during long development cycles. Feedback from customers and potential end-users can influence significant change in a product design during the development process as well. Therefore, failed experiments are anticipated. The product innovation process (see Chapter 4) is designed to link the company culture with a systematic approach to innovation such that early-stage exploration is less costly and that the business case is developed in parallel with experimentation that reduces overall financial risk. In addition to the product innovation process, leaders encourage innovation teams to take calculated risks for development. This means that mistakes and errors are not only tolerated but experiments designed for learning that fail are accepted as part of the overall innovation process.

Clarity of both team and individual criteria, as part of performance reviews, is essential to building and maintaining a positive culture.

Examples of team criteria include:

- Contribution to organization's financial health.
- Project milestones on time.
- Project on budget.
- Communications within the team.
- Communications outside the team.
- Individual feedback on team performance.

Examples of individual criteria include:

- Percent of team meetings attended.
- Communication with team members.
- Individuals' tasks on time and within budget.
- Contribution to team climate and culture.
- 360-degree feedback on performance from managers, team members, and other stakeholders.

7.7.2 Incentives for Innovation

Incentives are used to motivate employees to improve performance and increase productivity. Typical pay-for-performance incentive programs focus on short-term financial results and may include disincentives for failures, such as lower compensation and possible termination. While research shows that such pay-for-performance systems boost productivity in routine work, studies also indicate that open-ended, creative tasks call for different incentive systems (Manso, 2017). Staff must be encouraged to explore and exploit novel and unique opportunities with tolerance for, and perhaps expectation for, early failure.

Incentives should be chosen to be meaningful to the team or individual. These include:

For the team:

- Team financial bonus.
- Celebratory outing, dinner, etc.

- Recognition on honors board.
- High-performance team award.

For an individual team member:
- Financial bonus.
- Stock options.
- Promotion.
- Flight upgrades for a conference.
- Dinner for the family.
- All-expense paid holiday.

7.8 IN SUMMARY

- While strategy and process are essential to product development and product management, these alone will not lead to the sustained success of the organization. People are ultimately what makes an organization successful. It is culture and climate that provide the context within which strategy and process can be positively and successfully implemented.
- Every organization has a unique culture. However, there are key elements that underpin the innovation culture of most successful organizations:
  - A clear sense of goals and direction that are communicated, understood, and shared across the organization.
  - Experimentation is encouraged and tolerated.
  - Performance metrics for innovation teams are tied to strategic objectives, and team metrics outweigh individual performance metrics.
  - Hiring criteria includes the candidate's fit with the organization's innovation culture.
  - Constructive conflict and debate of ideas are encouraged.
  - Innovation work is designed in such a way that rewards reflect team, individual, professional, and personal accomplishments.
  - Ongoing training is integrated to build and support teams and leadership in the innovation journey.
- Cross-functional representation is the most effective model for product innovation teams. Specific team structures are selected based on the scope, scale, and complexity of the specific project and include:
  - Functional teams,
  - Lightweight teams,
  - Heavyweight teams, and
  - Autonomous teams.
- Teams grow and develop through a variety of processes. Understanding different work styles enhances the effectiveness of team communications and everyone's preferred way of working. Work style assessments are used in kickoff meetings and throughout the innovation project work to build a common vocabulary for team communications.
- A typical innovation project life cycle follows phases of idea creation, gaining support for the concept, planning the new product development project, and executing the work. Different team members have strengths and competencies that will enhance their capabilities during these life cycle phases. When the team lacks a specific skill set that is best suited for a specific phase of the innovation life cycle, the other team members must set aside their own preferred working styles to complete the project work.
- Conflict occurs on all projects due to different approaches to the technology or marketing of the new product as well as challenges in scope, schedule, and budget. Healthy conflict supports new ideas and problem-solving. The Thomas–Kilmann conflict model offers five categories of conflict management based on the level of assertiveness and cooperativeness. The ideal conflict management solution for each situation is different depending on the circumstances, especially the risk of the decision resulting from the various perspectives driving the conflict.
- Leaders and team members engage in communication within the team and with stakeholders external to the team. Communications include those with the functional organizations, such as legal and logistics, as well as the customer and senior management who fund the project. In all cases, the core team should provide a consistent message regarding the product innovation project.
- The best leaders score high on emotional intelligence (EQ), which involves factors of self-awareness, self-regulation, motivation, empathy, and social skills.
- Many organizations today use virtual or dispersed teams for innovation work. The benefits of a virtual team include access to specifically skilled team members, different cultural perspectives and insights into local markets.

- The VTM includes five elements and 16 practices to overcome the natural barriers of communication when the normal means of communicating are electronic rather than face-to-face. The five elements include:
  - Structure and initiation,
  - Communication modes and methods,
  - Meetings,
  - Knowledge management, and
  - Leadership practices.
- Individual and team performance metrics, together with appropriate recognition and rewards, are essential to the development and sustaining of a high-performing team.

A. CULTURE AND TEAMS PRACTICE QUESTIONS

1. Ensuring that the right team principles are in place to promote high performance is an aspect of a management role and is called?
   A. Team motivation.
   B. Team development.
   C. Project management.
   D. Senior leadership.
2. The role responsible for establishing the product innovation process and for ensuring its quality and consistency and on-going training in its application is called:
   A. Process manager.
   B. Process champion.
   C. Process owner.
   D. Process developer.
3. You are hired as a new product development consultant by an established company that is having challenges with its new product development innovation initiatives. Upon observing the work practices across the company, you note an emphasis on individuals working alone, very little social interaction either at or outside of work, limited recognition of staff performance, and frequent criticism of failure. What specific area would you recommend as the primary focus for the company to improve its product innovation?
   A. Better tools to assist in all aspects of product innovation.
   B. An improved product innovation process, used by all staff.
   C. Developing a culture of innovation.
   D. Encouraging senior managers to get more involved in product innovation.
4. The set of properties of the work environment, perceived directly or indirectly by employees to have a major effect on employee behavior, is termed:
   A. Culture.
   B. Environment.
   C. Climate.
   D. Organizational structure.
5. Ensuring, over time, that a product (or group of products) or services meets the needs of customers by continuously monitoring and modifying elements of the marketing mix is generally the role of a. . .
   A. Product manager.
   B. Project manager.
   C. General manager.
   D. Marketing manager.
6. Which of the following best describes a team?
   A. A group of individuals who meet regularly.
   B. People who enjoy doing the same things.
   C. A group of people with a common purpose for which they hold themselves mutually accountable.
   D. A group of people appointed to complete a specific task.
7. Tuckman identified five stages of team development. What does the forming stage involve?
   A. Deciding on team direction.
   B. Bringing the team together and getting to know each other.
   C. Appointing a team leader.
   D. Resolving differences among team members.

8. A functional team is best suited for which type of project.
   A. Highly complex, requiring strong cross-disciplinary collaboration.
   B. Product development that is totally new to the company.
   C. A project that demands total commitment from team members.
   D. Relatively simple product line extensions or improvements.
9. In a professional setting, what type of assessment helps a team achieve increased communication during the work of the project?
   A. Personality assessment, like the Myers-Briggs Type Indicator.
   B. Work style assessment, like DiSC.
   C. Functional resource assessment as used in creating the project budget.
   D. Project portfolio management assessment.
10. A company has decided to embark on a high-risk project that is focused on the development of a product outside the traditional boundaries of company operation. What type of team structure would be most suitable for this type of project?
    A. Autonomous.
    B. Lightweight.
    C. Functional.
    D. Agile.

Answers to practice questions

1. B
2. B
3. C
4. C
5. A
6. C
7. B
8. D
9. B
10. A

REFERENCES

Booz & Company. (2011). *The global innovation 1000: Why culture is key*. https://www.strategy-business.com/article/11404

Christensen, C.M. and Raynor, M.E. (2003). *The innovator's solution: Creating and sustaining successful growth*. Harvard Business School Press.

Denning, S. (2015). *Why U.S. firms are dying: Failure to innovate*. Forbes.

Gallup. (2023). *The 34 Clifton strengths themes explain your talent DNA*. https://www.gallup.com/cliftonstrengths/en/253715/34-cliftonstrengths themes.aspx

Goleman, D. (1998). *Working with emotional intelligence*. Bloomsbury.

Griffin, A., Price, R., and Vojak, B. (2012). *Serial innovators: How individuals create and deliver breakthrough innovations in mature firms*. Stanford Business Books.

Hardenbrook, D. and Jurgens-Kowal, T. (2018). Bridging communication gaps in virtual teams. In: *Leveraging constraints for innovation*, vol. 3 (ed. S. Gurtner, J. Spanjol, and A. Griffin), 95–117. Wiley https://doi.org/10.1002/9781119390299.ch6.

Harter, J. (2022). *U.S. employee engagement slump continues*. Gallup https://www.gallup.com/workplace/391922/employee-engagement-slump-continues.aspx.

Horowitz, B. (2019). *What you do is who you are: How to create your business culture*. HarperCollins.

Katzenbach, J.R. and Smith, D.K. (1993). The discipline of teams. *Harvard Business Review 71* (2): 111–120.

Kerzner, H.R. (2013). *Project management: A systems approach to planning, scheduling, and controlling*, 11e. Wiley.

Kilmann, R.H. and Thomas, K.W. (1978). Four perspectives on conflict management: An attributional framework for organizing descriptive and normative theory. *The Academy of Management Review* 3 (1): 59–68. https://doi.org/10.5465/amr.1978.4296368.

Lorenzo, R. and Reeves, M. (2018). *How and where diversity drives financial performance*. Harvard Business Review.

Manso, G. (2017). Creating incentives for innovation. *California Management Review 60* (1): 18–32. https://doi.org/10.1177/0008125617725287.

Markham, S.K. and Lee, H. (2013). Product development and management association's 2012 comparative performance assessment study. *Journal of Product Innovation Management 30* (3): 408–429. https://doi.org/10.1111/jpim.12025.

Olson, J.D., Appunn, F.D., McAllister, C.A. et al. (2014). Webcams and virtual teams: An impact model. *Team Performance Management 20* (3/4): 148–177. https://doi.org/10.1108/TPM-05-2013-0013.

Peart, N. (2020). *Meet the best company workplace culture awards 2020*. Forbes https://www.forbes.com/sites/nataliapeart/2020/12/14/best-company-workplace-culture-awards-2020.

Rosenfeld, R.B., Wilhelmi, G.J., and Harrison, A. (2011). *The invisible element*. Idea Connection Systems.

Ryan, R. (2023). *Here's what's happening to remote work in 2023*. Forbes https://www.forbes.com/sites/robinryan/2023/01/10/heres-whats-happening-to-remote-work-in-2023.

Scullard, M. and Baum, D. (2015). *Everything DiSC manual*. Wiley.

Strap, (2020). https://strapaccounting.com/blog/f/remote-work-statistics-shifting-norms-and-expectations.

Tuckman, B.W. (2001). Developmental sequence in small groups (reprint). *Group Facilitation: A Research and Applications Journal 3*: 66–81.

Wheelwright, S.C. and Clark, K.B. (1992). *Revolutionizing product development: Quantum leaps in speed, efficiency, and quality*. The Free Press.

Appendix A: Glossary of Terms

A/B testing: A form of multivariate research designed to test and compare two samples or variables. Other forms of multivariate testing, such as conjoint analysis, involve two or more variations and variables.

Agile product development: An iterative approach to product development that is performed in a collaborative environment by self-organizing teams.

Agile Stage-Gate: Combines the structure (stages and gates) of the classic Stage-Gate with the self-organized teams and short cycle iterations of Agile methodologies.

Alliance: Formal arrangement with a separate organization for purposes of development, and involving exchange of information, hardware, intellectual property, or enabling technology. Alliances involve shared risk and reward (e.g., co-development projects).

Alpha test: Pre-production product testing to find and eliminate the most obvious design defects or deficiencies, usually in a laboratory setting or in some part of the developing organization's regular operations, although in some cases it may be done in controlled settings with lead customers. See also beta test and gamma test.

Analyzer: An organization that follows an imitative innovation strategy, where the goal is to get to market with an equivalent or slightly better product very quickly once someone else opens up the market, rather than to be first to market with new products or technologies. Sometimes called an imitator or a "fast follower."

Applications development: The iterative process through which software is designed and written to meet the needs and requirements of the user base, or the process of enhancing or developing new products.

Acquisition effort: The extent to which your product or service is accessible to your customer.

Adjourning: The stage of a project team's work on the project is complete. In product innovation projects, the product is launched and turned over to standard business operations.

Architectural innovation: Combines technological and business disruptions. A well-quoted example is digital photography, which caused significant disruption for companies such as Kodak and Polaroid.

Architecture: See product architecture.

ATAR (Awareness-Trial-Availability-Repeat): A forecasting tool that attempts to mathematically model the diffusion of an innovation or new product.

Attribute testing: A quantitative market research technique in which respondents are asked to rate a detailed list of product or category attributes on one or more types of scales (such as relative importance, current performance, and current satisfaction with a particular product or service) for the purpose of ascertaining customer preferences for some attributes over others, to help guide the design and development process.

Product Development and Management Body of Knowledge: A Guidebook for Product Innovation Training and Certification, Third Edition.
Allan Anderson, Chad McAllister, and Ernie Harris.

Audit: When applied to new product development, an audit is an appraisal of the effectiveness of the processes by which the new product was developed and brought to market.

Augmented product: The core product, plus all other sources of product benefits, such as service, warranty, and image.

Augmented reality (AR): Similar to VR, whereas VR replaces the participant's real world with an entirely separate reality, AR overlays elements of a new reality into the participant's present environment.

Autonomous team: A completely self-sufficient project team with very little, if any, link to the funding organization. Frequently used as an organizational model to bring a radical innovation to the marketplace. Sometimes called a tiger team.

Awareness: A measure of the percent of target customers who are aware that the new product exists. Awareness is variously defined, including recall of brand, recognition of brand, recall of key features or positioning.

Balanced portfolio: A collection of projects where the proportion of projects in specific categories is selected according to strategic priorities.

Balanced Scorecard: A strategic management performance metric used to identify and improve various internal business functions and their resulting external outcomes.

Bass model: A tool used to forecast sales of an innovation, new technology, or a durable good.

Benchmarking: A process of collecting process performance data, generally in a confidential, blinded fashion, from a number of organizations to allow them to assess their performance individually and as a whole.

Benefit: A product attribute expressed in terms of what the user gets from the product rather than its physical characteristics or features. Benefits are often paired with specific features, but they need not be.

Best practice: Methods, tools, or techniques that are associated with improved performance. In new product development, no one tool, or technique assures success; however, a number of them are associated with higher probabilities of achieving success. Best practices likely are at least somewhat context specific. Sometimes called effective practice.

Best practice study: A process of studying successful organizations and selecting the best of their actions or processes for emulation. In new product development, it means finding the best process practices, adapting them, and adopting them for internal use.

Bang for the buck: Used in financial analysis. An idiom meaning the *worth of one's money or exertion*. The phrase originated from the slang usage of the words *bang* which means *excitement* and *buck* which means *money*.

Beta test: A more extensive test than the alpha test, performed by real users and customers. The purpose of beta testing is to determine how the product performs in an actual user environment.

Big data: A collection of large and complex data from different instruments at all stages of the process which go from acquisition, storage, and sharing, to analysis and visualization.

Biometrics: Relies on specialized tools, biometric technologies, and applications to study biological, cognitive, and emotional responses toward various products and services without actually asking questions or interfering with the experience.

Bottom-up portfolio selection: Starts first with a list of individual projects, and through a process of strict project evaluation and screening, ends up with a portfolio of strategically aligned projects.

Brainstorming: A group method of creative problem-solving frequently used in product concept generation. There are many modifications in format, each variation with its own name. The basis of all of these methods uses a group of people to creatively generate a list of ideas related to a particular topic. As many ideas as possible are listed before any critical evaluation is performed.

Brain writing: Participants are asked to write down ideas pertaining to a specific problem or question. Each participant then passes their ideas over to someone else. This person then adds to the list, and so the process continues.

Brand: A name, term, design, symbol, or any other feature that identifies one seller's good or service as distinct from those of other sellers. The legal term for brand is trademark. A brand may identify one item, a family of items, or all items of that seller.

Brand development index: Sales of your brand compared with its average performance in all markets.

Break-even point: The point in the commercial life of a product when cumulative development costs are recovered through accrued profits from sales.

Breakthrough projects: These projects strive to bring a new product to the market with new technologies, depart significantly from existing organizational practices, and have a high level of risk.

Bubble diagram: Visual representation of a product portfolio. Typically, a bubble diagram shows projects on a two-dimensional *X–Y* plot. The *X* and *Y* dimensions relate to specific criteria of interest (for example, risk and reward).

Burndown chart: A graphical representation of work left to do vs. time. The outstanding work (or backlog) is often on the vertical axis, with time along the horizontal. Burndown charts are a run chart of outstanding work. It is useful for predicting when all of the work will be completed.

Business analysis: An analysis of the business situation surrounding a proposed project. Usually includes financial forecasts in terms of discounted cash flows, net present values, or internal rates of returns.

Business case: The results of the market, technical, and financial analyses, or up-front homework. Ideally defined just prior to the "go to development" decision (gate), the case defines the product and project, including the project justification and the action or business plan.

Business Model Canvas (BMC): A strategic management and Lean startup template for developing new or documenting existing business models. It is a visual chart with elements describing an organization's or product's value proposition, infrastructure, customers, and finances.

Business-to-business: Transactions with non-consumer purchasers such as manufacturers, resellers (distributors, wholesalers, jobbers, and retailers, for example), institutional, professional, and governmental organizations. Frequently referred to as industrial businesses in the past.

Buyer: The purchaser of a product, whether or not they will be the ultimate user. Especially in business-to-business markets, a purchasing agent may contract for the actual purchase of a good or service, yet never benefit from the function(s) purchased.

Cannibalization: That portion of the demand for a new product that comes from the erosion of the demand for (sales of) a current product the organization markets.

Capacity planning: A forward-looking activity that monitors the skill sets and effective resource capacity of the organization. For product development, the objective is to manage the flow of projects through development such that none of the functions (skill sets) creates a bottleneck to timely completion. Necessary in optimizing the project portfolio.

Cash cows: Products that have a high share of a market and low overall growth.

Centers of excellence: A geographic or organizational group with an acknowledged technical, business, or competitive competency.

Certification: A process for formally acknowledging that someone has mastered a body of knowledge on a subject. In new product development, the PDMA has created and manages a certification process to become a New Product Development Professional (NPDP).

Champion: A person who takes a passionate interest in seeing that a particular process or product is fully developed and marketed. This informal role varies from situations calling for little more than stimulating awareness of the opportunity to extreme cases where the champion tries to force a project past the strongly entrenched internal resistance of organization policy or that of objecting parties.

Charter: A project team document defining the context, specific details, and plans of a project.

It includes the initial business case, problem and goal statements, constraints and assumptions, and preliminary plan and scope. Periodic reviews with the sponsor ensure alignment with business strategies. See also Product Innovation Charter.

Chasm: A critical part of the product life cycle occurs near the beginning – after introduction, but before growth has fully kicked in.

Checklist: A list of items used to remind an analyst to think of all relevant aspects. It finds frequent use as a tool of creativity in concept generation, as a factor consideration list in concept screening, and to ensure that all appropriate tasks have been completed in any stage of the product development process.

Circular economy: An economy that is restorative and regenerative by design, and which aims to keep products, components, and materials at their highest utility and value at all times, distinguishing between technical and biological cycles.

Climate: The set of properties of the work environment, perceived directly or indirectly by employees, that has a major effect on employee behavior.

Cluster sampling: The population is divided into clusters and a sample of clusters is taken.

Collaborative product development: When two organizations work together to develop and commercialize a specialized product.

Co-location: Physically locating project personnel in one area, enabling more rapid and frequent decision-making and communication among them.

Commercialization: The process of taking a new product from development to market. It generally includes production launch and ramp-up, marketing materials and program development, supply chain development, sales channel development, training development, training, and service and support development.

Competitive intelligence: Methods and activities for transforming disaggregated public competitor information into relevant and strategic knowledge about competitors' position, size, efforts, and trends. The term refers to the broad practice of collecting, analyzing, and communicating the best available information on competitive trends occurring outside one's own organization.

Concept: A clearly written and possibly visual description of a new product idea that includes its primary features and consumer benefits, combined with a broad understanding of the technology needed.

Concept generation: The processes by which new concepts, or product ideas, are generated. Sometimes also called idea generation or ideation.

Concept screening: The evaluation of potential new product concepts during the discovery phase of a product development project. Potential concepts.

Concept statement: A verbal or pictorial statement of a concept that is prepared for presentation to consumers to get their reaction prior to development.

Concept engineering: A customer-centered process that clarifies the "fuzzy front end" of the product development process with the purpose of developing product concepts. The method determines the customer's key requirements to be included in the design and proposes several alternative product concepts that satisfy these requirements.

Concept scenarios: Scenarios developed as a way to learn how product concepts would work in real-life situations.

Concept testing: The process by which a concept statement is presented to consumers for their reactions. These reactions can either be used to permit the developer to estimate the sales value of the concept or to make changes to the concept to enhance its potential sales value.

Concurrent engineering (CE): When product design and manufacturing process development occur concurrently in an integrated fashion, using a cross-functional team, rather than sequentially by separate functions. CE is intended to cause the development team to consider all elements of the product life cycle from conception through disposal, including quality, cost, and maintenance, from the project's outset. Also called simultaneous engineering.

Conjoint analysis: A market research technique in which respondents are systematically presented with a rotating set of product descriptions, each containing a rotating set of attributes and levels of those attributes. By asking respondents to choose their preferred product and/or to indicate their degree of preference from within each set of options, conjoint analysis can determine the relative contribution to overall preference of each variable and each level.

C-Conscientious: Team members who need to have a complete dataset to evaluate before making a rational decision. They are often perceived as unemotional but pride themselves on accurate, detailed work. Others may view their attention to detail as a hindrance to project work. (DiSC work assessment tool).

Consumer: The most generic and all-encompassing term for an organization's targets. The term is used in either the business-to-business or household context and may refer to the organization's current customers, competitors' customers, or current non-purchasers with similar needs or demographic characteristics. The term does not differentiate between whether the person is a buyer or a user target. Only a fraction of consumers will become customers.

Consumer market: The purchasing of goods and services by individuals and for household use (rather than for use in business settings). Consumer purchases are generally made by individual decision-makers, either for themselves or others in the family.

Consumer need: A problem the consumer would like to have solved. What a consumer would like a product to do for them.

Consumer panels: Groups of consumers in specific sectors, recruited by research companies and agencies, who are used as respondents to answer specific research questions relating to product testing, taste testing, or other areas. Most often, they are a specialist panel who take part in numerous projects. Consumer panels are particularly useful for short, quick surveys, where the emphasis is on a sample of those with specialist knowledge rather than a representative sample of the general population.

Contingency plan: A plan to cope with events whose occurrence, timing, and severity cannot be predicted.

Continuous improvement: The review, analysis, and rework directed at incrementally improving practices and processes. Also called Kaizen.

Continuous innovation: A product alteration that allows improved performance and benefits without changing either consumption patterns or behavior. The product's general appearance and basic performance do not functionally change. Examples include fluoride toothpaste and higher computer speeds.

Convergent thinking: Associated with analysis, judgement, and decision-making. It is the process.

Cooperation (team cooperation): The extent to which team members actively work together in reaching team level objectives.

Copyright: The exclusive and assignable legal right, given to the originator for a fixed number of years, to print, publish, perform, film, or record literary, artistic, or musical material.

Core Benefit Proposition (CBP): The central benefit or purpose for which a consumer buys a product. The CBP may come either from the physical good or service, or from augmented dimensions of the product. See also value proposition.

Core competence: That capability at which an organization does better than other organizations, which provides them with a distinctive competitive advantage and contributes to acquiring and retaining customers. Something that an organization does better than other organizations. The purest definition adds, "and is also the lowest cost provider."

Core product: The benefits that the target market will derive from the product.

Corporate culture: The "feel" of an organization. Culture arises from the belief system through which an organization operates. Corporate cultures are variously described as being authoritative, bureaucratic, and entrepreneurial. The organization's culture frequently impacts the organizational appropriateness for getting things done.

Corporate strategy: The overarching strategy of a diversified organization. It answers the questions of "in which businesses should we compete?" and "how does bringing in these businesses create synergy and/or add to the competitive advantage of the organization as a whole?"

Creativity: Creativity is defined as the tendency to generate or recognize ideas, alternatives, or possibilities that may be useful in solving problems.

Criteria: Statements of standards used by decision-makers at decision gates. The dimensions of performance necessary to achieve or surpass for product development projects to continue in development. In the aggregate, these criteria reflect a business unit's new product strategy.

Critical path: The set of interrelated activities that must be completed for the project to be finished successfully can be mapped into a chart showing how long each task takes, and which tasks cannot be started before which other tasks are completed. The critical path is the set of linkages through the chart that is the longest. It determines how long a project will take.

Critical path scheduling: A project management technique, frequently incorporated into various software programs, which puts all important steps of a given new product project into a sequential network based on task interdependencies.

Critical success factors: Those critical few factors that are necessary for, but don't guarantee, commercial success.

Cross-functional team: A team consisting of representatives from the various functions involved in product development, usually including members from all key functions required to deliver a successful product, typically including marketing, engineering, manufacturing/operations, finance, purchasing, customer support, and quality. The team is empowered by the departments to represent each function's perspective in the development process.

Crossing the chasm: Making the transition to a mainstream market from an early market dominated by a few visionary customers (sometimes also called innovators or lead adopters). This concept typically applies to the adoption of new, market-creating, technology-based products and services.

Crowdsourcing: The practice and use of a collection of tools for obtaining information, goods, services, ideas, funding, or other input into a specific task or project from a large and relatively open group of people, either paid or unpaid, most commonly via technology platforms, social media channels, or the Internet.

Culture: The shared beliefs, core values, assumptions, and expectations of people in the organization.

Customer: One who purchases or uses an organization's products or services.

Customer adoption model: Five distinct categories based on the degree of risk that customers are willing to take: *innovators, early adopters, early majority, late majority,* and *laggards*.

Customer needs: Problems to be solved. These needs, either expressed or yet to be articulated, provide new product development opportunities for the organization.

Customer site visits: A qualitative market research technique for uncovering customer needs. The method involves going to a customer's work site, watching as a person performs functions associated with the customer needs your organization wants to solve, and then debriefing that person about what they did, why they did those things, the problems encountered as they were trying to perform the function, and what worked well.

Cycle time: The length of time for any operation, from start to completion. In the new product development sense, it is the length of time to develop a new product from an early initial idea for a new product to initial market sales. Precise definitions of the start and end point vary from one organization to another and may vary from one project to another within the organization.

D-Dominance: Individuals with a "D" work style tend to prefer a fast pace of work. They are quick to make decisions and may be perceived as overly demanding. These team members are action-oriented. (DiSC work assessment tool).

Dashboard: A typically color-coded graphical presentation of a project's status or a portfolio's status by project resembling a vehicle's dashboard. Typically, red is used to flag urgent problems, yellow to flag impending problems, and green to signal projects on track.

Data: Measurements taken at the source of a business process.

Database: An electronic gathering of information organized in some way to make it easy to search, discover, analyze, and manipulate.

Decision tree: A diagram used for making decisions in business or computer programming. The "branches" of the tree diagram represent choices with associated risks, costs, results, and outcome probabilities. By calculating outcomes (profits) for each of the branches, the best decision for the organization can be determined.

Decline stage: The fourth and last stage of the product life cycle. Entry into this stage is generally caused by technology advancements, consumer or user preference changes, global competition, or environmental or regulatory changes.

Defenders: Organizations that stake out a product turf and protect it by whatever means, not necessarily through developing new products.

Deliverable: The output (such as test reports, regulatory approvals, working prototypes, or marketing research reports) that shows a project has achieved a result. Deliverables may be specified for the commercial launch of the product or at the end of a development stage.

Delphi: A technique that uses iterative rounds of consensus development across a group of experts to arrive at a forecast of the most probable outcome for some future state.

Demographic: The statistical description of a human population. Characteristics included in the description may include gender, age, education level, and marital status, as well as various behavioral and psychological characteristics.

Derivative projects: Spin-offs from other existing products or platforms. They may fill a gap in an existing product line, offer more cost-competitive manufacturing, or offer enhancements and features based on core organization technology. Generally, they are relatively low risk.

Design for assembly (DFA): Simplifies the product design to reduce the cost of assembly in the manufacturing process. An assembly is defined as a combination of parts and components needed to manufacture a product. This view includes all the working activities during and after production.

Design for the Environment (DFE): The systematic consideration of environmental safety and health issues over the product's projected life cycle in the design and development process.

Design for Excellence (DFX): The systematic consideration of all relevant life cycle factors, such as manufacturability, reliability, maintainability, affordability, and testability, in the design and development process. It allows for the intended behavior of the elements of design or their combination. DFF implies considerations such as design for safety (coffeemakers), design for simplicity (platform design), and design for redesign (product variants or derivatives).

Design for functionality: Functional design is the process of responding to the needs or desires of the people who will be using the product in a way that allows those needs and desires to be met.

Design for maintenance (DFM): Design decisions regarding the selection of materials, assemblies, parts, devices, and components determine the maintainability or the capacity of the system to be inspected, restored, and serviced when components fail as they reach their operational life. During the design stages, design for maintenance should facilitate executing corrective and preventive maintenance.

Design for manufacturing and assembly (DfMA): Design for Manufacturing and Assembly (DfMA) is an amalgam of two processes, design for manufacturing (DfM) and design for assembly (DfA). DfM is a design method to reduce the complexity of manufacturing operations and the overall cost of production, including the cost of raw materials. DfA is a design method to facilitate or reduce the assembly operations of parts or components of a product.

Design for production: Aims to minimize product costs and manufacture times while maintaining specified quality standards. A successful manufacturing process.

Design for recycling (DFR): DFR accounts for the use of materials that allow for reusing or reprocessing production waste, products, and parts of products. The methods for DFR center on reusing products and reprocessing products.

Design for Six Sigma (DFSS): The aim of DFSS is to create designs that are resource efficient, capable of exceptionally high yields, and are robust to process variations.

Design specifications: Where the concept statement provides a qualitative presentation of the product concept's benefits and features, the product design specifications provide the quantitative basis for further design and manufacture.

Design thinking: A creative solving approach – or more completely, a systematic and collaborative approach to identify and creatively solve problems.

Design for usability: Evaluates the functionality, serviceability, maintainability, ease of operation, reliability, safety, aesthetics, operating context and environment, and customizability of the concept system. Overall, DFU should be intrinsically connected to the product and manufacture design process.

Design for serviceability: Focuses on the ability to diagnose, remove, or replace any part, component, assembly, or subassembly of a product while performing service repairs and troubleshooting.

Design of experiments (DOE): Used to find cause-and-effect relationships. It is a systematic method to determine the relationship between factors affecting a process and the output of that process. DOE can be applied to both a product and its manufacturing process.

Design validation: Product tests to ensure that the product or service conforms to defined user needs and requirements. These may be performed on working prototypes or using computer simulations of the finished product.

Development: The functional part of the organization responsible for converting product requirements into a working product. Also, a phase in the overall concept-to-market cycle where the new product or service is developed for the first time.

Development teams: Teams formed to take one or more new products from concept through development, testing, and launch.

Digitization: Refers to creating a digital representation of physical objects or attributes. The goal of digitization is to make information more easily accessible, storable, maintained, and shared.

Digitalization: Refers to enabling or improving processes by leveraging digital technologies and digitized data. The purpose of digitalization is to enable, improve, and transform business operations through the use of digitized data and technologies in order to transform how organizations conduct business and improve productivity.

Digital Transformation. The Institute for Digital Transformation defines digital transformation as "The integration of digital technologies into a business resulting in the reshaping of an organization that reorients it around the customer experience, business value, and constant change."

Discontinuous innovation: Previously unknown products that establish new consumption patterns and behavior changes. Examples include microwave ovens and cellular phones.

Discounted Cash-Flow (DCF) analysis: One method for providing an estimate of the current value of future incomes and expenses projected for a project. Future cash flows for a number of years are estimated for the project, and then discounted back to the present using forecast interest rates.

Dispersed teams: Product development teams that have members working at different locations, across time zones, and perhaps even in different countries.

Disruptive innovation: Requires a new business model but not necessarily new technology. So, for example, Google's Android operating system potentially disrupts companies like Apple.

Distribution (physical and channels): The method and partners used to get the product (or service) from where it is produced to where the end user can buy it.

Divergent thinking: The process of coming up with new ideas and possibilities without judgement, without analysis, and without discussion. It is the type of thinking that allows for free-association, "stretching the boundaries," and thinking of new ways to solve difficult challenges that have no single, right, or known answer.

DMAIC: An acronym for the 5 phases that make up the process: Define, Measure, Analyze, Improve, Control.

DMASV: Is focused on the process of designing a new product, service, or process. The five phases of DMADV are: *Define, Measure, Analyze, Design, Verify*.

DMEDI: A creative approach to designing new robust processes, products, and services. Its 5 stages are: *Define, Measure, Explore, Develop, Implement*.

Double diamond framework: Used in Design Thinking, capturing the different stages and includes the notions of convergent and divergent thinking, as well as the problem space.

Early adopters: The second group of people to try a new product after the *innovators*. They are often opinion leaders in their field and are typically more risk-averse than innovators.

Early majority: The third group of people to try a new product. They are typically more risk-averse than early adopters and only adopt a new product once it has been validated by their peers.

Embodiment design: The stage of the design process that starts from the concept definition and continues to develop the design based on technical and economic criteria to reach the detail design stage, which leads to manufacturability.

Emergent emotions: Uses neural network and other non-linear dynamic modeling within the context of artificial intelligence to explain consumers' emotional processes.

Emotional design: Is based on eliciting the moods and feelings of consumers that allow for designing positive emotional associations and a feeling of trust in the product, and thus improve its usability.

Emotional intelligence: Comprised of self-management components and of elements directed toward managing relationships.

Empathy analysis: Involves the capacity to connect with and understand customers deeply and have a direct emotional connection with them.

Endurance testing: Is a type of testing that determines a product's performance, such as durability, speed, scalability, handling, and stability. The testing process includes manual and mechanical endurance tests designed to reveal faults and ensure your product is fit for market.

Enhanced new product: A form of derivative product. Enhanced products include additional features not previously found on the base platform, which provide increased value to consumers.

Entrepreneur: A person who initiates, organizes, operates, assumes the risk, and reaps the potential reward for a new business venture.

Ethnography: Observing customers and their environment to acquire a deep understanding of the life-styles or cultures as a basis for better understanding their needs and problems.

Expected Commercial Value (ECV): A financial evaluation method which seeks to maximize the expected commercial value, or worth, of a project.

Externalities: The effects of a product on people or the environment that are not reflected in the market price of the product.

Eye tracking: A specialized form of sensory testing that uses specialized tools, including connected headsets or goggles, measuring where people look and for how long. The equipment tracks and reports where participants look first, second, third, etc., and provides a visual scan overlaid on an image of the object tested. It is used to answer questions on consumers' reactions to various stimuli, online products and services, websites, apps, images of products, packaging, and messaging.

Factor analysis: A process in which the values of observed data are expressed as functions of a number of possible causes in order to find which are the most important.

Factory cost: The cost of producing the product in the production location including materials, labor, and overhead.

Failure mode and effects analysis (FMEA): A step-by-step approach for identifying possible failures in a design, a manufacturing or assembly process, or a product. ***Failure modes*** means the ways, or modes, in which something might fail.

Failure rate: The percentage of an organization's new products that make it to full market commercialization, but which fail to achieve the objectives set for them.

Feasibility analysis: The process of analyzing the likely success of a project or a new product.

Feature: The solution to a consumer need or problem. Features provide benefits to consumers. A handle (feature) allows a laptop computer to be carried easily (benefit). Usually, any one of several different features will be chosen to meet a customer need. For example, a carrying case with shoulder straps is another feature that allows a laptop computer to be carried easily.

Feature creep: The tendency for designers or engineers to add more capability, functions, and features to a product as development proceeds than were originally intended. These additions frequently cause schedule slip, development cost increases, and product cost increases.

Feature roadmap: The evolution over time of the performance attributes associated with a product. Defines the specific features associated with each iteration/generation of a product over its lifetime, grouped into releases (sets of features that are commercialized).

Field testing: Product use testing with users from the target market in the actual context in which the product will be used.

Financial success: The extent to which a new product meets its profit, margin, and return on investment goals.

First-to-market: The first product to create a new product category or a substantial subdivision of a category.

Five hexagon model: A methodology commonly used in Design Thinking.

Focus groups: A qualitative market research technique where 8–12 market participants are gathered in one room for a discussion under the leadership of a trained moderator. Discussion focuses on a consumer problem, product, or potential solution to a problem. The results of these discussions are not projectable to the general market.

Forecast: A prediction, over some defined time, of the success or failure of implementing a business plan's decisions derived from an existing strategy.

Front end of innovation: The starting point of the innovation process where opportunities are identified, and concepts are developed prior to entering the formal product development process. Often referred to as the *fuzzy from end*.

Forming: The first stage in team formation, where most team members are positive and polite. Some are anxious, as they haven't fully understood what the team will do.

Function: (1) An abstracted description of work that a product must perform to meet customer needs. A function is something the product or service must do. (2) Term describing an internal group within which resides a basic business capability such as engineering.

Function Analysis System Technique: (FAST) is a technique that builds on the results of a functional analysis. The purpose of FAST is to illustrate and provide insights on how the product system works in order to identify malfunctions, incoherence in the sequencing of operations, or operational flaws. The technique allows visualization of the cause–effect relationship among the functions in a product to enhance understanding of how it works.

Functional team: The project is divided into functional components with each component assigned to its own appropriate functional manager. Coordination is either handled by the functional manager or by senior management. Preceding the more formal product development process, it generally consists of three tasks: strategic planning, concept generation, and, especially, pre-technical evaluation.

Gamma test: A product use test in which the developers measure the extent to which the item meets the needs of the target customers, solves the problem(s) targeted during development, and leaves the customer satisfied.

Gantt chart: A horizontal bar chart used in project scheduling and management that shows the start date, end date, and duration of tasks within the project.

Gap analysis: Carried out in product development to determine the difference between expected and desired revenues or profits from currently planned new products if the corporation is to meet its objectives.

Gate: The point at which a management decision is made to allow the product development project to proceed to the next stage, to recycle back into the current stage to better complete some of the tasks, or to terminate. The number of gates varies by organization.

Gatekeepers: The group of managers who serve as advisors, decision-makers, and investors in a Stage-Gate® process. Using established business criteria, this multifunctional group reviews new product opportunities and project progress and allocates resources accordingly at each gate. This group is also commonly called a product approval committee or portfolio management team.

Greenwashing: When an organization or organization spends more time and money claiming to be "green" through advertising and marketing than actually implementing business practices that minimize environmental impact.

Growth stage: The second stage of the product life cycle, marked by a rapid surge in sales and market acceptance for the good or service. Products that reach the growth stage have successfully "crossed the chasm."

Guerrilla testing: Guerrilla testing is the simplest form of usability testing. Basically, guerrilla testing means going into a public place such as a supermarket or a park to ask people about a product prototype. They are asked to perform a quick usability test.

Heavyweight team: An empowered project team with adequate resourcing to complete the project. Personnel report to the team leader and are co-located as practical.

Hurdle rate: The minimum return on investment or internal rate of return percentage a new product must meet or exceed as it goes through development.

Idea: The most embryonic form of a new product or service. It often consists of a high-level view of the envisioned solution needed to solve the problem identified by a person, team, or organization.

Idea generation (ideation): All of those activities and processes that lead to creating broad sets of solutions to consumer problems. These techniques may be used in the early stages of product development to generate initial product concepts, in the intermediate stages for overcoming implementation issues, in the later stages for planning launch, and in the post-mortem stage to better understand success and failure in the marketplace.

IDOV: A specific methodology for designing new products and services to meet Six Sigma standards. It has 4 stages: *Identify, Design, Optimize, Validate*.

Implementation team: A team that converts the concepts and good intentions of the "should-be" process into practical reality.

Implicit product requirement: What the customer expects in a product, but does not ask for, and may not even be able to articulate.

Incremental improvement: A small change made to an existing product that serves to keep the product fresh in the eyes of customers. The research gathers detailed insights, perspectives, attitudes, thoughts, behaviors, and viewpoints on a problem, idea, program, situation, etc.

In-depth interview: Involves conducting longer intensive interviews probing and exploring a specific topic, one on one, with individual participants.

I-Influence: Team members who build social relationships easily and seek new people to engage with and add enthusiasm to the work. (DiSC work assessment tool).

Information: Knowledge and insight, often gained by examining data.

Initial screening: The first decision to spend resources (time or money) on a project. The project is born at this point. Sometimes called idea screening.

In-licensed: The acquisition from external sources of novel product concepts or technologies for inclusion in the aggregate NPD portfolio.

Innovation: A new idea, method, or device. The act of creating a new product or process. The act includes invention as well as the work required to bring an idea or concept into final form.

Innovation landscape map: A map of types of innovation divided into four quadrants – *routine, disruptive, radical, and architectural*.

Innovators: The first group of people to try a new product. They are typically early adopters who are willing to take risks.

Innovation-based culture: A corporate culture where senior management teams and employees work habitually to reinforce best practices that systematically and continuously churn out valued new products to customers.

Innovation steering committee: The senior management team or a subset of it responsible for gaining alignment on the strategic and financial goals for new product development, as well as setting expectations for portfolio and development teams.

Innovation strategy: Provides the goals, direction, and framework for innovation across the organization. Individual business units and functions may have their own strategies to achieve specific innovation goals, but it is imperative that these individual strategies are tightly connected with the overarching organizational innovation strategy.

Integrated Product Development (IPD): A philosophy that systematically employs an integrated team effort from multiple functional disciplines to effectively and efficiently develop new products that satisfy customer needs.

Intellectual property (IP): Information, including proprietary knowledge, technical competencies, and design information, which provides commercially exploitable competitive benefit to an organization.

Internal rate of return (IRR): The discount rate at which the present value of the future cash flows of an investment equals the cost of the investment. The discount rate with a net present value of 0.

Internet of Things (IoT): The interconnection via the Internet of computing devices embedded in everyday objects, enabling them to send and receive data.

Intrapreneur: The large-organization equivalent of an entrepreneur. Someone who develops new enterprises within the confines of a large corporation.

Introduction stage: The first stage of a product's commercial launch and the product life cycle. This stage is generally seen as the point of market entry, user trial, and product adoption.

ISO 9000: A set of five auditable standards of the International Organization for Standardization that establishes the role of a quality system in an organization and which is used to assess whether the organization can be certified as compliant to the standards. ISO 9001 deals specifically with new products.

Jobs to be done (JTBD): Provides a framework for categorizing, defining, capturing, and organizing the inputs that are required to support innovation.

Journal of Product Innovation Management: The premier academic journal in the field of innovation, new product development, and management of technology. The journal, which is owned by the PDMA, is dedicated to the advancement of management practice in all of the functions involved in the total process of product innovation.

Journey maps: A representation as a flowchart of all the actions and behaviors consumers take when interacting with the product or service.

Kaizen: A Japanese term meaning "change for the better" or "continuous improvement." It is a Japanese business philosophy regarding the processes that continuously improve operations and involve all employees.

Kanban: (English: signboard) started as a visual scheduling system and is best known as part of the Toyota production system. The Kanban board has become one of the most useful Agile project management tools, used widely by Agile teams and often referred to as Agile task boards.

Kano method: Used to identify customer needs and latent demands, determining functional requirements, developing concepts as candidates for further product definition, and analyzing competitive product or services within a product category.

Kansei engineering: Used to identify the relevant design elements (color, size, and shape) embedded in a product as determinant of user preference.

Laggards: The last group of people to try a new product. They are typically the most risk-averse and only adopt a new product out of absolute need with no other choices.

Late majority: The fourth group of people to try a new product. They are more risk-averse than the early majority and only adopt a new product once it is widely accepted.

Lead users: Users for whom finding a solution to one of their consumer needs is so important that they have modified a current product or invented a new product to solve the need themselves because they have not found a supplier who can solve it for them. When these consumers' needs are portents of needs that the center of the market will have in the future, their solutions are new product opportunities.

Lean product innovation: The Lean approach to meet the challenges of product innovation. Lean product innovation is founded on the fundamental Lean methodology initially developed by Toyota (the Toyota Production System, or TPS).

Lean startup: An approach to building new businesses based on the belief that entrepreneurs must investigate, experiment, test, and iterate as they develop products.

Learning organization: An organization that continuously tests and updates the experience of those in the organization and transforms that experience into improved work processes and knowledge that is accessible to the whole organization and relevant to its core purpose.

Life cycle assessment: A scientific method for analysis of environmental impacts (CO_2 footprint, Water footprint, etc.).

Lightweight team: New product team charged with successfully developing a product concept and delivering to the marketplace. Resources are, for the most part, not dedicated, and the team depends on the technical functions for resources necessary to get the work accomplished.

Line extension: A form of derivative product that adds or modifies features without significantly changing the product functionality.

Manufacturability: The extent to which a new product can be easily and effectively manufactured at minimum cost and with maximum reliability.

Manufacturing design: The process of determining the manufacturing process that will be used to make a new product.

Manufacturing test specification and procedure: Documents prepared by development and manufacturing personnel that describe the performance specifications of a component, subassembly, or system that will be met during the manufacturing process, and that describe the procedure by which the specifications will be assessed.

Market penetration: The percentage of your target market that you have reached at least once in a specific period of time.

Market research: Information about the organization's customers, competitors, or markets. Information may be from secondary sources (already published and publicly available) or primary sources (from customers themselves). Market research may be qualitative in nature, or quantitative. See entries for these two types of market research.

Market segmentation: Market segmentation is defined as a framework by which to subdivide a larger heterogeneous market into smaller, more homogeneous parts. These segments can be defined in many ways: demographic (men vs. women, young vs. old, or richer vs. poorer), behavioral (those who buy on the phone vs. the Internet vs. retail, or those who pay with cash vs. credit cards), or attitudinal (those who believe that store brands are just as good as national brands vs. those who don't).

Market share: An organization's sales in a product area as a percent of the total market sales in that area. Sales may be for the organization, a brand, a product, etc.

Market testing: Focuses on an *existing product in a new market* or market segment to provide customer feedback from specific target markets at later stages of product development. (Refer *test marketing*)

Marketing mix: Comprises the basic tools that are available to market a product. The market mix is often referred to as the 4 Ps – *Product, Price, Promotion, and Place.*

Marketing strategy: A process or model to allow an organization to focus limited resources on the best opportunities to increase sales and thereby achieve a unique competitive advantage.

Maturity stage: The third stage of the product life cycle, where sales begin to level off due to market saturation. It is a time when heavy competition, alternative product options, and (possibly) changing buyer or user preferences start to make it difficult to achieve profitability.

Metrics: A set of measurements to track product development and allow an organization to measure the impact of process improvements over time. These measures generally vary by organization but may include measures characterizing both aspects of the process, such as time to market and duration of particular process stages, as well as outcomes from product development such as the number of products commercialized per year and percentage of sales due to new products.

Mind-mapping: A graphical technique for imagining connections between various pieces of information or ideas. The participant starts with a key phrase or word in the middle of a page, then works out from this point to connect to new ideas in multiple directions – building a web of relationships.

Mission: The statement of an organization's creed, philosophy, purpose, business principles, and corporate beliefs. The purpose of the mission is to focus the energy and resources of the organization.

Mock-up: A model of what your final product will look like. Product mock-ups are frequently used to present a final product in a real-life context.

Morphological analysis: The approach generates a system-level solution that meets the needs and expectations of potential users. It aims to identify possible "elements" common to several possible "solutions" known as design parameters.

Multidimensional scaling (MDS): A means of visualizing the level of similarity of individual cases of a dataset (for example, products or markets).

Multifunctional team: A group of individuals brought together from the different functional areas of a business to work on a problem or process that requires the knowledge, training, and capabilities across the areas to successfully complete the work. See also cross-functional team.

Multiple regression analysis: Often used in product innovation to analyze survey-based data. It provides detailed insight that can be applied to new products or improve products or services when there are any number of factors, key drivers, and product attributes that can impact the product's value proposition from the customer's point of view. It can be used to identify which variables have an impact on the topic of interest and is used to predict the value of a variable based on the known value of two or more other variables (predictors).

Multivariate analysis: Explores the association between one outcome variable (referred to as the dependent variable) and one or more predictor variables (referred to as independent variables).

Needs statement: Summary of consumer needs and wants, described in customer terms, to be addressed by a new product.

Net present value (NPV): The difference between the present value of cash inflows and the present value of cash outflows. NPV is used in capital budgeting to analyze the profitability of a projected investment or project.

Net promoter score: The likelihood someone would recommend your product or service to a friend.

Network diagram: A graphical diagram with boxes connected by lines that shows the sequence of development activities and the interrelationship of each task with another. Often used in conjunction with a Gantt chart.

Neural networks: This approach creates non-linear models to examine the complex relationship between input variables (product features) and output variables (user perceptions).

New product: A term of many opinions and practices, but most generally defined as a product (either a good or service) new to the organization marketing it. Excludes products that are only changed in promotion.

New Product Development (NPD): The overall process of strategy, organization, concept generation, product and marketing plan creation and evaluation, and commercialization of a new product. Also frequently referred to as product development.

New product introduction (NPI): The launch or commercialization of a new product into the marketplace. Takes place at the end of a successful product development project.

New Product Development Professional (NPDP): A New Product Development Professional is certified by the PDMA as having mastered the body of knowledge in new product development, as proven by performance on the certification test. To qualify for the NPDP certification examination, a candidate must hold a bachelor's or higher university degree (or an equivalent degree) from an accredited institution and have spent a minimum of two years working in the new product development field.

New-to-the-world product: A good or service that has never before been available to either consumers or producers. The automobile was new-to-the-world when it was introduced, as were microwave ovens and pet rocks.

Non-product advantage: Elements of the marketing mix that create competitive advantage other than the product itself. These elements can include marketing communications, distribution, organization reputation, technical support, and associated services.

Norming: The third stage of team formation, when people start to resolve their differences, appreciate colleagues' strengths, and respect the leader's authority.

Open Innovation (OI): The strategy adopted by an organization whereby it actively seeks knowledge from external sources, through alliances, partnerships, and contractual arrangements, to complement and enhance its internal capability in pursuit of improved innovation outcomes. These innovation outcomes may be commercialized internally, through new business entities, or through external licensing arrangements.

Operations: A term that includes manufacturing but is much broader, usually including procurement, physical distribution, and, for services, management of the offices or other areas where the services are provided.

Opportunity: A business or technology gap that an organization or individual realizes, by design or accident, exists between the current situation and an envisioned future in order to capture competitive advantage, respond to a threat, solve a problem, or ameliorate a difficulty.

Option pricing theory: Estimates a value of an options contract by assigning a price, known as a premium, based on the calculated probability that the contract will be successful.

Organizational identity: Fundamental to the long-term success of an organization is a clear definition and understanding of what the organization stands for, why it exists.

Outsourcing: The process of procuring a good or service from someone else, rather than the organization producing it themselves.

Outstanding Corporate Innovator Award: An annual PDMA award given to organizations acknowledged through a formal vetting process as being outstanding innovators. The basic requirements for receiving this award are: (1) Sustained success in launching new products over a five-year time frame; (2) Significant organization growth from new product success; (3) A defined new product development process, that can be described to others; (4) Distinctive innovative characteristics and intangibles.

Patent: A government authority or license conferring a right or title for a set period, especially the sole right to exclude others from making, or selling an invention.

Payback: The time, usually in years, from some point in the development process until the commercialized product or service has recovered its costs of development and marketing. While some organizations take the point of full-scale market introduction of a new product as the starting point, others begin the clock at the start of development expense.

Perceptual mapping: A quantitative market research tool used to understand how customers think of current and future products. Perceptual maps are visual representations of the positions that sets of products hold in consumers' minds.

Performance measurement system: The system that enables an organization to monitor the relevant performance indicators of new products in the appropriate time frame.

Performance metrics: A set of measurements to track product development and to allow an organization to measure the impact of process improvement over time.

Performing: The fourth stage of team formation when hard work leads, without friction, to the achievement of the team's goals. The team structures and processes, established by the leader, are working well.

Personas: Fictional characters built based on objective and direct observations of groups of users. These characters become "typical" users or archetypes, enabling developers to envision specific attitudes and behaviors toward product features.

PERT (Program Evaluation and Review Technique): An event-oriented network analysis technique used to estimate project duration when there is a high degree of uncertainty in estimates of duration times for individual activities.

PESTLE: A structured tool based on the analysis of Political, Economic, Social, Technological, Legal, and Environmental factors. It is particularly useful as a strategic framework for seeking a better understanding of trends in factors that will directly influence the future of an organization – such as demographics, political barriers, disruptive technologies, and competitive pressures.

Phase review process: A staged product development process in which first one function completes a set of tasks, then passes the information generated sequentially to another function, which in turn completes the next set of tasks, and then passes everything along to the next function.

Pipeline (product pipeline): The scheduled stream of products in development for release to the market.

Pipeline management: A process that integrates product strategy, project management, and functional management to continually optimize the cross-project management of all development-related activities.

Plant variety rights: An exclusive right to produce for sale and sell propagating material of a plant variety.

Platform product: The design and components that are shared by a set of products in a product family. From this platform, numerous derivative products can be designed. See also product platform.

Portfolio: Commonly referred to as a set of projects or products that an organization is investing in and making strategic trade-offs against. See also project portfolio and product portfolio.

Portfolio criteria: The set of criteria against which the business judges both proposed and currently active product development projects to create a balanced and diverse mix of ongoing efforts.

Portfolio management: A business process by which a business unit decides on the mix of active projects, staffing, and dollar budget allocated to each project currently being undertaken. See also pipeline management.

Portfolio management team: See gatekeeper.

Portfolio rollout scenarios: Hypothetical illustrations of the number and magnitude of new products that would need to be launched over a certain time frame to reach the desired financial goals; accounts for success/failure rates and considers organization and competitive benchmarks.

Primary market research: Original research conducted by you (or someone you hire) to collect data specifically for your current objective. They are also responsible for the ongoing training, innovation input, and continuous improvement of the process.

Principal components analysis (PCA): Identifies the smallest number of factors or components necessary to explain as much (or all) of the variance, in a data set, as possible. It reduces the unnecessary features present in the data by creating or deriving new dimensions (or also called components).

Process managers: The operational managers responsible for ensuring the orderly and timely flow of ideas and projects through the process.

Process champion: A senior executive responsible for establishing the new product innovation process. Works to ensure the quality and consistency of the implementation of product innovation processes.

Process owner: The executive manager responsible for the strategic results of the NPD process. This includes process throughput, quality of output, and participation within the organization.

Product: All goods, services, or knowledge sold. Products are bundles of attributes (features, functions, benefits, and uses) and can be tangible, as in the case of physical goods, intangible, as in the case of those associated with service benefits; or can be a combination of the two.

Product and process performance success: The extent to which a new product meets its technical performance and product development process performance criteria.

Product approval committee: See gatekeeper.

Product architecture: The way in which functional elements are assigned to the physical chunks of a product and the way in which those physical chunks interact to perform the overall function of the product.

Product backlog: A basis of Agile product development. The requirements for a system, expressed as a prioritized list of product backlog items. These include both functional and non-functional customer requirements, as well as technical team-generated requirements.

Product definition: Defines the product, including the target market, product concept, benefits to be delivered, positioning strategy, price point, and even product requirements and design specifications.

Product design specifications: All necessary drawings, dimensions, environmental factors, ergonomic factors, aesthetic factors, cost, maintenance that will be needed, quality, safety, documentation, and description. It also gives specific examples of how the design of the project should be executed, helping others work properly.

Product development: The overall process of strategy, organization, concept generation, product and marketing plan creation and evaluation, and commercialization of a new product.

Product Development & Management Association (PDMA): A not-for-profit professional organization whose purpose is to seek out, develop, organize, and disseminate leading-edge information on the theory and practice of product development and product development processes. Website: www.pdma.org.

Product development portfolio: The collection of new product concepts and projects that are within the organization's ability to develop, are most attractive to the organization's customers, and deliver short- and long-term corporate objectives, spreading risk and diversifying investments.

Product development process: A disciplined and defined set of tasks, steps, and phases that describe the normal means by which an organization repetitively converts embryonic ideas into salable products or services.

Product development team: That group of persons who participate in a product development project. Together they represent the full set of capabilities needed to complete the project.

Product discontinuation: A product or service that is withdrawn or removed from the market because it no longer provides an economic, strategic, or competitive advantage in the organization's portfolio of offerings.

Product failure: A product that does not meet the objective of its charter or marketplace.

Product family: The set of products that have been derived from a common product platform. Members of a product family normally have many common parts and assemblies.

Product innovation: The creation and subsequent introduction of a good or service that is new or an improved version of previous goods or services.

Product Innovation Charter (PIC): A critical strategic document, the Product Innovation Charter (PIC), is the heart of any organized effort to commercialize a new product. It contains the reasons the project has been started, the goals, objectives, guidelines, and boundaries of the project.

Product life cycle: The six stages that a new product is thought to go through from birth to death: development, introduction, growth, maturity, decline, and retirement. This is often reduced to four stages with deletion of development and retirement. Controversy surrounds whether products go through this cycle in any predictable way.

Product life cycle management: Changing the features and benefits of the product, elements of the marketing mix, and manufacturing operations, over time to maximize the profits obtainable from the product over its life cycle.

Product line: A group of products marketed by an organization to one general market. The products have some characteristics, customers, and uses in common and may also share technologies, distribution channels, prices, services, and other elements of the marketing mix.

Product management: Ensuring over time that a product or service profitably meets the needs of customers by continually monitoring and modifying the elements of the marketing mix, including: the product and its features, the communications strategy, distribution channels, and price.

Product manager: The person assigned responsibility for overseeing all of the various activities that concern a particular product. Sometimes called a brand manager in consumer-packaged goods organizations.

Product owner: Commonly used in Agile product development. The product owner is the single person who must have final authority representing the customer's interests in backlog prioritization and requirements questions.

Product platform: Underlying structures or basic architectures that are common across a group of products or that will be the basis of a series of products commercialized over a number of years.

Product portfolio: The set of products and product lines the organization has placed in the market.

Product positioning: How a product will be marketed to customers. Product positioning refers to the set of features and benefits that are valued by (and therefore defined by) the target customer audience, relative to competing products.

Product rejuvenation: The process by which a mature or declining product is altered, updated, repackaged, or redesigned to lengthen the product life cycle and in turn extend sales demand through testing).

Product roadmap: Illustrates high-level product strategy and demonstrates how a product will evolve over time. It is essential to organizational alignment.

Product superiority: Differentiation of an organization's products from those of competitors, achieved by providing consumers with greater benefits and value. This is one of the critical success factors in commercializing new products.

Productivity index: Gauges the economic well-being of a project in real-time. For example, by dividing the Ney Present Value by the money remaining to be spent.

Program manager: The organizational leader charged with responsibility of executing a portfolio of NPD projects.

Project: A temporary endeavor undertaken to create a unique product, service, or result (PMI-PMBOK Guide).

Project decision-making & reviews: A series of Go/No-Go decisions about the viability of a project that ensure the completion of the project provides a product that meets the marketing and financial objectives of the organization. Includes a systematic review of the viability of a project as it moves through the various phase Stage-Gates in the development process. These periodic checks validate that the project is still close enough to the original plan to deliver against the business case.

Project leader: The person responsible for managing an individual new product development project through to completion. They are responsible for ensuring that milestones and deliverables are achieved and that resources are utilized effectively. See also team leader.

Project management: The set of people, tools, techniques, and processes used to define the project's goal, plan all the work necessary to reach that goal, lead the project and support teams, monitor progress, and ensure that the project is completed in a satisfactory way.

Project pipeline management: Fine-tuning resource deployment smoothly for projects during ramp-up, ramp-down, and mid-course adjustments.

Project plan: A formal, approved document used to guide both project execution and control. Documents planning assumptions and decisions, facilitates communication among stakeholders, and documents approved scope, cost, and schedule deadlines.

Project portfolio: The set of projects in development at any point in time. These will vary in the extent of newness or innovativeness.

Project resource estimation: This activity provides one of the major contributions to the project cost calculation. Turning functional requirements into a realistic cost estimate is a key factor in the success of a product delivering against the business plan.

Project sponsor: The authorization and funding source of the project. The person who defines the project goals and to whom the final results are presented. Typically, a senior manager.

Project strategy: The goals and objectives for an individual product development project. It includes how that project fits into the organization's product portfolio, who the target market is, and what problems the product will solve for those customers.

Project team: A multifunctional group of individuals chartered to plan and execute a new product development project.

Prospectors: Organizations that lead in technology, product and market development, and commercialization, even though an individual product may not lead to profits. Their general goal is to be first to market with any particular innovation.

Prototype: A physical model of the new product concept. Depending upon the purpose, prototypes may be non-working, functionally working, or both functionally and aesthetically complete.

Psychographics: Characteristics of consumers that, rather than being purely demographic, measure their attitudes, interests, opinions, and lifestyles. Frequently used to gather initial consumer needs and obtain initial reactions to ideas and concepts. Results are not representative of the market in general nor projectable. Qualitative marketing research is used to show why people buy a particular product, whereas quantitative marketing research reveals how many people buy it.

Qualitative Market Research: Research conducted with a very small number of respondents, either in groups or individually, to gain an impression of their beliefs, motivations, perceptions, and opinions. Frequently used to gather initial consumer needs and obtain initial reactions to ideas and concepts. Results are not representative of the market in general or projectable to that market.

Quality: The collection of attributes, which when present in a product, means a product has conformed to or exceeded customer expectations.

Quality assurance/compliance: Function responsible for monitoring and evaluating development policies and practices, to ensure they meet organization and applicable regulatory standards.

Quality-by-design: The process used to design quality into the product, service, or process from the inception of product development.

Quality control specification and procedure: Documents that describe the specifications and the procedures by which they will be measured which a finished subassembly or system must meet before judged ready for shipment.

Quality Function Deployment (QFD): A structured method employing matrix analysis for linking what the market requires to how it will be accomplished in the development effort. This method is most frequently used during the stage of development when a multifunctional team agrees on how customer needs relate to product specifications and the features that deliver those needs. By explicitly linking these aspects of product design, QFD minimizes the possibility of omitting important design characteristics or interactions across design characteristics. QFD is also an important mechanism in promoting multifunctional teamwork. Developed and introduced by Japanese auto manufacturers, QFD is widely used in the automotive industry.

Promotion: A entire set of activities, which communicate the product, brand, or service to the user.

Quantitative market research: Consumer research, often surveys, conducted with a large enough sample of consumers to produce statistically reliable results that can be used to project outcomes to the general consumer population. Used to determine importance levels of different customer needs, performance ratings of and satisfaction with current products, probability of trial, repurchase rate, and product preferences. These techniques are used to reduce the uncertainty associated with many other aspects of product development.

Quartz Open Framework: A framework for managing product innovation based on six key elements – connect, discover, commit, describe, create, and deliver.

Radical innovation: A new product, generally containing new technologies, that significantly changes behaviors and consumption patterns in the marketplace.

Random sample: A subset of a statistical population in which each member of the subset has an equal probability of being chosen, e.g., 3D printing.

Rapid prototyping: The physical modeling of a design using machine technology in an additive and/or subtractive manufacturing process.

Reactors: Organizations that have no coherent innovation strategy. They only develop new products when absolutely forced to by the competitive situation.

Release plan: A tactical document designed to capture and track the features planned for an upcoming release. A release plan usually spans only a few months and is typically an internal working document for product and development teams.

Reposition: To change the position of the product in the minds of customers, either on failure of the original positioning or to react to changes in the marketplace. Most frequently accomplished through changing the marketing mix rather than redeveloping the product.

Resource matrix: An array that shows the percentage of each non-managerial person's time that is to be devoted to each of the current projects in the organization's portfolio.

Resource plan: Detailed summary of all forms of resources required to complete a product development project, including personnel, equipment, time, and finances.

Return on investment (ROI): A standard measure of project profitability, this is the discounted profits over the life of the project expressed as a percentage of initial investment.

Reverse engineering: The implementation of value analysis (VA) tear-down processes to formulate ideas for product improvement.

Risk: An event or condition that may or may not occur, but if it does occur will impact the ability to achieve a project's objectives. In new product development, risks may take the form of market.

Risk acceptance: An uncertain event or condition for which the project team has decided not to change the project plan. A team may be forced to accept an identified risk when they are unable to identify any other suitable response to the risk.

Risk avoidance: Changing the project plan to eliminate a risk or to protect the project objectives from any potential impact due to the risk.

Risk management: The process of identifying, measuring, and mitigating the business risk in a product development project.

Risk mitigation: Actions taken to reduce the probability and/or impact of a risk to below some threshold of acceptability.

Risk tolerance: The level of risk that a project stakeholder is willing to accept. Tolerance levels are context specific. That is, stakeholders may be willing to accept different levels of risk for different types of risk, such as risks of project delay, price realization, and technical potential.

Risk transference: Actions taken to shift the impact of a risk and the ownership of the risk response actions to a third party.

Road-mapping: A graphical multistep process to forecast future market and/or technology changes, and then plan the products to address these changes.

Routine innovation: Builds on an organization's existing technological competencies and fits with its existing business models. Innovation is focused on feature improvement and new versions or models.

S-Curve (Technology S-Curve): Technology performance improvements tend to progress over time in the form of an "S" curve. When first invented, technology performance improves slowly and incrementally. Then, as experience with a new technology accrues, the rate of performance increase grows and technology performance increases by leaps and bounds. Finally, some of the performance limits of a new technology start to be reached and performance growth slows.

Sales forecasting: Predicting the sales potential for a new product using techniques such as the ATAR (Awareness-Trial-Availability-Repeat) model.

Sales wave research: Customers who are initially offered the product at no cost are re-offered it, or a competitor's product, at slightly reduced prices. The offer may be made as many as five times. The number of customers continuing to select the product and their level of satisfaction is recorded.

Scamper: An ideation tool that utilizes actions verbs as stimuli. S – Substitute; C – Combine; A – Adapt; M – Modify; P – Put to another use; E – Eliminate; R – Reverse.

Scenario analysis: A tool for envisioning alternate futures so that a strategy can be formulated to respond to future opportunities and challenges.

Screening: The process of evaluating and selecting new ideas or concepts to put into the project portfolio. Most organizations now use a formal screening process with evaluation criteria that span customer, strategy, market, profitability, and feasibility dimensions.

Scrum: A term used in Agile product development. Arguably it is the most popular framework for implementing Agile. With scrum, the product is built in a series of fixed-length iterations, giving teams a framework for shipping software on a regular cadence.

Scrum-master: Commonly used in Agile product development. The facilitator for the team and product owner. Rather than manage the team, the scrum-master works to assist both the team and the product owner.

Scrum team: Commonly used in Agile product development. Usually made up of seven, plus or minus two team members. The team usually comprises a mix of functions or disciplines required to successfully complete the sprint goals (cross-functional team).

Secondary market research: Research that involves searching for existing data originally collected by someone else.

Segmentation: The process of dividing a large and heterogeneous market into more homogeneous subgroups. Each subgroup, or segment, holds similar views about the product, and values, purchases, and uses the product in similar ways.

Senior management: That level of executive or operational management above the product development team that has approval authority or controls resources important to the development effort.

Sensitivity analysis: A calculation of the impact that an uncertainty might have on the new product business case. It is conducted by setting upper and lower ranges on the assumptions involved and calculating the expected outcomes.

Sensory testing: A quantitative research method that evaluates products in terms of the human sensory response (sight, taste, smell, touch, hearing) to the products tested.

Sentiment analysis: Used to classify and understand people's opinions in product review blogs or social networks.

Services: Products, such as an airline flight or insurance policy, which are intangible or at least substantially so. If totally intangible, they are exchanged directly from producer to user, cannot be transported or stored, and are instantly perishable. Service delivery usually involves customer participation in some important way. Services cannot be sold in the sense of ownership transfer, and they have no title of ownership.

Simulated test market: A form of quantitative market research and pre-test marketing in which consumers are exposed to new products and to their claims in a staged advertising and purchase situation. Output of the test is an early forecast of expected sales or market share, based on mathematical forecasting models, management assumptions, and input of specific measurements from the simulation.

Six Sigma: A level of process performance that produces only 3.4 defects for every one million operations.

Six thinking hats: A tool developed by Edward de Bono which encourages team members to separate thinking into six clear functions and roles. Each role is identified with a color-symbolic "thinking hat."

Sketch: A freehand drawing. It is very cheap and fast and often done during brainstorming sessions. Its main goal is to illustrate the product or aspects of the product.

Social media: Computer-mediated tools that allow people, companies, and other organizations to create, share, or exchange information, ideas, and pictures/videos in virtual communities and networks.

Specification: A detailed description of the features and performance characteristics of a product. For example, a laptop computer's specification may read as a 90-megahertz Pentium, with 16 megabytes of RAM and 720 megabytes of hard disk space, 3.5 hours of battery life, weight of 4.5 pounds, with an active matrix 256 color screen.

Sponsor: An informal role in a product development project, usually performed by a higher-ranking person in the organization who is not directly involved in the project, but who is ready to extend a helping hand if needed or provide a barrier to interference by others.

Sprint: A term used in Agile product development. A set period of time during which specific work has to be completed and made ready for review.

Stage: One group of concurrently accomplished tasks, with specified outcomes and deliverables, of the overall product development process.

Stage-Gate® process: A widely employed product development process that divides the effort into distinct time-sequenced stages separated by management decision gates. Multifunctional teams must successfully complete a prescribed set of related cross-functional tasks in each stage prior to obtaining management approval to proceed to the next stage of product development. The framework of the Stage-Gate® process includes work-flow and decision-flow paths and defines the supporting systems and practices necessary to ensure the process's ongoing smooth operation.

Staged product development activity: The set of product development tasks commencing when it is believed there are no major unknowns and that result in initial production of salable product, carried out in stages.

Standard cost: See factory cost.

Star products: Products that command a significant market share in a growing overall market.

Storming: The stage in team formation where people start to push against the boundaries established. This is where many teams fail. Storming often starts where there is a conflict between team members' natural working styles.

Storyboarding: Focuses on the development of a story, possibly about a consumer's use of a product, to better understand the problems or issues that may lead to specific product design attributes.

S-Steady: Team members who are considered even-tempered and calm, valuable traits when the new product innovation work is chaotic or unstructured. These team members are more accommodating than many other people and will easily show empathy for others (DiSC work assessment tool).

Strategic balance: Balancing the portfolio of development projects along one or more of many dimensions such as focus vs. diversification, short vs. long term, high vs. low risk, extending platforms vs. development of new platforms.

Strategic fit: Ensures projects are consistent with the articulated strategy. For example, if certain technologies or markets are specified as areas of strategic focus, do the projects fit into these areas?

Strategic partnering: An alliance or partnership between two organizations (frequently one large corporation and one smaller, entrepreneurial organization) to create a specialized new product. Typically, the large organization supplies capital and the necessary product development, marketing, manufacturing, and distribution capabilities, while the small organization supplies specialized technical or creative expertise.

Strategic priorities: Ensures the investment across the portfolio reflects the strategic priorities. For example, if the organization is seeking technology leadership, then the balance of projects in the portfolio should reflect this focus.

Strategy: The organization's vision, mission, and values. One subset of the organization's overall strategy is its innovation strategy.

Stratified sampling: The population is divided into strata according to some variables that are thought to be related to the variables that we are interested in. A sample is taken from each stratum.

Support projects: Can be incremental improvements in existing products or improvements in manufacturing efficiency of an existing product. Generally they are low risk.

Sustainable development: Development which meets the needs of current generations without compromising the ability of future generations to meet their own needs.

Sustainable innovation: The process in which new products or services are developed and brought to commercialization and in which the characteristics of sustainable development are respected from the economic, environmental, and social angle, in the sourcing, production, use, and end-of-service stages of the product life cycle.

Sustaining innovation: Does not create new markets or value networks but only develops existing ones with better value, allowing companies to compete against each other's sustaining improvements.

SWOT Analysis: Strengths, Weaknesses, Opportunities, and Threats Analysis. A SWOT analysis evaluates an organization in terms of its advantages and disadvantages vs. competitors, customer requirements, and market/economic environmental conditions.

Systems engineering (SE): Combine the concepts of systems thinking and SE process models to take a *problem* through a systematic and integrated process of design and project management tools and methods into a *solution*.

Tangible product: The physical and aesthetic design features that give the product its appearance and functionality.

Target market: The group of consumers or potential customers selected for marketing. This market segment is most likely to buy the products within a given category. These are sometimes called prime prospects.

Task: The smallest describable unit of accomplishment in completing a deliverable.

Target market: A particular group of consumers at which a product or service is aimed.

Team leader: The person leading the new product team. Responsible for ensuring that milestones and deliverables are achieved but may not have any authority over project participants.

Technology-driven: A new product or new product strategy based on the strength of a technical capability. Sometimes called solutions in search of problems.

Technology foresighting: A process for looking into the future to predict technology trends and the potential impact on an organization.

Technology roadmap: A graphic representation of technology evolution or technology plans mapped against time. It is used to guide new technology development or technology selection in developing new products.

Technology S-curve: The life cycle that applies to most technologies — embryonic, growth, and maturity stage.

Technology strategy: A plan for the maintenance and development of technologies that supports the future growth of the organization and aids the achievement of its strategic goals.

Technology transfer: The process of converting scientific findings from research laboratories into useful products by the commercial sector. May also be referred to as the process of transferring technology between alliance partners.

Test marketing: Focuses on a *new product in a market* to provide customer feedback from specific target markets at later stages of product development.

Time to market: The length of time it takes to develop a new product from an early initial idea for a new product to initial market sales. Precise definitions of the start and end point vary from one organization to another and may vary from one project to another within the organization. (Also called Speed to Market).

Top-down portfolio selection: Also known as the strategic bucket method, relies on starting with strategy and placing significant emphasis on project selection according to this strategy.

Total Quality Management (TQM): A business improvement philosophy that comprehensively and continuously involves all of an organization's functions in improvement activities.

Trade secrets: Information related to IP that is retained confidentially within an organization.

Trademark: A symbol, word, or words legally registered or established by use as representing an organization or product.

Triple constraint: The combination of the three most significant restrictions on any project: scope, schedule, and cost. The triple constraint is sometimes referred to as the project management triangle or the iron triangle.

Triple bottom line: Reports an organization's performance against three dimensions: Financial, social, environmental.

TRIZ: The acronym for the Theory of Inventive Problem Solving, which is a Russian systematic method of solving problems and creating multiple-alternative solutions. It is based on an analysis and codification of technology solutions from millions of patents. The method enhances creativity by getting individuals to think beyond their own experience and to reach across disciplines to solve problems using solutions from other areas of science.

Unarticulated customer needs: Those needs that a customer is either unwilling or unable to explain.

Usage and purchase intent: The extent to which someone says they will use or purchase your product or service.

User: Any person who uses a product or service to solve a problem or obtain a benefit, whether or not they purchase it.

User experience (UX): In current vernacular, UX is often associated with interface design, human factor design, etc., and while those are definitely a part of the user experience, UX ultimately comes down to understanding the customer.

User interface (UI): Refers to the interfaces with which users engage.

Value: Any principle to which a person or organization adheres with some degree of emotion. It is one of the elements that enter into formulating a strategy.

Value-added: The act or process by which tangible product features or intangible service attributes are bundled, combined, or packaged with other features and attributes to create a competitive advantage, reposition a product, or increase sales.

Value proposition: A short, clear, and simple statement of how and on what dimensions a product concept will deliver value to prospective customers. The essence of "value" is embedded in the trade-off between the benefits a customer receives from a new product and the price a customer pays for it.

Virtual reality (VR) testing: A growing segment of the market research field, conducted using specialized equipment including a headset and/or gloves with tracking sensors that create three-dimensional (3D) simulations and enable participants to interact in a realistic environment.

Virtual team: Dispersed teams that communicate and work primarily electronically may be called virtual teams.

Vision: An act of imagining, guided by both foresight and informed discernment, that reveals the possibilities as well as the practical limits in new product development. It depicts the most desirable future state of a product or organization.

Vitality index: The sales of new products as defined by the business divided by the sales of all products for a given product line or department during a designated period.

Voice of the Customer (VOC): A process for eliciting needs from consumers that uses structured in-depth interviews to lead interviewees through a series of situations in which they have experienced and found solutions to the set of problems being investigated. Needs are obtained through indirect questioning by coming to understand how the consumers found ways to meet their needs, and, more important, why they chose the particular solutions they found.

Waste: Any activity that utilizes equipment, materials, parts, space, employee time, or other corporate resource beyond minimum amount required for value-added operations to ensure manufacturability. These activities could include waiting, accumulating semi-processed parts, reloading, passing materials from one hand to the other, and other non-productive processes. The seven basic categories of waste that a business should strive to eliminate are overproduction, waiting for machines, transportation time, process time, excess inventory, excess motion, and defects.

Waterfall method: Widely applied within the software industry. The five phases of the classical waterfall process are *requirement gathering, design, implementation, verification, and maintenance*.

Whole product: A product definition concept that emphasizes delivering all aspects of a product which are required for it to deliver its full value. This would include training materials, support systems, cables, how-to recipes, additional hardware/software, standards and procedures, implementation, applications consulting – any constitutive elements necessary to assure the customer will have a successful experience and achieve at least minimum required value from the product.

Willingness to pay: The highest price a customer says they will definitely buy your product or service.

Wireframe: A detailed black-and-white layout of the product. They provide specifics about how the product is laid out or constructed. Wireframes are commonly used for software products and websites.

Workplan: Detailed plan for executing the project, identifying each phase of the project, the major steps associated with them, and the specific tasks to be performed along the way. Best practice workplans identify the specific functional resources assigned to each task, the planned task duration, and the dependencies between tasks. See also Gantt chart.

Appendix B: The BoK in Practice Cases

Following is a collection of cases demonstrating application of various aspects of PDMA's BoK.

B.1 CASE 1

Global DairyCo (not the actual name of the company).

B.1.1 Purpose

This case study provides an overview of how a large multinational food company used its business goals to provide a direction for its product innovation strategy and application of its R&D resources. It demonstrates the importance of the linkage between corporate and innovation strategies.

B.1.2 Corporate Strategy

The role of corporate strategy in product innovation is to provide the context, goals, and direction for product innovation portfolio development, prioritization of projects, and management of individual projects.

B.1.3 Company Background

Global DairyCo is a multinational dairy product manufacturing and marketing company, with a wide portfolio of products across commodity, ingredients, and consumer categories, in 120 countries worldwide. Characteristics of the company include:

- One of the largest dairy R&D capabilities in the world with extensive research, development, and pilot plant facilities
- Connections with over 100 research institutions worldwide
- R&D investment of over $100m annually
- Established customer base with several large infant formula and sports food manufacturers

B.1.4 The Challenge

Global DairyCo recognized the need for a clearer vision and strategy on which to base its future growth. It was concerned that the return on its R&D investment was not being maximized, mainly due to a lack of clear focus, based on the overall company direction and goals. Significant resources and expenditures were devoted to projects that lacked any apparent link to the company's current or projected future business.

Product Development and Management Body of Knowledge: A Guidebook for Product Innovation Training and Certification, Third Edition.
Allan Anderson, Chad McAllister, and Ernie Harris.

B.1.5 The Approach

A cross-functional team comprising global representatives of marketing, R&D, manufacturing, and finance was established with the objective "to define the key business areas for R&D focus, thus providing criteria for R&D resource allocation and expenditure."

B.1.6 The Process

1. ***Define the key themes (areas of strategic focus) that will provide the basis for further company sustenance and growth.***

Seven key strategic themes were identified (see Figure B.1):

i. Lowest cost commodity supplier
ii. Leading price and inventory manager
iii. Developer of dairy ingredients partnerships
iv. Specialty milk components innovator
v. Leading nutritional milks marketer
vi. Leading dairy marketer to food-service
vii. Integrated regional strategies

Themes 1 and 2: Aim to protect two of the current competitive advantages – maintaining the company's leading position as a low-cost commodity product supplier and through global inventory control, thus providing the potential to significantly control prices.

Theme 3, 4, 5, and 6: Aim to underpin the "growth engines" for higher profit margins from value-added businesses (products and services).

Theme 7: Aims to reinforce the importance of cross-regional cooperation and integration in achieving optimal leverage from the strategic efforts.

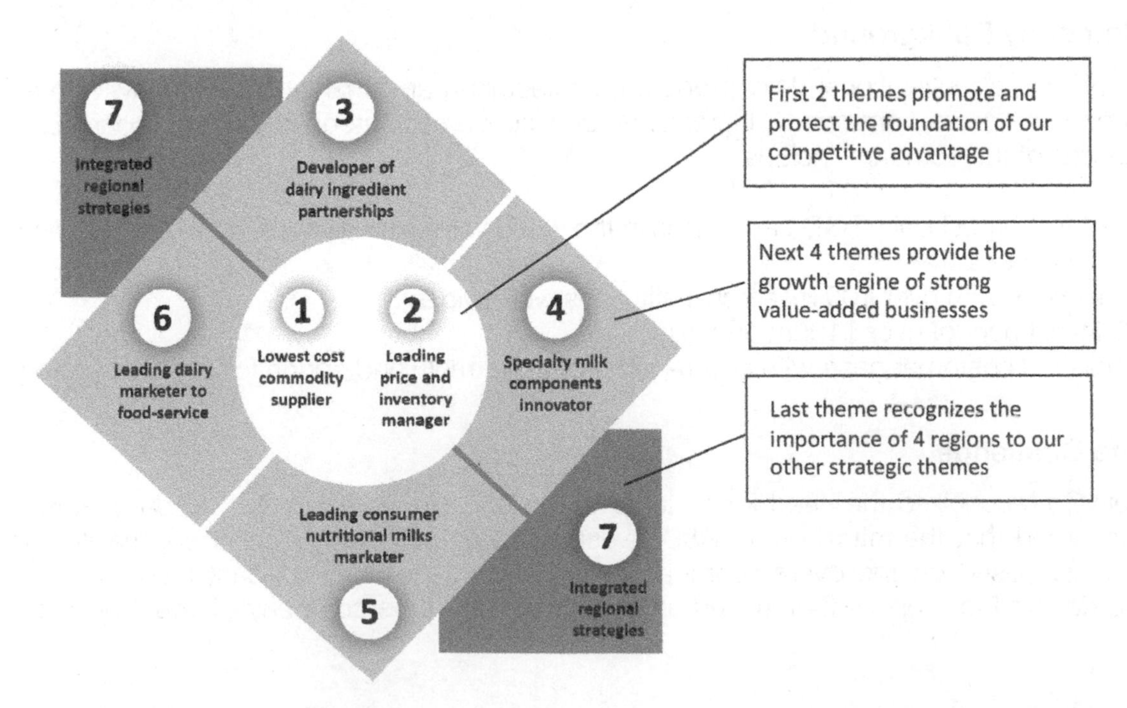

FIGURE B.1 Seven strategic themes

2. ***Identify which of the 7 strategic themes have the greatest potential for company growth from R&D investment** – **where should the R&D funding and resources be focused?***

Four of the strategic themes were prioritized as having the greatest potential from R&D investment. Following were specific initiatives to address each theme:

i. **Developer of dairy ingredient partnerships**
Global DairyCo already had several partnerships with well-established international brands (Figure B.2). A key to future growth was seen through further leveraging these partnerships and developing new partnerships focused on R&D and product innovation. Specific actions included:
- Improving communications with all major partners through the regular exchange of staff and information.
- Identifying and working closely with key lead customers to develop new products based on their specific needs. Specifically, getting closer to lead customers, understanding what they want, and working with them to meet their requirements, often requiring Global Dairy R&D and partner companies to work collaboratively at each others' facilities for extended periods.
- Improving intellectual property practices to ensure maximum return from product innovation through direct product sales, licensing, or sale of IP.

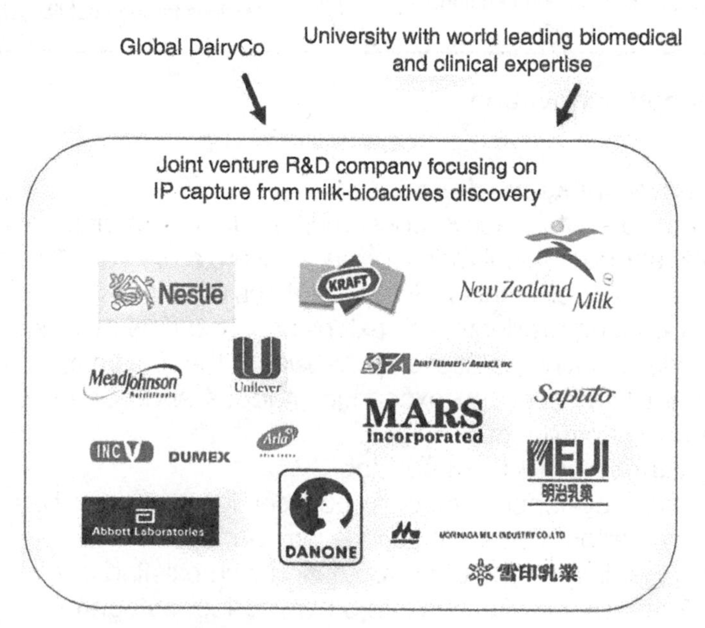

FIGURE B.2 Global Ingredients partnerships

ii. **Specialty milk components innovator**
Global DairyCo had, for several years, carried out research into the separation, manufacture, and application of "*minor milk components*", many of which had far higher market value than traditional products. However, specialized capabilities were lacking within the company. Specific actions included (Figure B.3):
- Identification of core capability to be developed within the company and capability that was better outsourced – either through individual contracts or partnerships.
- A joint venture with a high-profile university with world-leading expertise in biomedical research and clinical testing. The joint venture focused on milk bioactives discovery leading to IP value capture.
- Refocusing on bioactive ingredients that could form a platform across a range of ingredient and consumer products.

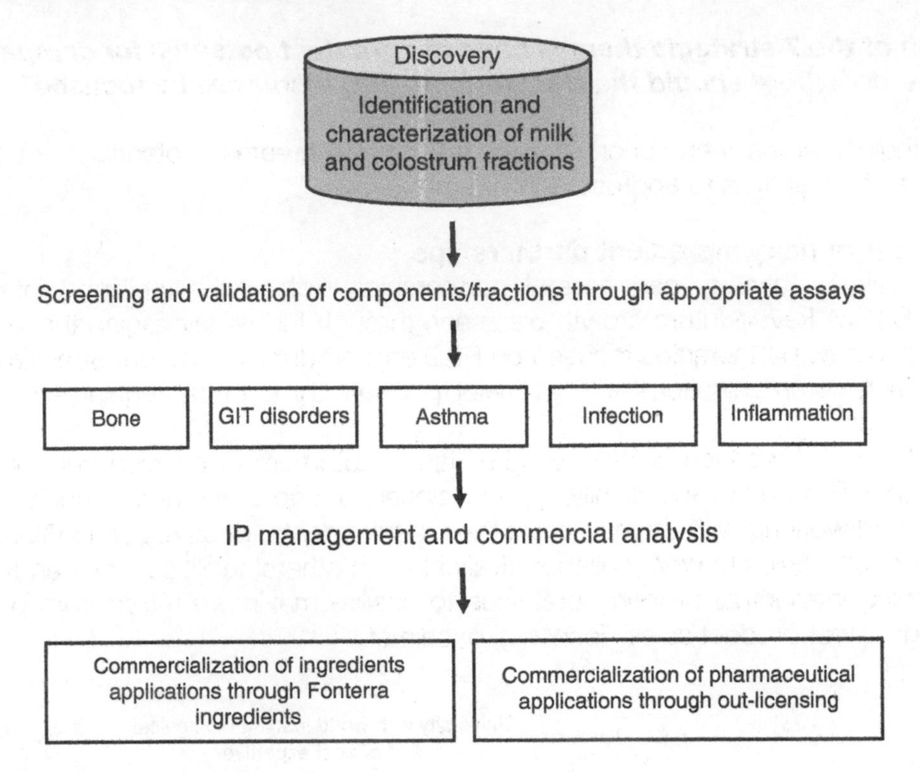

FIGURE B.3 Milk component innovation

iii. **Leading consumer milks marketer**

Global DairyCo had over 118 active consumer products – mainly focused on the Asian region. Although relatively strong, Global DairyCo brands were a distant 3rd behind two other major international consumer dairy companies. Specific actions included:

- Revitalization of existing products, line extensions, and new products.
- Ensuring a close working relationship between the scientists working on the bioactives research program to ensure the exchange of ideas and speed up the commercialization of new bioactive-based products.

iv. **Leading dairy marketer to food service**

Global DairyCo had developed a strong food service business based on its well-established cheese technology platform. Specific actions included:

- Maintaining and building on the good reputation with existing customers.
- Extending the food service offerings into other product categories.
- Leveraging of the bioactives research program and "*platform ingredients*" that might add value to food service customers.

B.1.7 The Outcomes

In the two years after the initiation of this process, the following outcomes were achieved:

- A key outcome from the process was the ongoing impact of forming the cross-functional team. Collaboration across the company functions was enhanced, resulting in strong cooperation across the company on product innovation and other company initiatives.
- An annual product innovation exhibition was hosted by the R&D center, sharing research and new products in the pipeline with marketing colleagues from around the world. This, in turn, promoted greater collaboration between R&D and marketing functions and a greater understanding of each other's roles.

- The bioactives research program received significantly more funding with a defined portfolio of research projects forming a platform of applications across a targeted range of consumer and ingredient products.
- R&D partnerships at various levels were established with over 100 institutions globally.
- Small cross-functional teams were established to focus on servicing key customers – working closely with lead customers to develop new products with well-founded customer requirements.
- The increase in returns from the launch of value-added consumer and ingredient products, together with forward projections of significant growth, resulted in a 20% increase in R&D funding.

Case Questions

1. What was the fundamental first step required by the company to provide the direction for product innovation?
2. What are the important features of the 7 strategic themes identified by Global DairyCo?
3. Would you describe the Global DairyCo strategy as a business or a corporate strategy?
4. How would you define Global DairyCo's strategy in terms of the Porter Framework?
5. What have you identified as the key areas of focus for product innovation in response to the company strategy?

Answers to Case Questions

1. What was the fundamental first step required by the company to provide the direction for product innovation? What would you expect the company to have done to take this first step?

   The first step was to identify the key business areas for the company to focus on to enable future growth. Although not explicitly explained in the case study, Global DairyCo had:
   - Carried out extensive study of its current businesses – products and markets.
   - Sought market research on future trends and growth opportunities.
   - Analyzed current capability – internally and externally with respect to technology and marketing.
   - Identified growth targets across a balanced portfolio of businesses to minimize risk.
2. What are the important features of the 7 strategic themes identified by Global DairyCo?
   - They are based on current business operations and capability.
   - They provide a balanced portfolio of business – some quite traditional and low risk, while others rely on a high level of research with significant risk.
3. Would you describe the Global DairyCo strategy as a business or a corporate strategy?

   The range of themes and the description of Global DairyCo implies that this is more a corporate than a business strategy. In fact, it could be described as an integration of several business strategies with interlinking components. For example, the potential for research from the "*milk components innovator*" theme provides development opportunities in food service or consumer products. In fact, this would almost certainly provide the basis for a platform strategy.
4. How would you define Global DairyCo's strategy in terms of the Porter Framework?
   - Lowest cost commodity supplier – cost leadership.
   - Leading price and inventory manager – cost leadership.
   - Developer of dairy ingredients partnerships – segmentation.
   - Specialty milk components innovator – differentiation.
   - Leading nutritional milks marketer – potentially segmentation or differentiation.
   - Leading dairy marketer to food-service – segmentation.
   - Integrated regional strategies – cost leadership.

   Overall, we can see from this case study that a company does not have to focus on what specific type of strategy. Different strategy types can be applied to different business or product offerings.

5. What have you identified as the key areas of focus for product innovation in response to the company strategy?

   Some of the key areas identified by the company were:

   - Get closer to our customers.
   - Get better at prioritizing and directing our R&D efforts.
   - Get better integration across R&D, manufacturing, and marketing.
   - Get better at accessing external IP.
   - Get more efficient in capturing value from our internal IP.
   - Develop and leverage off science and technology platforms, e.g., bioactives from milk and cheese technology.

B.2 CASE 2

LeanMed, Medical Technology Start-up

B.2.1 Purpose

This case provides an overview of how LeanMed, a medical technology start-up applied Lean start-up principles in their innovation process. It demonstrates the importance of using a structured methodology in innovation strategy and process, particularly as applied in start-up situations. The case study is provided with the permission of LeanMed founder Mark Adkins.

B.2.2 Company Background

LeanMed is a medical technology company founded in 2018. It is dedicated to bringing essential medical treatment to underserved regions of the world through innovative technologies.

B.2.3 Lean Start-Up Methodology

This case study highlights the application of lean start-up methodology based on the following elements:

1. A 3-stage process.
2. The business model canvas (BMC).
3. The build/measure/learn cycle.
4. The minimum valuable product (MVP).

In establishing the company and identifying its first product, LeanMed followed the three-stage process shown in Figure B.4

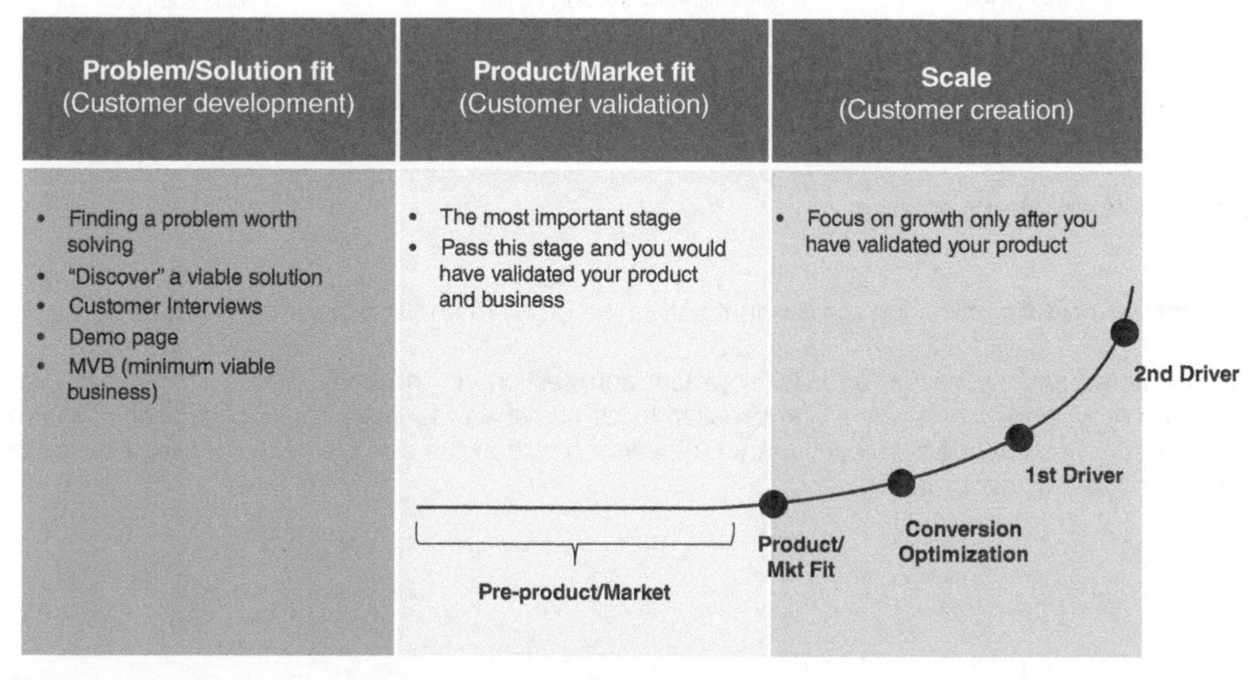

FIGURE B.4 LeanMed's 3-stage start-up process

B.2.4 Applying the 3-Stage Process

1. ***Identifying a problem:*** Finding a problem worth solving and discovering a viable solution.

Pediatric pneumonia is the #1 killer of children in the world, averaging 800,000 deaths per year. 99% of these deaths occur in developing countries.
Oxygen is required to treat children with pneumonia yet it is in short supply in the developing world where poor infrastructure, environmental, and financial challenges limit their ability to access life-saving O_2.

LeanMed's solution

LeanMed's first product, the O_2 Cube, delivers vital medical oxygen to the 1.2 billion patients of rural hospitals and health clinics that lack grid electricity, as shown in Figure B.5.

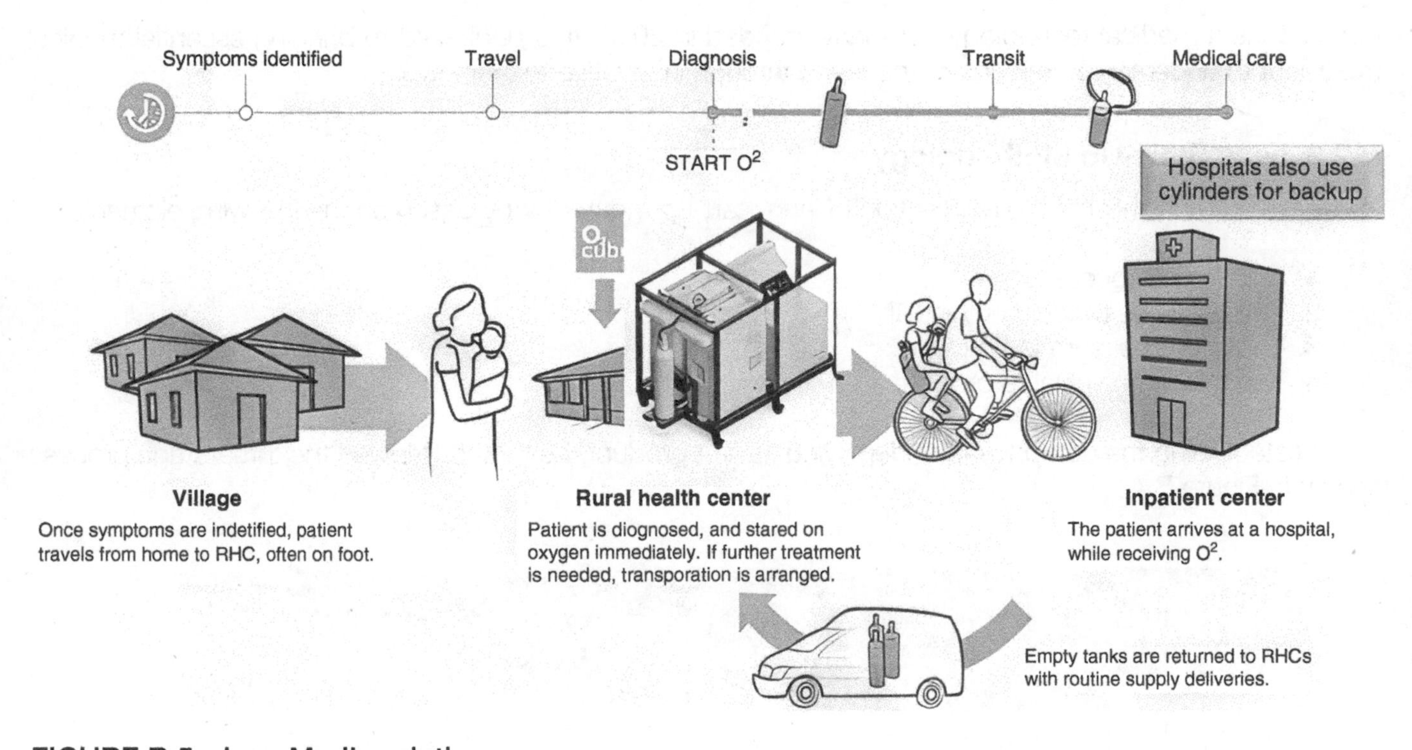

FIGURE B.5 LeanMed's solution

2. ***Product/market fit:*** validating the product solves the problem and is a fit for the market.

LeanMed entered various design and development competitions in the early stages of concept development, prototyping, and preliminary market research, as shown in Figure B.6. Success in these competitions and positive indications from early market research provided the confidence to continue development to a minimum viable product (MVP).

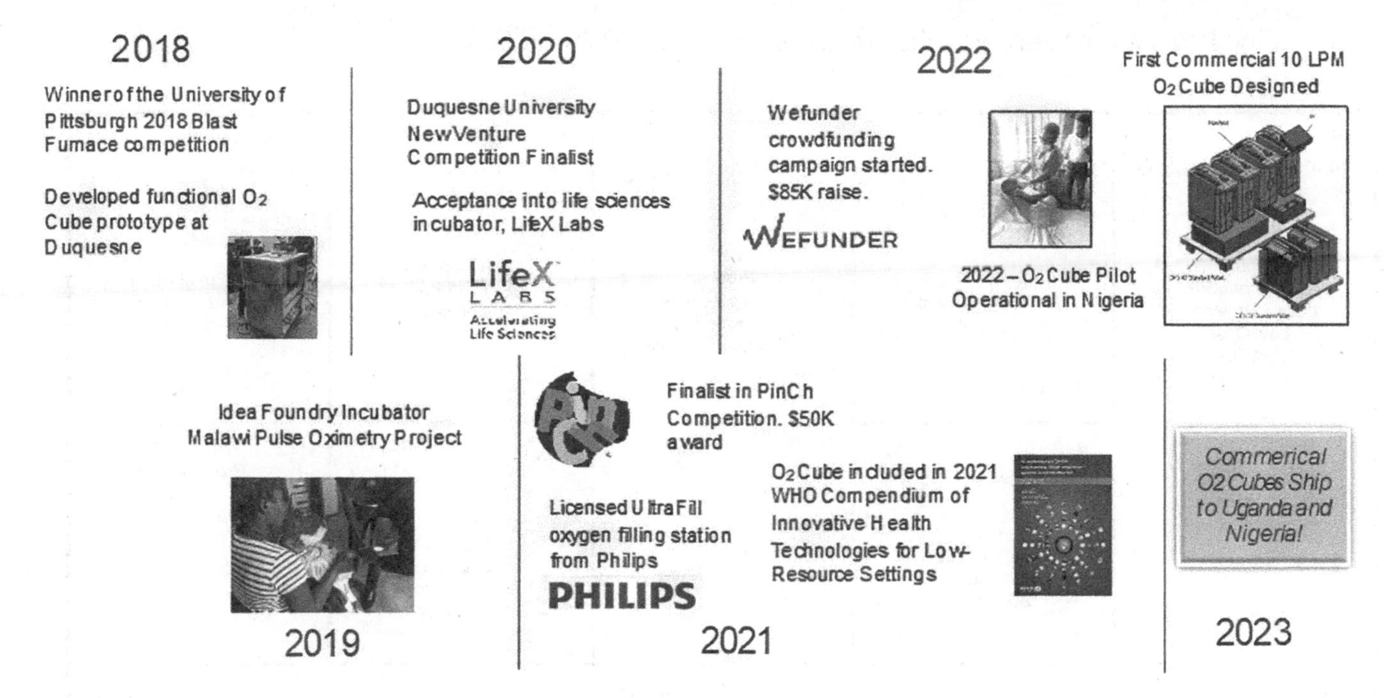

FIGURE B.6 Product/market fit validation process

The Business Model Canvas (BMC)

LeanMed's BMC is shown in Figure B.7. The BMC was used to create alignment in nine areas that contribute to business success.

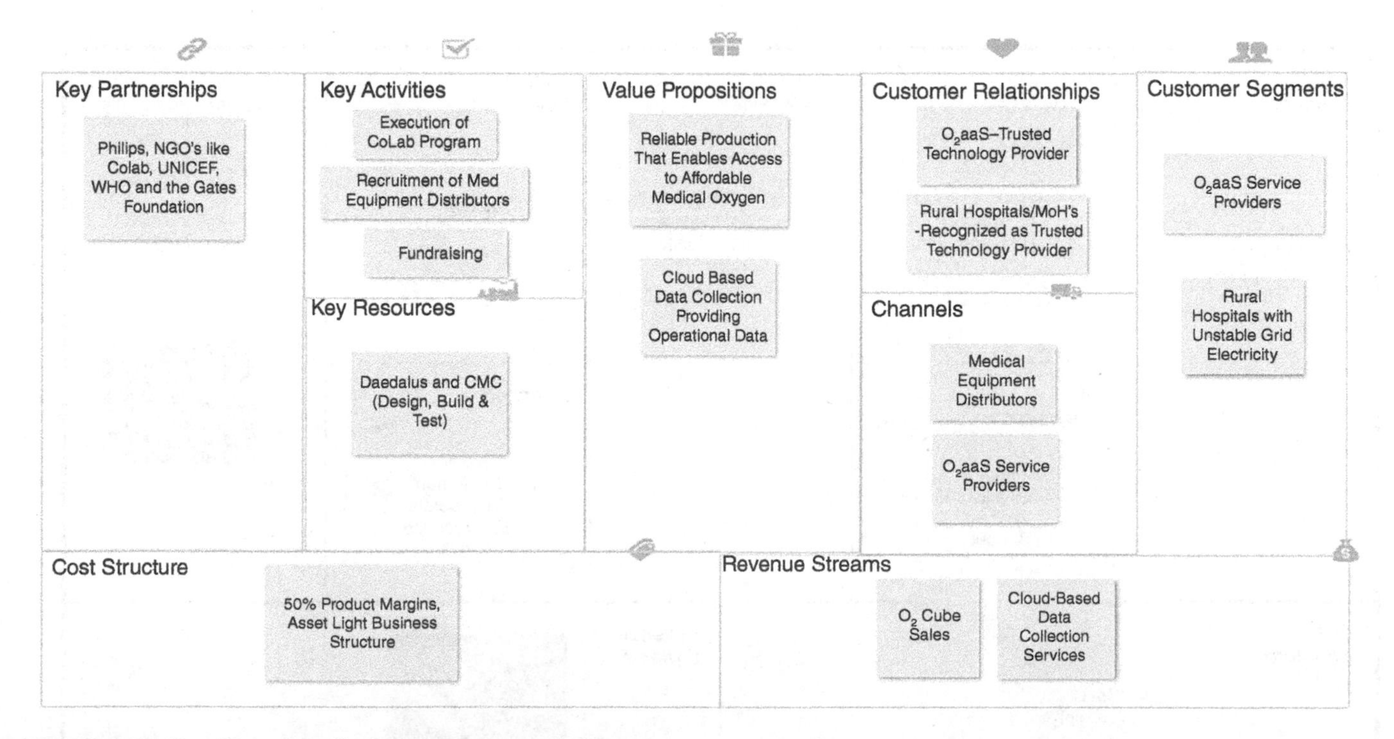

FIGURE B.7 LeanMed's business model canvas

The key hypotheses and risks are shown in Figures B.8 and B.9.

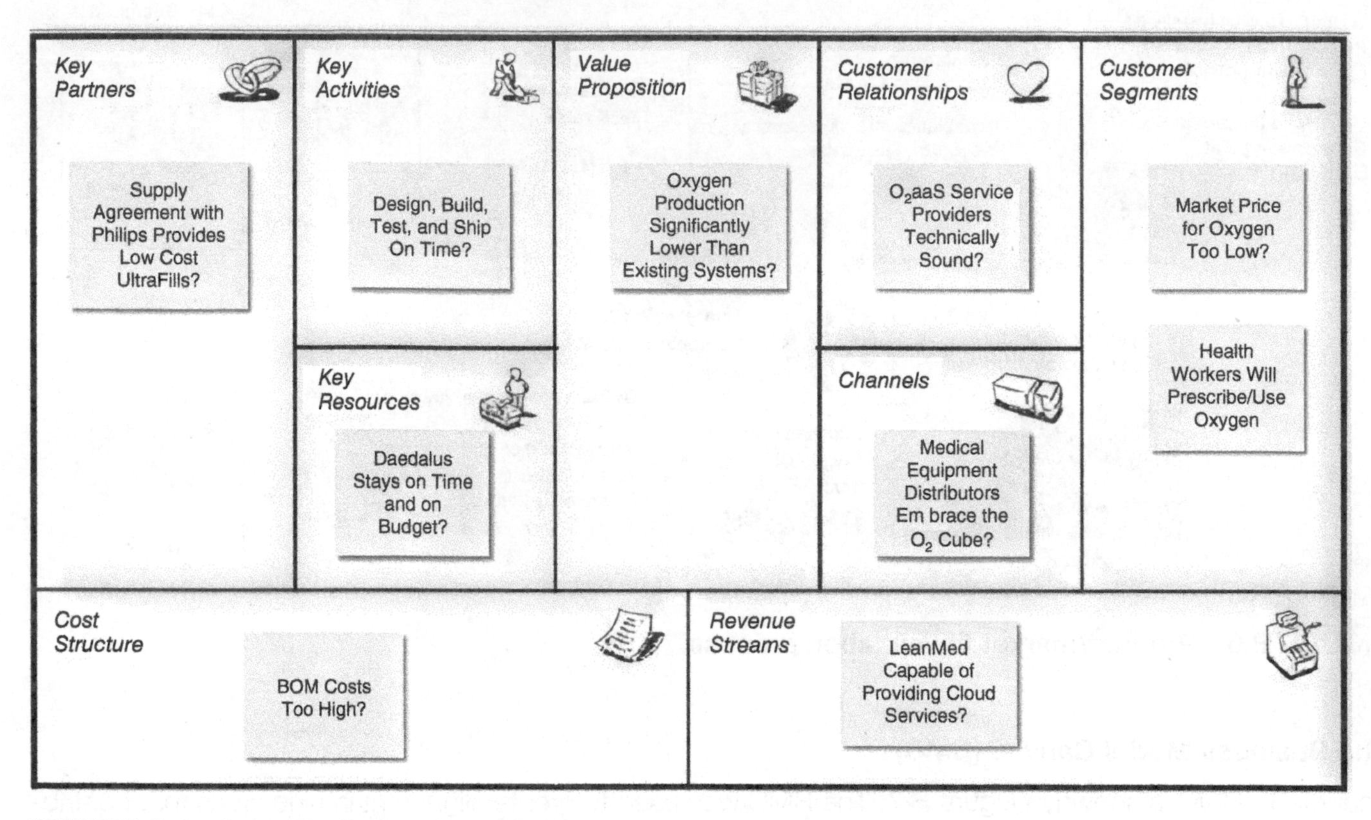

FIGURE B.8 LeanMed's key hypotheses

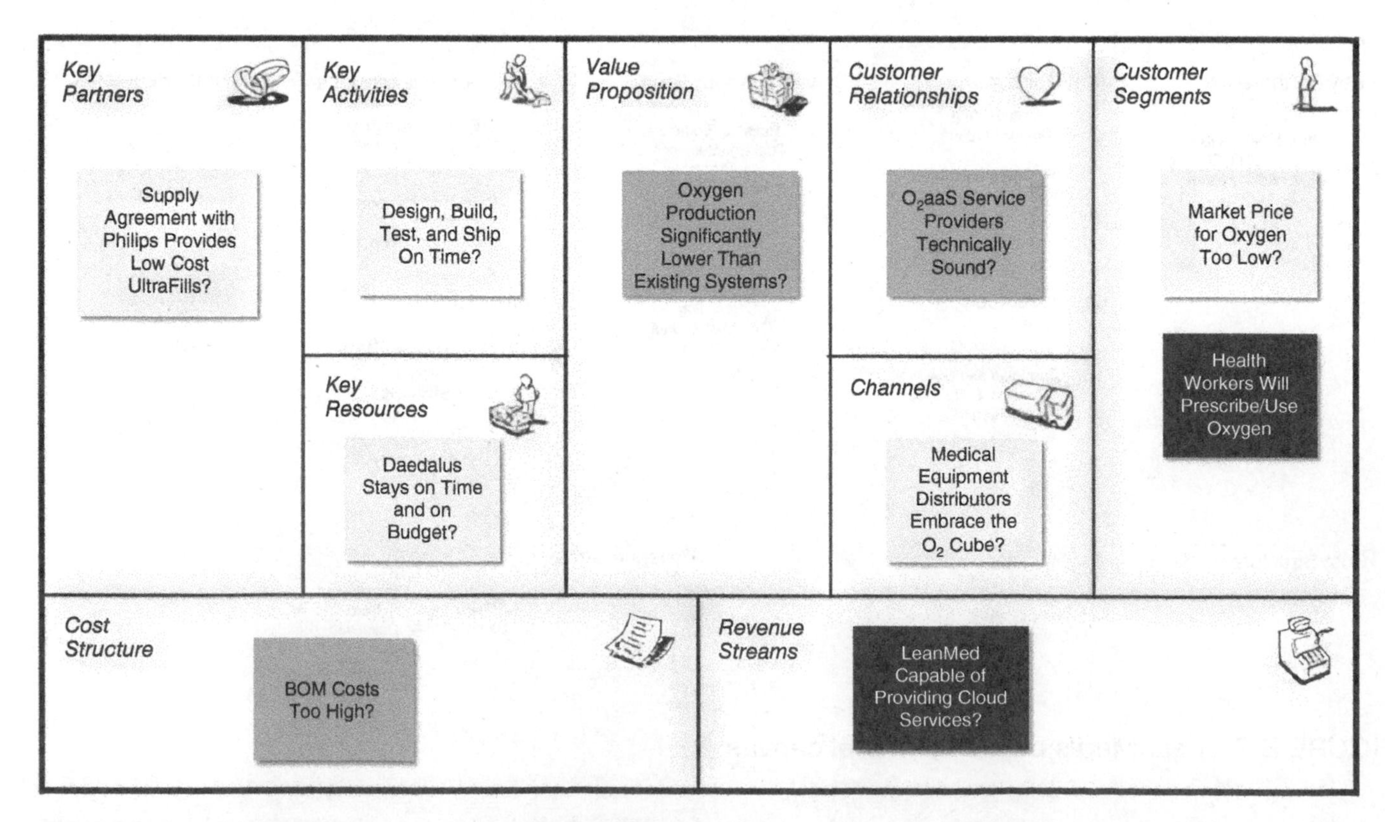

FIGURE B.9 Identifying and prioritizing the risks.

Fast track as an MVP

A minimum viable product (MVP) is a version of a product with just enough features to be valuable to early customers who can then provide feedback for future product development.

Using the MVP

A pilot program was initiated to better understand the delivery of oxygen and learn how versatile the O_2 Cube is by implementing variations of the product in multiple locations with different needs. The Fast Track O_2 Cube is the minimum viable product version of the O_2 Cube that uses already FDA-approved components donated by various organizations, including Philips, Goal Zero, and Masimo.
The MVP was tested in a 26 Bed Pediatric Ward Ota in Nigeria. Feedback was used to learn what improvements can be made to the product to enable greater usability and market acceptance. Results are shown in Figure B.10.

FIGURE B.10 Results from MVP market test

3. **Scale up to commercialization and future business growth.**

LeanMed's commercialization and growth strategy is shown in Figure B.11.

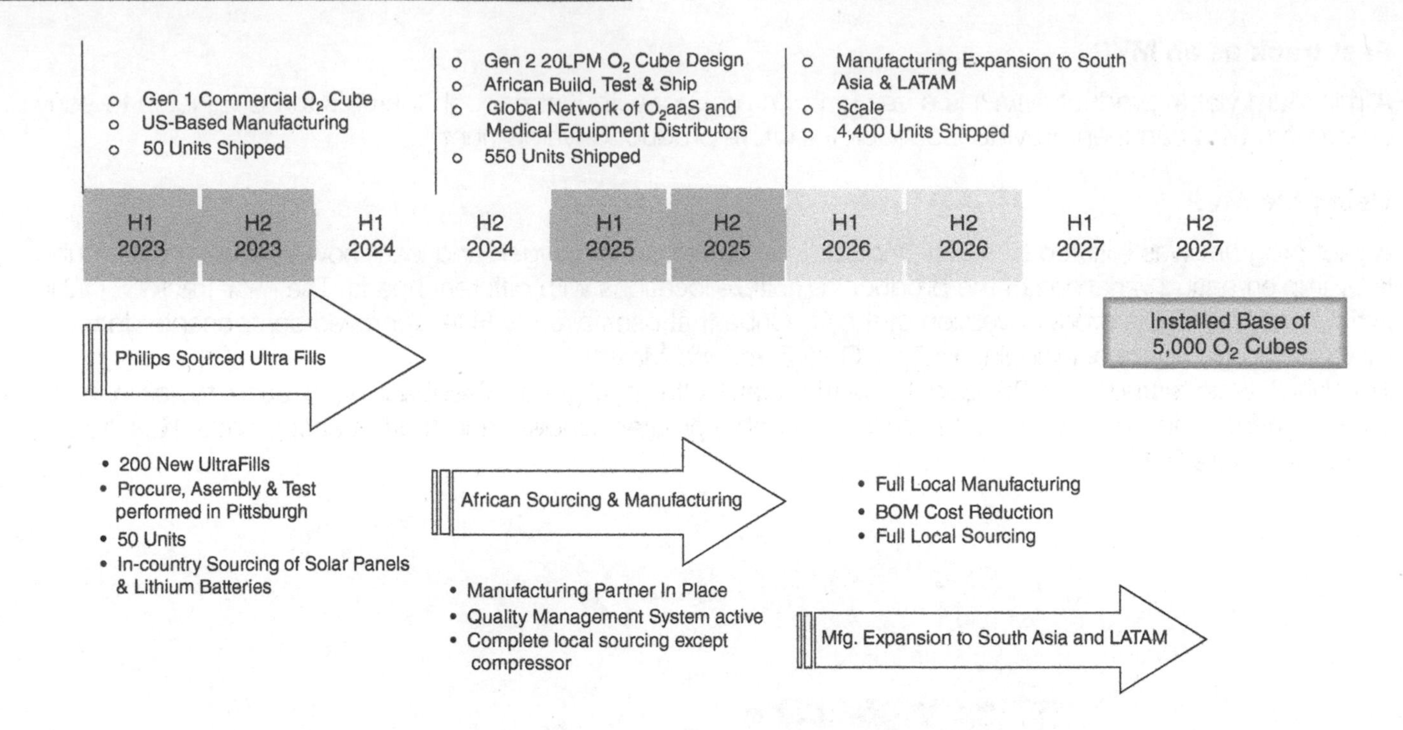

FIGURE B.11 LeanMed's business growth plan

Case Questions

1. What are the key elements in lean start-up methodology used by LeanMed?
2. What are the basic components of a Business Model Canvas (BMC)?
3. How does a BMC help in strategic planning?
4. What ways can be used to validate a product concept for further development?
5. Why is product concept validation so important?
6. What is a minimum viable product, and why is it such an important part of the product innovation process?

Answers to Case Questions

1. The key elements of the lean start-up methodology used by LeanMed are:
   a. Problem/solution fit (customer development)
   b. Product/market fit (customer validation)
   c. Scale (customer creation)
2. The basic components of the BMC are:
   a. Key partners
   b. Key activities
   c. Value proposition
   d. Customer relationships
   e. Customer segments
   f. Key resources
   g. Channels
   h. Cost structure
   i. Revenue streams

3. The Business Model Canvas provides a clear structure and specific issues that need to be addressed in strategic planning. It is also an excellent tool for communication and discussion.
4. There are several ways of validating a concept for further development. These include:
   a. Stakeholder research focused on the target market, either the end-user or specific stakeholders in the distribution channels, for example, medical practitioners or distributors.
   b. Experts in the product and/or market, for example, commercialization partners or academics.
   c. In-market testing of a prototype, for example, a minimum viable product (MVP) version of a product with just enough features, provides end-users and other stakeholders a good appreciation of what the final product could offer.
5. Throughout this Body of Knowledge, we have emphasized how cost and risk increase significantly as product innovation progresses. It is critically important to ensure that the decisions to continue with further development are based on sound information. Validating the product concept using a clear concept description or, preferably, a well-developed prototype, can provide critical information for decision-making, either for modifying the product or "killing" the project.
6. The minimum viable product is a prototype of the final product with features and functionality that provide the user with a good appreciation of how the final product will perform.

B.3 CASE 3

Stingray. Firefighter's Communication Device

This case study was prepared by students at the School of Engineering and Advanced Technology, Massey University, New Zealand: Jessica Braddon-Parsons, Errol Chrystal, Afnan Kayed, and Jacob Patrick.

B.3.1 Purpose

To provide an example of a product concept and design specifications based on the development of a product called the Stingray. The purpose was to address a number of the current problems faced by emergency services workers (specifically firefighters). It comprises an earpiece (or Bluetooth receiver to work with firefighters' headsets) and a main radio unit, which can be repositioned on uniforms for optimal accessibility.

B.3.2 The Core Product

The Stingray enables firefighters to communicate with other emergency services over large distances with clarity, ease, and reliability. This communication will be secure, safe, and reliable. The movement of the user will not be impeded, nor will the radio pose a safety risk.

B.3.3 The Tangible Product

The Stingray has two pieces: the main radio unit and the earpiece. Descriptions of design requirements for each follows along with augmented product capabilities.

The radio

- The radio is housed in a smaller and more compact unit than the older models. A curved profile helps mold the unit to the wearer and a lower, flatter design will stop any twisting motion on the hip while moving.
- A clip on the back of the radio unit allows the radio to be attached directly to a belt. This clip can be removed so the radio can be placed in the pockets of uniforms.
- The upgrade to a digital network will ensure improved audio quality.
- The case will be constructed from a high-strength plastic polymer, polyether ether ketone (PEEK). This helps protect internal componentry from damage due to rough treatment, heat, or water. The corners will also be surrounded by a soft rubber edge that helps prevent damage if the unit is dropped.
- A GPS will provide real-time tracking of personnel.
- The gyroscope and accelerometer can detect if an officer has fallen and is no longer moving. If this happens, the radio will send an emergency signal with their name or ID and geographic position.
- A sheet of copper wire will be placed in the officers' vests to act as an aerial. For emergency personnel who do not wear these vests, a smaller aerial will be provided to plug into the unit, like the current screw-in aerials. These plug into the coaxial connector on top of the radio.
- An encrypted Bluetooth signal will connect the radio with the wireless earpiece. The volume and channel control dials located at the top of the radio will be kept the same as on previous radios. This was a feature that the emergency services requested as it is easy to use in all situations. However, the dials now have an added lock-in feature to prevent accidental changes, which can be activated by pressing down on the top of either dial.

- A speaker and microphone are built into the radio unit for use if the earpiece fails to work for any reason. The speaker is activated when the Bluetooth connection to the earpiece is lost. The microphone can be activated by pressing the large "talk button" on the top of the radio.
- An emergency panic button is located on the side of the unit. If this button is pressed, the location and name or ID of the user will be sent to the operator at the base station, and emergency procedures will be followed.
- A chip (SD memory card) will have to be inserted into the radio to make it operational. This is an added security feature, and the chip will also retain information about the user to be sent out with GPS locations so operators know which person is using the radio unit. It will also store voice commands to activate or disable audio transmission through the earpiece.
- A removable lithium ion (Li-ion) battery will be used in place of the old nickel metal hydride battery. This decreases weight and increases battery life.
- The radio can be docked and charged through a mini-USB port located on the bottom of the unit. This port can also be used to program the unit (such as setting the channels) with a provided USB cable.

The earpiece

- Using a wireless earpiece replaces the old, wired speaker that was a large hindrance to firefighting personnel.
- A power button is used to turn the earpiece on or off.
- The earpiece will connect to the main radio unit via an encrypted Bluetooth signal.
- The ear grip will be removable so everyone can attach a fitted ear grip. This will ensure a snug fit on each user to prevent the earpiece falling off.
- A large "talk button" is located on the side of the earpiece to activate the microphone. Voice activation will also be available to the user for situations where both hands are occupied.
- An extended microphone comes down the side of the user's face to get the best possible audio quality through proximity to their mouth.
- Audio quality is improved by placing the audio source close to the ear, and by using a small surface transducer.
- A large battery indicator around the edge of the "talk button" will indicate how much battery life is left.
- The earpiece can be charged via a mini-USB connection on the underside. A USB cable is provided with every earpiece for this purpose. This connection can also be used to update the earpiece's firmware.

Augmented product capabilities

- A deployment service including technical training provided to users.
- Ongoing support services including a 24/7 service desk and online service portal.
- A USB cable.
- A pocket to hold the SD memory card (which can then be kept in a personal storage space such as a locker).
- A 10-port docking station that charges up to 10 units.

B.3.4 The Design Specifications

The translation of the concept description to design specifications is outlined in Tables B.1 and B.2.

Step 1: Define the user needs and their relative importance. This was derived from user research including focus groups and observational studies. The list of user needs is shown in Table B.1.

TABLE B.1 The user needs and priorities

#	Needs		Importance
1	Case	Must be strong	5
2	Case	Must be waterproof, shockproof and heat resistant	5
3	Volume and channel knobs	Needs to be lockable	2
4	Earpiece	Must have a personally fitted ear grip	4
5	Radio	Must be lightweight and small	4
6	Radio	Needs to have a longer battery life	3
7	Radio	Needs to be digital	5
8	Radio	Must display battery levels for both parts	3
9	Radio	Must be personalized to identify the user	1
10	Radio	Must have GPS	3
11	Radio	Must have a kinetic energy charger	1
12	Radio	Needs an emergency microphone and speaker	4
13	Radio	Transmission range greater than current product	5

Step 2: Translate the consumer needs into engineering design specifications with specific units and target values. These target values were derived from a combination of user research and competitor benchmarking and are shown in Table B.2. Further examples of design specifications and their development are provided in Ulrich and Eppinger (2016).

TABLE B.2 The design metrics, units, and targets

Metric #	Needs #	Metric	Importance	Units	Target value
1	1,2,6	High strength polymer casing	5	Nm^2 or Pascal	>1000
2	6	Total mass	4	G	<700
3	7,12	Battery life	3	mAh	>2500
4	13	Effective transmission range	5	Km	>1.5
5	3,4	Easy to use buttons	4	Subjective	>3 (on a 1–5 scale)
6	6	Size of radio unit	4	Cm^3	<400
7	2	Water resistance	5	X-bar	>5
8	2,3	Effective life of unit	5	Years	>6

Case Questions

1. What is the *core product* and why is this so important in defining product specifications?
2. What is the *tangible product* and how does its definition aid in the development of design specifications?
3. What are the two basic steps in defining product specifications?

Answers to Case Questions

1. The core product defines the key benefits the target market will derive from the product. These are critical elements that must be included in the design specifications.
2. The tangible product is the physical and aesthetic design features that give the product its appearance and functionality. Clear definition of these features provides the basis for the product design specifications and for translation of these specifications into a physical product.
3. The two basic steps in defining product specifications are:
   a. Define the user needs and their relative importance. This requires user research including focus groups and observational studies.
   b. Translate the consumer needs into engineering design specifications with specific units and target values.

REFERENCE

Ulrich, K.T. and Eppinger, S.D. (2016). *Product design and development*, 6e. McGraw-Hill Education.

B.4 CASE 4

OBO. Field Hockey Equipment

Presented with permission from Simon Barnett, Chief Executive, OBO, Field Hockey Equipment, New Zealand.

B.4.1 Purpose

To demonstrate the application of market research in product innovation. This case describes the use of social media as a market research tool to inform new product innovation and product improvement.

B.4.2 Company Background

OBO is a New Zealand-based company that design and manufacture field hockey equipment for goalkeepers. They hold around 65% of this niche market globally, exporting products to 62 countries. The company is an active product innovator, continually seeking improvement to existing products and the addition of new products to its range.

Field hockey is played globally and is an Olympic sport. A team comprises 11 players,10 field players, and a goalie. The game is played with a hard plastic ball which can reach speeds of up to 100 mph presenting significant danger to all players, especially the goalie. OBO markets a range of hockey equipment, specializing in equipment required by goalies. It's designed not only to enable goalkeepers to move freely, perform at their best, and look good but also to provide optimal protection for all parts of the body: head, throat, chest, hands, legs, and feet. See Figure B.12.

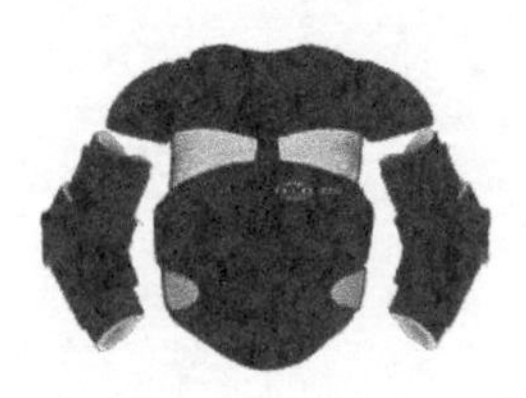

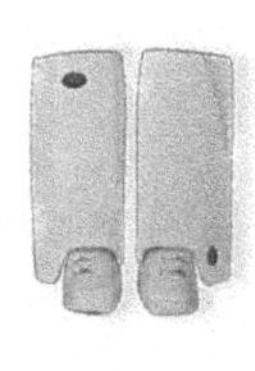

FIGURE B.12 **A sample of OBO's product range**

Over recent years, OBO has embraced social media as a key tool for communication with its customers and for underpinning its product innovation. Recognizing the need for a clear strategy to optimize the value of its social media effort, OBO developed social media guidelines, and the specific use of four key social media tools: (1) the company website, (2) Facebook, (3) YouTube, and (4) Instagram.

B.4.3 OBO Social Media Guidelines (from OBO website)

"As times are changing, we all need to keep up with them. Social media is still ever evolving, and you are all evolving with it. Locally based OBO pages are becoming more and more common, so these guidelines are designed to provide insight into how we in New Zealand utilize Social Media and how your own pages can work with OBO HQ content and augment each other."

"If executed well our social media channels will not only highlight our products but help build brand love and trust. Goalkeepers are our hero's and goalkeepers are unique. They deserve somewhere to go that they can own and totally relate to. Our Social Media channels are designed to make the goalkeeper the hero while still having a product/sales presence. It is designed to bring us closer to the goalkeeper and the goalkeeper closer to us. Social Media enables us to have two-way conversations about products, training drills, news and communicate brand.

- Everything needs to be well executed and exactly on brand and through the goalkeepers eyes
- We would rather do fewer things but do them really well
- Everything needs to have a why and overall goal for each different media. This ensures we are doing it for the right reason

We have found that having different goals/activities for each media tool means people are more likely to engage in all our medias as opposed to just one. It also means that our work is more planned and targeted."

B.4.4 OBO'S Four Key Media and Their Goals

1. ***Website:*** Product Education and enhancing choosing experience. 70% choosing experience/30% brand experience.
2. ***Facebook:*** Big focus on two-way communication between brand and goalkeepers. Builds community and is designed to bring the goalkeeper closer to the brand.
3. ***Instagram:*** Highlighting the world of OBO (products, goalkeeper, and brand) through photos. 70% brand/30% choosing experience and product education.
4. ***YouTube:*** Use of videos to allow OBO to add extra detail that OBO can't communicate as easily in writing.

The key focus is on product information tips and tricks and training videos.

Things that are important to OBO

- Goalkeepers
- Language – positive, fun, genuine/real, humor, no bullshit
- Engagement – two-way conversations
- Content – useful and beneficial
- Brand – feel and execution
- Original – messaging, content, and executions
- Details matter
- Feedback

OBO content feel. . .

- All about the goalkeeper
- Fun
- Serious
- Color
- Feedback, input
- Random, crazy, unexpected

How OBO uses the various social media tools is shown in Figure B.13.

	Photos	Videos	Two Way Comms	Product Launch	Product Updates	Competitions	Feedback	Training Videos	Product Info/Details
Instagram	Frequently	Sometimes	Sometimes	Frequently	Sometimes	Sometimes	Sometimes	Sometimes	Frequently
YouTube	Never	Frequently	Never	Sometimes	Never	Never	Never	Frequently	Never
Facebook	Frequently	Frequently	Frequently	Frequently	Frequently	Frequently	Frequently	Frequently	Frequently

Frequently	Sometimes	Never

FIGURE B.13 The application and emphasis of the 4 social media tools

Facebook is the primary social media platform used for player product research as it enables more in-depth and nuanced two-way conversations with keepers. OBO uses Facebook to understand, identify, and quantify consumer needs through questions, discussions, and polls.

B.4.5 A Product Innovation Example: The Left-hand Protector

Field hockey goalkeepers use their left-hand protector not only to protect their vulnerable hands, but to defend the goal and deflect the ball away from attacking players. While OBO had a successful range of left-hand protectors for various skill levels, player feedback from international players was that an evolving game presented new opportunities to utilize the left hand, and a new product would be required to achieve this. When Facebook members were asked about their needs, OBO received similar comments.

While international goalkeepers were intimately involved in the detailed development of the new product, social media was used to verify and quantify needs and features.

The design brief for the left-hand protector contains:

Qualifiers

- Comfort = At least the same.
- Protection = At least the same.
- Rebound = Increase.
- Weight = Similar would be ideal but can afford to be slightly heavier.

Differentiators

- Improve/Aid left-hand side diving.
- Improve stability and control of hand protector through more purposeful thumb and pinky placement.
- Increase rebound/clearance predictability – flatter/squarer/:
  - Front saving face.
  - Side Edges.
- Align with other Plus Range styling.
- Increase overall saving area size and usefulness of saving area:
  - Within International Hockey Federation rules.
  - Hand placement.
- Improve security on hand while still ensuring comfort and wrist movement.

Figure B.14 shows the design stages through to production and the interactions with goalkeepers at each stage. The final product, as described in social media promotion, is shown in Figure B.15.

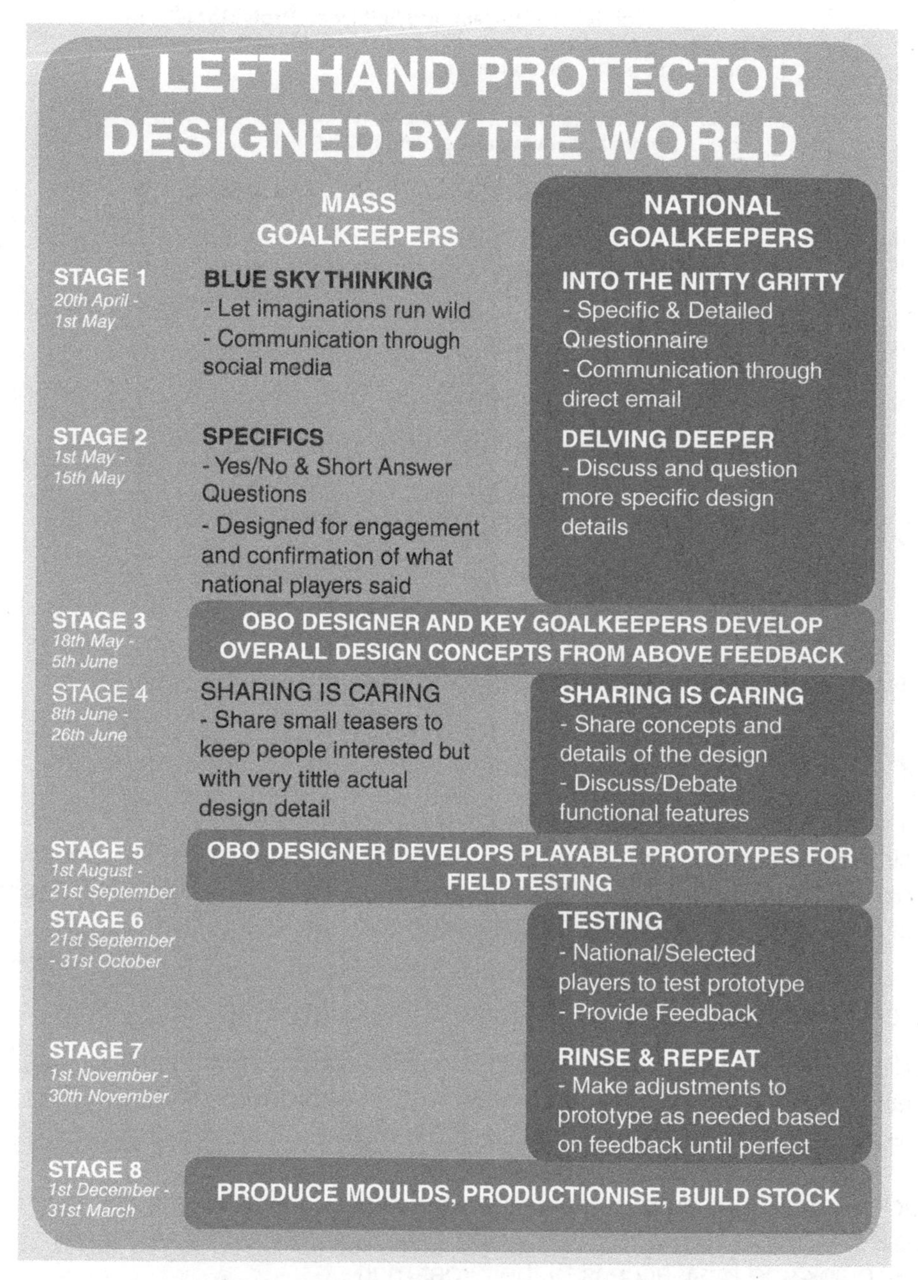

FIGURE B.14 Left hand protector design stages.

The ROBO Hi-Rebound Left Hand Protector is flat, with loads of rebound and great control. It also has a soft inner for increased comfort and fit.

It is sized so that the protector is stable on the hand, achieving more ball control and greater power.

It has a fairly flat face but still shaped to control the ball and direct downwards.

It has slightly square edges which can be used to deflect the ball and an almost flat bottom edge when placed on the ground.

The anatomically shaped inner, results in a natural hand position and good grip, especially in the thumb area.

The wrist protector is fully shaped for maximum coverage.

The heavy elastic strap keeps hand protector secure on the hand at all times, but it doesn't inhibit the movement.

There is a soft pad under the wrist area for increased comfort and better ball control.

FIGURE B.15 Left hand protector final product description

Case Questions

1. Where can social media contribute to product innovation?
2. Why is social media such a valuable market research tool in product innovation?
3. Lead users are important to many categories of product innovation. How has social media increased the value of lead user groups to product innovation?

Answers to Case Questions

1. Social media can contribute at various stages of the product innovation process:
   a. In the early stages of idea generation – either for new products or for product improvement.
   b. In concept development and product design specifications through contribution of specific user needs and iterative response to design improvements.
   c. Feedback on near-to-market-ready prototypes in real-use situations.
2. Social media is valuable to product innovation in several ways:
   a. Relatively quick and easy access to the target market.
   b. Engaging potential consumers throughout the innovation process.
   c. Identifying and engaging individuals who may become early adopters and influencers in purchase of the new product. For example, international goalies in the field hockey case.
3. A lead user is a type of consumer or user who recognize product challenges and who has a high level of expertise and experience in a specific field or market. Lead users often anticipate future trends and demands and create solutions that can benefit not only themselves but also other users and potential customers. In certain targets, for example, the field hockey case, lead users form on-line communities to share their ideas and experiences. These communities provide an excellent forum for insights and direction through product innovation.

B.5 CASE 5

Developing a Culture of Innovation: Adobe Kickbox and rready®

B.5.1 Purpose

This case study provides an overview of how a large multinational software company developed and implemented a program to kickstart its innovation culture. Further, as they opensourced their approach, examples from other companies are addressed as well. The examples also show impacts to management of innovation portfolios and processes.

B.5.2 Body of Knowledge Areas

This case study primarily relates to Culture & Teams and secondarily addresses Process as well as Portfolio.

B.5.3 The Challenge

Adobe is well known for its software products used and loved by companies and individual creatives for photo and video editing, drawing, stock imagery, and more. Photoshop, Illustrator, Premiere Pro, and the Creative Cloud are among their flagship products. Since its beginning in 1982, innovation has been a characteristic of the organization and remains one of its four core values today. With nearly 30,000 employees, the scale of Adobe provides resources and abilities to tackle many types of opportunities. However, with such scale, innovation can slow as well. At Adobe, slowing innovation was reflected in its stock price. Like many stocks after the financial collapse in 2008, Adobe lost half of its market value. By 2012, the price was recovering, but at a slow pace. Adobe's approach to innovation was limiting growth. Using a top-down approach, they would identify a few projects each year and, through a business case process, fund a handful that appeared most promising. It's a standard approach but not one that makes innovation accessible to many employees. Adobe was potentially missing out on creative ideas as only a few employees shared ideas that received funding.

Adobe was not alone, as most organizations don't make it easy for innovation to occur. While CEOs and other senior leaders share the importance of innovation, few of the same leaders say they have effective innovation practices in their organizations. The Center for Creative Leadership puts the gap at about 80%, with 94% of senior leaders expressing that innovation is a key driver of success, but only 14% are confident in their organization's ability to drive innovation effectively (Horth and Vehar, 2015). There are many challenges that help explain the gap, including:

- ***Resistance to change:*** Innovation often requires breaking away from the status quo, which can be met with resistance from employees, managers, and even the organizational culture itself.
- ***Risk aversion:*** Organizations that prioritize stability and predictability may be hesitant to take risks associated with new ideas or approaches. Fear of failure and the potential negative consequences can stifle innovation.
- ***Lack of resources:*** Innovation requires dedicated resources such as time, funding, and people. Other operational priorities and short-term goals may take precedence, leaving little room for innovation.
- ***Siloed mentality:*** In organizations with departmental silos, collaboration and cross-functional communication can be hindered. If there is a lack of collaboration and information sharing, it becomes difficult to generate and implement innovative ideas.
- ***Lack of a supportive culture:*** A culture that discourages risk-taking, punishes failure, or emphasizes strict adherence to rules and procedures can stifle innovation. In such a culture, employees may hesitate to propose new ideas or take risks.

- ***Short-term focus:*** Organizations that prioritize short-term results and immediate financial returns may struggle to invest in long-term innovation projects. Balancing short-term goals with long-term innovation can be challenging.
- ***Leadership and management practices:*** An organization's leadership and management style can greatly influence its ability to innovate. Leaders who are resistant to change, fail to provide clear vision and support for innovation, or stifle creativity through excessive control can hinder the organization's innovative potential.

Overcoming these challenges requires a commitment to fostering a culture of innovation, providing resources and support for innovation initiatives, promoting collaboration and cross-functional communication, and cultivating leadership that encourages and empowers employees to think creatively and take risks. Making these changes is an overwhelming responsibility, yet that is exactly what organizations need to be innovative and what innovation leaders are responsible for accomplishing.

B.5.4 The Approach

In 2012, Mark Randall, their Vice President of Creativity, had started applying Lean Start-up-type practices. He was seeing success pursuing ideas using experimentation, iteration, and small amounts of funding. Randall's work attracted the attention of the head of Adobe's largest business unit, who wanted to expand Randall's approach to innovation (Dann, 2023).

Randall surveyed employees to gain insights about what led them to not share ideas. The most frequent response was "convincing management" (Burkus, 2015). Instead of needing to convince management to pursue an idea, Randall wanted to empower employees to begin innovating without the need for approval. After all, it was likely that many of the most creative people at Adobe were uninterested in dealing with the bureaucratic structure to pitch an idea. Randall thought instead of Adobe funding a $1 million project, he could fund 1000 projects at $1000 each. If only one of those 1000 projects became a new growth engine, it would pay for the innovation cost.

Randall assembled the innovation resources he had collected into a training approach, and coupled with a few other items, he created a physical box titled RedBox. The innovation program was named Kickbox, since it was intended to kickstart the Adobe innovation culture. The Kickbox process is intended to be time-bound and follow a simple process constructed around answering four questions:

1. ***Problem:*** Is the problem worth solving? (2 weeks)
2. ***Solution:*** Does the solution create value for customers? (4 weeks)
3. ***Concept:*** Do I have a concept for the next phase? (1 week)
4. ***Convince:*** Can I convince others of my idea? (1 week)

Thankfully, we don't have to provide the details of Kickbox and the fascinating contents of the RedBox (which includes chocolate and coffee) in this case study as Adobe open-sourced the RedBox program. This also means that 1000s of other organizations have deployed Kickbox to foster innovation culture. Find the open-sourced material at www.Kickbox.org.

Further, a small team at Swisscom, the leading Swiss IT- and Telco-Company, was one of the companies that implemented Kickbox based on the open-sourced material. After several successes and years of optimizing and enriching the program within Swisscom, the team decided to spin-off and make the program available to any organization that would like to create a culture of bottom-up innovation using the 360 approach of their KICKBOX program. The spin-off is the company rready®, which has helped numerous organizations successfully implement Kickbox and related software solutions to get even greater value from Kickbox. rready® is accelerating the learning that comes from the numerous organizations using Kickbox and rolling those lessons into improvements to the open-sourced materials as well as the services they provide organizations. Below are a few examples of organizations, based on information provided by rready® (2023), that have seen Kickbox used to overcome the challenges this case study started with.

B.5.5 How Organizations Created Innovation using Kickbox

After open-sourcing Kickbox, it has spread to over 1000 companies, including US-based tech companies, large enterprises such as 3M, Caterpillar, MasterCard, and P&G, and unique organizations like US Peace Corps, the Gates Foundation, and DARPA. Highlights of how innovation culture has changed at non-US-based companies ZF, Swisscom, Implenia, and CSS are shared below, based on rready® case studies and used with permission (rready®, 2023).

ZF

ZF is an automotive systems business with more than 100 years of experience. Most of ZF's customers are big manufacturers, but innovation leader Manuel Leichtle expects that customer base to become more distributed in the future. "We are a tier 1 supplier, and we are very familiar with receiving orders from the original equipment manufacturer (OEM). And, from my point of view, that will change in the near future. It will not be exclusively the big customer who will continue to pay for our products," he says. "And this also requires the engineers to change. In the past, it was common to receive a technical specification, and based on that, we develop a product. And in my opinion, this won't happen anymore."

This shift in the design process means that engineers will have to adopt a broader view of product requirements to ensure they understand and address the customer's problem. Manuel says that "In the future, you'll need to discover more about your customer—and I mean the end customer, the one who really uses [the product]. I think whoever manages this best in the future will survive. And Kickbox is part of it because it trains you to take a different look at your potential customer." Kickbox played a major role in educating ZF employees on customer-centricity in an experiential learning environment. Another strength of using the Kickbox approach to influence innovation culture is that employees practice methodologies like Human Centered Design, Lean Start-up, and Design Thinking without even realizing it because the program is simple, gamified, and fun.

While the cultural changes have been welcomed, the Kickbox method has led to even more positive impacts at ZF than Manuel expected. In particular, the innovation ecosystem has been a valuable resource for cultivating ideas, allowing project leaders to find the external support they need easily. Additionally, by working with innovators in other industries, ZF has been able to leverage their experiences and learnings to improve the likelihood of success of its own innovation projects. "We didn't need to make the same errors again," Manuel says. "You're talking to someone who had the same challenges."

Swisscom

Swisscom is Switzerland's number-one telecommunications firm. They generate the largest part of their revenue with products that did not exist a few years ago. This trend is expected to accelerate, requiring Swisscom to foster an innovation culture at all levels of the organization. They compete in a fast-changing and highly dynamic market. At the same time, Switzerland is one of the leading innovation countries, with universities producing top talent and attractive employers such as Google, Apple, Microsoft, and Facebook gobbling them up. How can Swisscom keep attracting talent to stay ahead in the market and continue to innovate at an ever-quickening pace?

Upon hearing a keynote presentation by Mark Randall about the organizational benefits Kickbox creates, Dave Hengartner, Swisscom's Innovation Lab Manager, sought to adopt Kickbox. After rolling out Kickbox to a few volunteer employees, Hengartner was energized. He had found a way to build an innovation culture and increase the speed of innovation. He shared, "I've discovered an approach that I believe will bring entrepreneurial ownership into a big organization." Over time, as Kickbox reached more employees at Swisscom, Hengartner found ways to simplify the Kickbox implementation and improve participants' experience. Swisscom created a software platform and a services marketplace to accompany their use of Kickbox. Today, Kickbox is the leading entrepreneurship program at Swisscom, shaping the culture for innovation.

Implenia

Implenia is a leading multinational integrated construction and real estate services company. In addition to building construction and civil engineering, Implenia is a key player in the infrastructure sectors of its home markets and a successful real estate developer.

In 2019, the company officially made innovation one of the four key pillars of its business strategy. As part of this, Implenia formed the Innovation Hub, a team dedicated to accelerating innovation internally and working with external partners to make Implenia fit for the future.

The Innovation Hub was looking for a structured and centralized solution to foster innovation and intrapreneurship that is accessible to the entire workforce. While the primary aim was to encourage ideation for new, innovative business models, the Innovation Hub also sought to promote a cultural shift from solution-based thinking to a stronger focus on desirability. The specific challenge faced by Implenia was mobilizing the entire employee base across all functions and sites – from accountants to project managers to construction workers. This required an approach to promoting innovation that was accessible to all employees.

Implenia rolled out Kickbox to all employees regardless of their job title or location. By providing a standardized structure, Kickbox gave employees an accessible entry point to express their ideas and the tools to structure and develop new exciting concepts that translate to viable business models. In addition, the Innovation Hub recognized the importance of removing any obstacles to participation – organizational or cultural – so they got top management involved for a big launch event and promoted Kickbox during site visits where they could answer questions directly. The culture across the organization has evolved to support innovation and use it to deliver on Implenia's business strategy.

CSS

CSS is one of Switzerland's leading health insurance companies. Innovation is a core value, both to differentiate themselves from the competition as well as to ensure they provide their customers with the most convenient, quality care possible. They're continuously investing in innovative new offerings and using Kickbox to drive innovation throughout the company.

As with many industries, insurance is undergoing a digital transformation. Self-service options like online services and apps allow customers to manage their healthcare without reaching out to their insurers or providers directly. This digital transformation is making health insurance services more convenient and accessible. It's also helping to contain rising healthcare costs.

"One of the reasons why we focus on automating services is to stem those rising costs," says Daniel Alzer, Innovation Manager. It's not a simple process, though. Insurance companies like CSS rely on a lot of legacy systems, and it's not easy to just add new features to these systems. "It's not like updating Office on your PC," Daniel explains. "All the company's core processes are based on that system." That means creating a new digital offering is a long-term project, which can be hard to get through the pipeline when other projects with more immediate value take priority. As Daniel explains, "You have to convince the stakeholders the idea is really worth investing in. Kickbox makes you validate the idea from the beginning by talking to real people in the market. If you can show that customers are interested in it, that makes it much easier to convince them."

Daniel also had the objective of decentralizing innovation. "We had a central innovation team and an innovation board, but we thought it would be great to include all the employees as well—because every employee has great ideas." Most importantly, decentralizing innovation by using Kickbox has helped CSS streamline its innovation process by putting employees in the driver's seats of their own projects from the beginning. "Before Kickbox, you'd need to create a project approval document, form a team, and make a budget. . .now, individuals can start working on projects themselves."

While Daniel believes in the importance of innovation, in his experience, it can be hard to get people engaged with innovation programs. "They have to leave their comfort zone, and not everyone is able or wants to do that," he explains. When he first introduced Kickbox to CSS, Daniel wanted to foster a culture of innovation that would empower all employees to be a part of the innovation process and allow them to work on their own ideas. His colleagues see that occurring, with Kickbox impacting the culture at CSS and helping their people grow personally and professionally. Daniel recalls that one manager came to him with glowing feedback after seeing the effects on one member of his team. This individual, who had led a Kickbox project involving an online pharmacy platform, was so motivated to bring his idea to life that he took it upon himself to teach himself new programming skills—opening up new opportunities both for himself and for his team.

B.5.6 Summary

The best innovating companies recognize the importance of senior leader support for innovation (Knudsen et al., 2023). They often form an innovation group or lab to be responsible for innovation. This model, while helpful, is also restricted on the number of projects they can start and manage at any one time. This can lead to bureaucratic systems for pitching and selecting ideas for innovation projects. While innovation is occurring, some employees with the most creative and valuable ideas are not being involved in innovation. A bottom-up approach is needed to involve all employees in innovation; to democratize innovation in an organization. Doing so impacts the culture in several meaningful ways, mitigating factors that limit innovation:

- Fostering a culture of innovation
- Providing resources and support for innovative initiatives
- Promoting collaboration and cross-functional communication
- Cultivating leadership that encourages and empowers employees to think creatively and take risks.

Kickbox is a free, open-source system to create bottom-up innovation. Over 1000 organizations have adopted it and have seen their innovation culture change and become much more inclusive.

Case Questions

1. What are some factors that limit innovation effectiveness in organizations?
2. What initiatives mitigate factors that limit innovation effectiveness?
3. Why is a top-down approach to innovation insufficient?
4. How does Kickbox provide a bottom-up approach to innovation?

Answers to Case Questions

1. What are some factors that limit innovation effectiveness in organizations?

   Factors have been introduced throughout the BoK. The ones highlighted in this case are resistance to change, risk aversion, lack of resources, siloed mentality, lack of a supportive culture, short-term focus, and misaligned leadership and management practices.
2. What initiatives mitigate factors that limit innovation effectiveness?

   The BoK highlights many practices common to organizations that innovate well. The examples shared in this case address fostering a culture of innovation, providing resources and support for innovative initiatives, promoting collaboration and cross-functional communication, and cultivating leadership that encourages and empowers employees to think creatively and take risks.
3. Why is a top-down approach to innovation insufficient?

   A top-down approach to innovation implies that senior leaders are involved in selecting and funding a few innovation projects each year. As seen in the Portfolio chapter, selecting innovation projects is essentially placing bets where to commit resources. A top-down approach favors safer bets on opportunities that are better understood and contain fewer unknowns. These projects are identified and promoted throughout the organization by only a few employees. Consequently, the experience, insights, and creativity of most employees is not leveraged for innovation.
4. How does Kickbox provide a bottom-up approach to innovation?

   Kickbox equips any employee for innovation. They are provided instructions, motivation, and resources to pursue an idea, test the importance of the idea, consider solutions, and validate solutions with customers. They learn to conduct fast inexpensive experiments to explore the problem/opportunity and test their assumptions. For the cost of one traditional top-down innovation project, numerous innovation projects can be pursued. In Adobe's case, instead of one top-down project, they could fund a 1000 Kickbox projects. Only one needs to become a growth leader to pay for the cost of operating the Kickbox bottom-up innovation program.

REFERENCES

Burkus, D. (2015). *Inside adobe's innovation kit*. Harvard Business Review. https://hbr.org/2015/02/inside-adobes-innovation-kit.

Dann, J.B. (2023). *"Kickboxing" around the world: An intrapreneurship revolution?* USC-Marshall/Lloyd Greif Center for Entrepreneurial Studies.

Horth, D.M. and Vehar, J. (2015). *Innovation: How leadership makes the difference*. Center for Creative Leadership.

Knudsen, M.P., Zedtwitz, M., Griffin, A., and Barczak, G. (2023). Best practices in new product development and innovation: Results from PDMA's 2021 global survey. *Journal of Product Innovation Management* 40 (3): 257–275.

rready® (2023). *Corporate innovation success stories*. rready® Case Studies https://www.rready.com/case-studies.

Appendix C: Examples of Product Innovation Best Practice from PDMA's Outstanding Corporate Innovators' (OCI) Awards

C.1 BACKGROUND TO THE OCI AWARDS

The OCI Award is the only global innovation award that recognizes sustained and quantifiable innovation success AND the practices which have contributed to that success.

Criteria include:

- Sustained record of success launching new products/services over a 5-year period.
- Significant and quantifiable business results delivered by new products/services.
- Consistent use of a set of new product development practices from which others can learn.

PDMA employs a rigorous selection process for the OCI Award which is managed by volunteers with extensive and diverse innovation backgrounds.

Since 1987, winners have shared their specific practices at the PDMA Annual Conference providing a unique "behind the scenes" look at successful corporate innovators and providing lessons learned for attendees. OCI Winners reflect diversity in industry, size, and global location.

Product Development and Management Body of Knowledge: A Guidebook for Product Innovation Training and Certification, Third Edition.
Allan Anderson, Chad McAllister, and Ernie Harris.

C.2 ANALYSES AND LEARNINGS FROM OCI WINNERS

Throughout the history of OCI, PDMA has conducted analyses of the practices of all the winners. Key takeaways from those analyses are:

- Common practices have contributed to the innovation success of the winners within each timeframe despite the diversity of industry, size, and geography.
- Evolution of the practices which have contributed to the winners' success over time.
- Increasing complexity of the practices for achieving innovation success.

A summary of those practices, common to most winners over recent years is shown in Figure C.1.

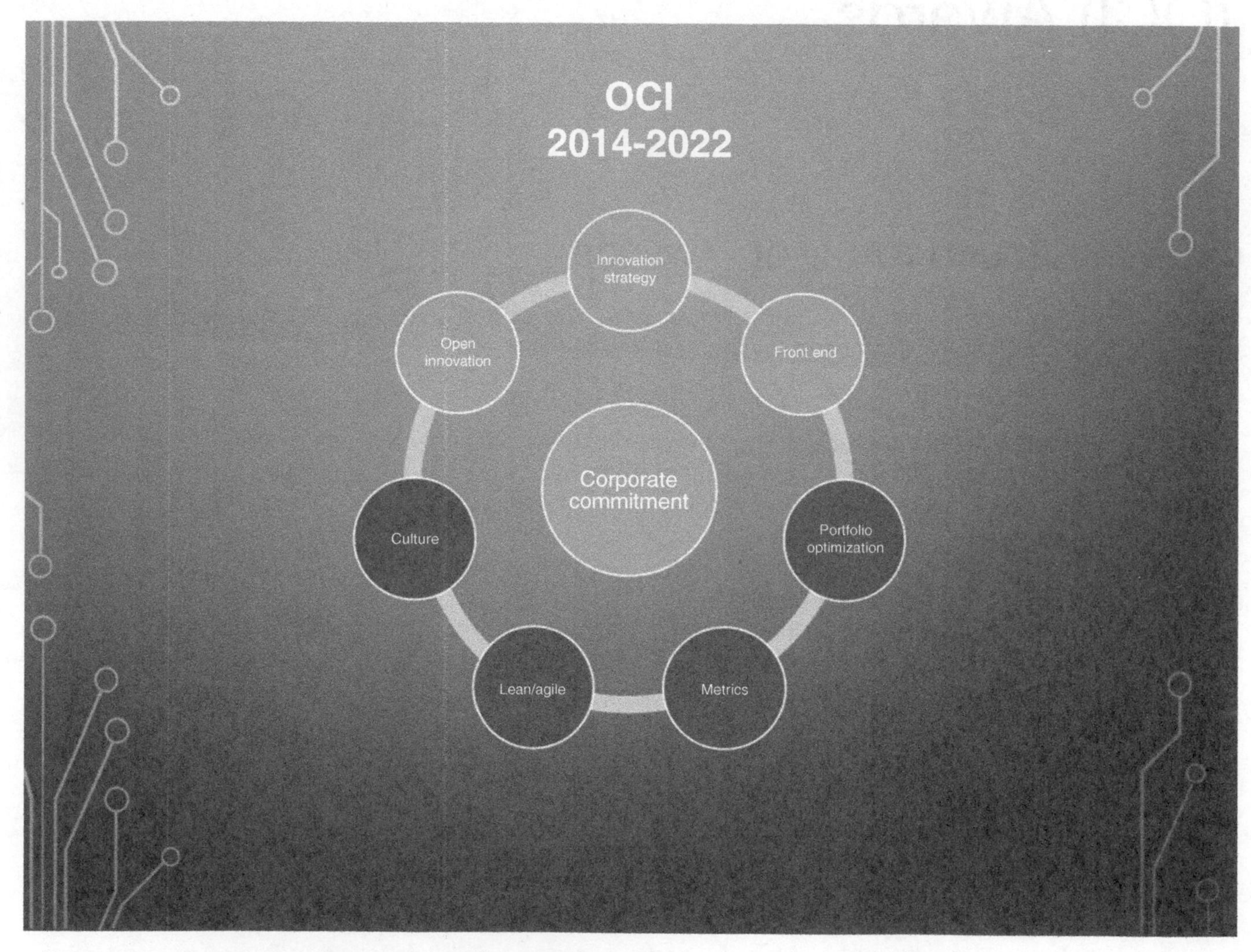

FIGURE C.1 Summary of learnings from OCI awards

Following are insights that PDMA gained about these practices as well as specific examples how some of the Winners executed against these practices.

Corporate Commitment: For the winners in recent years, innovation has become a corporate strategic core value which is reflected in senior management involvement and support, a dedicated innovation function, and the resources needed to consistently deliver significant innovations to the market.

An example of that commitment is a quote during an economic downturn from the VP of Seating for Faurecia, a global automotive equipment supplier headquartered in France---"A crisis is a terrible thing to waste; I will double your innovation budget for next year."

Innovation Strategy: Recent OCI winners have shown that having a well-defined innovation strategy is critical to success. The strategy clearly defines for the organization the key factors for achieving innovation success. The innovation strategy provides focus and internal alignment and drives resource allocation.

Corning, a leading global innovator in materials science, calls its innovation strategy its Innovation Recipe, which is depicted in Figure C.2.

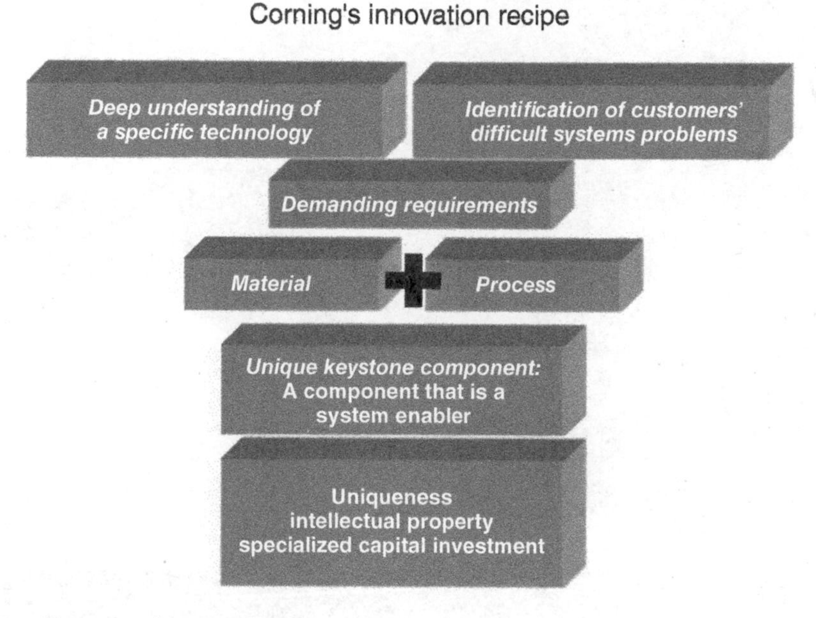

FIGURE C.2 Corning "innovation recipe"

Focus on the Front End: Recent OCI winners all conduct extensive market and customer discovery at the front end of the innovation process to identify, screen, and validate potential new opportunities before they start the development process. The example of a front-end process in Figure C.3 comes from Church & Dwight.

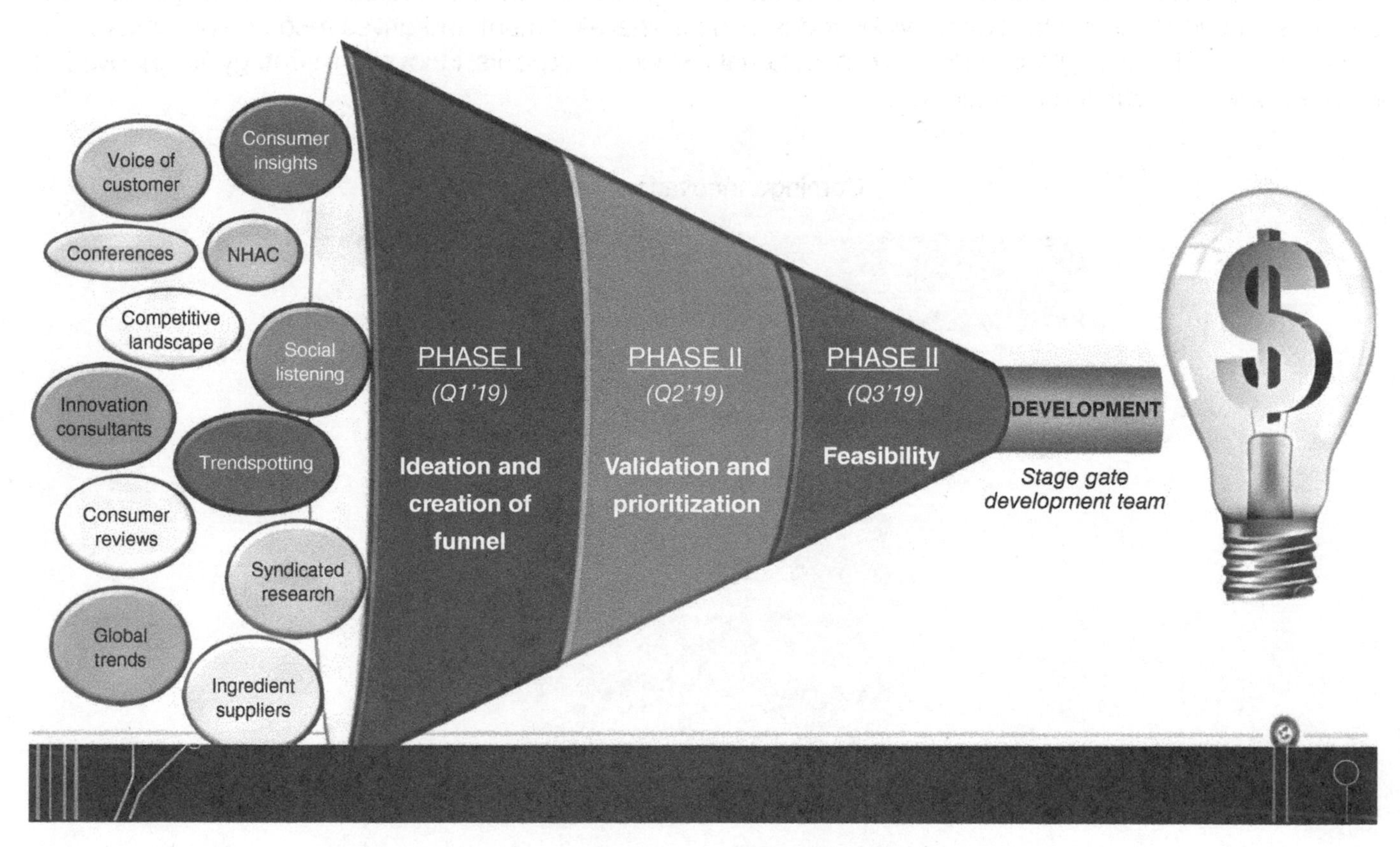

FIGURE C.3 Church and Dwight front-end of innovation process

Figures C.4 and C.5 depict how UnitedHealth Group systematizes the front end to identify unarticulated needs and develop unique solutions.

Systematizing the "Front End"

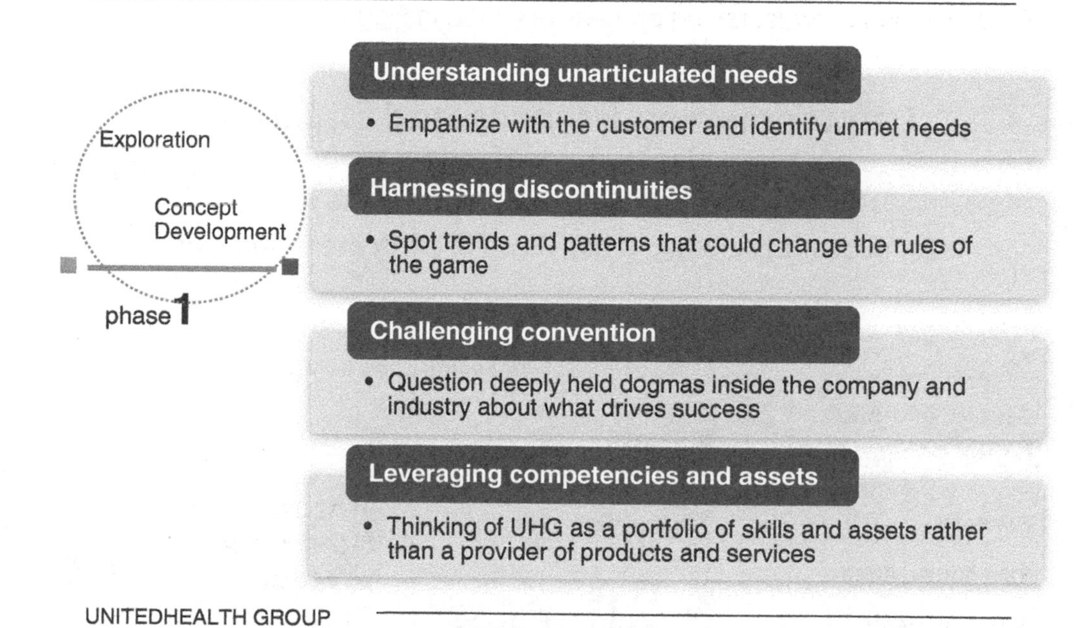

UNITEDHEALTH GROUP

FIGURE C.4 United-Health front-end of innovation system

Harnessing unarticulated needs: design thinking in health care

- A new way to understand and solve problems
  - Common to CPG – Revolutionary in health
- Developing competency for:
  - Empathy
  - Contextual understanding
  - Seeing beyond the survey
- Creates rich insights that spur creativity
- Inspires solutions with meaningful impact
  - My Health Sherpa

UNITEDHEALTH GROUP

FIGURE C.5 United-Health application of design thinking

Recent winners are also developing new ways to generate ideas for potential new products/services. An example of this is provided by Novozymes, a global biotechnology company headquartered in Denmark, with their internally developed crowdsourcing process depicted in Figure C.6.

Crowdsourcing Process and Outcome

Our pilot study created a high level of excitement and trust

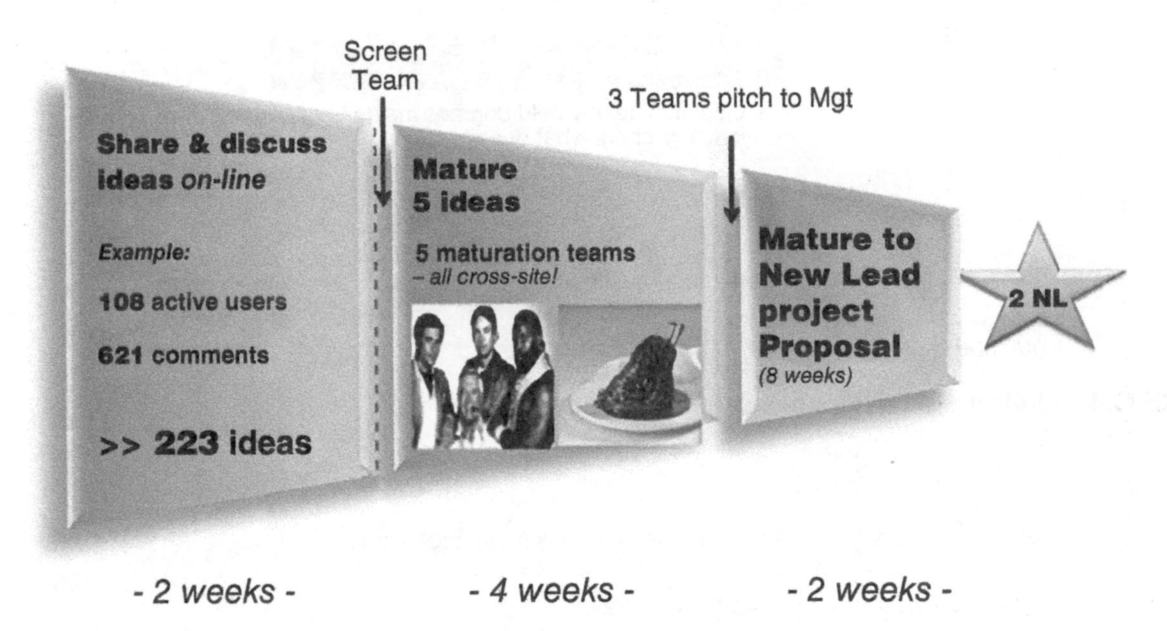

FIGURE C.6 Novozymes application of crowdsourcing

Portfolio Optimization: For recent OCI winners, achieving innovation success involves working on the right projects the right way at the right time and solving the right problems. It also means achieving a balance of incremental, adjacent, and disruptive opportunities in the portfolio. For some of the winners, it involves leveraging technology platforms across business units and across the globe. Frequently, portfolio optimization may involve looking outside the company's current markets and even looking at new business models. All these activities are done using a disciplined process, as shown in Figure C.7 from BD (Becton, Dickinson & Co.), a global medical technology company.

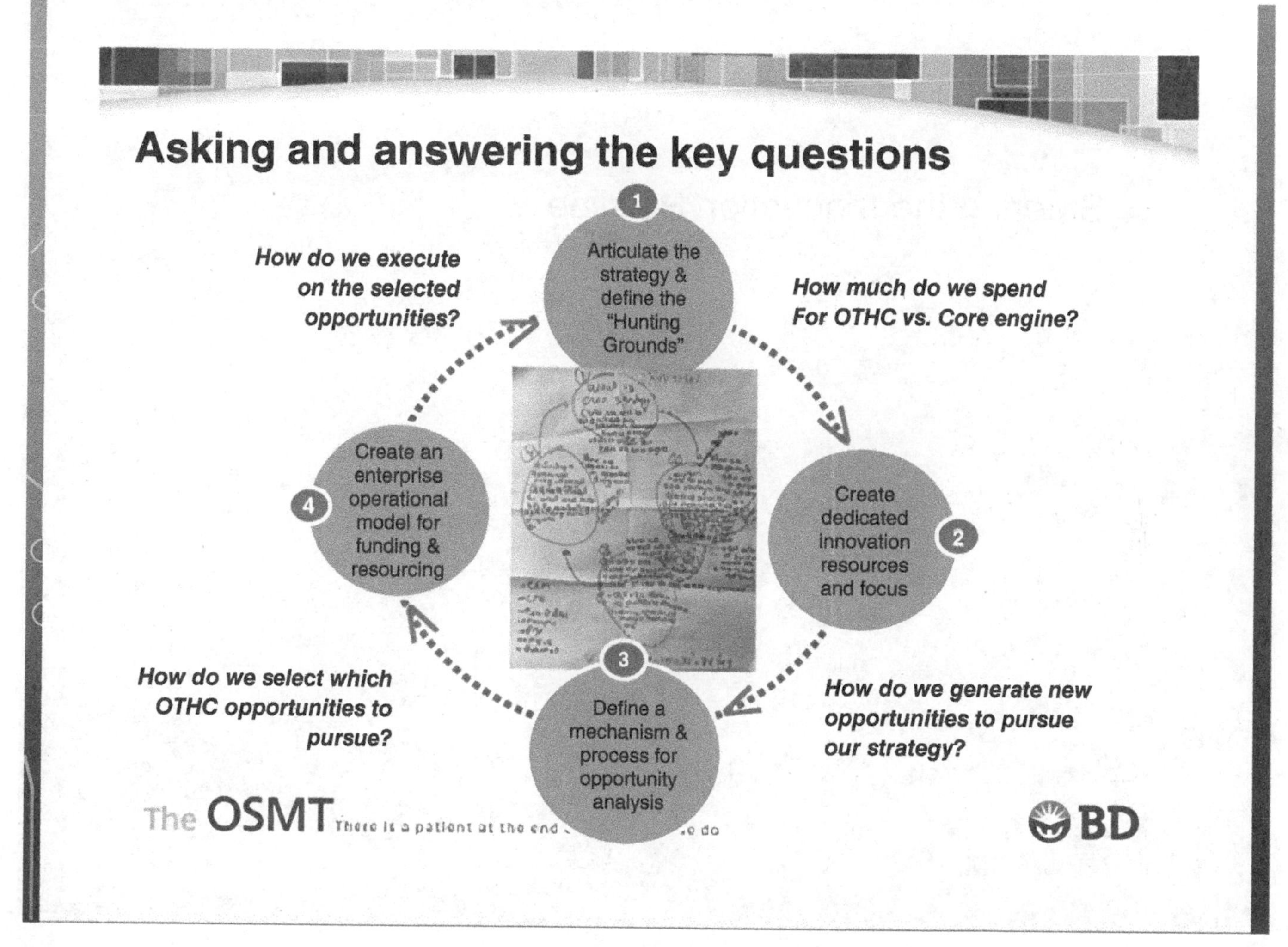

FIGURE C.7 Becton, Dickson Co. process for portfolio optimization

Open Innovation: For all the recent OCI winners, open innovation/collaborative development has become a key corporate strategy. As depicted in Figure C.8 from DSM, a global leader in health and nutrition headquartered in The Netherlands, open innovation is used throughout the innovation process. Like DSM, many of the OCI Winners have also created their own venture funds to invest in university research and in startups.

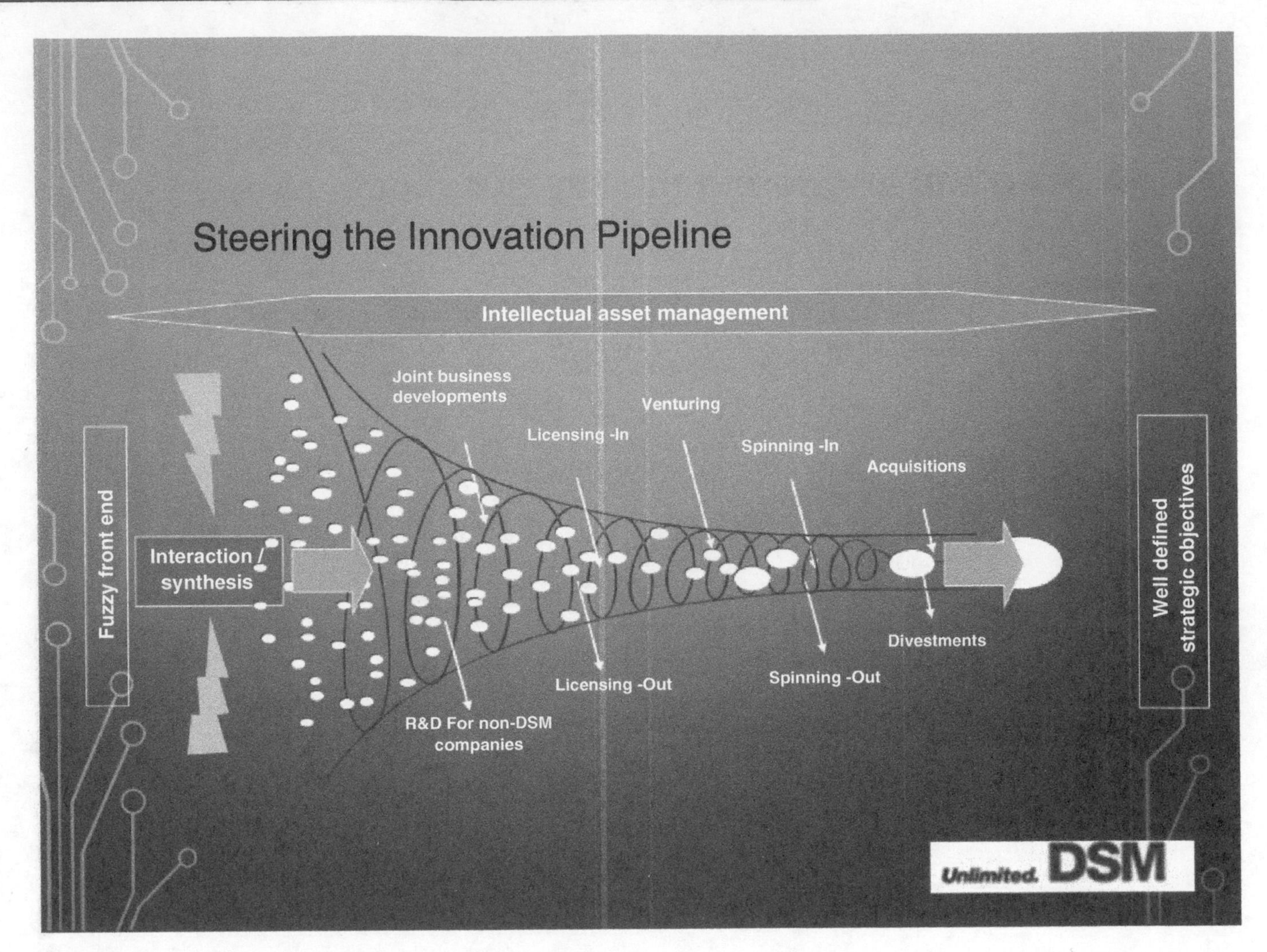

FIGURE C.8 DSM application of open innovation

Culture: Culture creates the foundation for successful corporate innovation. Senior management creates a culture which values and respects the diversity of all employees and supports an environment where people are not afraid to take risks, think boldly or even to fail. Other key elements of culture which are important to achieving and sustaining innovation success are creating a collaborative environment and one where leaders are committed to team success by building innovation with them.

Lean and Agile: In recent years with rapidly changing market dynamics and customer needs in order to succeed, companies have had to employ lean and agile processes. Past winners agree that these practices have helped to improve transparency and alignment on innovation activities within the company. One of the benefits of an agile test and learn approach is to get quicker feedback from customers on new products and services. Also, agile practices allow companies to adapt more quickly to market disrupters by applying MVP's, pivots, and fast learning cycles to the innovation process. Figure C.9 shows an agile process used by Porex, a global leader in porous polymer solutions.

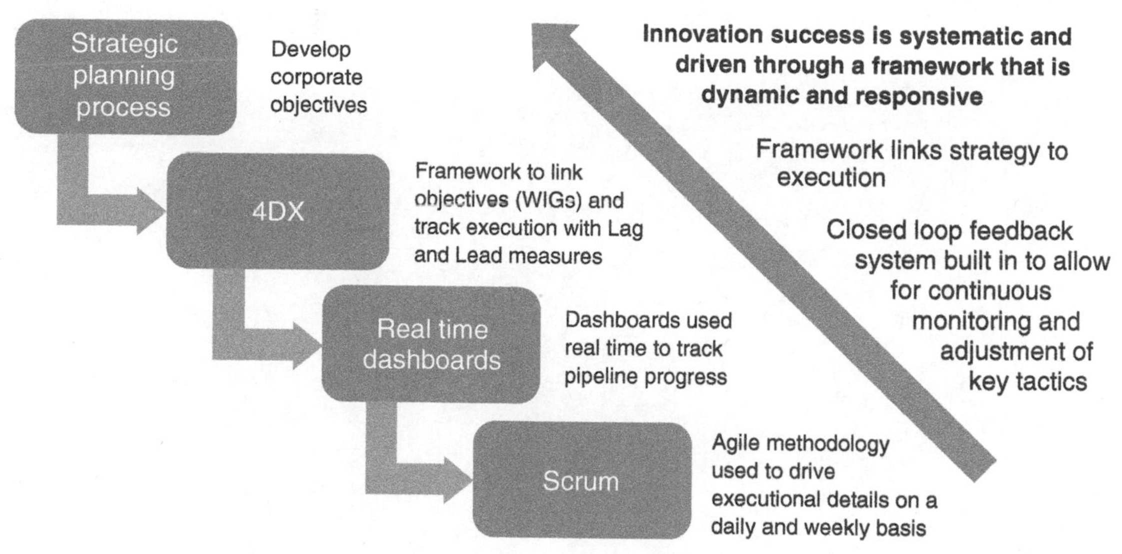

FIGURE C.9 **Porex Agile Process**

Metrics: In order to sustain innovation success, OCI winners look not at just financial measures of success such as sales from new products, vitality index, and profit margins but also measure the efficiency and effectiveness of their innovation process. They track metrics on a number of ideas generated and evaluated, milestone achievements, and time to market. They also capture lessons learned from each project and make certain these are reflected in process improvements.

Index

Product Development and Management Body of Knowledge: A Guidebook for Product Innovation Training and Certification, Third Edition.
Allan Anderson, Chad McAllister, and Ernie Harris.

D

Q

T